Human-AI Cognitive Collaboration for Industrial Evolution

Raul Villamarin Rodriguez
https://orcid.org/0000-0001-9831-905X
Woxsen University, India

Hemachandran Kannan
https://orcid.org/0000-0002-7424-5022
Woxsen University, India

Mark Esposito
Northeastern University, USA & Harvard University, USA

IGI Global Scientific Publishing
Publishing Tomorrow's Research Today

Vice President of Editorial Melissa Wagner
Director of Acquisitions Mikaela Felty
Director of Book Development Jocelynn Hessler
Production Manager Mike Brehm
Cover Design Jose Rosado

Published in the United States of America by
IGI Global Scientific Publishing
701 East Chocolate Avenue
Hershey, PA, 17033, USA
Tel: 717-533-8845 | Fax: 717-533-7115
Website: https://www.igi-global.com E-mail: cust@igi-global.com

Library of Congress Cataloging-in-Publication Data

Names: Rodriguez, Raul Villamarin editor | Kannan, Hemachandran, 1985- editor | Esposito, Mark, 1974- editor
Title: Human-AI cognitive collaboration for industrial evolution / edited by: Raul Villamarin Rodriguez, Hemachandran Kannan, and Mark Esposito.
Description: Hershey PA : IGI Global Scientific Publishing, [2027] | Includes bibliographical references. | Summary: "This book explores how human intelligence and artificial intelligence can co-evolve as complementary forces driving industrial transformation"-- Provided by publisher.
Identifiers: LCCN 2025052878 (print) | LCCN 2025052879 (ebook) | ISBN 9798337370361 hardcover | ISBN 9798337370378 paperback | ISBN 9798337370385 ebook
Subjects: LCSH: Human-computer interaction | Artificial intelligence--Industrial applications | Cognition
Classification: LCC QA76.9.H85 H836 2026 (print) | LCC QA76.9.H85 (ebook)
LC record available at https://lccn.loc.gov/2025052878
LC ebook record available at https://lccn.loc.gov/2025052879

British Cataloguing in Publication Data
A Cataloguing in Publication record for this book is available from the British Library.

Table of Contents

Detailed Table of Contents

"Industry 5.0" is the most recent industrial revolution, and it seeks to transform the ways in which the manufacturing processes under "Industry 4.0" and its previous generations were handled. Industry 5.0 was introduced and promoted by the European Commission, and it mainly focuses on humans and autonomous machines collaboration. It refers to people working together with the smart machines rather than machines taking over the jobs of the people. This particular industrial revolution adds a human touch to the pillars of the latest technological advancements. The present study focuses on the society's perspective towards the introduction of Industry 5.0 and how they think it can aid in the better efficiency of manufacturing processes. Also, what is their opinion towards the question of whether Industry 5.0 will bring broader benefits to the marketplacethan Industry 4.0 and how industries are preparing themselves to embrace this new change?

The role of human intelligence brings many changes in the artificial intelligence (AI) which some features common in factories. The study will focus on developing interaction between "human expertise and AI technology" within the framework of industrial AI implementation. AI to become more resourceful In: manufacturing, energy, logistics, and healthcare" sectors. However, for AI to work well, it needs to be based on both the technological maturity of the industry and a people-centered approach to AI integration. This is when a human expert helps retrain and reskill the workforce so that people and AI can work together well. It's also very important for people to be able to understand what AI output means make decisions when things aren't clear, and stand up for ethics. Case studies and real-world evidence back this up. The talk has made it clear that we need to find a way to make AI systems that work with human thinking clear, easy to understand, and clear. It also means that people might see AI as a partner instead of a replacement technology.

creating resilient workforces, able to continuously disruption-proof and to capitalize on the opportunities of the human-AI collaboration.

Major transformation industrial design has undergone was not limited to tools but also brought about a shift in the design paradigm. This study brings wheelchairs into focus and investigates new perspectives introduced by digital technologies, utilising a literature review and case studies. A wheelchair is an exemplary product to represent universal design principles and allow full observation of concrete effects of digital transformation with its sensitivity towards individual needs. Study uses qualitative analysis techniques supported by the literature review and case studies. A comparative analysis was conducted using the criteria obtained from this literature review. With the help of this transformation, design processes have become a more innovative and holistic structure by integrating technology. In this context, the digitalisation process in industrial design does not only refer to a technical phase but also a social transformation representing barrier-free and universal design.

The increasing complexity of agricultural and food systems necessitates human intelligence to engage in a successful collaboration with the artificial intelligence (AI) to transform agricultural systems into more productive, sustainable, and accurate in their judgments. The given chapter examines the idea of human-AI cognitive collaboration in agriculture with a specific focus on the complementary nature of human expertise and machine intelligence. It discusses the ways in which the state-of-the-art AI methods, such as machine learning, deep learning, and computer vision, contribute to the agricultural processes, such as crop monitoring, disease detection, yield prediction, and food quality evaluation. The chapter also talks about the human-in-the-loop models that combine the farmer knowledge and feedback with the AI-driven insights to enhance the trust and usability. . The chapter ends by giving future directions of research and industrial implications of human-machine synergy in sustainable agriculture and food systems.

Education is entering a new era where Artificial Intelligence (AI) is no longer seen merely as a tool for automating tasks but as a partner in shaping meaningful learning experiences. This chapter moves beyond the traditional view of AI as a replacement for routine processes, to focus on how humans and machines can co-create knowledge, foster critical thinking, and build collaborative intelligence in education. The chapter begins by tracing the shift from automation-driven applications such as grading, scheduling, and basic tutoring to more dynamic, interactive forms of human AI collaboration. It highlights how AI can act as a learning companion that adapts to students' needs, supports teachers in identifying hidden learning patterns, and creates personalized pathways for skill development. Importantly, it also underscores the irreplaceable role of human educators in nurturing empathy, ethical reasoning, and creativity areas where machines remain limited.

This chapter develops a conceptual architecture for Human-AI Collaborative Intelligence and Hybrid Cognition in post-industrial learning ecosystems. Drawing on bounded rationality, distributed cognition, joint cognitive systems, and sociotechnical alignment, it shows how collaboration reliability depends on interaction loops, interface regimes, role ecologies, and audit-ready accountability. It maps sectoral task ecologies to modality stacks that build uncertainty articulation, provenance discipline, escalation competence, and recovery-from-error performance. A multilevel capability model aligns individual, team, organizational, and ecosystem strata with curriculum design, assessment validity, and integrity-by-design institutions, concluding with polycentric governance and metric families grounded in procedural justice, capability expansion, ethics of care, privacy, and contestability for globally portable praxis.

With the introduction of AI agents, more data, faster computer processing hardware, and more interconnected third party developers, the ability for humans to manage or interfere with AI autonomous decisions is decreasing. Human agency is a key criteria for accountability. Previously, for oversight, agency was provided broadly by the concept of human-in-the-loop (HITL), however it is unclear if this is still relevant. This chapter briefly outline the concept of HITL which originated from a military context. In this paper we argue that HITL is difficult to implement, and it is often used as AI Ethics Washing, where the term is used to protect AI companies and stakeholders and blame the users. We introduce a framework for HITL where it moves away from being discretionary where AI and human agency are dependent on the transparency accountability of the power, trust and complexity to prevent AI Ethics Washing.

This chapter explores how policy, regulation, and governance can align artificial intelligence with the public good in the era of industrial evolution. Drawing on public value theory, the responsible innovation framework, and stakeholder governance theory, it conceptualises a Policy–Regulation–Governance Compass to guide ethical and inclusive AI development. Through a comparative analysis of global approaches, from the European Union's AI Act and the U.S. AI Bill of Rights to emerging African and Asian strategies, the chapter reveals asymmetries in regulatory capacity and institutional readiness. It highlights persistent challenges, including regulatory lag, power concentration, and global inequality. The chapter proposes actionable policy pathways that emphasise adaptive frameworks, institutional capacity, inclusivity, and international cooperation. It argues that governance is not a constraint on innovation

Preface

OVERVIEW OF THE SUBJECT MATTER

The relationship between human intelligence and artificial intelligence stands as one of the defining features of contemporary industrial and social life. Once imagined primarily as a contest a zero-sum competition in which machines would inevitably displace human workers this relationship has grown considerably more nuanced, productive, and promising. The emergence of collaborative intelligence, in which humans and AI systems work in deliberate complementarity to achieve outcomes neither could accomplish alone, represents a profound shift in how we understand both the nature of work and the potential of technology.

This volume, Human-AI Cognitive Collaboration for Industrial Evolution, is conceived as a rigorous scholarly and practical examination of this shift. It brings together seventeen chapters authored by researchers and practitioners spanning management science, cognitive psychology, educational theory, agricultural science, industrial design, ethics, and public policy. Together, these contributions map the contours of an evolving human-AI relationship that is reshaping industries, redefining workforce competencies, transforming educational systems, and raising urgent questions about governance, equity, and the public good.

The subject matter of this book sits at the confluence of several powerful intellectual currents: the ongoing maturation of artificial intelligence as a technological capability; the transition from Industry 4.0's automation-first paradigm to Industry 5.0's human-centric vision; growing scholarly and public interest in the ethical dimensions of AI deployment; and practical organizational demand for frameworks that help leaders, managers, and workers navigate the complexity of human-AI integration. This volume responds to all of these currents, offering both theoretical depth and actionable insight.

THE TOPIC IN THE WORLD TODAY

We are living through a period of extraordinary technological transformation. Artificial intelligence has moved rapidly from the research laboratory to the factory floor, the classroom, the hospital, the farm, and the corporate boardroom. Machine learning systems now make consequential decisions in hiring, lending, medical diagnosis, and criminal justice. Robotic systems collaborate with human workers in manufacturing and logistics. AI-powered tools are reshaping creative professions from graphic design to content creation. And generative AI systems are beginning to transform knowledge work in ways that challenge longstanding assumptions about which human capabilities are irreplaceable.

In this context, the question of how humans and AI systems should relate to one another has become one of the most pressing questions of our time. Industry 5.0, as conceptualized by the European Commission and adopted as a guiding framework by policymakers and business leaders globally, insists that the answer must center the human: AI should serve human flourishing, not supplant it. Workers should be empowered by AI collaboration, not rendered redundant by automation. The industrial future should be characterized by resilience, sustainability, and human dignity, not merely by efficiency and output maximization.

This book intervenes in these conversations at a moment when their stakes could not be higher. The choices that organizations, policymakers, educators, and individuals make today about how to design, deploy, and govern human-AI collaborative systems will shape the trajectory of industrial evolution for generations. By providing rigorous, multidisciplinary analysis of human-AI cognitive collaboration across diverse domains and contexts, this volume aims to inform those choices with the depth and nuance they deserve.

TARGET AUDIENCE

This volume is designed to serve a diverse and intellectually engaged readership united by a shared interest in understanding and shaping the human-AI collaborative future. Its primary audiences include:

Academic researchers and graduate students in fields including management science, organizational behavior, artificial intelligence, cognitive science, educational technology, agricultural science, industrial design, ethics, and public policy will find in this volume a rich compendium of theoretical frameworks, empirical findings, and conceptual contributions that advance scholarly understanding of human-AI collaboration across disciplinary boundaries.

Business leaders, managers, and organizational practitioners will benefit from the volume's practical frameworks, case analyses, and strategic recommendations for implementing human-AI collaborative systems in ways that are effective, ethical, and sustainable. The book speaks directly to the challenges that executives and managers face as they navigate AI adoption in their organizations.

Policymakers, regulators, and governance professionals will find the volume's treatment of AI ethics, responsible deployment, human-in-the-loop governance, and public policy frameworks particularly valuable as they design regulatory environments that can keep pace with rapidly evolving AI capabilities while protecting human rights and the public interest.

Educators and education technology professionals will be engaged by the volume's examination of AI's transformative impact on learning environments, pedagogical approaches, and the cultivation of collaborative intelligence as an educational outcome.

Finally, this book will be of value to informed general readers workers, citizens, and anyone concerned with the social implications of AI who seek a rigorous and accessible guide to understanding how artificial intelligence is reshaping the human world of work, learning, and governance.

ORGANIZATION OF THE BOOK

The seventeen chapters of this volume are organized to guide readers from foundational theoretical frameworks through domain-specific applications and culminating reflections on ethics, governance, and policy. The following synopses provide an orientation to the volume's intellectual architecture.

Chapter 1. Industry 5.0 and its Societal Perspective: The Empirical Aspects of Human-Centric Revolution

Moving beyond the automation-first paradigm of Industry 4.0, this chapter provides an empirical examination of Industry 5.0's human-centric revolution. Drawing on real-world data and case studies, it investigates how the fifth industrial revolution repositions the human worker not as a redundant element but as a valued collaborator with AI systems. The chapter explores societal implications including shifts in labor markets, evolving social contracts between employers and employees, and the broader economic consequences of human-AI co-production. It presents empirical evidence demonstrating how organizations that embrace human-centric design in their AI strategies achieve superior outcomes in productivity, worker satisfaction, and societal value creation.

Chapter 2. The Evolving Role of Human Intelligence in Industrial AI Adoption

This chapter examines how the adoption of AI in industrial settings reshapes—rather than replaces—human intelligence. It investigates the cognitive and professional adaptations required of workers as AI systems assume routine tasks, exploring what uniquely human competencies become more valuable in AI-augmented workplaces. The chapter addresses how critical thinking, emotional intelligence, ethical judgment, and creative problem-solving emerge as the most prized human contributions when AI handles data processing and pattern recognition. Through organizational case studies and cognitive science perspectives, the authors chart a roadmap for cultivating and preserving human intellectual capital in an era of accelerating AI adoption.

Chapter 3. Human AI Cognitive Collaboration in Industrial Contexts: A Managerial Framework for Assessment, Execution, and Governance—From Theory to Sustainable Practice

This chapter bridges theory and practice by presenting a comprehensive managerial framework for implementing human-AI cognitive collaboration in industrial settings. It addresses three interrelated dimensions: assessment of organizational readiness and capability gaps, execution strategies for deploying collaborative AI systems, and governance structures that ensure accountability and sustainability. The authors draw on management science, organizational behavior, and AI research to construct a practical toolkit for business leaders navigating the complexity of human-AI integration. Special attention is paid to governance mechanisms that protect human agency while maximizing the productive potential of AI collaboration, ensuring that implementations remain ethically grounded and strategically sustainable over time.

Chapter 4. Strategic Decision-Making Frameworks with Human-AI Systems: Transforming Workforce Dynamics and Future Synergies

Focusing on the executive and strategic dimensions of human-AI collaboration, this chapter explores how AI augments and transforms decision-making processes in modern organizations. It presents frameworks for integrating AI-generated insights into strategic planning while preserving the irreplaceable value of human judgment in ambiguous and high-stakes situations. The chapter examines how human-AI decision-making partnerships alter organizational hierarchies, shift workforce dynamics, and create new synergies at every level of the enterprise. Through case analyses of leading organizations, the authors illustrate how strategic deployment of human-AI systems creates competitive advantages while also demanding new leadership competencies and organizational cultures.

Chapter 5. Workforce Transformation and AI-Driven Skill Development in Business and Management

Complementing the previous chapter with a sharper focus on business and management contexts, this contribution examines how AI-driven skill development is reshaping professional competencies in organizational settings. It explores the specific skills that AI adoption demands from managers, analysts, and business professionals, including data literacy, AI system oversight, and the ability to collaborate effectively with algorithmic decision-support tools. The chapter presents empirical evidence from business environments undergoing AI-driven transformation and offers strategic recommendations for organizations seeking to build management capabilities commensurate with their AI investments. It also addresses the cultural dimensions of skill development, emphasizing the role of organizational learning cultures in sustaining workforce transformation.

Chapter 6. Future-Proofing the Workforce in the Age of Human-AI Synergy

This chapter takes a forward-looking perspective on workforce resilience, exploring strategies for future-proofing human talent in an era of accelerating AI advancement. It addresses the existential challenge of preparing workers for roles that do not yet exist while equipping them to remain relevant as AI capabilities expand. The authors examine macro-level trends in AI development and their projected impacts on labor markets across sectors, offering evidence-based recommendations for individuals, organizations, and policymakers. The chapter emphasizes adaptability, continuous learning, and the cultivation of distinctly human capacities as the cornerstones of workforce resilience, arguing that future-proofing is not about resisting AI change but about proactively shaping a human-AI synergy that sustains meaningful work.

Chapter 7. Digitalisation in Industrial Design: A Holistic Review of Wheelchair Design

Offering a compelling domain-specific illustration of human-AI collaboration, this chapter presents a holistic review of how digitalization and AI are transforming industrial design through the lens of wheelchair design. It demonstrates how AI-assisted design processes, digital simulation, and human-centered design principles converge to produce mobility solutions that more precisely meet the needs of users with disabilities. The chapter illustrates broader lessons about how collaborative intelligence between human designers and AI tools can accelerate innovation, improve ergonomic outcomes, and expand access to assistive technologies. It serves as a powerful case study in the humanistic potential of industrial AI when guided by genuine concern for human welfare and inclusion.

Chapter 8. Human and Machine Synergy in Agriculture and Food Systems

This chapter extends the volume's analytical lens to agriculture and food systems, examining how human-machine synergy is transforming one of humanity's oldest industries. It explores how AI, robotics, precision agriculture technologies, and human expertise combine to address global food security challenges, optimize resource use, and reduce environmental impact. The authors investigate the specific collaborative dynamics between farmers, agricultural scientists, and AI systems, highlighting how local knowledge and human judgment remain indispensable complements to algorithmic optimization. The chapter also addresses equity dimensions, considering how the benefits of agricultural AI can be made accessible to smallholder farmers in developing economies.

Chapter 9. Collaborative AI for Smart and Adaptive Educational Environments

Turning to the educational domain, this chapter examines how collaborative AI is enabling the creation of smart, adaptive learning environments that respond dynamically to individual student needs. It explores AI-powered personalization, intelligent tutoring systems, and data-driven pedagogical adaptation as tools that augment—rather than replace—the irreplaceable human dimensions of teaching and mentorship. The authors present evidence of how human-AI collaboration in education can improve learning outcomes, reduce achievement gaps, and make quality education more accessible. The chapter also considers the professional implications for educators, arguing that AI collaboration enhances rather than diminishes the value of skilled, empathetic human teachers.

Chapter 10. Human-AI Collaborative Intelligence and Pedagogies of Hybrid Cognition in Learning Ecosystems: Infrastructure in Post-Industrial Education Design

Building on the previous chapter's foundation, this contribution presents a more theoretical exploration of hybrid cognition pedagogies in post-industrial educational design. It examines how learning ecosystems must be redesigned to cultivate collaborative intelligence the capacity to think effectively in partnership with AI systems as a core educational outcome. The chapter explores the infrastructural requirements of post-industrial education, including digital platforms, data architectures, and institutional cultures that support human-AI collaborative learning at scale. It proposes a new pedagogical paradigm in which developing the capacity for effective human-AI collaboration becomes as fundamental an educational goal as literacy and numeracy in the knowledge economy.

Chapter 11. AI-Powered Branding Strategies: Gravitational Pull of Innovation through AI's Impact on Advertising and Design

This chapter examines how AI is revolutionizing branding, advertising, and design, creating new gravitational forces of innovation in the creative economy. It explores how human-AI collaboration in creative processes generates branding strategies and advertising content that are simultaneously more data-driven and more creatively ambitious than those produced by human teams alone. The authors investigate how AI tools augment human creative intuition, accelerate design iteration, and enable hyper-personalized consumer engagement at scale. The chapter also addresses the authenticity challenge how brands maintain genuine human connections with consumers in an era of AI-generated content arguing that the most successful AI-powered branding strategies are those that keep human creativity and values at their center.

Chapter 12. Responsible and Ethical AI in Industrial Contexts

This chapter confronts the ethical imperatives of AI deployment in industrial settings, presenting a comprehensive framework for responsible AI that goes beyond compliance to embrace genuine ethical commitment. It examines the ethical challenges most pressing in industrial AI contexts: algorithmic bias in hiring and promotion systems, worker surveillance and privacy erosion, the displacement of vulnerable workers, and the accountability gaps that emerge when AI systems make consequential decisions. The authors argue that responsible AI requires proactive ethical design—embedding fairness, transparency, and human dignity into AI systems from their inception and present practical frameworks for organizations seeking to operationalize ethical principles in their AI strategies.

Chapter 13. Human-in-the-Loop for Artificial Intelligence: Avoiding AI Ethics Washing

Directly addressing a critical challenge in AI governance, this chapter examines the human-in-the-loop principle as both a technical design choice and an ethical commitment, while exposing the risk of AI ethics washing the superficial adoption of ethical language without substantive changes in AI system design or deployment. It explores how genuine human oversight of AI systems differs from performative compliance, presenting criteria for evaluating whether human-in-the-loop implementations provide real accountability or merely the appearance of it. The chapter offers organizational and regulatory recommendations for ensuring that human oversight mechanisms are meaningful, that the humans in the loop are appropriately empowered, and that ethics commitments translate into measurable improvements in AI system behavior and outcomes.

Chapter 14. Policy, Regulation, and the Public Good of Artificial Intelligence

The concluding chapter situates the book's discussions within the broader landscape of AI policy and regulation, examining how governance frameworks can ensure that AI serves the public good in industrial and social contexts. It reviews emerging regulatory approaches to AI governance across major jurisdictions, assessing their strengths and limitations in addressing the challenges documented throughout the volume. The authors argue for a policy framework that balances innovation with protection, that is adaptive to the rapid pace of AI development, and that centers human well-being as the ultimate measure of AI's social value. The chapter concludes with a call for inclusive, participatory AI governance that brings workers, citizens, and affected communities into the policy conversations that will shape the future of human-AI collaboration in industry and society.

CONCLUSION: THE BOOK'S CONTRIBUTION TO THE FIELD

Human-AI Cognitive Collaboration for Industrial Evolution makes a distinctive and substantial contribution to an increasingly important field of scholarly inquiry and practical concern. By bringing together seventeen rigorous contributions spanning management, cognitive science, education, agriculture, design, ethics, and policy, the volume achieves a breadth and depth of coverage that no single-discipline treatment of this subject can match.

The book's most fundamental contribution lies in its reframing of the human-AI relationship. Against narratives of replacement and displacement, it marshals evidence and argument for a vision of collaborative intelligence in which human capabilities and AI capabilities are deliberately designed to complement and reinforce one another. This reframing has implications not only for how organizations design and deploy AI systems, but for how educational institutions prepare future workers, how policymakers design governance frameworks, and how individuals understand their own roles and value in an AI-augmented world.

The volume also makes important contributions to understanding the sectoral diversity of human-AI collaboration. By examining collaborative intelligence in manufacturing, agriculture, education, healthcare design, branding, and public administration, it demonstrates that while the principles of effective human-AI collaboration are generalizable, their application must be sensitive to the specific cognitive demands, ethical stakes, and organizational contexts of particular domains. This sectoral breadth enriches both theoretical understanding and practical guidance.

Perhaps most importantly, this book insists on the irreducible importance of the ethical and governance dimensions of human-AI collaboration. The contributions on responsible AI, human-in-the-loop governance, ethics washing, and public policy collectively argue that technological capability must be matched by ethical commitment and robust governance if human-AI collaboration is to serve human flourishing rather than undermine it. This ethical insistence is not an add-on to the volume's technical and organizational analyses but integral to them: the book consistently treats the question of how humans and AI should collaborate as inseparable from the question of what that collaboration should be for.

As artificial intelligence continues to evolve in capability and expand in reach, the need for rigorous, multidisciplinary scholarship on human-AI cognitive collaboration will only grow. We hope that this volume serves not only as a resource for current researchers, practitioners, and policymakers, but as a catalyst for the next generation of inquiry one that keeps human dignity, equity, and flourishing at the center of the conversation about our shared technological future.

The Editors

Chapter 1 Industry 5.0 and Its Societal Perspective: The Empirical Aspects of Human-Centric Revolution

Sneha Verma
http://orcid.org/0000-0003-1927-0062
Bharatiya Vidya Bhavan College, New Delhi, India

ABSTRACT

"Industry 5.0" is the most recent industrial revolution, and it seeks to transform the ways in which the manufacturing processes under "Industry 4.0" and its previous generations were handled. Industry 5.0 was introduced and promoted by the European Commission, and it mainly focuses on humans and autonomous machines collaboration. It refers to people working together with the smart machines rather than machines taking over the jobs of the people. This particular industrial revolution adds a human touch to the pillars of the latest technological advancements. The present study focuses on the society's perspective towards the introduction of Industry 5.0 and how they think it can aid in the better efficiency of manufacturing processes. Also, what is their opinion towards the question of whether Industry 5.0 will bring broader benefits to the marketplacethan Industry 4.0 and how industries are preparing themselves to embrace this new change?

1. INTRODUCTION

The first industrial revolution started in England and moved to the United States in the late 18th century, and continued till the 19th century. Mechanisation of industries, or simply put, 'machines', was introduced in the first industrial revolution for

DOI: 10.4018/979-8-3373-7036-1.ch001

the very first time. This contributed enormously to the growth and development of the economies around the globe. The Industrial Revolution had an impact on industries such as textiles, agriculture, mining, etc. Since then, human civilization has witnessed various industrial revolutions, as can be seen in Figure 1.

Figure 1. Stages of Industrial Revolutions

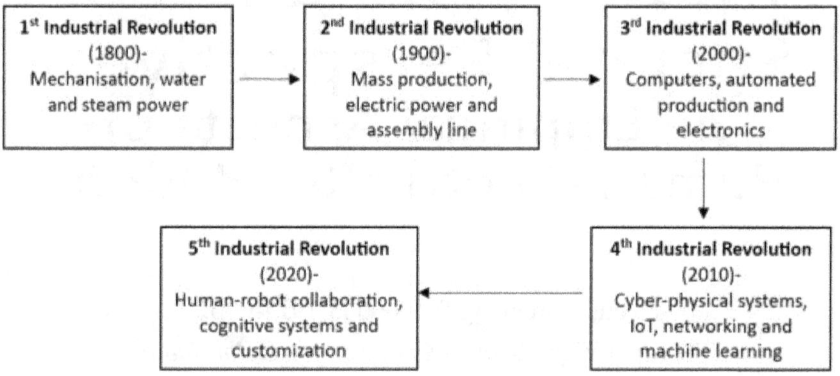

The world today is experiencing the fourth stage of the industrial revolution, or simply put, '*Industry 4.0*'. Industry 4.0 refers to a revolutionary change in production and manufacturing brought about by the integration of cutting-edge automation systems and digital technology. Adoption of technologies like big data analytics, artificial intelligence (AI), the Internet of Things (IoT), and cyber-physical systems, which together generate connected environments known as "smart factories," signifies this shift. These technologies facilitate real-time data analysis and interchange, which promotes better product quality, increased operational efficiency, and data-driven decision-making.

In contrast to Industry 4.0, *'Industry 5.0'* has emerged as a new concept. The ultimate goal of Industry 5.0 is the reinforcement of the role and contribution of industries towards society, rather than focusing only on efficiency and mass production. Industry 5.0 restores the human element to industrial processes, encouraging a more resilient, human-centric, and sustainable approach to production and innovation (*European Commission: Directorate-General for Research and Innovation, 2021*). It is concerned more with the people at work and puts them at the centre of the production process. In other words, 'Industry 5.0' adopts an approach where human beings are kept at the center and technology revolves around them. It believes that humans and machines should work together, contributing towards the sustainable development process and also keeping in mind the resource limit of the very earth we live on. This changing paradigm believes that societal demands, such as worker welfare, environmental sustainability, and ethical issues, must be well-balanced with technological growth. Industry 5.0 aims to integrate human intelligence with technologies like artificial intelligence (AI), collaborative robots (cobots), and the Internet of Things (IoT) by developing mechanisms that are not only more efficient but also more flexible to human values and societal objectives (Nahavandi, 2019). Also, Industry 5.0 deals with important global issues pertaining to sustainability and climate change by emphasizing resource efficiency and environmental stewardship. It seeks to reduce waste and carbon footprints across a variety of sectors, including supply chain management, manufacturing, healthcare, and agriculture.

Under Industry 4.0, human beings are an essential part of the manufacturing process, but all kinds of everyday and monotonous duties are performed by robots, i.e., the machines and automated systems. While on the other hand, the main aim of 'Industry 5.0' is encouraging and supporting the role of workforce (human beings) in the process of manufacturing and production. This lifts the self-confidence of the workers and improves human abilities that are unique to humans only. As a result, this can add value to the production process, such as decision-making, creativity and solving complex problems.

1.1 Key aspects of Industry 5.0:

The following outlines the key aspects of Industry 5.0.

1. **Human Centricity:**
    - Collaboration between Humans and Machines: Unlike Industry 4.0, where automation often led to the replacement of human labour, Industry 5.0 emphasises the collaboration between humans and machines. The goal is to enhance human capabilities rather than replace them. This paradigm emphasises that humans are vital for the effectiveness and flexibility of manufacturing processes and are not merely resources.
    - Personalization and Customization: Industry 5.0 seeks to tailor products and services to individual needs and preferences, enhancing consumer satisfaction and experience.
    - Ethical AI and Technology: This phase promotes the development and use of technologies that respect human values and ethical standards. It advocates for AI systems that are transparent, fair, and accountable.
2. **Sustainability:**
    - Environmental Impact: Industry 5.0 puts a great emphasis on the creation of sustainable industrial practices that minimize environmental

impact. This includes reducing waste, conserving energy, and adopting circular economy principles.
- Social Responsibility: Companies are encouraged to be mindful of the societal impacts of their operations to promote equitable economic growth and social well-being.

3. **Societal Impact:**
   - Job Transformation: While automation and AI may reduce the need for certain types of labour, Industry 5.0 emphasizes reskilling and upskilling the labour workforce to adapt to new roles that require creativity, problem-solving, and emotional intelligence.
   - Quality of Life: By integrating advanced technologies with human needs, Industry 5.0 aims to improve the quality of life, promoting work-life balance, and enhancing health and safety in the workplace.
   - Inclusion and Accessibility: Ensuring that technological advancements are accessible to all segments of society, including marginalized and disadvantaged groups, is a core principle of Industry 5.0.
4. **Resilience:** In Industry 5.0, resilience is the capacity of manufacturing systems to quickly adjust to modifications, interruptions, and new demands from the market. Businesses can determine the best alternate routes in the event of disruptions by employing digital technologies like AI-enhanced modelling and simulations. This maintains operational efficiency while taking cost, quality, and logistics into account. Such adaptability is essential in an era when technology is evolving rapidly and customer needs are changing accordingly.

1.2 Empirical Aspects of the Human-Centric Revolution:

1. **Case Studies and Real-World Examples:**
   - Collaborative Robots (Cobots): Companies like Universal Robots have developed collaborative robots designed to work alongside humans, taking on repetitive tasks while allowing workers to focus on more complex, value-added activities.
   - Sustainable Manufacturing Practices: Organizations are increasingly adopting sustainable practices. For instance, the use of digital twins and AI in manufacturing to optimize processes, reduce waste, and improve energy efficiency.
2. **Surveys and Data:**
   - Worker Satisfaction: Empirical studies show that human-centric approaches in Industry 5.0 can lead to increased job satisfaction, higher productivity, and better employee retention.

- Consumer Trends: Data indicate a growing consumer demand for personalized products, which aligns with the human-centric and customization focus of Industry 5.0.

3. **Policy and Regulation:**
    - Governance Models: Governments and international organizations are beginning to develop frameworks to guide the ethical development and deployment of Industry 5.0 technologies, ensuring they benefit society as a whole.
    - Standardization: Efforts are being made to standardize the practices and technologies associated with Industry 5.0, facilitating global adoption and ensuring that the benefits are widely shared.

2. RELATED LITERATURE

Xu et al. (2021) examined the inception, conception and perception of 'Industry 4.0' and 'Industry 5.0'. The results state that different stages of the industrial revolution so far have changed the way an industry works and operates. 'Industry 4.0' is guided by technology like its ancestors. And, 'Industry 5.0' is human- centric and value- driven. However, they both are like the two-sides of the same coin. They both need each other to function smoothly. Potočan, et al. (2020) stated that 'Society 5.0' is meant to solve significant social problems along with creating innovative technological advancements. Industries are working on their CSR activities keeping in mind the human- centric approach and sustainable development.

The results (Stentoft et al. 2019) concluded that the readiness for Industry 4.0 of SMEs in Denmark is relatively low as compared to large companies. The possible reason for this could be that big companies have access to resources and can easily exploit new technologies. Another reason could be zero awareness about the latest technology. Demir & Cicibaş (2019) stated that 'Industry 4.0' and 'Industry 5.0' are not mutually exclusive. Mass production and sustainable development should go hand in hand. Sudibjo et al. (2019) studied the characteristics of learning and education in this age of 'Industry 5.0'. They concluded that in this era of digital age, learning should be student- centric. Students or learners now have access to internet which has obviously made teaching and learning quicker and easier. Collaborative learning and project- based learning should be given more preference.

Cetinkaya & Unsacar (2021) stated that some industries do not fully understand the concept of Industry 4.0. Also, some are afraid that their manufacturing and business operations can be affected by the recent technological developments. Thus, training can help prepare these industries to easily adopt the changes in technology. Ghobakhloo et al. (2023) concluded that 'Industry 5.0' is the lucid extension of the fourth industrial revolution, which is mainly concerned with giving more importance to the workers working in industries and attaining sustainability. 'Industry 5.0' is focused towards bringing a harmonious alliance between machines and humans and also managing the digital industrial revolution. The result (Zizic, et al., 2022) concluded that humans are the main drivers of all the activities. They have created the production systems and an effort is made to create such autonomous robots that collaborate with humans and work together.

Skobelev & Borovik (2017) found out that the growing prevalence of digitalization of economies around the world and its practical application have formed the basis for Industry 4.0 and are also working for Society 5.0. Yadav et al. (2023) stated that Industry 5.0 is the change that connects humans and machines. Industry 5.0 also aids environmental sustainability as industries are becoming more responsible towards the discard of waste and other harmful materials. Barata & Kayser (2023) conducted a bibliometric analysis which stated that Industry 5.0 envisages a societal-driven and future oriented transformation of the manufacturing processes. Ramu (2020) concluded that this particular revolution is related to interaction between human beings and digital machines. It will transform the production processes by incorporating collaboration between humans and robots on a large scale and providing customer-oriented products. Özdemir & Hekim (2018) concluded that 'Industry 5.0' is related to introducing complex digital networks without incriminating the sustainability factor.

3. RESEARCH OBJECTIVES

The research objectives of the study are:

1. To determine whether organisations are ready to embrace the new industrial revolution.
2. To examine the human-centric approach of Industry 5.0.
3. To assess whether the fifth industrial revolution will lead to greater benefits for the marketplace than Industry 4.0.

4. RESEARCH HYPOTHESES

The hypotheses that have been developed for the study are:

$H0_1$: There is no correlation between an organisation's readiness to embrace 'Industry 5.0' and benefits related to the adoption of Industry 5.0.

$H0_2$: There is no correlation between the human-centric approach of 'Industry 5.0' and benefits related to the adoption of Industry 5.0.
$H0_3$: There is no significant impact of the human-centered approach of Industry 5.0 and the organisation's readiness to embrace Industry 5.0 on the benefits related to the adoption of Industry 5.0.

5. RESEARCH METHODOLOGY

The present study is exploratory and empirical in nature. For carrying out the study, primary data have been collected. An online questionnaire was sent through Google Forms to 150 individuals working as students, as academicians, individuals in the corporate sector, in the government sector, and also individuals working on their own. A total of 105 individuals responded to the questionnaire, out of 150 sent. Cronbach's alpha is used to test the reliability of the data and to check the internal consistency among the items in the questionnaire.

For analysis, statistical tests like Pearson's correlation and regression have been used, and results have been interpreted accordingly. Also, charts have been employed for the presentation of all the demographic variables, and interpretations have been made accordingly.

6. DATA ANALYSIS AND INTERPRETATION

Figure 2.

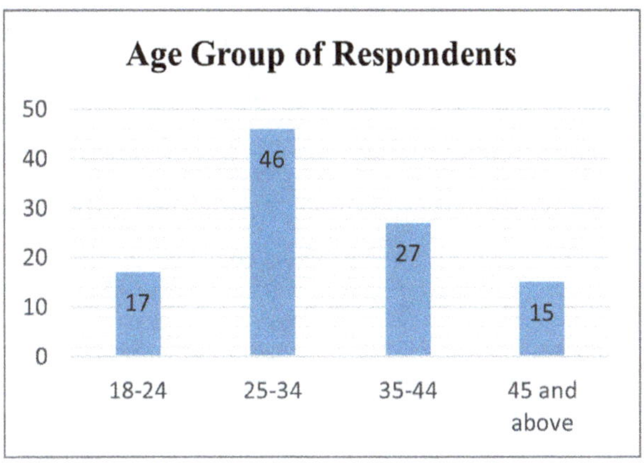

Figure 3.

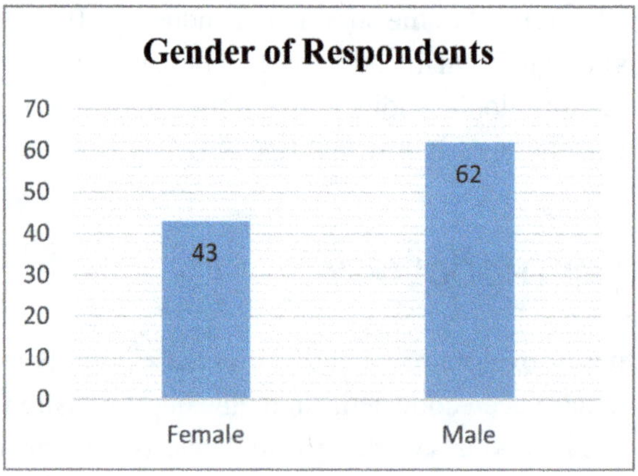

Figure 4.

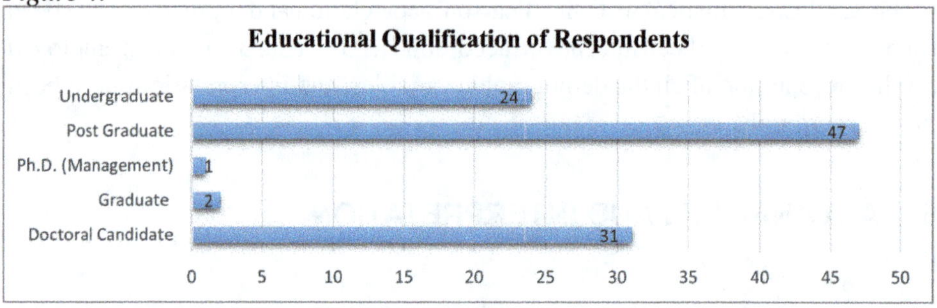

Figure 5.

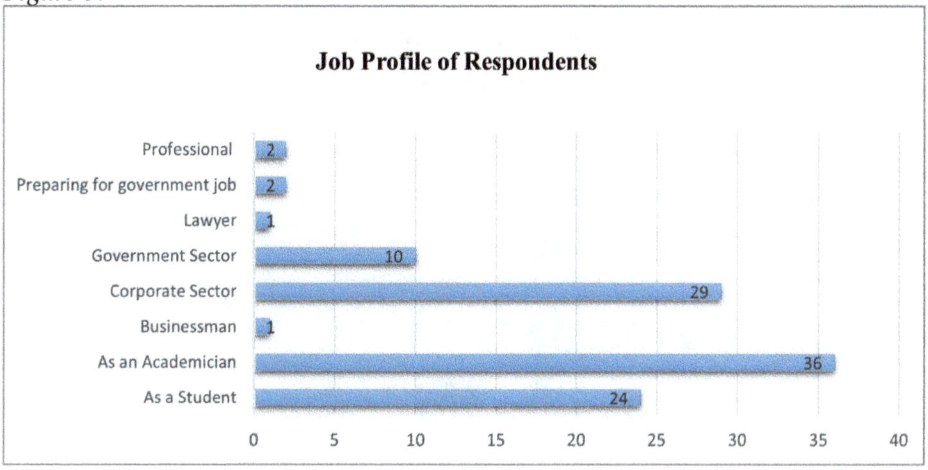

Figure 6.

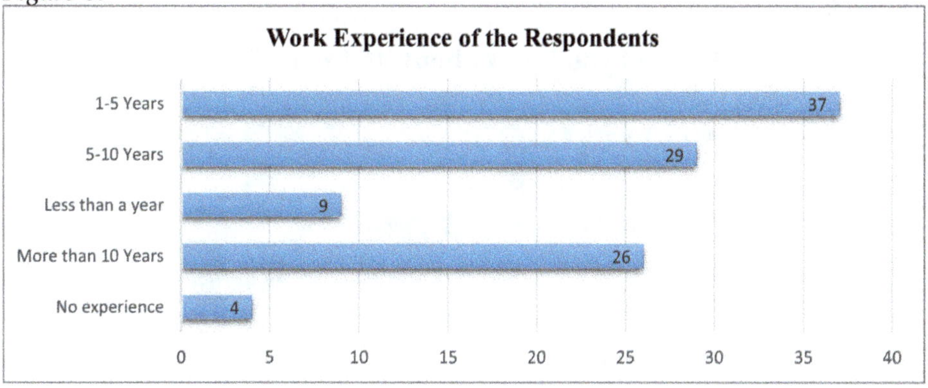

Figure 7.

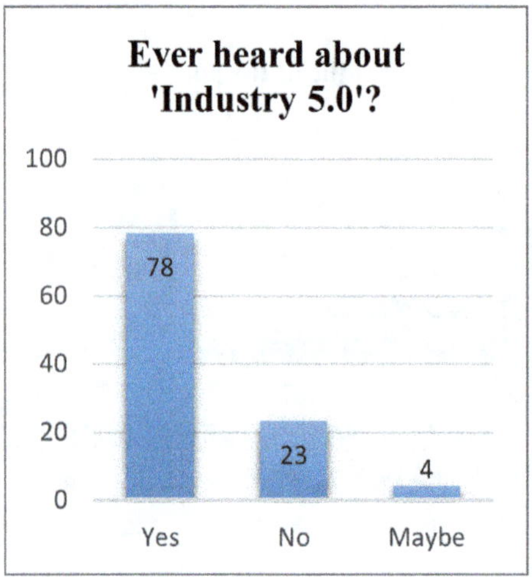

Figure 8.

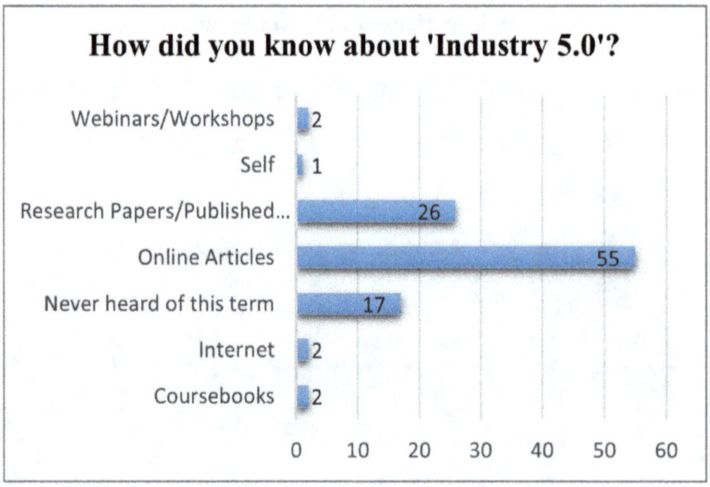

Figure 9.

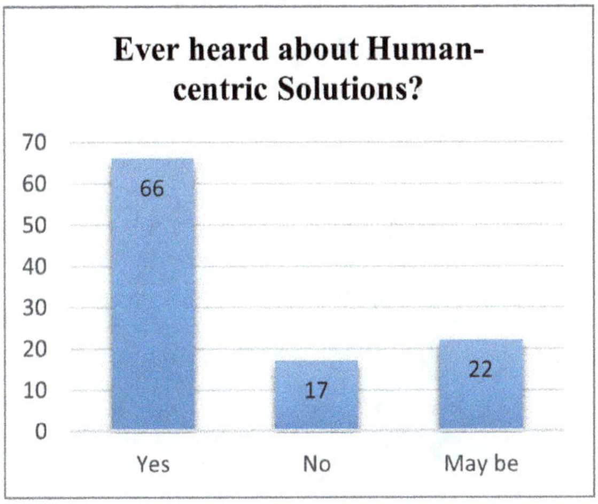

Figure 10.

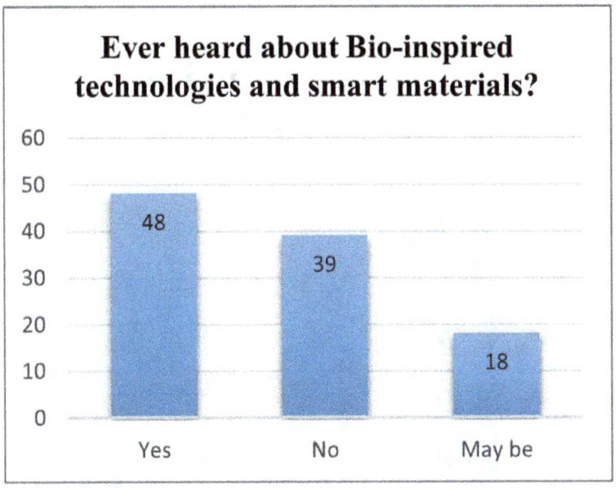

Figure 11.

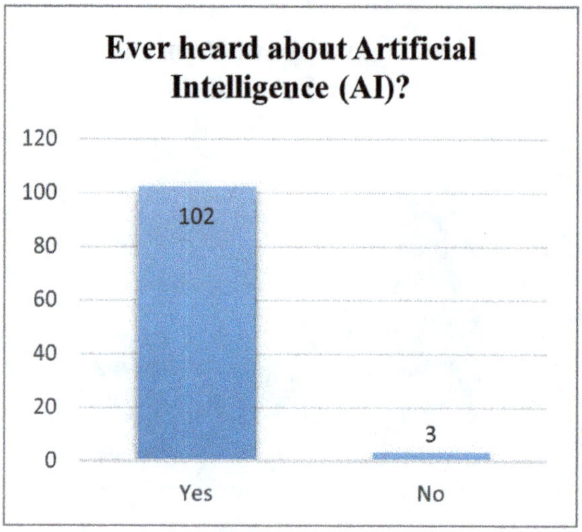

Figure 12.

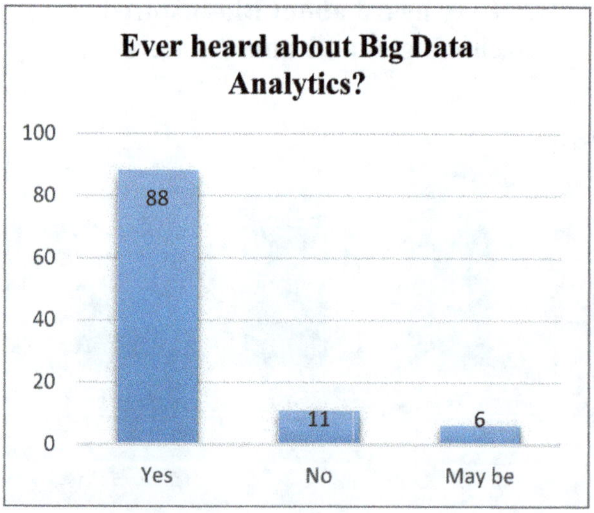

Figure 13.

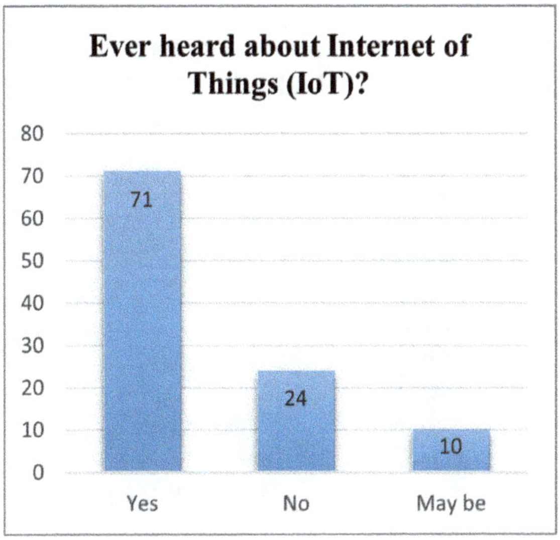

Figure 14.

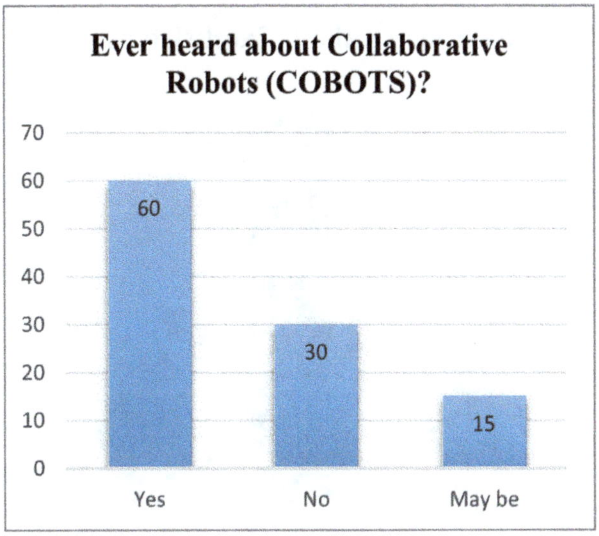

Figure 15.

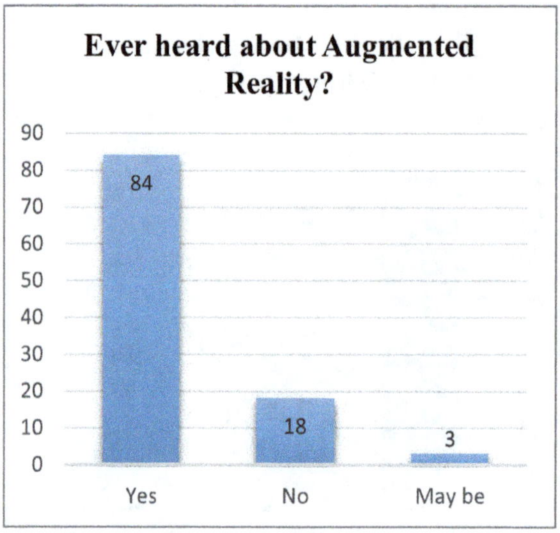

Figure 16.

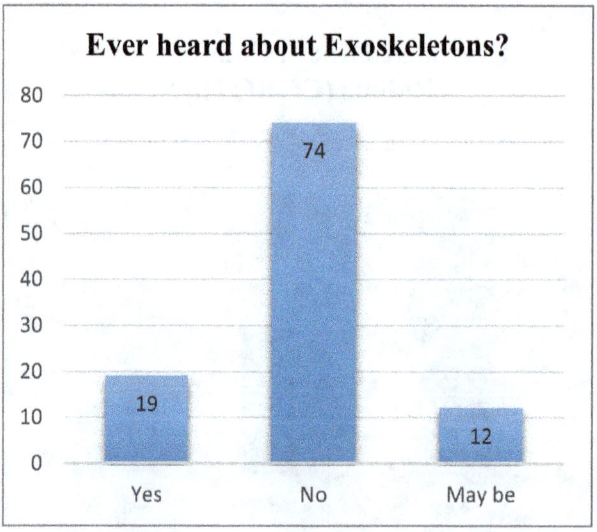

As can be seen in the above figures, out of 105 responses received, 62 are from males and 43 from females. The maximum number of respondents (46) belong to the age group of 25-34 years and are post-graduates and doctorate. Around 36 of them work as academicians, followed by 29 working in the corporate sector. They have a work experience of around 1-5 years (37) and only 78 of them have heard about Industry 5.0 through online articles.

In respect of terminologies associated with Industry 5.0, the following provides detailed information:

Table 1.

Ever heard about?	Yes	No	May be
Human- centric solution	66	17	22
Bio- inspired technologies and smart materials	48	39	18
Artificial Intelligence	102	03	00
Big data analytics	88	11	06
Internet of Things (IoT)	71	24	10
Collaborative Robots (COBOTS)	60	30	15
Augmented Reality	84	18	03
Exoskeletons	19	74	12

Source: Compiled by the author

Table 2. Reliability Statistics

Cronbach's Alpha	Cronbach's Alpha Based on Standardized Items	N of Items
.788	.797	10

Source: Compiled by the author

The Cronbach's alpha value for all the variables used for the purpose of analysis is more than 0.7. This exhibits that the items in the questionnaire are sufficiently consistent and indicates that the measure is reliable.

Table 3. Statistics

	N		Mode
	Valid	Missing	
Do you agree that Industry 5.0 is similar to Industry 4.0?	105	0	4
Do you agree that Industry 5.0 can enhance the efficiency of manufacturing processes as compared to Industry 4.0?	105	0	5
Do you agree that organisations can easily embrace the concept of Industry 5.0?	105	0	4

continued on following page

Table 3. Continued

	N		Mode
	Valid	Missing	
Do you agree that Industry 5.0 will lead to better automation of manufacturing processes as compared to Industry 4.0?	105	0	4
Do you agree that the interaction between humans and computerised machinery will drastically improve the optimisation and automation of manufacturing enterprises?	105	0	4
Do you agree that the collaboration between the two industrial processes entities will offer new techniques and ideas for managing a workforce that includes both people and software robots?	105	0	4
Do you agree that robots are not supposed to supersede humans but co-op with them?	105	0	5
Do you agree that Industry 5.0 will bring broader benefits to the marketplace than industry 4.0?	105	0	4
Do you agree that Industry 5.0 will create more jobs than eliminate?	105	0	4
Do you agree that Industry 5.0 can support sustainable manufacturing practices by optimizing processes, reducing waste, and enabling energy-efficient operations?	105	0	4

The above table shows the *mode* of all the values of the questionnaire received in the form of responses. As can be seen, 4 is the value with the highest frequency for eight questions out of ten. For the rest two, 5 is the mode value. This represents that respondents have either somewhat agreed or totally agreed with the questions asked of them in the questionnaire.

In case of question regarding the similarity between Industry 4.0 and Industry 5.0, most of the respondents have agreed that there exists a similarity between the two. Also, the respondents have agreed to the fact that Industry 5.0 can enhance the efficiency of the manufacturing processes as compared to Industry 4.0. It will also lead to better automation of manufacturing processes and industries can easily embrace the new concept of Industry 5.0. The synergy and cooperation between humans and digital machineries will exceptionally upgrade the working of manufacturing industries. Almost all of them have agreed to the fact that robots are not supposed to replace/remove humans but are meant to work with them in collaboration. ‘Industry 5.0’ will lead to greater benefits for the industrial marketplace as compared to ‘Industry 4.0’. Respondents have agreed that Industry 5.0 will generate more jobs and has the capacity to support sustainable manufacturing processes.

Table 4.

Correlations		**Organisation's readiness to embrace Industry 5.0**	**Benefits of adopting Industry 5.0**	**Human-centric approach of Industry 5.0**
Organisation's readiness to embrace Industry 5.0	Pearson Correlation	1	.253**	.147
	Sig. (2-tailed)		.009	.135
	N	105	105	105
Benefits of adopting Industry 5.0	Pearson Correlation	.253**	1	.350**
	Sig. (2-tailed)	.009		.000
	N	105	105	105
Human-centric approach of Industry 5.0	Pearson Correlation	.147	.350**	1
	Sig. (2-tailed)	.135	.000	
	N	105	105	105

***. Correlation is significant at the 0.01 level (2-tailed).*

As can be seen in table 3, a Pearson correlation coefficient was computed to find out the relationship between an organisation's readiness to embrace Industry 5.0 and the benefits of adopting Industry 5.0. There was a significant but very weak positive relationship between an organisation's readiness to embrace Industry 5.0 and the benefits related to adoption of Industry 5.0, r (103) = .253, p = .009. Also, a Pearson correlation coefficient was computed to find out the relationship between the human-centric approach of 'Industry 5.0' and benefits related to the adoption of Industry 5.0. There was again a significant but moderate positive relationship between the human-centric approach of 'Industry 5.0' and the benefits related to adoption of Industry 5.0, r (103) = .350, p= < .001. Thus, the null hypotheses $H0_1$ and $H0_2$ tend to be rejected.

Table 5. Model Summary

Model	**R**	**R Square**	**Adjusted R Square**	**Std. Error of the Estimate**	**Change Statistics**				
					R Square Change	**F Change**	**df1**	**df2**	**Sig. F Change**
1	.501[a]	.251	.228	.714	.251	11.257	3	101	.000

a. Predictors: (Constant), Human-centric approach of Industry 5.0, Organisation's readiness to embrace Industry 5.0

Table 6. ANOVA[a]

Model		Sum of Squares	df	Mean Square	F	Sig.
1	Regression	17.230	3	5.743	11.257	.000[b]
	Residual	51.532	101	.510		
	Total	68.762	104			

a. Dependent Variable: Benefits of adopting Industry 5.0

b. Predictors: (Constant), Human-centric approach of Industry 5.0, Organisation's readiness to embrace Industry 5.0

Table 4 & 5 show the results of the regression for the null hypothesis $H0_{3.}$ A regression analysis was computed to determine whether the human-centred approach of 'Industry 5.0' and an organisation's readiness to embrace this new industrial revolution predict the benefits of adopting it. The results of ANOVA were significant, $F(3,101) = 11.257$, $p = .000$. Therefore, we reject the null hypothesis and conclude that the new human-centred approach of 'Industry 5.0' and the organisation's readiness to embrace it predict the benefits related to adoption of the fifth industrial revolution.

7. FINDINGS AND CONCLUSIONS

Businesses, particularly small ones, still face difficulty and are attempting to digitize their operations in an attempt to compete in the rapidly evolving technological world. They are actively looking for methods to establish the critical link with the Industry 4.0 model that will increase production. With everything going on, Industry 5.0—the next phase of the industrial revolution—is rapidly emerging. Many of the top and well-known academic institutions and policy-makers like the European Union Commission, think that technological changes and advancements are inevitable and cannot be avoided. However, the societal and environmental needs should not be ignored anymore and must be combined with the latest techno-industrial revolution.

Due to time constraints, the number of responses collected for the purpose of the study is less. The study found that people and society at large are of the opinion that Industry 5.0 is better than Industry 4.0 in terms of the human-centric approach of the former. In other words, robotic machineries are not meant to replace humans/ or workers in the manufacturing processes. Instead, robots and humans must work jointly (COBOTS). This is the essence that makes this fifth industrial revolution more relevant than the fourth one. 'Industry 5.0' can enhance efficiency and lead to better automation of manufacturing processes. The synergy between human beings and digital machineries will inordinately enhance the working and automation of manufacturing industries. This will prove to be beneficial for the marketplace and for the industry as a whole.

The fourth and the fifth industrial revolutions can be said as the two sides of the same coin. 'Industry 4.0' is guided by technology, whereas 'Industry 5.0' is value-oriented. The former requires the latter to recall and prompt the important societal needs, values and responsibilities towards the society as the fundamental goal to be achieved. And also, the latter needs the former for all kinds of technological advancements. The main aim of 'Industry 4.0' is production in bulk, and that of 'Industry 5.0', the ultimate goal is achieving sustainability. Mass/bulk production and sustainability are not mutually exclusive. In that sense, Industry 4.0 and Industry 5.0 are insufficient in themselves. Thus, they both need each other to achieve the goal of sustainable mass production.

Industry 5.0 represents a shift towards a more human-centric, sustainable, and inclusive industrial paradigm. Its societal perspective emphasizes the importance of integrating technology with human needs and values, ensuring that technological advancements contribute to the overall well-being of society. The empirical aspects of this revolution highlight the potential for Industry 5.0 to transform industries in ways that prioritize people and the planet, offering a more balanced and ethical approach to progress.

However, Industry 5.0 does not come without limitations. The highly automated manufacturing systems demand high skill development for the workforce. Industries, especially those operating at small levels, find it really difficult and challenging to cope up with the ever-changing technical needs. Training the workforce according to the changing technologies proves to be a ginormous task, requiring both time and money. Also, there are chances of increased cybersecurity threats in industries due to the use of standard communication protocols and internet connectivity. Thus, measures must be taken to reap out the maximum benefits from it and minimise the possible threats that can occur.

REFERENCES

Barata, J., & Kayser, I. (2023). Industry 5.0–Past, Present, and Near Future. *Procedia Computer Science*, *219*, 778–788. DOI: 10.1016/j.procs.2023.01.351

Cetinkaya, A. S., & Unsacar, M. K. (2021). Assessment of industry 4.0 on manufacturing enterprises: Demographic perspective. University of South Florida (USF) M3 Publishing, 5(2021), 107. https://www.doi.org/10.5038/9781955833035

Demir, K. A., & Cicibaş, H. (2019). The next industrial revolution: industry 5.0 and discussions on industry 4.0. In 4th International Management Information Systems Conference "Industry (Vol. 4, pp. 17-20). https://www.researchgate.net/publication/336653504_The_Next_Industrial_Revolution_Industry_50_and_Discussions_on_Industry_40

European Commission. Directorate-General for Research and Innovation. (2021). Industry 5.0: towards a sustainable, human-centric and resilient European industry. Publications Office of the European Union. https://data.europa.eu/doi/10.2777/308407

Ghobakhloo, M., Iranmanesh, M., Tseng, M. L., Grybauskas, A., Stefanini, A., & Amran, A. (2023). Behind the definition of Industry 5.0: a systematic review of technologies, principles, components, and values. Journal of Industrial and Production Engineering, 1-16. DOI: 10.1080/21681015.2023.2216701

Nahavandi, S. (2019). Industry 5.0—A human-centric solution. *Sustainability (Basel)*, *11*(16), 4371. DOI: 10.3390/su11164371

Özdemir, V., & Hekim, N. (2018). Birth of industry 5.0: Making sense of big data with artificial intelligence, "the internet of things" and next-generation technology policy. *OMICS: A Journal of Integrative Biology*, *22*(1), 65–76. DOI: 10.1089/omi.2017.0194 PMID: 29293405

Potočan, V., Mulej, M., & Nedelko, Z. (2020). Society 5.0: Balancing of Industry 4.0, economic advancement and social problems. *Kybernetes*, *50*(3), 794–811. DOI: 10.1108/K-12-2019-0858

Ramu, N. B. (2020). Management Challenges and Opportunities of Industry 5.0. https://www.researchgate.net/publication/341932600_Management_Challenges_and_Opportunities_of_Industry_50

Skobelev, P. O., & Borovik, S. Y. (2017). On the way from Industry 4.0 to Industry 5.0: From digital manufacturing to digital society. Industry 4.0, 2(6), 307-311. https://stumejournals.com/journals/i4/2017/6/307

Stentoft, J., Jensen, K. W., Philipsen, K., & Haug, A. (2019). Drivers and barriers for Industry 4.0 readiness and practice: a SME perspective with empirical evidence. https://doi.org/DOI: 10.1080/09537287.2020.1768318

Sudibjo, N., Idawati, L., & Harsanti, H. R. (2019, December). Characteristics of Learning in the Era of Industry 4.0 and Society 5.0. In *International Conference on Education Technology (ICoET 2019)* (pp. 276-278). Atlantis Press. DOI: 10.2991/assehr.k.250122.052

Xu, X., Lu, Y., Vogel-Heuser, B., & Wang, L. (2021). Industry 4.0 and Industry 5.0—Inception, conception and perception. *Journal of Manufacturing Systems*, *61*, 530–535. DOI: 10.1016/j.jmsy.2021.10.006

Yadav, M., Vardhan, A., Chauhan, A. S., & Saini, S. (2023, February). A Study on Creation of Industry 5.0: New Innovations using big data through artificial intelligence, Internet of Things and next-origination technology policy. In 2023 IEEE International Students' Conference on Electrical, Electronics and Computer Science (SCEECS) (pp. 1-12). IEEE. DOI: 10.1109/SCEECS57921.2023.10063069

Zizic, M. C., Mladineo, M., Gjeldum, N., & Celent, L. (2022). From industry 4.0 towards industry 5.0: A review and analysis of paradigm shift for the people, organization and technology. *Energies*, *15*(14), 5221. DOI: 10.3390/en15145221

Chapter 2
The Evolving Role of Human Intelligence in Industrial AI Adoption

Sanjeev Kumar
http://orcid.org/0000-0002-7375-7341
Lovely Professional University, India

ABSTRACT

The role of human intelligence brings many changes in the artificial intelligence (AI) which some features common in factories. The study will focus on developing interaction between "human expertise and AI technology" within the framework of industrial AI implementation. AI to become more resourceful In: manufacturing, energy, logistics, and healthcare" sectors. However, for AI to work well, it needs to be based on both the technological maturity of the industry and a people-centered approach to AI integration. This is when a human expert helps retrain and reskill the workforce so that people and AI can work together well. It's also very important for people to be able to understand what AI output means make decisions when things aren't clear, and stand up for ethics. Case studies and real-world evidence back this up. The talk has made it clear that we need to find a way to make AI systems that work with human thinking clear, easy to understand, and clear. It also means that people might see AI as a partner instead of a replacement technology.

INTRODUCTION

Technological improvements have always had an impact on the history of industrial growth. The first industrial revolution used steam and machines, the second used electricity and mass production, and the third used computers and automation. A lot

DOI: 10.4018/979-8-3373-7036-1.ch002

of people call the fourth industrial revolution (Industry 4.0) "intelligence artificial" (IA) these days. This time is different from others because it has machines that can work and systems that can learn, change, and make choices without much help from people. Intelligence Artificial is when businesses use machine learning, computer vision, natural language processing, robotics, and data-driven optimization. It is also necessary because a modern production environment makes a lot of data, like data from the supply chain, machine sensors, maintenance logs, and quality checks (Dutta, 2018). IA can help businesses get useful information from this data, guess when equipment will break down, improve operations, and even change procedures automatically when they need to. The rise of computer-aided manufacturing and early robots in the late 20th century marked the beginning of industrial artificial intelligence (Martini et al., 2024). The real reasons were "the Internet of Things (IoT), cloud computing, and better processing power." Advanced algorithms and the real-time flow of data between connected sensors and devices made it possible for businesses to move away from traditional automation and toward smart, dynamic systems. For instance, IA -based predictive maintenance is replacing planned maintenance, which saves money and cuts down on downtime. Also, machines that use IA for quality control are better at finding small flaws in products than people (AL-Shboul, 2024). But you should know that efficiency isn't the most important thing in the IA industrial age. It means that the relationship between "technology and human intelligence" is changing in a bigger way. In the past, when there were industrial revolutions, people didn't need as many manual skills, but they often needed more technical skills. In the age of IA -based industry, people need to learn how to work with smart systems. Employees are no longer just robots; they are now managers, decision-makers, and creative problem solvers who help IA meet the needs of the organization and uphold human values. This change is hard. People are worried about moral "decision-making, algorithmic bias, security threats, and the replacement of human judgment" when they use IA (Heimberger et al., 2024). To make sure that machines make people smarter instead of replacing them, businesses need to find a balance between automation and human oversight. The industrial age of IA shows how cognitive things took the place of mechanization. So, machines shouldn't just be able to do what they're told anymore. They should be able to learn, adapt, and engage in substantive conversations with individuals (Ahmed et al., 2022). Companies that use IA don't ask if robots will ever be better than people; instead, they ask how human and artificial intelligence can work together to make industrial ecosystems that are more creative, moral, and strong. This necessitates the analysis of alterations to human intelligence concerning the application of IA (Gupta et al., 2024).

UNDERSTANDING INDUSTRIAL IA

People often think of intelligence artificial (IA) as a technology of the future, but it is already being used in real life as a useful tool that has changed how businesses work today. Industrial IA is the use of IA technology in specific areas of industry, such as "supply chains, manufacturing, energy, logistics, and healthcare." Industrial IA is different from consumer-focused AI, like chatbots or recommendation engines, in that it is supervised to work in complicated physical environments, with large amounts of data, and is focused on safety, dependability, and efficiency (Peres et al., 2020). Industrial artificial intelligence (IA) is the main combination of data-driven algorithms with industrial knowledge. This speeds up decision-making and makes processes more efficient. The combination of technologies and digitalization works together with the collaboration of "Industrial Internet of Things" (IoT) and machine learning (Lee & Su, 2025). Real-time analysis helps in businesses to move from proactive and predictive and solving problems. Using sensor data like; "vibration, temperature, and sound to predict" before it is made is called predictive analysis based on artificial intelligence. The another important aspect in IA algorithms make things new and work along with supply chain to fix the problems and provide the suggestions on time (Bawack et al., 2021). In the smart factories the uses of IA brings more new changes and increases the production and fulfil the demand on time as per the availability of raw materials. IA also helps to optimize the effective planning and warehouse management which cutting the cost and tie. Industrial IA closely ties up with digital technology and increase the IA powered which simulate and test different scenarios before implementing changes in the real world (Li et al., 2025). It also reduces the risks and increases the efficacy of the work and accelerates innovation in term of; "energy companies use digital twins of power plants to test efficiency" improvements or emission-reduction strategies without disrupting actual operations (Lee et al., 2019). The benefits of industrial IA have vast implementation which also have some common challenges in term of; "data quality, interoperability of systems, cybersecurity risks", and IA integrating. On the other hand industrial IA cannot function effectively without human support. Human intelligence plays a major role in contextualizing "IA's outputs, validating predictions, and ensuring that ethical and safety standards". Industrial IA recognizing with the collaborative of "intelligence system rather than a replacement for human workers. It increases the "efficiency, reliability, and innovation" in industrial processes while relying on humans to provide "oversight, creativity, and contextual judgment" (Zhang et al., 2019).

THE HUMAN IA RELATIONSHIP AND HUMAN INTELLIGENCE VS. INTELLIGENCE ARTIFICIAL

Artificial Intelligence, however, does not consist of biological processes. It is not made up of anything else; it is made up of algorithms, data, and processing power. IA is good at sorting through a lot of information, finding patterns, and making things better. For instance, machine learning algorithms can find patterns in huge datasets that are too big for people to understand. IA can quickly and accurately look at sensor data from thousands of machines in factories at the same time (Jarrahi et al., 2022). It can also tell when maintenance will be needed or make production schedules better. But IA doesn't have self-awareness, consciousness, or an innate understanding. It can only do what it's programming and the data it gets allow it to do (Saghiri et al., 2022). The differences between human and artificial intelligence show how they can help each other (Yau et al., 2021). IA is good at math and getting things done quickly, but people are good at being creative, flexible, and honest. IA can tell that a factory machine part is likely to break down in a week, but a human engineer has to decide what to do next: stop production right away, order spare parts, or change the schedule to cause the least amount of trouble. IA can't fully weigh the economic, safety, and social trade-offs that this decision involves without human judgment (Korteling et al., 2021).

In business, the relationship between people and IA is changing from one of replacement to one of working together. "Human-in-the-loop" (HITL) systems are what this balance is called. IA gives data-driven insights in HITL frameworks, and people check them, put them in context, and make the final decisions. This is very important in fields like aviation, healthcare, and energy where mistakes could kill people (Tai, 2020). IA suggestions should follow the rules of ethics, the law, and the organization's goals. IA also changes how people act by making them smarter and helping them makes better decisions. IA -powered forecasting systems, for example, let supply chain managers see more data than they could on their own, such as global trade flows and weather patterns. Instead of having to handle data processing tasks, people can now focus on strategy, new ideas, and building relationships (Pomerol, 1997). Despite these benefits, there are still issues with people's relationship with the IA. One potential issue is that people might place too much trust in the IA without taking into account potential biases or algorithmic limitations. However, employees may be reluctant to use it if they believe that IA could replace them or if they don't trust the computers' decision-making. Understanding of technology and an atmosphere that values "accountability, transparency, and teamwork to strike' the correct balance (Ferrario et al., 2020). Instead of being in a power dynamic, humans and IA work together to improve each other. Artificial intelligence and human intelligence are very different. Artificial intelligence is based on calculation, accuracy, and size, while human intelligence depends on morality, meaning, and flexibility as mention in the table 01.

Table 1. Human Intelligence vs. Artificial Intelligence: Source: Author Contributions

Aspect	Human Intelligence	Intelligence Artificial
Origin	Evolved biologically over millions of years	Created by humans through algorithms, data, and computational systems
Nature	Conscious, self-aware, emotional, and context-driven	Unconscious, lacks emotions and self-awareness; functions based on programmed logic
Learning Ability	Learns from experience, intuition, emotions, and social interaction	Learns from data patterns, algorithms, and training sets
Creativity	Can think abstractly, imagine new possibilities, and innovate beyond known patterns	Limited to recombining existing data; cannot generate true originality
Decision-Making	Considers context, ethics, morality, and long-term consequences	Focuses on efficiency, optimization, and data-driven outcomes
Adaptability	Highly adaptable to new, uncertain, or ambiguous environments	Limited to pre-defined rules and the scope of its training data
Processing Power	Slower at handling large-scale data but excels in reasoning with incomplete information	Extremely fast at analyzing vast datasets and identifying complex patterns
Error Handling	Can make errors due to bias, fatigue, or emotions	Can make errors if data is biased or incomplete; mistakes can be systematic
Ethical/Moral Sense	Possesses empathy, values, and ethical judgment	Lacks intrinsic ethics; decisions depend on human-defined rules
Scope of Application	Broad and general across all aspects of life and industries	Narrow and specialized; excels in specific, well-defined tasks
Dependency	Independent thinking and reasoning; not reliant on external programming	Dependent on human programming, data quality, and maintenance
Social Skills	Uses language, culture, and empathy for complex social interactions	Limited to natural language processing and programmed responses without genuine understanding

INDUSTRIAL IA ADOPTION: OPPORTUNITIES AND CHALLENGES

The way businesses operate, develop new ideas, and compete has changed significantly as a result of artificial intelligence (IA). Artificial intelligence has become a transformative force as businesses strive to do more, make better use of their resources, and provide higher-quality goods and services (Hassani et al.,

2020). However, during the adoption process, both positive and negative things may happen. Companies that want to use IA safely and effectively need to be aware of these two factors.

THE ADVANTAGES OF IA IN BUSINESS

1. **Better Productivity and Efficiency:** IA has advantages for businesses since it may improve operational efficiency and streamline procedures. IA systems are able to keep an eye on manufacturing lines, spot regions where activity stops, and suggest real-time process enhancements. Predictive maintenance systems, for instance, use machine sensor data to anticipate when a machine may break down (Rana & Rathore, 2023). It saves time and money on repairs and downtime. IA helps logistics with planning routes and keeping track of inventory. This speeds up delivery and cuts down on waste.
2. **Making Choices Based on Data:** Industrial IA lets companies make choices based on data that is up to date, not just gut feelings or past trends. Advanced analytics can find patterns in large datasets that aren't easy to see. This can help with planning for demand, forecasting, and making sure the quality is good. For instance, manufacturers can change their production schedules on the fly based on things like how well machines are working, how much demand there is in the market, or problems in the supply chain. This level of accuracy reduces waste and makes the whole system work better (Bhambri et al., 2023).
3. **New Business Models and New Ideas:** Using IA lets you come up with new products, services, and business models, which is how innovation happens. IA and IoT make smart factories possible. These factories can try out different ways of making things, such as flexible production, mass customization, and digital twins. Digital twins are virtual copies of real systems that let companies try out changes to processes or equipment before they happen in the real world. This makes things safer and speeds up the process of coming up with new ideas.
4. **Safety and Risk Management:** IA systems make things safer by watching dangerous areas and guessing when things might go wrong. IA can find problems in chemical plants, power plants, or mines that people might not see. This protects workers and stops accidents from happening. IA can also model different risk scenarios, which helps businesses, get ready for things like broken equipment, supply chain problems, or changes in the market.

CHALLENGES OF INDUSTRIAL IA ADOPTION

1. Displacement of Workers and Skills Gap: IA can make things run more smoothly, but it also makes people worry that they will lose their jobs. Workers may be worried that more and more routine and manual tasks are being automated. There is also a growing need for people who can use IA, analyze data, and work with IA. Businesses should pay for programs that help their employees learn new skills and get better at the ones (Park, 2025).
2. **Data Quality and Integration:** IA works best when the data is accurate. In factories, data often comes from a lot of different places, like old machines, IoT sensors, and business systems. Bad or incomplete data can lead to wrong predictions, operational issues, and costly mistakes (Wang et al., 2021). From a technical point of view, it's also hard to add IA to existing systems because it needs careful planning and a strong IT architecture.
3. **Ethical and Regulatory Concerns:** The use of IA brings up moral questions, especially when it comes to being honest, responsible, and making choices. If, for example, predictive models are trained on historical data that shows past inequalities, they may unintentionally reinforce biases. There may be very strict rules about data privacy, safety standards, and reporting in some parts of the world. Businesses have to deal with these problems to make sure they follow the law and do the right thing.
4. **Over-Reliance and Systemic Risks:** Putting too much faith in IA systems can make the whole system weaker. The activities may encounter significant challenges in the event of algorithmic defects, cyberattacks, or programming problems. People that rely too much on IA results could not regularly review them, which could lead to a loss of confidence in such results (Jagatheesaperumal et al., 2022). It is essential to implement a "human-in-the-loop" approach, which entails human supervision and evaluation of IA choices, in order to guarantee resilience and accountability.
5. **Cost and How to Make It Happen Difficult:** Artificial intelligence-based solutions may be quite expensive, especially for small or medium-sized businesses. The expenses cover the purchase of technology, its integration with current systems, staff training, and the assurance of smooth operation. Organizations may not fully utilize the potential of artificial intelligence if they do not carefully prepare and do not employ a phase-by-phase implementation strategy. The application of artificial intelligence in industry makes it possible to discover new ways to work more productively, develop innovations, and increase security. At the same time, it supports systemic risk management, data management, staff changes, and ethical norms. Finding a reasonable balance between taking advantage of the IA 's benefits and demonstrating alertness, investing in employee training,

ensuring adherence to ethical standards, and maintaining a strong infrastructure are all crucial to getting the most out of it. In the industrial era dominated by IA, the sectors that manage these risks and opportunities well will emerge as the best.

THE ROLE OF HUMAN INTELLIGENCE IN IA-DRIVEN INDUSTRIES

As artificial intelligence becomes more prevalent in factories, manual labor is becoming obsolete; instead, intelligent systems are being used for "guidance, supervision, and collaboration". Artificial intelligence is very useful for managing enormous amounts of data, improving work flow, and providing predictions. However, it lacks a natural understanding of concepts, esoteric reasoning, and the ability to make "situational decisions. Intuition, human creativity, and moral consciousness" are essential for ensuring that artificial intelligence programs are safe, effective, and aligned with commercial goals. People in the loop systems: One of the best ways that human intelligence helps IA -driven enterprises is through "Human-in-the-loop" (HITL) systems. People are responsible for making sure everything is done right, reviewing the work, and making decisions. IA handles the laborious tasks of these systems. IA, for instance, can simultaneously monitor hundreds of machines in a plant to identify problems. On the other hand, it is up to individuals to decide whether these problems need to be fixed right away or if they can wait until the next scheduled maintenance. This partnership reduces risks, ensures accuracy, and avoids relying too much on automated procedures. People can also add context to IA -driven tasks with HITL systems. IA can find patterns in production data, but it doesn't always get what's going on outside the company, like when suppliers are late, there are labor disputes, or the market changes suddenly. People use their brains to figure out what these contextual clues mean and make smart choices that go beyond just looking at the numbers. This helps the business stay on track with its goals and keeps things running smoothly.

IA systems are very good at finding patterns in structured datasets, but they have a hard time with situations that are new or not clear. People can change the way they think thanks to their intelligence, which lets them come up with new ways to solve problems they didn't see coming. In sectors like "energy, aircraft, and pharmaceuticals", this flexibility is essential. Situations that deviate from accepted paradigms are common for managers and engineers. IA's computational powers are improved by its capacities to synthesize knowledge, apply reasoning across domains, and execute creative solutions. Although IA can suggest improvements to procedures based on their past performance, it is unable to forecast how supply chains will be impacted by abrupt regulatory changes or geopolitical events. To keep the business strong and competitive, staff members evaluate the IA 's suggestions and make the necessary strategic changes.

1. **Making choices and keeping an eye on morals:** The tasks specified by their programmers and training data are all that the IA systems can accomplish. The data collections may contain bias or other unintended influences. The involvement of individuals is essential to ensuring that the IA's decisions are fair, open, and consistent with legal and social norms. In fields like; "health", where IA can help with "diagnosis or treatment planning", human oversight is crucial to preventing ethical mistakes that could harm patients or undermine trust. Human education helps us hold people accountable (Al Badi et al., 2022).
2. **Working together and improving skills:** Artificial intelligence helps to improve human intelligence by adding skills rather than removing them. Because of their understanding of the information based on data, people are able to perform more complex mental tasks, such as planning, designing, and solving problems. For instance, the IA may discover people's desires and identify the best routes for the transportation of goods. However, human managers use this information to establish relationships, reach agreements, and make decisions that artificial intelligence cannot measure. In order to achieve better results, this collaboration combines the accuracy of the calculations with human judgment in IA sociologists.
3. **Changes in the workforce:** Companies that spend money on training their employees not only make it easier for IA to be used, but they also give their workers the skills they need to make a difference in a workplace that uses IA. Human intelligence is still important in fields that use IA, and it's even more important now. IA can't do things like keeping an eye on things, understanding context, making moral decisions, or coming up with creative solutions to problems. People work with IA to make sure that operations are correct, strategies are strong, and morals are upheld (Ware & Grantham, 2003). The best industrial ecosystems will be those that see IA and human intelligence as two sides of the same coin. They will use computers to their full potential and also make the most of what makes humans unique. IA handles the hard parts of this partnership, like size and complexity. Human intelligence makes sure that everything is relevant, moral, and flexible. This will make the future of industry strong, creative, and responsible (Bard & Purnomo, 2006).

WORKFORCE AND SKILLS TRANSFORMATION

The use of intelligence artificial (IA) in factories brings the major changes in the technology and the people who work together. IA makes the routine tasks effectively and increases the efficiency in the operations and also improves the skill in human which brings the changing in the operations.

1. **New Skills for Industrial Workers:** The usage of IA in Industries trained the employee their technical skill, analytical, and cognitive tasks (Goglio, 2025). To understanding the new IA systems and robotics technologies, and IoT integration are all examples of technical skills where all the employees learn the fundamental of IA tools and IA outputs where fix the different problems with automated systems. To apply the analytical skills is another important task where workers evaluate IA -generated insights, spot outliers, and make decisions based on both quantitative and qualitative data. IA applies the better tools and techniques when it works with the collaboration of human beings. IA can also bring the improvement in the routes and inventory management system in supply chain management.
2. **Strategies for lifelong learning and retraining:** Reskilling programs are especially important for workers whose jobs are greatly affected by automation. For instance, people who work on production lines might become IA supervisors, maintenance analysts, or data interpreters. Upskilling programs not only give workers the technical skills they need, but they also boost morale and make people less likely to resist IA adoption by showing them a clear path to career growth (Arutyunova et al., 2022).
3. **Changing Roles in the Organization:** As IA takes over tasks that are repetitive and require a lot of math, people's jobs are changing to include overseeing, working with others, and making strategic decisions. More and more, employees are working with IA by guiding algorithms, checking outputs, and making sure that decisions are ethical and based on the right information. This change turns businesses into human– IA ecosystems, where machines and people must work together smoothly for the business to be successful. Leadership is also very important for changing the workforce. Leaders need to share their vision for how IA will be used, explain what new skills employees will need, and show employees how to adapt to IA as a helpful tool instead of a threat (Dess & Picken, 2000). Policies within organizations should encourage collaboration between different fields, bringing together ideas from engineering, data science, operations, and human resources to get the most out of IA while still allowing people to make their own decisions.

4. **Problems with Changing the Workforce:** Changing the workforce can be hard, even though there are opportunities. People who don't want to change, who don't have the right skills, and who don't have equal access to training materials can all slow down the adoption of IA. Also, IA technologies may change so quickly that employees can't keep up with new skills, which means that both businesses and governments will need to keep spending money on them. To successfully use industrial IA, you need to change the skills and the workforce. Companies need to encourage continuous learning, reskilling programs, and adaptive leadership, while employees need to learn new technical, analytical, and cognitive skills to work with IA systems (Ozkan-Ozen & Kazancoglu, 2022). In IA -driven workplaces, the industrial workforce is moving away from doing the same tasks over and over again and toward working together, supervising, and making strategic decisions. Industries that accept this change will not only be able to use AI to its fullest potential, but they will also build workforces that are strong, creative, and ready for the future, so they can thrive in the changing industrial landscape.

CASE STUDIES MANUFACTURING: HUMAN-IA COLLABORATION ON THE FACTORY FLOOR

1. Priestley's Gourmet Delights (PGD) is an Australian bakery that uses IA:

Priestley's Gourmet Delights is a bakery in Queensland that uses IA and robots to change the way it makes things. The company spent $53 million on a state-of-the-art smart factory with self-driving cars and robots that can work together. These technologies have not only doubled the bakery's production capacity, but they have also freed up people to do more skilled tasks like quality control and recipe development. This change has led to more jobs and a big drop in greenhouse gas emissions. This is because solar panels and electric ovens that use less energy have been installed.

2. BMW and Figure Robotics: Improving the Production line:

BMW has worked with Figure Robotics, a California-based company, to put a group of humanoid robots on their production lines. These robots that look real have gotten a lot better; their speed has gone up by 400%. They can now put 1,000 car parts in the right place every day with amazing accuracy. The goal of the partnership is to make a fleet of fully autonomous robots that can do hard factory work. This will allow human workers to do less demanding tasks. The first fleet was successfully tested in Munich, Germany, in August. There are plans for more tests with BMW in

January (Kochan, 2005). The industrial landscape is changing a lot as intelligence artificial (IA) becomes more common in areas like production, logistics, energy, healthcare, and more. Most of the time, IA is used to make things run more smoothly, guess when things will break, and improve processes. In the future, industrial IA will focus on collaborating with human intelligence, adhering to ethical standards, and operating on a global scale. These directions are important for businesses that want to stay strong and competitive in the changing industrial age.

1. **Moving toward smart systems that can work together**: The next step for industrial IA is to make symbiotic intelligence systems, where people and IA work together to make decisions, solve problems, and come up with new ideas. Symbiotic intelligence emphasizes co-creation, contrasting with existing systems that regard individuals as supervisors or validators. IA gives us the ability to process data, find patterns, and make predictions. People give us creativity, intuition, and an understanding of the situation (Giese et al., 2015). This partnership will help businesses solve very hard industrial problems that neither people nor machines could solve on their own. For example, IA could change production schedules in smart factories in real time, while people change their plans based on new rules, changes in the market, or things that happen that they didn't expect (Horváth, 2021).
2. **Responsible and Ethical** IA **Use:** As IA becomes more important in business, making sure it is used in a responsible and ethical way will be very important. Fairness, openness, and responsibility must be built into IA systems of the future. This means that choices must follow the rules of society and the law. People will still need to keep an eye on things, especially in areas where a lot is at stake, like healthcare, energy, and defense. Companies will probably use standard ethical frameworks, IA auditing practices, and ways to make sure they are following the rules to stop bias, discrimination, and misuse (Peters et al., 2020). Also, making sure that IA systems can be explained will help people understand, check, and trust what IA does, which will make it easier to work together and lower the risks of running a business.
3. **Evolution of the Workforce and Human-Centric Skills:** The future industrial workforce will have human-centric skills that make IA work better. You will still need to know a lot about IA, data analysis, and robotics. But you'll also need to have soft skills like creativity, critical thinking, being able to change, and being able to make moral decisions (Flores et al., 2020). It will be very important to keep teaching and retraining workers so they can move from jobs that are simple or repetitive to jobs that need strategic oversight, working together to solve problems, and IA supervision. Human intelligence will be very important

to make sure that IA is used in a way that is effective, moral, and in line with the organization's goals.

4. **Global Views and Scalability:** How widely industrial IA is used will also depend on global views and scalability. Because each part of the world has its own problems with technology, rules, and workers, IA will be used in different ways. For example, developing countries might focus on IA to get the most out of their resources and save money. On the other hand, developed countries might focus on ethical integration, new ideas, and better ways for people and IA to work together. IA solutions greater while still being "safe, reliable, and ethical", people from different countries will need to work together, share information, and set global standards for IA.
5. **New ideas through combining:** Instead of relying on isolated installations, the future of artificial intelligence in industry will involve connecting all of the facilities inside the industrial ecosystem. Artificial intelligence will collaborate with blockchain, digital assets, augmented reality, and the Internet of Things (IoT) to develop industrial networks that are completely connected, robust, and flexible. With this kind of integration, people can keep guiding, overseeing, and coming up with new ideas for these complex systems. Additionally, they will be able to make decisions instantly, use predictive analysis, and demonstrate flexibility in their work style. The use of artificial intelligence in industry will be more ethical, human-centered, and globally connected in the future. The next generation of artificial intelligence companies will be built on simbiotic intelligence systems, a transformational labor force, ethical supervision, and integrated industrial ecosystems. The most successful companies will be those who are able to integrate the new technology with the existing knowledge.

CONCLUSION

A significant shift in business operations, innovation, and competition can be seen in the ever-evolving role of human intelligence in the industrial application of IA. This study examined the impact of IA technologies, such as automated learning, robotics, the Internet of Things (IoT), and digital tools, on the evolution of industrial operations. It has also emphasized the significance of human intelligence in ensuring an efficient, moral, and flexible use of IA.

The findings demonstrate that while IA augments human capabilities, it does not take the place of humans in tasks that require human supervision, judgment, and creativity—qualities that machines cannot supply on their own. The study found that human-in-the-loop (HITL) systems are very important. In sectors including manufacturing, logistics, healthcare, and energy, IA functions best when humans maintain decision-making authority, according to HITL frameworks. Although IA can process enormous volumes of data, streamline processes, and forecast future events, it is unable to understand the broader context, think through moral quandaries, or adjust when errors are made. These gaps are filled by human intelligence, which bestows creativity, critical thinking skills, the ability to make moral decisions, and subject-matter expertise. The human-IA relationship creates a cooperative ecosystem in which each party's forces are mutually reinforced. It also illustrates how artificial intelligence is changing the jobs and abilities people need to do in their line of work. The need for people with high cognitive, technological, and ethical abilities has grown as repetitive tasks have been automated. People are leaving traditional operational positions to work with IA, make strategic decisions, and solve problems. This change illustrates the need for workers to continue education throughout their life, gain new skills, and adapt to workplaces where IA is used. Businesses are better at using IA in a way that preserves employee engagement and happiness when they take the initiative to adapt their workforces.

Numerous opportunities are presented by the use of KI in industry, including the creation of novel ideas, the execution of safer and higher-quality work, increased productivity, and proactive warfare. However, it also creates challenges including poor data quality, cyber threats, esoteric challenges, lack of use, and implementation issues. For adoption to be successful, it is essential to take into account both technology and people. In order to reduce risks and increase benefits, policies that prioritize ethical supervision, transparency, and responsibility are needed, along with staff development programs. The study emphasizes that one of the key characteristics of the industries driven by artificial intelligence in the future will be intelligence.

REFERENCES

Ahmed, I., Jeon, G., & Piccialli, F. (2022). From Artificial Intelligence to Explainable Artificial Intelligence in Industry 4.0: A Survey on What, How, and Where. *IEEE Transactions on Industrial Informatics*, *18*(8), 5031–5042. DOI: 10.1109/TII.2022.3146552

Al Badi, F. K., Alhosani, K. A., Jabeen, F., Stachowicz-Stanusch, A., Shehzad, N., & Amann, W. (2022). Challenges of AI Adoption in the UAE Healthcare. *Vision: The Journal of Business Perspective*, *26*(2), 193–207. DOI: 10.1177/0972262920988398

AL-Shboul, M. A.AL-Shboul. (2024). Artificial intelligence drivers' effect on willingness to adopt the human capital supply chain in manufacturing firms: An empirical investigation from developing countries – a mediation model. *Industrial Management & Data Systems*, *124*(10), 2919–2938. DOI: 10.1108/IMDS-12-2023-0977

Arutyunova, A. E., Khanseviarov, R. I., Okunkova, E. A., Alieva, P. R., & Makareva, E. A. (2022). Youth, Work and Skills: A Map of the Transformation of the Workforce and Employment. In Ostrovskaya, V. N., & Bogoviz, A. V. (Eds.), *Big Data in the GovTech System* (Vol. 110, pp. 149–160). Springer International Publishing., DOI: 10.1007/978-3-031-04903-3_18

Bard, J. F., & Purnomo, H. W. (2006). Incremental changes in the workforce to accommodate changes in demand. *Health Care Management Science*, *9*(1), 71–85. DOI: 10.1007/s10729-006-6281-y PMID: 16613018

Bawack, R. E., Fosso Wamba, S., & Carillo, K. D. A. (2021). A framework for understanding artificial intelligence research: Insights from practice. *Journal of Enterprise Information Management*, *34*(2), 645–678. DOI: 10.1108/JEIM-07-2020-0284

Bhambri, P., Rani, S., Balas, V. E., & Elngar, A. A. (2023). *Integration of AI-Based Manufacturing and Industrial Engineering Systems with the Internet of Things* (1st ed.). CRC Press., DOI: 10.1201/9781003383505

Dess, G. G., & Picken, J. C. (2000). Changing roles: Leadership in the 21st century. *Organizational Dynamics*, *28*(3), 18–34. DOI: 10.1016/S0090-2616(00)88447-8

Dutta, S. (2018). An overview on the evolution and adoption of deep learning applications used in the industry. *Wiley Interdisciplinary Reviews. Data Mining and Knowledge Discovery*, *8*(4), e1257. DOI: 10.1002/widm.1257

Ferrario, A., Loi, M., & Viganò, E. (2020). In AI We Trust Incrementally: A Multi-layer Model of Trust to Analyze Human-Artificial Intelligence Interactions. *Philosophy & Technology*, *33*(3), 523–539. DOI: 10.1007/s13347-019-00378-3

Flores, E., Xu, X., & Lu, Y. (2020). A Reference Human-centric Architecture Model: A skill-based approach for education of future workforce. *Procedia Manufacturing*, *48*, 1094–1101. DOI: 10.1016/j.promfg.2020.05.150

Giese, H., Vogel, T., & Wätzoldt, S. (2015). Towards Smart Systems of Systems. In Dastani, M., & Sirjani, M. (Eds.), *Fundamentals of Software Engineering* (Vol. 9392, pp. 1–29). Springer International Publishing., DOI: 10.1007/978-3-319-24644-4_1

Goglio, V. (2025). A Better Skilled Workforce for the Digital Transformation: What Role for Platform Education? In Goglio, V., & Biancalana, C. (Eds.), *Mapping the Evolution of Platform Society* (pp. 27–46). Emerald Publishing Limited., DOI: 10.1108/978-1-83608-028-220251003

Gupta, R., Nair, K., Mishra, M., Ibrahim, B., & Bhardwaj, S. (2024). Adoption and impacts of generative artificial intelligence: Theoretical underpinnings and research agenda. *International Journal of Information Management Data Insights*, *4*(1), 100232. DOI: 10.1016/j.jjimei.2024.100232

Hassani, H., Silva, E. S., Unger, S., TajMazinani, M., & Mac Feely, S. (2020). Artificial Intelligence (AI) or Intelligence Augmentation (IA): What Is the Future? *AI*, *1*(2), 143–155. DOI: 10.3390/ai1020008

Heimberger, H., Horvat, D., & Schultmann, F. (2024). Exploring the factors driving AI adoption in production: A systematic literature review and future research agenda. *Information Technology Management*. Advance online publication. DOI: 10.1007/s10799-024-00436-z

Horváth, I. (2021). Connectors of smart design and smart systems. *Artificial Intelligence for Engineering Design, Analysis and Manufacturing*, *35*(2), 132–150. DOI: 10.1017/S0890060421000068

Jagatheesaperumal, S. K., Rahouti, M., Ahmad, K., Al-Fuqaha, A., & Guizani, M. (2022). The Duo of Artificial Intelligence and Big Data for Industry 4.0: Applications, Techniques, Challenges, and Future Research Directions. *IEEE Internet of Things Journal*, *9*(15), 12861–12885. DOI: 10.1109/JIOT.2021.3139827

Jarrahi, M. H., Lutz, C., & Newlands, G. (2022). Artificial intelligence, human intelligence and hybrid intelligence based on mutual augmentation. *Big Data & Society*, *9*(2), 20539517221142824. DOI: 10.1177/20539517221142824

Kochan, A. (2005). BMW uses even more robots for both flexibility and quality. *The Industrial Robot*, *32*(4), 318–320. DOI: 10.1108/01439910510600173

Korteling, J. E., van de Boer-Visschedijk, G. C., Blankendaal, R. A. M., Boonekamp, R. C., & Eikelboom, A. R. (2021). Human- versus Artificial Intelligence. *Frontiers in Artificial Intelligence*, *4*, 622364. DOI: 10.3389/frai.2021.622364 PMID: 33981990

Lee, J., Singh, J., & Azamfar, M. (2019). *Industrial Artificial Intelligence* (Version 3). arXiv. DOI: 10.48550/ARXIV.1908.02150

Lee, J., & Su, H. (2025). Rethinking industrial artificial intelligence: A unified foundation framework. *International Journal of AI for Materials and Design*, *2*(2), 56. DOI: 10.36922/IJAMD025080006

Li, S., Meng, W., Liu, C., Long, C., & He, S. (2025). S[4] FD: Self-Supervision-Enhanced Semisupervised Fault Diagnosis for Complex Industrial Processes. *IEEE Transactions on Industrial Informatics*, *21*(5), 3585–3594. DOI: 10.1109/TII.2024.3523590

Martini, B., Bellisario, D., & Coletti, P. (2024). Human-Centered and Sustainable Artificial Intelligence in Industry 5.0: Challenges and Perspectives. *Sustainability (Basel)*, *16*(13), 5448. DOI: 10.3390/su16135448

Ozkan-Ozen, Y. D., & Kazancoglu, Y. (2022). Analysing workforce development challenges in the Industry 4.0. *International Journal of Manpower*, *43*(2), 310–333. DOI: 10.1108/IJM-03-2021-0167

Park, C. (2025). Addressing Challenges for the Effective Adoption of Artificial Intelligence in the Energy Sector. *Sustainability (Basel)*, *17*(13), 5764. DOI: 10.3390/su17135764

Peres, R. S., Jia, X., Lee, J., Sun, K., Colombo, A. W., & Barata, J. (2020). Industrial Artificial Intelligence in Industry 4.0—Systematic Review, Challenges and Outlook. *IEEE Access : Practical Innovations, Open Solutions*, *8*, 220121–220139. DOI: 10.1109/ACCESS.2020.3042874

Peters, D., Vold, K., Robinson, D., & Calvo, R. A. (2020). Responsible AI—Two Frameworks for Ethical Design Practice. *IEEE Transactions on Technology and Society*, *1*(1), 34–47. DOI: 10.1109/TTS.2020.2974991

Pomerol, J.-C. (1997). Artificial intelligence and human decision making. *European Journal of Operational Research*, *99*(1), 3–25. DOI: 10.1016/S0377-2217(96)00378-5

Rana, B., & Rathore, S. S. (2023). Industry 4.0 – Applications, challenges and opportunities in industries and academia: A review. *Materials Today: Proceedings*, *79*, 389–394. DOI: 10.1016/j.matpr.2022.12.162

Saghiri, A. M., Vahidipour, S. M., Jabbarpour, M. R., Sookhak, M., & Forestiero, A. (2022). A Survey of Artificial Intelligence Challenges: Analyzing the Definitions, Relationships, and Evolutions. *Applied Sciences (Basel, Switzerland)*, *12*(8), 4054. DOI: 10.3390/app12084054

Tai, M.-T. (2020). The impact of artificial intelligence on human society and bioethics. *Tzu-Chi Medical Journal*, *32*(4), 339. DOI: 10.4103/tcmj.tcmj_71_20 PMID: 33163378

Wang, K., Zhao, Y., Gangadhari, R. K., & Li, Z. (2021). Analyzing the Adoption Challenges of the Internet of Things (IoT) and Artificial Intelligence (AI) for Smart Cities in China. *Sustainability (Basel)*, *13*(19), 10983. DOI: 10.3390/su131910983

Ware, J., & Grantham, C. (2003). The future of work: Changing patterns of workforce management and their impact on the workplace. *Journal of Facilities Management*, *2*(2), 142–159. DOI: 10.1108/14725960410808177

Yau, K.-L. A., Lee, H. J., Chong, Y.-W., Ling, M. H., Syed, A. R., Wu, C., & Goh, H. G. (2021). Augmented Intelligence: Surveys of Literature and Expert Opinion to Understand Relations Between Human Intelligence and Artificial Intelligence. *IEEE Access : Practical Innovations, Open Solutions*, *9*, 136744–136761. DOI: 10.1109/ACCESS.2021.3115494

Zhang, X., Ming, X., Liu, Z., Yin, D., Chen, Z., & Chang, Y. (2019). A reference framework and overall planning of industrial artificial intelligence (I-AI) for new application scenarios. *International Journal of Advanced Manufacturing Technology*, *101*(9–12), 2367–2389. DOI: 10.1007/s00170-018-3106-3

Chapter 3
Human-AI Cognitive Collaboration in Industrial Contexts:
A Managerial Framework for Assessment, Execution, and Governance - From Theory to Sustainable Practice

Tamrisha Patnaik
http://orcid.org/0000-0002-6961-2203
Woxsen University, India

Nitya Sundar Nanda
VJIM, India

ABSTRACT

TThis chapter provides a comprehensive, theory-driven framework for understanding and executing cognitive collaboration between humans and artificial intelligence in industrial contexts. Grounded in socio-technical systems theory, cognitive augmentation theory, and organizational learning theory.The chapter rethinks AI as a cognitive partner that helps people do their jobs better and changes how organizations work, rather than as a technology that takes over. It is founded on socio-technical systems theory, cognitive augmentation theory, and organizational learning theory. The framework identifies three essential pillars in critical sectors such as healthcare, manufacturing, banking, and logistics: collaborative decision-making ecosystems, adaptive learning systems, and cognitive complementarity in decision-making. The chapter presents comprehensive assessment methodologies that

DOI: 10.4018/979-8-3373-7036-1.ch003

utilize both qualitative indicators (including employee engagement, psychological safety, and innovation culture) and quantitative indicators (such as productivity gains and return on investment).

1. INTRODUCTION

Artificial intelligence has quickly moved from supporting roles to being central in today's industries (McKinsey & Company, 2023; Russell & Norvig, 2021). In manufacturing, predictive systems help with maintenance and production. In healthcare, AI-assisted diagnostics aid clinical evaluation (Google Health, 2024). In finance, algorithms support real-time trading and risk analysis (JPMorgan Chase & Co., 2023). In logistics, optimization platforms manage complex supply chains (FourKites, 2024). In all these areas, AI now plays a key part in both daily operations and long-term planning.

Even with these advances, bringing AI into daily work creates management challenges that traditional performance measures do not fully capture (Davenport & Ronanki, 2018). Organizations often judge AI by productivity, efficiency, speed, and cost savings. While these show clear benefits, they do not tell us if working with AI actually improves decision quality (Jarrahi, 2018; Shneiderman, 2022). AI can speed up decisions, but it might also weaken the human judgment needed to review and interpret its suggestions (Raisch & Krakowski, 2021). As a result, organizations may gain short-term efficiency but lose long-term expertise, adaptability, and learning. Balancing productivity with lasting decision quality is a key management issue in today's rapidly changing environment.

Working with AI is more than just using new technology or software (Wilson & Daugherty, 2018). It is an organizational challenge that involves combining human judgment, machine intelligence, trust, and governance (Makarius et al., 2020). Successful AI integration relies not only on technology but also on how well organizations encourage people and AI systems to work together (Amershi et al., 2019; Xu et al., 2021). In most workplaces, AI does not replace human decision-making. Instead, people and AI act as partners in shared decision-making. AI brings speed, data processing, predictions, and pattern recognition, while humans provide context, ethics, creativity, and flexible judgment (Jarrahi, 2018).

The main management challenge is not choosing between humans or AI making decisions alone, but designing ways for both to work together and improve each other's strengths. This chapter looks at how to make human–AI collaboration effective by exploring three related problems.

The first challenge is finding the right balance between making good decisions and working quickly. If employees always follow AI recommendations without checking them, they may slowly lose the expertise needed to judge those decisions well (Raisch & Krakowski, 2021). Over time, this can make organizations more efficient but also more dependent on automation and less skilled in human judgment. The key question is how organizations can make sure that working with AI actually improves decision quality, not just speeds things up.

The second challenge is about trust between people and AI systems. For collaboration to work well, trust should align with the AI's reliability (Lee & See, 2004; Siau & Wang, 2018). Too much trust can lead to unquestioning acceptance of AI results, while too little trust can prevent employees from using AI effectively. The challenge is how organizations can keep adjusting the level of trust people have in AI as they work together.

The third challenge is about accountability when people and AI make decisions together. As more results come from both employees and AI systems working together, it gets harder to know who is responsible for mistakes or failures. This uncertainty poses ethical, legal, and management risks, especially in critical sectors such as healthcare, finance, and government (European Commission, 2021; US NIST, 2023). The challenge is how organizations can establish clear rules for accountability while maintaining strong human oversight.

To tackle these challenges, this chapter presents a single framework for managing how people and AI work together. It uses ideas from socio-technical systems theory (Baxter & Sommerville, 2011; Trist & Bamforth, 1951), distributed cognition (Hutchins, 1995), and governance research (Floridi & Cowls, 2019; OECD, 2019). The chapter sees AI as a tool that can boost human judgment when used wisely, not as a replacement for people. It shows that successful collaboration depends more on how organizations are set up than on the technology itself. This is shown through examples of measurement systems, evaluation models, implementation plans, and governance structures.

1.2 Theoretical Foundations: Three Lenses, One Framework

The framework uses three main theories to address the key challenges of human–AI collaboration. Socio-Technical Systems (STS) theory says that focusing only on technology and ignoring human roles, skills, and decision-making can cause AI projects to fail (Baxter & Sommerville, 2011; Trist & Bamforth, 1951). This theory stresses the need to redesign workflows and responsibilities alongside AI adoption (Makarius et al., 2020; Venkatesh et al., 2010). Distributed Cognition and Cognitive Augmentation theory sees intelligence as shared between people and AI, so good collaboration depends on the right balance of trust and skills (Hutchins, 1995). The centaur model is an example of this, combining human judgment with machine intelligence (Wilson & Daugherty, 2018). Managerial Governance theory emphasizes the importance of accountability, transparency, and ongoing learning in environments where AI supports decision-making (Floridi & Cowls, 2019; OECD, 2019).

Collectively, these theories emphasize the importance of a chapter's main idea: inman–AI collaboration decision-making to succeed, organizations need to combine complementary skills, adjust trust as needed, and build strong governance systems.

1.3 Chapter Architecture: A Unified Argument in Eleven Sections

The eleven sections of this chapter work together as one integrated argument. The discussion starts by identifying the problem, then moves through theory, practical use, measurement, evaluation, implementation, validation, and governance. Table 1 shows how each section addresses a specific challenge and introduces a part of the framework.

Table 1. Chapter Architecture: How Each Section Connects to the Unified Framework

Section	Core Question Answered	Challenge(s) Addressed	Framework Element Introduced	Output / Deliverable
1 Introduction	Why does this matter?	Identifies Challenges 1, 2, and 3	Problem framing and theoretical grounding	Research agenda
2.Centaur Model	What is the ideal partnership?	Provides the answer to all three challenges through the 3 pillars	Cognitive Collaboration Architecture	Operational principles
3.Industry Frameworks	How do the pillars work in practice?	Ch. 1: shows complementarity; Ch. 2: calibrates trust; Ch. 3: defines decision rights	Sector-specific collaboration maps	4 application tables
4. Measurement	How do we know it is working?	Quantitative: Ch.1 & 2; Qualitative: Ch.2 & 3	KPI framework (Table 7)	12 diagnostic KPIs
5.IMPACT+ & BSC	How do we evaluate holistically?	Integrates all KPI data; links governance to strategy	IMPACT+ model & 360° BSC	Strategic scorecard
6.Readiness	Are we organisationally prepared?	Prerequisites for resolving Ch. 3 (governance) and Ch. 1 & 2 (capability)	4-Dimensional Readiness Diagnostic	Readiness checklist
7. Roadmap	How do we sequence implementation?	Phase 1→Ch.3; Phase 2→Ch.1&2; Phase 3→sustained resolution	12–24 month phased pathway	Implementation plan

continued on following page

Table 1. Continued

Section	Core Question Answered	Challenge(s) Addressed	Framework Element Introduced	Output / Deliverable
8. Case Studies	What does success look like?	Validates framework resolution of all three challenges	Three industry evidence studies	Validated outcomes
9. Ethics & Governance	How do we safeguard the partnership?	Formally resolves the accountability gap	Governance architecture	Ethical safeguards
10–11. Future & Conclusion	Where must scholarship and practice go next?	Identifies remaining gaps in framework coverage	Research agenda & integrated synthesis	Unified framework map

Source: Author's Compilation

Table 1 helps readers follow three main themes. The automation bias problem is explored in theory in section 2, applied in section 3, measured in section 4, evaluated in section 5, prepared for in section 6, implemented in section 7, and validated in section 8. Trust calibration follows the same steps. The accountability gap is mainly resolved in section 9, but is also prepared for in section 6 and built into section 5. Section 11 brings all three themes together. With this structure in place, the chapter introduces the conceptual model that enables collaboration.

2. THE CENTAUR MODEL: THE ARCHITECTURE OF EFFECTIVE COLLABORATION

2.1 Origins, Evidence, and Core Insight

The centaur model emerged from freestyle chess competitions where human players collaborated with chess engines. Instead of replacing human expertise, hybrid human–AI teams outperformed both grandmasters and AI systems working independently (Wilson & Daugherty, 2018). Humans contributed contextual reasoning, strategic judgment, and intuition, while AI systems provided computational speed, tactical analysis, and pattern recognition. These findings demonstrated that collaboration between human and machine intelligence produces superior outcomes when both contribute complementary strengths.

The model has since been validated across industries. In healthcare, radiologists supported by AI achieve higher diagnostic accuracy than either humans or AI alone (Topol, 2019). In finance, trader–algorithm partnerships improve risk-adjusted returns, while in research and drug discovery, human–AI teams accelerate innovation and identify insights more effectively (Wilson & Daugherty, 2018). These examples establish the centaur model as an evidence-based framework for collaborative intelligence (Jarrahi, 2018). The model is built on three interconnected pillars that address the challenges of decision quality, trust calibration, and accountability.

- Complementary Intelligence ensures an effective division of cognitive labor, with AI handling analytical tasks and humans providing contextual and ethical judgment.
- Adaptive Learning Systems enable continuous improvement through human feedback, corrections, and AI-assisted capability development.
- Shared Decision Ecosystems establish transparency, explainability, human override authority, and auditable decision processes, ensuring accountability remains embedded within collaboration.

Together, these pillars form the theoretical foundation for the sector frameworks, measurement systems, and governance structures discussed in subsequent sections. (Refer Table-2)

Table 2. The Centaur Collaboration Framework:Three Pillars, Three Challenge Resolutions

Pillar	Meaning	AI Contribution	Human Contribution	Collaboration Advantage	Challenge Resolved
Complementary Intelligence	Combine asymmetric strengths for superior outcomes	Pattern recognition at scale; computational consistency; speed	Creative synthesis; contextual reasoning; ethical judgment	Results exceed working alone by 20–40%	Challenge 1: Over-reliance is prevented when each partner does what only they can do
Adaptive Learning Systems	Mutual improvement through continuous feedback loops	Self-corrects based on human override decisions and outcome data	Identifies AI blind spots; develops AI literacy and critical thinking	Compounding performance gains; prevents expertise decay	Challenge 2: Trust calibration improves as humans learn AI strengths and limits
Shared Decision Ecosystems	Transparent, auditable, jointly accountable decision architecture	Surfaces ranked options with confidence scores and reasoning traces	Validates against ethical, contextual, and practical standards	Faster, smarter, and more accountable decisions	Challenge 3: Accountability is embedded in the decision architecture itself

Source: Author's Compilation

With the three pillars established, the next step is to see how they work across different industries and which collaboration designs actually succeed. Section 3 examines four detailed sector frameworks, showing how the pillars connect to the human and AI skills, decision-making, and governance required in manufacturing, healthcare, finance, and logistics.

3. INTEGRATED HUMAN–AI COLLABORATION FRAMEWORK ACROSS INDUSTRIAL CONTEXTS

The centaur model manifests differently across industries due to operational risk, decision complexity, and regulatory requirements. Human and AI collaboration in manufacturing, healthcare, finance, and logistics illustrates how AI supports analytical and computational tasks while humans provide contextual judgment and oversight (Jarrahi, 2018; Wilson & Daugherty, 2018).

In manufacturing, AI-driven predictive maintenance systems analyze sensor data to spot equipment failures before breakdowns happen (Baur et al., 2024; Siemens AG, 2024). Technicians interpret operational conditions and validate recommendations. In healthcare, AI-assisted diagnostic systems quickly process medical data (IBM Watson Health, 2023; Topol, 2019). Clinicians offer ethical reasoning, evaluate patient context, and hold final decision-making authority. In finance, AI systems analyze real-time market data and predictive signals (JPMorgan Chase & Co., 2023). Analysts and traders interpret geopolitical uncertainty, market behavior, and strategic implications. In logistics, AI systems optimize routing, scheduling, and inventory coordination while human planners handle disruptions, local adaptation, and strategic prioritization (FourKites, 2024; project44, 2024).

Common collaboration patterns appear across these sectors. AI systems excel at computational and data-heavy tasks, while humans add contextual reasoning, ethical judgment, and adaptability (Jarrahi, 2018). Human oversight and the ability to override decisions are crucial for maintaining trust, supporting organizational learning, and ensuring accountability (Shneiderman, 2022). These applications show that human and AI collaboration is not a one-size-fits-all model but a design process sensitive to the specific needs of each sector and the expertise of humans (Amershi et al., 2019).

Table 3. Integrated Human–AI Collaboration Framework Across Sectors

Sector	Primary AI Role	Primary Human Role	Collaboration Structure	Key Organizational Benefit	Major Collaboration Risk
Manufacturing	Predictive analytics, maintenance forecasting, process optimization	Contextual equipment evaluation, anomaly interpretation, and intervention decisions	AI identifies operational patterns while technicians validate and act on recommendations	Reduced downtime and improved operational efficiency	Overdependence on automated maintenance recommendations
Healthcare	Diagnostic analysis, medical imaging evaluation, patient-data processing	Clinical judgment, ethical reasoning, patient-context interpretation	AI supports diagnostic assessment while clinicians retain final decision authority	Improved diagnostic accuracy and reduced clinician workload	Automation bias and reduced critical evaluation
Finance	Real-time data analysis, market prediction, risk modeling	Strategic interpretation, geopolitical assessment, risk oversight	AI generates analytical recommendations while analysts interpret broader market implications	Faster analytical capability and improved risk management	Opaque algorithmic decisions and accountability ambiguity
Logistics	Route optimization, scheduling, and inventory coordination	Contingency planning, local adaptation, and customer coordination	AI optimizes operational variables while planners manage disruptions and strategic priorities	Improved supply chain efficiency and responsiveness	Reduced adaptability during unexpected disruptions

Source: Author's Compilation

The logical next question is: How does an organization know whether these collaboration designs are actually working? Section 4 answers this through a comprehensive measurement framework.

4. MEASURING WHAT MATTERS: A FRAMEWORK FOR TRACKING COLLABORATION QUALITY

The sector frameworks in Section 3 explain how human–AI collaboration should be designed. This section explains how organizations can evaluate whether collaboration is functioning effectively or drifting toward failure. The framework combines quantitative, qualitative, and governance-based indicators to assess decision quality, trust calibration, and accountability within AI-supported work systems.

4.1 The Strategic Need for Measurement Redesign

Traditional performance metrics such as output, efficiency, and cost reduction are insufficient for evaluating human–AI collaboration because they focus mainly on speed and volume. A productivity increase achieved through meaningful human augmentation differs significantly from one created through excessive automation and declining human judgment (Raisch & Krakowski, 2021). Without broader evaluation systems, organizations may appear efficient while gradually weakening the expertise required to supervise and improve AI systems (Shneiderman, 2022). A redesigned measurement framework, therefore, evaluates both system performance and human capability development. This distinction helps organizations identify whether AI is strengthening collaboration or encouraging passive dependence on automated recommendations (Amershi et al., 2019).

4.2 Quantitative Metrics for Monitoring Decision Quality

Quantitative indicators assess whether AI collaboration improves operational outcomes while preserving decision quality. Productivity measures such as throughput, processing speed, and operational efficiency should be evaluated alongside quality indicators, including first-pass yield, error rates, audit outcomes, and retrospective decision accuracy. Rising efficiency combined with declining quality may indicate automation bias and reduced human oversight (Raisch & Krakowski, 2021).

Innovation indicators also distinguish augmentation from automation. While automation improves existing processes, augmentation supports new capabilities and creative problem-solving (Wilson & Daugherty, 2018). Relevant indicators include patent filings, new product ideas, process improvements, and the speed of prototype development. These measures are best evaluated over time because collaborative innovation develops gradually. Financial performance remains important and includes ROI, cost savings, and revenue growth. However, sustainable evaluation must also consider investments in infrastructure, workforce training, governance systems, cybersecurity, and organizational change management.

4.3 Qualitative Indicators for Monitoring Trust Calibration

Qualitative indicators assess how employees experience AI-supported work systems. Cognitive load measures, workload surveys, NASA-TLX assessments (Hart & Staveland, 1988), behavioral observations, and employee feedback help organizations identify whether AI reduces analytical burden or creates overdependence and distrust (Lee & See, 2004).

Trust calibration can be evaluated through override rates, manual verification frequency, and acceptance-rejection patterns. Extremely low override rates may signal automation bias, while excessively high rates may indicate poor AI quality or insufficient trust (Lee & See, 2004; Siau & Wang, 2018). Organizations should also assess whether employees continue developing analytical capabilities through interaction with AI systems. Knowledge-sharing systems, annotations, and decision trails help preserve institutional expertise and support long-term organizational learning (Jarrahi, 2018).

4.4 Cross-Industry Benchmarks and KPI Integration

Table 4 provides industry benchmarks for productivity, override rates, and trust calibration. Healthy override ranges help organizations detect over-reliance or under-trust, while low trust-calibration scores indicate the need for governance intervention and workflow redesign (Lee & See, 2004).

Table 4. Performance Benchmarks: Human–AI Collaboration Across Industries

Metric	Manufacturing	Healthcare	Finance	Logistics
Productivity Gain	20–40%	15–25%	10–20%	15–30%
Error / Risk Reduction	First-pass yield +18%	Misdiagnosis −12%	Execution errors −25%	Late deliveries −20%
ROI Payback	18–24 months	12–20 months	6–14 months	14–22 months
Employee Satisfaction	+25%	+40% (fatigue ↓)	+18%	+30%
Healthy Override Rate	15–25%	20–35%	10–20%	8–15%
Trust Calibration Score	7.2 / 10	7.8 / 10	6.9 / 10	7.5 / 10

Note: Healthy override rates represent calibrated collaboration. Below-range rates signal automation bias; above-range rates signal under-trust or AI quality issues. Source: Author's Compilation

Table 5 presents the integrated KPI framework—twelve measurement categories each explicitly linked to the challenge it monitors. This linkage is intentional: measurement should be diagnostic as well as evaluative. When an organization reviews its KPI dashboard, it should be able to identify not merely which metrics are declining, but which challenge that decline signals and which framework element requires attention.

Table 5. **Integrated KPI Framework: Twelve Categories Linked to Three Challenges**

KPI Category	Sample Metric	Measurement Method	Challenge Monitored
1. Output & Productivity	Throughput per hour; task completion rate	Workflow analytics; time-and-motion studies	Challenge 1
2. Quality & Error Rate	First-pass yield; rework %; defect density	QA audit logs; process control charts	Challenge 1
3. Time Efficiency	Cycle time; time-to-decision; time-to-market	Process mining; decision log timestamps	Challenge 1
4. Resource Optimization	Energy per unit; material waste %; asset uptime	IoT telemetry; ERP data integration	Challenge 1 & 3
5. Employee Experience	Job satisfaction score; meaningful work index	Pulse surveys; engagement platforms	Challenge 1 & 2
6. Cognitive Load	Self-reported mental effort; NASA-TLX score	NASA-TLX scale; biometric sensors	Challenge 2
7. Innovation Output	Ideas generated per quarter; patent filings	R&D databases; innovation platforms	Challenge 1
8. Decision Quality	Outcome accuracy vs benchmark; audit score	Decision logs; retrospective outcome analysis	Challenge 1 & 2
9. AI Override Rate	% of recommendations modified by humans	Decision system logs; governance dashboards	Challenge 2
10. Bias & Fairness	Bias detection score; demographic disparity index	Algorithmic audit tools; fairness dashboards	Challenge 3
11.Transparency Score	User-rated AI explanation clarity; traceability	User surveys; audit trail completeness review	Challenge 3
12.Stakeholder Value	NPS; customer retention; ESG score; partner sentiment	CRM analytics; ESG reporting frameworks	Challenge 3

Source: Author's Compilation

The twelve KPIs in Table 7 provide the operational measurement layer of the framework. But tracking individual KPIs is insufficient for strategic management: organizations need an evaluation system that integrates these KPIs into a holistic assessment of collaboration health and connects that assessment to strategic decision-making. Section 5 provides this integration through the IMPACT+ model and the 360° Balanced Scorecard.

5. STRATEGIC EVALUATION AND ORGANIZATIONAL INTEGRATION

5.1 Why Strategic Evaluation is Needed

Operational metrics like productivity, speed, and efficiency help assess immediate performance results, but they do not provide a complete view of whether human–AI collaboration leads to lasting organizational value. Organizations need broader evaluation systems that assess workforce capabilities, governance quality, adaptability, innovation, and long-term organizational strength, in addition to operational performance. The IMPACT+ model combines the KPI framework from Table 5 into seven strategic evaluation areas that together measure collaboration effectiveness, governance quality, workforce ability, and long-term organizational sustainability.

Table 6. Strategic Evaluation and Organizational Integration

Dimension	Strategic Focus	Organisational Objective
Innovation	Creativity and organizational learning	Strengthen innovation capability
Manpower	Workforce capability and wellbeing	Improve employee adaptability
Performance	Productivity and operational quality	Enhance efficiency and decision quality
Adaptability	Organisational resilience	Support continuous learning
Customer	Stakeholder and service value	Improve customer trust and satisfaction
Trust & Ethics	Governance and accountability	Preserve transparency and fairness
Sustainability	Long-term organizational impact	Ensure responsible and sustainable growth

Source: Author's Compilation

5.2 IMPACT+ Strategic Dimensions

The IMPACT+ framework provides a broad framework for evaluating how people and AI work together, moving beyond basic operational measures. Each dimension highlights an important organizational skill needed for lasting collaboration. The Innovation dimension assesses whether working with AI encourages creativity, experimentation, and learning rather than merely automating simple tasks. The Manpower dimension looks at how well employees develop new skills, adapt, and work together in AI-supported environments (World Economic Forum, 2023). The Performance dimension focuses on raising productivity, efficiency, and the quality of decisions. The Adaptability dimension is about how well an organization can stay resilient and adjust its workflows and decision-making when things change. The Customer dimension examines how collaboration affects service quality, stakeholder trust, and customer satisfaction. The Trust & Ethics dimension reviews transparency, accountability, fairness, and how well decisions can be explained in systems where people and AI work together (Floridi & Cowls, 2019). The Sustainability dimension looks at long-term responsibility, including social impact, keeping the workforce strong, and using AI responsibly. Together, these dimensions form a strategic way to see if human–AI collaboration makes an organization stronger over time.

5.3. 360° Balanced Scorecard Integration

The IMPACT+ framework gives a broad strategic view, but the 360° Balanced Scorecard ties these ideas to overall organizational performance by linking day-to-day results with long-term strengths (Kaplan & Norton, 1996). The Balanced Scorecard examines collaboration from four angles: financial results, customer value, how well internal processes work, and learning and growth. Human–AI collaboration influences all these areas at the same time, which shows that real success depends on aligning strategy, not just focusing on single results.

Table 7. Balanced Scorecard Perspectives for Human–AI Collaboration

Balanced Scorecard Perspective	Human–AI Collaboration Focus	Strategic Outcome
Financial Perspective	Productivity, operational efficiency, cost optimization	Improved organizational performance
Customer Perspective	Service quality, responsiveness, stakeholder trust	Enhanced customer satisfaction
Internal Process Perspective	Workflow optimization, decision support, coordination	Better process effectiveness
Learning & Growth Perspective	Workforce capability, adaptability, and AI literacy	Sustainable organizational learning

Source: Author's Compilation

5.4 Strategic Organizational Alignment

Strategic evaluation should ensure that human–AI collaboration remains aligned with broader organizational objectives rather than functioning solely as a technological efficiency initiative. Sustainable collaboration requires balancing operational performance with workforce capability development, governance quality, organizational learning, and stakeholder trust. Organizations that evaluate collaboration strategically are more likely to achieve long-term resilience, innovation, and responsible AI integration (Raisch & Krakowski, 2021). Human–AI collaboration, therefore, becomes not only a technological capability but also a strategic organizational design process supporting sustainable competitive advantage.

6. ORGANIZATIONAL READINESS: THE PREREQUISITES FOR SUCCESSFUL COLLABORATION

The measurement framework in Section 4 and the evaluation instruments in Section 5 define what effective collaboration looks like and how to assess it. This section addresses the prior question: what must be in place before effective collaboration is possible? Organizational readiness is not a binary state—an organization is either ready or not—but a multidimensional profile of strengths and gaps that determines where implementation will succeed, where it will struggle, and what investments must be sequenced before deployment begins (ATS Global, 2025). The four readiness dimensions presented here are directly linked to the three challenges: readiness gaps create the conditions in which challenges manifest.

6.1 The Four-Dimension Readiness Diagnostic

Table 8 presents the four-dimensional readiness diagnostic, structured to show how each dimension's profile connects to a specific failure vulnerability. An organization with mature technical infrastructure but immature governance is particularly exposed to accountability gaps: AI recommendations will be produced at scale without the structures to assign and document responsibility. An organization with strong workforce capability but weak technical infrastructure generates enthusiasm for collaboration that unreliable systems cannot sustain—producing trust miscalibration as employees oscillate between enthusiasm and frustration. Cultural readiness failures create the psychological conditions for both over-reliance on AI and resistance to transparency in governance.

Table 8. Organizational Readiness Diagnostic: Four Dimensions Linked to Three Challenges

Readiness Dimension	What Is Assessed	Indicators of Readiness	Challenge Link
Technical Infrastructure	Data quality; integration capability; system reliability; cybersecurity posture	Clean, integrated data pipelines; API connectivity; < 1% data latency; SOC 2 compliance	Ch. 1: Unreliable AI systems drive over-reliance when employees cannot critically evaluate outputs
Workforce Capability	AI literacy, analytical skill, adaptability, and psychological readiness for change	A 70%+ workforce can interpret AI confidence scores; critical override behavior is observable in pilots	Ch. 1 & 2: capabilities determine whether humans can correct AI errors and calibrate trust
Governance Maturity	Decision rights frameworks; accountability structures; ethics review capacity	Decision logging systems live; ethics board constituted; escalation pathways documented.	Ch. 3: mature governance prevents accountability voids when AI and human decisions interact
Cultural Readiness	Leadership attitudes, experimentation tolerance, commitment to workforce development	Leadership publicly uses measurement insights; employees co-design AI workflows; failure is treated as learning	Ch. 2 & 3: cultural openness is a prerequisite for the trust-building and transparency that governance requires

Source: Author's Compilation

6.2 Leadership Commitment: Strategic Direction and Accountability

Successful human–AI collaboration depends heavily on visible and sustained leadership commitment. When senior leaders actively use collaboration metrics in strategic decisions, participate in governance reviews, support workforce development, and acknowledge implementation challenges openly, they signal that AI collaboration is a long-term organizational priority. This visibility encourages managerial alignment and reduces resistance to workflow redesign (Davenport & Ronanki, 2018). Leadership commitment is especially important for accountability because governance systems require investment in ethics boards, decision logging, bias auditing, and transparent oversight mechanisms. Leaders who treat governance as a strategic responsibility rather than a compliance burden strengthen organizational trust and collaboration effectiveness (Shneiderman, 2022).

6.3 Governance Architecture: Institutional Infrastructure

Effective AI governance requires structured institutional systems rather than informal oversight. Dedicated governance boards involving operations, HR, legal, compliance, ethics, and technology teams should regularly review collaboration performance, ethical risks, and accountability outcomes across the IMPACT+ dimensions (European Commission, 2021; US NIST, 2023). Governance systems must also include decision-logging mechanisms that record AI recommendations, human responses, and final outcomes. These logs support organizational learning, regulatory compliance, auditability, and measurement of override behavior, linking governance directly to collaboration evaluation.

6.4 Workforce Upskilling and Change Management

Human–AI collaboration changes work roles from routine task execution to judgment-based decision-making. Employees, therefore, require AI literacy, critical evaluation skills, and the ability to make ethical decisions (World Economic Forum, 2023). Resistance to AI adoption often reflects concerns about autonomy, expertise, and job security (Venkatesh et al., 2010). Effective change management addresses these concerns through employee participation in workflow design, transparent communication, workforce wellbeing measures, and visible career pathways within AI-augmented environments. These initiatives help organizations strengthen trust, adaptability, and long-term collaboration readiness before implementation begins (Wilson & Daugherty, 2018).

7. IMPLEMENTATION ROADMAP: SEQUENCING THE RESOLUTION OF THREE CHALLENGES

Organizational readiness is a prerequisite; implementation is the process by which readiness is converted into operational human–AI collaboration. The roadmap sequences the resolution of all three concerns across a 24-month horizon. Accountability governance must be established first, as a precondition for all subsequent activities—without it, pilot learning cannot be captured, trust cannot be monitored, and accountability gaps immediately fill the space that collaboration creates. Automation bias and trust calibration are then actively managed in the pilot phase, where measurement data and feedback loops generate the evidence base for continuous improvement. The scaling phase consolidates these resolutions and ensures they are sustained rather than degraded under expansion pressure.

Table 9. ***Human–AI Collaboration Implementation Roadmap: Three Phases, Three Challenge Resolutions***

Phase	Timeline	Key Activities (linked to Readiness Table 9)	Success Milestones & Challenge Targets
Phase 1: Foundation	Months 1–4	Conduct readiness diagnostic (Table 9 dimensions); constitute AI governance board (Governance Maturity); select pilot cases; baseline all 12 KPIs (Table 7); launch AI literacy workforce program (Workforce Capability)	Ch. 3: governance charter signed; ethics framework live. Ch. 1 & 2: baseline KPIs documented; pilot scope defined with explicit trust-calibration targets
Phase 2: Pilot & Learn	Months 5–12	Deploy AI collaboration systems; activate IMPACT+ quarterly reviews (Table 8); implement feedback loops (Adaptive Learning pillar); role-specific upskilling; conduct first bias audit; refine trust calibration protocols using override rate data (KPI 9)	Ch. 1: pilot productivity >15%; employee adoption >70%. Ch. 2: override rate 10–30%; trust score >6.5/10. Ch. 3: first bias audit complete; accountability logs active
Phase 3: Scale & Optimize	Months 13–24	Expand to all departments; integrate 360° BSC into strategic planning; advance upskilling (Shared Decision pillar); deploy predictive analytics; publish stakeholder transparency reports; and review the annual measurement framework.	All Challenges: productivity >25%; positive ROI confirmed; innovation output doubled; satisfaction >7.5/10; ESG targets integrated; governance fully operational across all sites

Source: Author's Compilation

7.1 Phase 1: Foundation — Establishing Governance (Months 1–4)

The foundation phase builds the infrastructure required for accountable human–AI collaboration. Governance is established before deployment to prevent accountability gaps and unmanaged AI adoption. Key activities include conducting the readiness diagnostic (Table 8), forming a cross-functional AI governance board, identifying pilot use cases, establishing baseline KPIs, and launching AI literacy programs. The primary outcome is institutional readiness, as measured by governance charters, ethics frameworks, baseline reports, and clearly defined pilot objectives linked to decision quality and trust calibration.

7.2 Phase 2: Pilot and Learn — Monitoring Collaboration Quality (Months 5–12)

The pilot phase focuses on understanding how collaboration functions within the organization rather than only proving financial value. AI systems are deployed in selected operational areas while the IMPACT+ framework and governance mechanisms generate decision logs, override data, and trust assessments. Productivity, decision quality, override rates, and cognitive-load indicators are monitored to identify automation bias or trust miscalibration (Lee & See, 2004). Governance reviews support refinement before large-scale deployment.

7.3 Phase 3: Scale and Optimize — Sustaining Collaboration Capability (Months 13–24)

The scaling phase expands successful collaboration models while ensuring governance systems, workforce capabilities, and measurement frameworks develop alongside AI deployment. IMPACT+ and the 360° Balanced Scorecard become integrated into strategic planning and organizational reviews. As Adaptive Learning matures, AI accuracy improves through continuous feedback, employees become more effective collaborators, and governance systems strengthen oversight capability. Continuous improvement through governance reviews, KPI updates, and bias audits ensures sustainable long-term collaboration performance (Shneiderman, 2022).

8. CASE STUDY VALIDATION

Several organizational cases show how collaborative intelligence works in practice and support the framework developed in this chapter. GE Aviation implemented the Predix predictive maintenance platform to help with turbine maintenance by continuously analyzing sensor data (General Electric Aviation, 2024). Before this, maintenance systems relied mostly on scheduled inspections and reactive interventions. The AI-supported predictive maintenance improved failure forecasting and operational coordination. However, GE allowed technicians to have authority in the decision-making process. Employees could confirm, change, or disregard AI-generated recommendations. This collaborative setup reduced the risk of over-relying on automation while enhancing organizational learning. The behavior of overriding decisions and the outcomes of maintenance improved AI accuracy through adaptive learning (General Electric Aviation, 2024).

Mayo Clinic used AI-assisted diagnostic systems in breast cancer screening processes (Mayo Clinic, 2024). Even though the AI systems showed high diagnostic accuracy, organizational leaders recognized the risks of over-dependence on algorithmic recommendations. AI outputs were presented with confidence scores and supporting reasoning so that clinicians could critically assess recommendations. Governance safeguards like anonymization procedures and demographic bias audits strengthened accountability and trust within the institution. This collaboration enhanced diagnostic accuracy while lowering clinician fatigue (Mayo Clinic, 2024).

JPMorgan Chase expanded the COIN platform into a hybrid decision-support environment (JPMorgan Chase & Co., 2023). This combined algorithmic analysis with trader expertise. AI systems analyzed market information and provided recommendations, while traders kept authority over interpretation, risk assessment, and execution decisions. Explainable AI mechanisms offered confidence intervals and clear reasoning, allowing employees to evaluate recommendations in the broader market context. Decision logs ensured accountability and supported institutional learning (JPMorgan Chase & Co., 2023).

These cases highlight several important principles. First, collaboration works best when human judgment remains active. Second, adaptive learning systems enhance both algorithmic ability and employee expertise over time. Third, transparency and accountability mechanisms are vital for building trust and legitimacy within institutions (Shneiderman, 2022). The cases also support the chapter's main argument that human-AI collaboration is an organizational design challenge rather than a solely technological issue. Sustainable collaboration develops when governance structures, workforce capabilities, and adaptive learning systems grow alongside technological deployment.

Table 10. Case Study Validation of Human–AI Collaboration Framework

Organisation	Industry	AI Function	Human Role	Governance Mechanism	Strategic Outcome
GE Aviation	Manufacturing	Predictive maintenance	Technician validation and intervention	Override authority and monitoring	Reduced downtime and improved maintenance efficiency
Mayo Clinic	Healthcare	AI-assisted diagnostics	Clinical interpretation and ethical judgment	Explainability and bias auditing	Improved diagnostic quality and reduced fatigue
JPMorgan Chase	Finance	Market analytics and risk modeling	Strategic interpretation and execution oversight	Decision logging and explainable AI	Enhanced analytical capability and accountability

Source: Author's Compilation, based on GE Aviation (2024), Mayo Clinic (2024), and JPMorgan Chase (2023)

The three cases confirm that the framework's predictions are empirically grounded: Complementary Intelligence prevents automation bias; Adaptive Learning resolves trust miscalibration over time; and Shared Decision Ecosystems, backed by governance architecture, close accountability gaps. They also reveal a critical implementation lesson: the concern that receives the least attention in initial deployment most severely limits overall collaboration performance. Section 9 addresses the governance dimension that underpins accountability resolution across all three cases—and across all industrial contexts.

9. ETHICS AND GOVERNANCE: RESOLVING ACCOUNTABILITY IN HUMAN–AI COLLABORATION

The accountability gap in human–AI collaboration cannot be resolved through measurement systems alone. Sustainable collaboration requires governance structures that embed accountability, transparency, and ethical oversight into organizational processes. Governance, therefore, becomes a strategic capability that supports trust and responsible AI integration at scale (Floridi & Cowls, 2019; OECD, 2019).

9.1 Privacy and Data Protection

Human–AI collaboration depends on organizational data such as decision logs, interaction records, cognitive-load indicators, and employee feedback. While essential for evaluation, these systems also create privacy and surveillance concerns (Zuboff, 2019). Employees who feel excessively monitored may avoid overriding AI recommendations or suppress honest feedback, weakening organizational learning and reducing the reliability of measurement systems. Privacy governance is therefore both an ethical requirement and a functional necessity for effective collaboration. Organizations should implement: anonymization of employee-level data, controlled access protocols, informed consent procedures, secure data-retention policies, and periodic privacy assessments. These mechanisms ensure that measurement systems support learning rather than workplace surveillance while strengthening employee trust and psychological safety.

9.2 Algorithmic Fairness and Bias Auditing

AI systems can reproduce biases embedded in training data or historical organizational practices (Bender et al., 2021). In workforce settings, biased algorithms may influence recruitment, promotion, task allocation, or performance evaluation. The Bias and Fairness KPI and the Trust and Ethics dimension of the IMPACT+ framework help organizations monitor these risks. However, measurement alone is insufficient without institutional corrective mechanisms. Regular bias audits should examine whether AI-supported decisions create unequal outcomes across demographic groups. Governance teams must also have the authority to modify or suspend systems when fairness problems emerge. Override-rate analysis is especially important because patterns in override behavior may reveal hidden inequalities within collaboration systems. Diverse representation in governance boards and evaluation processes further improves the organization's ability to identify and address fairness risks across different operational contexts.

9.3 Transparency and Explainability

Transparency is essential for maintaining meaningful human authority in AI-supported environments (Shneiderman, 2022). Employees cannot exercise informed judgment if AI recommendations are opaque or difficult to interpret (Amershi et al., 2019). The Transparency Score (KPI 11) evaluates whether AI explanations are understandable and actionable for decision-makers. Effective systems should clearly communicate: why recommendations were generated, factors influencing outcomes, confidence levels, and situations where predictions may become unreliable. Transparency also strengthens accountability. In collaborative decision systems, employees remain responsible for final decisions, while organizations retain responsibility for AI quality, governance standards, and oversight mechanisms.

9.4 Governance as Strategic Capability

Ethical governance is not merely a compliance activity; it is a strategic capability that enables organizations to sustain trustworthy and effective collaboration over time. Important governance mechanisms include AI governance boards, decision-logging systems, transparency reporting, ethical impact assessments, and regular fairness audits. Together, the readiness framework, implementation roadmap, and governance architecture form an integrated system that embeds accountability, transparency, and fairness into organizational design. These governance mechanisms collectively resolve Challenge 3 by ensuring that responsibility remains visible, auditable, and institutionally supported throughout human–AI collaboration.

10. FUTURE RESEARCH DIRECTIONS: EXTENDING THE FRAMEWORK

Although the framework presented in this chapter is grounded in current evidence and managerial practice, several important research gaps remain. The first gap concerns longitudinal collaboration dynamics. Most existing studies examine human–AI collaboration over short observation periods, making it difficult to determine whether long-term AI integration strengthens or gradually weakens human expertise (Raisch & Krakowski, 2021). Longitudinal research tracking employee judgment, learning, and professional identity over several years is therefore essential for understanding whether Complementary Intelligence genuinely prevents expertise erosion.

A second research priority involves trust calibration measurement. Current instruments—including override rates, cognitive-load indicators, and trust indexes—remain insufficiently validated for large-scale organizational deployment. More reliable and context-sensitive tools are needed to evaluate whether employee trust remains appropriately aligned with actual AI performance (Lee & See, 2004).

A third gap concerns cross-cultural and cross-organizational variation. Most current evidence comes from large technology-intensive organizations in Western economies. Additional research is needed to examine how the centaur model, IMPACT+ framework, and readiness diagnostics apply in small and medium enterprises, public-sector institutions, low-resource environments, and non-Western organizational contexts.

Finally, future scholarship should examine the political and institutional dimensions of AI governance. Questions regarding who defines organizational metrics, whose interests AI systems optimize, and how workers challenge unfair algorithmic decisions remain underexplored (Zuboff, 2019). Addressing these issues requires interdisciplinary collaboration across management studies, organizational behavior, labor policy, and ethics research.

11. CONCLUSION: HUMAN–AI COLLABORATION AS ORGANIZATIONAL DESIGN

The central argument of this chapter is that the major challenges of human–AI collaboration—decision quality, trust calibration, and accountability—are interconnected dimensions of a single organizational problem rather than separate technological concerns. The centaur model in Section 2 provides the theoretical foundation through Complementary Intelligence, Adaptive Learning, and Shared Decision Ecosystems. Section 3 demonstrates how these principles operate across manufacturing, healthcare, finance, and logistics, while Section 4 introduces KPIs that measure productivity, trust, governance, and workforce capability. Section 5 extends these operational indicators into strategic evaluation through the IMPACT+ framework and the 360° Balanced Scorecard. Section 6 identifies the organizational conditions required before implementation, and Section 7 presents a phased roadmap for governance, workforce adaptation, and scaling. Section 8 validates the framework through evidence from GE Aviation, Mayo Clinic, and JPMorgan Chase, showing that sustainable collaboration depends on organizational learning, governance maturity, and transparency rather than technological capability alone.

Section 9 completes the framework by establishing accountability, fairness, and transparency as institutional requirements for long-term collaboration.

Across all three case studies, the evidence supports a common conclusion: effective human–AI collaboration is fundamentally an organizational design challenge (Makarius et al., 2020). GE Aviation preserved technician expertise within AI-supported maintenance systems, Mayo Clinic improved diagnostic safety through governance structures, and JPMorgan Chase strengthened trading performance through transparency mechanisms that enabled human judgment to complement algorithmic analysis. The broader objective of human–AI collaboration should therefore not be efficiency alone, but the creation of organizational environments in which AI strengthens human capability while operating within systems that preserve accountability, fairness, and trust (Shneiderman, 2022). The frameworks, KPIs, governance mechanisms, and implementation models presented in this chapter provide organizations with an integrated approach for achieving that objective. Ultimately, the success of human–AI collaboration depends less on machine intelligence and more on how organizations design the relationship between human and AI systems.

REFERENCES

Amershi, S., Weld, D., Vorvoreanu, M., Fourney, A., Nushi, B., Collisson, P., Suh, J., Iqbal, S., Bennett, P. N., Inkpen, K., Teevan, J., Kikin-Gil, R., & Horvitz, E. (2019). Guidelines for human-AI interaction. In *Proceedings of the 2019 CHI Conference on Human Factors in Computing Systems* (pp. 1–13). Association for Computing Machinery. DOI: 10.1145/3290605.3300233

Baur, C., Wee, D., & Deichmann, J. (2024). *Digital twins: The next frontier of factory optimization*. McKinsey & Company.

Baxter, G., & Sommerville, I. (2011). Socio-technical systems: From design methods to systems engineering. *Interacting with Computers, 23*(1), 4–17. DOI: 10.1016/j.intcom.2010.07.003

Bender, E. M., Gebru, T., McMillan-Major, A., & Shmitchell, S. (2021). On the dangers of stochastic parrots: Can language models be too big? In *Proceedings of the 2021 ACM Conference on Fairness, Accountability, and Transparency* (pp. 610–623). Association for Computing Machinery. DOI: 10.1145/3442188.3445922

Davenport, T. H., & Ronanki, R. (2018). Artificial intelligence for the real world. *Harvard Business Review, 96*(1), 108–116.

European Commission. (2021). Proposal for a regulation laying down harmonized rules on artificial intelligence (Artificial Intelligence Act). *European Union.*

Floridi, L., & Cowls, J. (2019). A unified framework of five principles for AI in society. *Harvard Data Science Review, 1*(1). Advance online publication. DOI: 10.1162/99608f92.8cd550d1

FourKites. (2024). *AI-powered supply chain visibility and logistics optimization report*. FourKites.

General Electric Aviation. (2024). *Predix predictive maintenance platform performance report*. GE Aviation.

Global, A. T. S. (2025). *The smart industry readiness index*. https://www.ats-global.com/resources/blogs/the-smart-industry-readiness-index/

Google Health. (2024). *Advances in AI-assisted medical diagnostics and clinical decision support*. Google Health Research.

Hart, S. G., & Staveland, L. E. (1988). Development of NASA-TLX (Task Load Index): Results of empirical and theoretical research. In Hancock, P. A., & Meshkati, N. (Eds.), *Human mental workload* (pp. 139–183). Elsevier.

Hutchins, E. (1995). *Cognition in the wild.* MIT Press. DOI: 10.7551/mitpress/1881.001.0001

IBM Watson Health. (2023). *AI-enabled healthcare transformation report.* IBM Corporation.

Jarrahi, M. H. (2018). Artificial intelligence and the future of work: Human-AI symbiosis in organizational decision making. *Business Horizons, 61*(4), 577–586. DOI: 10.1016/j.bushor.2018.03.007

JPMorgan Chase & Co. (2023). *COIN platform and AI-enabled financial intelligence systems*. JPMorgan Technology Review.

Kaplan, R. S., & Norton, D. P. (1996). *The balanced scorecard: Translating strategy into action*. Harvard Business School Press.

Lee, J. D., & See, K. A. (2004). Trust in automation: Designing for appropriate reliance. *Human Factors, 46*(1), 50–80. DOI: 10.1518/hfes.46.1.50.30392

Makarius, E. E., Mukherjee, D., Fox, J. D., & Fox, A. K. (2020). Rising with the machines: A sociotechnical framework for bringing artificial intelligence into the organization. *Journal of Business Research, 120*, 262–273. DOI: 10.1016/j.jbusres.2020.07.045

Mayo Clinic. (2024). *AI-assisted breast cancer screening pilot outcomes.* Mayo Clinic Digital Health Reports.

. McKinsey & Company. (2023). *The state of AI in 2023: Generative AI's breakout year*. McKinsey Global Survey.

Organization for Economic Co-operation and Development. (2019). *OECD principles on artificial intelligence*. OECD Publishing., DOI: 10.1787/eedfee77-

project44. (2024). *Real-time transportation visibility and AI logistics analytics.* project44.

Raisch, S., & Krakowski, S. (2021). Artificial intelligence and management: The automation-augmentation paradox. *Academy of Management Review, 46*(1), 192–210. DOI: 10.5465/amr.2018.0072

Russell, S., & Norvig, P. (2021). *Artificial intelligence: A modern approach* (4th ed.). Pearson.

Shneiderman, B. (2022). *Human-centered AI.* Oxford University Press. DOI: 10.1093/oso/9780192845290.001.0001

Siau, K., & Wang, W. (2018). Building trust in artificial intelligence, machine learning, and robotics. *Cutting-Edge Technologies in Higher Education, 17*, 47–59. DOI: 10.1108/S2044-996820180000017005

Siemens, A. G. (2024). *MindSphere industrial IoT platform overview*. Siemens Digital Industries.

Topol, E. (2019). *Deep medicine: How artificial intelligence can make healthcare human again*. Basic Books.

Trist, E. L., & Bamforth, K. W. (1951). Some social and psychological consequences of the longwall method of coal-getting. *Human Relations, 4*(1), 3–38. DOI: 10.1177/001872675100400101

United States National Institute of Standards and Technology. (2023). *Artificial intelligence risk management framework (AI RMF 1.0)*. U.S. Department of Commerce.

Venkatesh, V., Bala, H., & Sykes, T. A. (2010). Impacts of information and communication technology implementations on employees' jobs in service organizations in India: A multi-method longitudinal field study. *Production and Operations Management, 19*(5), 591–613. DOI: 10.1111/j.1937-5956.2010.01148.x

Wilson, H. J., & Daugherty, P. R. (2018). Collaborative intelligence: Humans and AI are joining forces. *Harvard Business Review, 96*(4), 114–123.

World Economic Forum. (2023). *The future of jobs report 2023*. World Economic Forum.

Xu, W., Dainoff, M. J., Ge, L., & Gao, Z. (2021). From human-computer interaction to human-AI interaction: New challenges and opportunities for enabling human-centered AI. *International Journal of Human-Computer Interaction, 37*(10), 945–955. DOI: 10.1080/10447318.2021.1881589

Zuboff, S. (2019). *The age of surveillance capitalism: The fight for a human future at the new frontier of power*. PublicAffairs.

KEY TERMS AND DEFINITIONS

Automation Bias: The tendency to over-rely on AI recommendations by accepting algorithmic outputs without exercising independent critical judgment, progressively eroding the contextual expertise that makes AI systems valuable—the primary manifestation of Challenge 1.

Centaur Model: A human–AI collaboration architecture, empirically validated across chess, radiology, finance, and drug discovery, in which human and machine intelligence are combined as genuine complements, consistently outperforming either working independently.

Cognitive Augmentation: The use of AI systems to expand human analytical capacity, access, and decision speed while preserving human decision authority, contextual judgment, and accountability—the defining characteristic of genuine collaboration as distinct from automation.

Complementary Intelligence: The first pillar of the centaur model: the asymmetric division of cognitive labor between AI systems (pattern recognition, computational consistency) and human intelligence (contextual reasoning, ethical judgment, creative synthesis) that makes collaboration more valuable than either alone.

Distributed Cognition: A theoretical framework conceptualizing intelligence as distributed across multiple agents—human and technological—forming the basis for the centaur model's empirical predictions about collaboration performance.

IMPACT+ Model: An integrative evaluation framework for human–AI collaboration encompassing seven dimensions (Innovation, Manpower, Performance, Adaptability, Customer, Trust and Ethics, Sustainability), each linked to specific KPIs and challenge resolutions.

Socio-Technical Systems Theory: An organizational theory holding that technology integration succeeds only when technical deployment and social system redesign—roles, decision rights, culture, capability development—are optimized simultaneously.

Trust Calibration: The alignment of human reliance on AI recommendations with actual AI system reliability within specific operational contexts—the central concern of Challenge 2, monitored through the AI Override Rate (KPI 9) and cognitive load instruments.

Chapter 4
Strategic Decision-Making Frameworks With Human-AI Systems:
Transforming Workforce Dynamics and Future Synergies

Göran Roos
http://orcid.org/0000-0003-0943-3585
Adelaide University, Australia

ABSTRACT

This chapter examines strategic decision-making frameworks integrating ten AI architectures, from deterministic expert systems to non-deterministic large language models. It develops a three-dimensional framework mapping AI characteristics, decision contexts, and human cognitive roles across 36 configuration scenarios. Effective human-AI collaboration requires matching AI types to decision environments through human-in-the-loop, human-on-the-loop, or human-out-of-the-loop configurations. Workforce transformation sees humans evolving from operators to orchestrators, with 1% of skills fully automatable whilst 25% of jobs face substantial reconfiguration. Governance must adapt to architectural heterogeneity, maintaining human accountability whilst enabling AI autonomy. The trajectory toward 2040 cognitive ecosystems promises emergent intelligence through multi-agent networks, demanding investment in interoperability, trust architectures, and education preparing workers for meta-cognitive orchestration roles.

DOI: 10.4018/979-8-3373-7036-1.ch004

INTRODUCTION

The emergence of hybrid human-artificial intelligence systems represents a fundamental shift in how organisations approach strategic decision-making, reshaping cognitive architectures through which decisions are formulated, evaluated, and executed across industrial ecosystems. This transformation extends beyond mere technological augmentation, reconceptualising how human and artificial agents interact within decision frameworks (Akata et al., 2020; Dellermann et al., 2019; Jarrahi, 2018; Jarrahi et al., 2022). The convergence of human cognitive capabilities with artificial intelligence systems creates hybrid intelligence wherein complementary strengths merge to achieve outcomes neither could accomplish independently (Akata et al., 2020; Vaccaro et al., 2024).

Contemporary industrial fields witness organisations embedding artificial intelligence into strategic decision processes through varied architectures. Early implementations such as rule-based expert systems offered deterministic logic for repetitive decisions, whilst modern deep-learning algorithms and large language models introduce probabilistic reasoning and generative capabilities (Raisch & Krakowski, 2021; Jarrahi, 2018). This examination develops frameworks for implementing hybrid human-AI decision systems across ten distinct AI types: expert systems employing rule-based reasoning (Buchanan & Shortliffe, 1984), supervised machine learning models including support vector machines and random forests (Breiman, 2001), unsupervised learning algorithms for pattern discovery, reinforcement learning systems exhibiting non-deterministic training behaviours, deep learning architectures comprising convolutional neural networks and recurrent neural networks (Dosovitskiy et al., 2020), large language models with inherent non-determinism (Vaswani et al., 2017), computer vision systems for visual analysis, generative AI for creative applications, and embodied robotics AI. Each technology is examined as part of an integrated ecosystem where human and artificial agents collaborate, compete, and co-evolve.

Table 1. Ten AI Types and Their Characteristics

AI Type	Determinism	Technical Approach	Primary Applications	Key Strengths	Key Limitations	Typical Collaboration Mode
Expert Systems	Deterministic	Rule-based inference using IF-THEN logic	Medical diagnosis, financial planning, legal compliance, troubleshooting	Transparent reasoning, explainable decisions, reliable in constrained domains	Inflexible, cannot learn from data, difficult to maintain with complex rule sets	Human-in-the-loop
Supervised ML (Decision Trees, Random Forests, SVM)	Deterministic (post-training)	Statistical pattern learning from labelled data	Credit scoring, spam detection, customer churn prediction, risk assessment	Accurate on specific tasks, fast inference, well-understood behaviour	Requires labelled data, limited to trained scenarios, potential overfitting	Human-in-the-loop
Unsupervised ML (Clustering, PCA)	Deterministic	Pattern discovery without labels	Customer segmentation, anomaly detection, data compression, market basket analysis	Identifies hidden patterns, no labels needed, useful for exploration	Results can be ambiguous, difficult to validate, parameter-sensitive	Human-in-the-loop
Reinforcement Learning	Non-deterministic (training); Deterministic (deployment)	Trial-and-error optimisation through environmental interaction	Robotics control, game playing, autonomous vehicles, resource optimisation	Learns optimal strategies, handles sequential decisions, environmental adaptation	Extensive training required, sample inefficient, difficult to debug	Human-on-the-loop
Deep Learning - CNNs	Deterministic	Hierarchical visual feature extraction	Image classification, object detection, facial recognition, medical imaging	Exceptional visual perception, automatic feature extraction, scales with data	Requires massive datasets, computationally expensive, limited explainability	Human-on-the-loop
Deep Learning - RNNs/Transformers	Deterministic	Sequential pattern processing with temporal dependencies	Time series forecasting, speech recognition, video analysis, sequence modelling	Handles temporal dependencies, captures long-range patterns, versatile	Computationally intensive, training complexity, gradient issues	Human-on-the-loop
Large Language Models	Non-deterministic (default); Deterministic (temperature=0)	Transformer-based text generation and understanding	Content generation, translation, chatbots, code generation, text analysis	Human-like understanding, versatile across tasks, few-shot learning	Hallucination risk, expensive computation, training bias, context limitations	Human-in-the-loop
Computer Vision Systems	Deterministic	Integrated visual processing pipelines	Quality control, autonomous driving, surveillance, retail analytics, AR/VR	High-speed processing, consistent performance, detects subtle patterns	Novel scenario struggles, adversarial vulnerability, privacy concerns	Human-on-the-loop

continued on following page

Table 1. Continued

AI Type	Determinism	Technical Approach	Primary Applications	Key Strengths	Key Limitations	Typical Collaboration Mode
Generative AI (Image/Video)	Non-deterministic (unless seed fixed)	Probabilistic content synthesis	Art creation, design prototyping, content production, data augmentation	Creative output, rapid iteration, diverse variations, fills data gaps	Unpredictable quality, copyright concerns, resource intensive, potential bias	Human-in-the-loop
Robotics AI	Mixed (deterministic perception, variable planning)	Embodied intelligence with sensorimotor integration	Warehouse automation, surgery, manufacturing, agriculture, delivery	Operates in hazardous environments, continuous operation, precise and consistent	High development cost, safety risks, unpredictable situation challenges	Human-on-the-loop

Note: Collaboration modes represent typical configurations but may vary based on specific decision contexts, stakes, and regulatory requirements.

Industrial sectors show increased adoption of IntelliFusion strategies, with organisations expecting gains in productivity and innovation as AI systems handle routine tasks, enabling human employees to focus on creative and strategic work (Shrestha et al., 2019). Through combination of human and AI capabilities, efficiency and profitability in decision-making are expected to increase (Jarrahi, 2018). Recent evidence indicates large language models can generate and evaluate business strategies comparable to experienced entrepreneurs and investors, suggesting human-AI teams might explore strategic options more efficiently than humans alone (Csaszar et al., 2024).

The problem space addressed covers understanding how distinct AI systems transform cognitive and procedural decision frameworks. Each AI typology introduces affordances and constraints that reshape attention allocation, sensemaking processes, trust calibration, and accountability mechanisms within decision-making architectures (Shen et al., 2021; Okamura & Yamada, 2020). Organisations deploying these diverse systems confront questions about appropriate division of cognitive labour, evolution of human roles from operators to orchestrators, and governance structures required to maintain strategic coherence and ethical alignment (World Economic Forum, 2025).

This examination addresses seven fundamental questions critical to understanding and implementing hybrid human-AI decision systems:

First, how do different AI architectures—from deterministic expert systems to non-deterministic large language models—transform cognitive and procedural decision frameworks within organisations? This question explores the distinctive impacts of each AI type on attention allocation, interpretation processes, and sensemaking activities.

Second, what frameworks and typologies effectively characterise human-AI collaboration configurations, and how should organisations match collaboration modes to AI architectural characteristics and decision contexts? This covers understanding when human-in-the-loop, human-on-the-loop, or human-out-of-the-loop configurations prove most appropriate.

Third, how does AI integration transform workforce roles, cognitive identities, and required competencies as workers transition from direct task execution toward oversight and orchestration of hybrid systems? This addresses the evolution from operators to orchestrators and the skills necessary for effective collaboration with diverse AI types.

Fourth, what governance structures, ethical frameworks, and accountability mechanisms are necessary to ensure AI systems operate within societal norms and organisational objectives whilst maintaining transparency and explainability? This examines regulatory requirements, algorithmic auditing, and bias mitigation strategies.

Fifth, what are the strategic and organisational implications of deploying diverse AI architectures for competitive dynamics, organisational structures, and value creation mechanisms? This explores how AI affects organisational agility, resilience, and dynamic capabilities.

Sixth, how will future cognitive ecosystems evolve from current isolated deployments toward interconnected networks of multiple AI architectures and human agents engaging in continuous co-evolution? This addresses emerging multi-agent systems, trust architectures, and human roles in 2040 scenarios.

Seventh, what integrated conceptual framework best characterises the architecture, dynamics, and implications of human-AI strategic decision-making systems across diverse organisational contexts? This synthesises insights into a comprehensive model mapping relationships between AI characteristics, decision contexts, and human cognitive roles.

This analysis adopts a multi-level, cross-sector, socio-technical approach. The multi-level perspective examines transformations across strategic, tactical, and operational organisational layers. Cross-sector analysis identifies patterns and variations in human-AI collaboration across manufacturing, defence, healthcare, and logistics domains. The socio-technical framework recognises that technological capabilities cannot be divorced from human cognitive processes, organisational structures, and institutional governance mechanisms (Perry & Macredie, 2000; Steyvers & Kumar, 2023). The AIeSA, as a hybrid AI, is designed as a partner-in-thought with the human user, built to ask questions and promote higher levels of cognition through dynamic partnering with access to authoritative data (United States Air Force Office of the Chief Scientist [USAF], 2015).

Building upon foundational theories of bounded rationality (Simon, 1955), distributed cognition (Hutchins, 1995), and organisational decision-making (Cyert & March,

1963), this examination develops comprehensive frameworks for understanding and implementing hybrid human-AI decision systems. The methodological approach integrates multiple theoretical perspectives and empirical domains, with classical decision theory providing scaffolding for understanding how limitations in cognitive capacity, information processing bandwidth, and computational resources constrain decision-making (Kahneman & Tversky, 1979). Distributed cognition theory offers insights into how cognitive labour becomes allocated across human actors, artificial agents, and mediating artefacts within socio-technical systems. Contemporary hybrid intelligence frameworks synthesise these perspectives, conceptualising human-AI collaboration as emergent cognitive architectures (Dellermann et al., 2019).

The argument developed here operates simultaneously at three levels. At the micro level, individual human operators, AI systems and the interaction design between them determine whether a decision is sound, auditable and justifiable in the moment. At the meso level, organisational structures, workflows and incentive regimes translate these micro interactions into reproducible patterns—what gets automated, what remains under human control, and where accountability sits when something goes wrong. At the macro level, regulatory regimes, labour-market institutions and cross-border standards shape which patterns become viable, which are foreclosed, and how quickly a given configuration diffuses across sectors. Cross-level feedback matters: a micro failure (an erroneous recommendation accepted without challenge) can trigger meso redesign (new oversight roles) and macro response (fresh regulatory guidance), while macro choices (the risk categories of the EU AI Act) in turn constrain which micro configurations are permitted. Keeping the three levels in view avoids the common error of treating human-AI systems as a purely technical or purely organisational phenomenon (Geels, 2002; Shrestha et al., 2019).

THEORETICAL AND CONCEPTUAL FOUNDATIONS

The theoretical foundations underpinning human-AI decision-making systems draw from multiple disciplinary traditions contributing insights into how cognitive processes can be distributed, augmented, and transformed through technological mediation. Herbert Simon's concept of bounded rationality challenged assumptions of perfect rationality, showing that decision-makers operate within cognitive constraints necessitating satisficing strategies rather than optimisation (Simon, 1955, 1997). Simon believed that agents face uncertainty about the future and costs in acquiring information in the present, factors limiting the extent to which agents may make fully rational decisions (Simon, 1955). Administrative behaviour reflects "a choice of appropriate means to reach designated ends" whilst acknowledging that "the task of rational decision involves the discovery of such means" (Simon, 1997).

These limitations manifest differently across the AI types examined. Expert systems with deterministic rule-based processing address human limitations in consistently applying complex rule sets but struggle with situations outside programmed parameters. Large language models exhibit flexibility in handling novel situations but introduce non-deterministic elements that can lead to hallucination and unpredictability (Augenstein et al., 2024). Standard economic theory assumed individuals are perfectly rational decision-makers, yet Simon emphasised limits to rationality that real-life administrators face regarding memory, attention, and capacity (Augier & March, 2001).

The behavioural theory of the firm developed by Cyert and March (1963) extended these insights to organisational contexts, conceptualising firms as coalitions of participants with divergent goals engaging in negotiated decision processes characterised by sequential attention to goals, problematic search, and organisational learning. Organisational decision-making reflects satisficing behaviour constrained by limited search, standard operating procedures, and sequential resolution of conflicting objectives. AI systems may become new organisational agents participating in decision coalitions, bringing capabilities that shift how problems are identified and solved. Algorithmic decision systems could hard-code certain decision premises, acting as de facto members of dominant coalitions setting decision criteria.

Daniel Kahneman and Amos Tversky's research programme on judgement and choice under uncertainty mapped systematic biases characterising human decision-making, showing that intuitive judgements deviate from normative rationality in predictable ways (Kahneman, 2011). Their dual-process theory distinguishes between System 1 processing (fast, intuitive, automatic) and System 2 processing (slow, deliberate, analytical). Tversky and Kahneman (1974) demonstrated reliance on mental shortcuts leading to predictable biases including anchoring, availability, and representativeness. Prospect theory revealed that individuals evaluate outcomes relative to reference points, exhibit loss aversion, and apply probability-weighting functions (Kahneman & Tversky, 1979).

The heuristics-and-biases programme has been remarkably fruitful, unveiling dozens of cognitive biases in decision-making (Baron, 2008). Integration of AI raises questions about whether AI systems might mitigate human biases or introduce new ones. AI can provide data-driven analysis free of emotion, potentially correcting human over-reliance on intuition (Kahneman, 2011). However, AI trained on human-generated data may learn and amplify human biases, as evidenced by studies showing LLMs reproduce moral or cognitive biases present in training corpora (Mukherjee et al., 2023).

Distributed cognition theory, pioneered by Edwin Hutchins, offers perspective on how cognitive processes extend beyond individual minds to encompass tools, representations, and social structures. Cognition involves not only the brain but also external artefacts, work teams, and cultural systems for interpreting reality (Hutchins, 1995). What distinguishes distributed cognition from other approaches is commitment to theoretical principles regarding boundaries of the unit of analysis for cognition and the range of mechanisms participating in cognitive processes (Hollan et al., 2000).

Application of distributed cognition to human-AI systems reveals how decision-making capabilities emerge from interaction between human agents, AI systems, and organisational structures. Computer vision systems transform visual data into representations humans can interpret and act upon, creating distributed perceptual systems (Zhang & Patel, 2006). Large language models function as external memory and reasoning systems, extending human cognitive capabilities in ways paralleling how navigation charts and instruments create distributed cognitive systems in maritime navigation (Hutchins & Palen, 1997).

Contemporary hybrid intelligence frameworks define this capability as "the ability to achieve complex goals by combining human and artificial intelligence, thereby reaching superior results to those each of them could have accomplished separately" (Dellermann et al., 2019). This definition emphasises complementarity and synergy rather than substitution or competition. Four key capabilities distinguish hybrid intelligence systems: ability to collaborate effectively through shared understanding, to continually learn and adapt through reciprocal interaction, to explain decisions comprehensibly to human partners, and to ensure appropriate human oversight and control (Akata et al., 2020).

The diverse AI architectures examined engage differently with these theoretical frameworks. Expert systems instantiate classical symbolic AI approaches, encoding domain expertise through explicit rule structures and logical inference mechanisms (Russell & Norvig, 2021). Their deterministic nature ensures predictable, explainable decisions but constrains adaptability. Supervised machine learning algorithms exemplify data-driven pattern recognition discovering predictive relationships through statistical optimisation rather than explicit programming (Hastie et al., 2009). Whilst deterministic in application after training, these systems learn implicit rules that may prove difficult to articulate.

Unsupervised learning approaches discover latent structures within data absent predefined categories, supporting exploratory analysis and anomaly detection where relevant patterns remain unknown (Hastie et al., 2009). Reinforcement learning systems optimise decision policies through trial-and-error interaction with dynamic environments, exhibiting non-deterministic exploration during training (Sutton & Barto, 2018). Deep learning architectures achieve performance through hierarchical feature learning across vast datasets, though opacity introduces explainability challenges (LeCun et al., 2015; Vaswani et al., 2017).

From a cognitive perspective, deterministic AI can be seen as extension of explicit knowledge, whereas non-deterministic AI captures tacit patterns experts might not articulate (Polanyi, 1966). Rule-based systems are easy to understand and explain because decisions follow predefined IF-THEN rules (Vale et al., 2022). A model's temperature refers to a scalar value used to adjust probability distribution predicted by the model (Pursnani et al., 2023). Temperature determines balance between adhering to learned data patterns and generating diverse or creative responses (Zhu et al., 2023).

Simon's pioneering research in AI included viewing computers as tools to extend human problem-solving capabilities. As early as the 1950s, Simon and colleagues predicted AI would become essential for management decision-making (Newell & Simon, 1956), developing some of the first expert systems and heuristics-based algorithms. March and Simon (1958) argued that organisations simplify decision environments for individuals; in the presence of AI, some simplifications may be delegated to algorithms, freeing human managers to focus on higher-order judgement and contextual interpretation.

HUMAN-AI COGNITIVE COLLABORATION AND DECISION ENVIRONMENTS

The cognitive collaboration between humans and artificial intelligence systems manifests through mechanisms of attention modulation, interpretation enhancement, and distributed sensemaking that vary across different AI architectures. In strategic decision environments characterised by uncertainty, complexity, and ambiguity, such collaboration has potential to enhance decision quality whilst reshaping the cognitive environment in which decisions are made (Klein et al., 2006; Weick, 1995). Understanding these transformations requires examining how each AI type influences human cognitive processes and how humans adapt decision-making strategies when working with diverse AI capabilities.

In contemporary decision environments, attention represents a scarce cognitive resource strategically allocated across multiple information streams and decision points. Different AI types modulate human attention in distinct ways. Expert systems direct attention through explicit rule chains, making transparent the logical pathways leading to recommendations (Buchanan & Shortliffe, 1984). This transparency enables humans to trace reasoning processes and identify potential gaps in rule bases. Deep learning systems such as convolutional neural networks process vast amounts of visual information and direct human attention to salient features that might otherwise be overlooked, though the basis for salience detection often remains opaque (Moawad et al., 2022).

Human decision-making processes can be enhanced by artificial intelligence; nevertheless, use of AI in critical domains has raised concerns regarding final users, those affected by undertaken decisions, and broader society (Akata et al., 2020). The process of interpretation in hybrid systems involves translating between human semantic understanding and various forms of AI representation. Supervised learning models such as random forests generate probabilistic outputs requiring interpretation within specific decision contexts (Liaw & Wiener, 2002). This interpretation involves judgement about relevance and reliability of model outputs given particular circumstances. Large language models present interpretation challenges, as natural language outputs can create illusion of understanding whilst potentially containing factual errors or biased reasoning (OpenAI, 2023).

Studies show that humans and AI possess complementary cognitive strengths, with effective collaboration arising when each party focuses on capabilities (Jarrahi, 2018; Dellermann et al., 2019). AI systems excel at data-driven analysis: aggregating historical data, detecting patterns or correlations, and performing computations exceeding human working-memory limits. An AI-driven analytics platform might evaluate millions of transactions to flag emerging market trends or anomalies humans might never notice (Gershman et al., 2015; Shrestha et al., 2019). AI's analytical prowess comes without limitations of fatigue or cognitive bias toward recent information. Computer vision systems might inspect every product on assembly lines for defects with consistent precision, supporting operational decisions about quality control.

Humans contribute judgement, domain experience, and intuition valuable in unstructured or novel situations. Humans excel at intuitive decision-making when faced with ambiguity, drawing on tacit knowledge, common sense, and understanding of context, ethics, or nuance that current AIs cannot replicate (Vincent, 2021; Jarrahi, 2018). Humans have imagination and creativity, generating hypotheses or strategic visions not directly extrapolated from past data. Humans weigh objectives and value considerations, balancing stakeholder interests or ethical principles where AI optimising defined metrics might falter. AI provides raw intelligence and speed, but "context is what holds the system together" – context remains a distinctly human contribution (Gill, 2020).

Trust calibration emerges as a determinant of effective human-AI collaboration. If humans are to accept algorithmic prescriptions, they need to trust them (Lee & See, 2004). Incompleteness in formal trust criteria is a barrier to optimisation (Lyons, 2013). Trust must be calibrated differently for each AI type based on respective strengths and limitations. Deterministic systems such as expert systems and supervised learning models enable straightforward trust calibration through consistent performance within defined parameters. Non-deterministic systems including reinforcement learning agents during training and large language models with temperature settings above zero require nuanced trust calibration accounting for variability in outputs (Bansal et al., 2019).

Insufficient trust can lead to under-utilisation of valuable AI insights, whereas over-trust can lead to over-reliance where humans accept outputs uncritically even when conditions have changed (Dzindolet et al., 2003). Trust in AI develops based on perceptions of system reliability, predictability, and helpfulness (McKnight et al., 2011). As AI agents become more anthropomorphic or autonomous, people may attribute to them qualities of interpersonal trust (Choung et al., 2023; de Visser et al., 2018). If predictive models provide not just output but explanation or rationale, human partners can better judge when to heed or override AI. This has led to research in explainable AI to support decision-makers' cognitive alignment with AI systems (Adadi & Berrada, 2018).

Cognitive load distribution between humans and AI systems varies across architectures and contexts. Rule-based expert systems reduce cognitive load by automating routine decision processes but may increase load when exceptions arise requiring human intervention (Chen & Barnes, 2015). Deep learning systems process complex perceptual information that would overwhelm human capacity but generate new demands in understanding and validating outputs. Integration of human cognitive strengths with machine computational capabilities proves important (Xiong et al., 2022).

Sensemaking in hybrid systems involves collaborative construction of understanding from ambiguous or incomplete information. Unsupervised learning algorithms contribute by identifying patterns and structures in data humans might not perceive directly. These patterns must be interpreted and contextualised by decision-makers bringing domain knowledge and understanding of broader organisational goals (Pea, 1993). Computer vision systems transform visual data into symbolic representations integrated with other information sources, enabling new forms of multimodal sensemaking (Zhang & Norman, 1994).

The dynamics of joint situational awareness in human-AI teams depend on characteristics of AI systems involved. Deterministic systems enable predictable coordination, as humans develop accurate mental models of system behaviour (Endsley, 2017). Non-deterministic systems, particularly those exhibiting emergent behaviours such as reinforcement learning agents or generative AI, require continuous updating of mental models and flexible coordination strategies. Hybrid intelligence emphasises collaboration between humans and AI, where machine learning applications support rather than replace human decision-making (Döppner et al., 2019). Machine learning augments human expertise by providing data-driven insights, enabling operators to make more informed decisions (Schneider & Leyer, 2019).

Cognitive collaboration involves managing different temporal scales at which humans and AI systems operate. Whilst AI systems process information and generate outputs at computational speeds, human understanding operates on biological timescales. This temporal mismatch creates challenges for real-time decision-making but also opportunities for AI systems to pre-process information and prepare it for human consumption at appropriate moments. Design of effective hybrid systems must account for these temporal dynamics, creating interfaces and workflows bridging different operational speeds (Shively et al., 2017).

Explanation and justification in cognitive collaboration varies across AI architectures. Expert systems provide explicit rule-based explanations aligning with human logical reasoning. Supervised learning models offer feature importance scores and decision boundaries requiring statistical interpretation (Caruana & Niculescu-Mizil, 2006). Deep learning systems present challenges for explanation, leading to development of post-hoc explainability techniques attempting to make decision processes more transparent (Miller, 2019). Large language models can generate natural language explanations, but these may not accurately reflect actual processing, raising concerns about reliability of self-explanation.

Adaptation and learning in hybrid systems occur bidirectionally, with humans and AI systems modifying behaviour based on interaction experiences. Humans develop new cognitive strategies for working with different AI types, learning to draw on strengths whilst compensating for weaknesses (Rastogi et al., 2022). Reinforcement learning systems explicitly adapt based on feedback, potentially learning from human corrections and preferences (Lample & Chaplot, 2017). This co-adaptation creates dynamic hybrid systems whose capabilities evolve through interaction.

Empirical work by Logg et al. (2019) found people exhibit bias to follow algorithmic advice, sometimes even when suboptimal – a phenomenon dubbed algorithm appreciation. Design of decision environments can encourage positive effects by prompting users to critically evaluate AI suggestions or providing diverse AI outputs to avoid single-answer anchoring (Mukherjee et al., 2023). The principle is cognitive calibration: aligning human confidence and understanding with AI's actual reliability. When well calibrated, humans correct AI when it errs and defer when it has superior information or analysis (Bansal et al., 2021).

The ideal human-AI collaboration involves division of cognitive labour: AI handles data processing and option generation whilst humans guide processes, set objectives, and make final judgements infused with experience and values (Gill, 2020; Wang et al., 2021). This could mean AI systems evaluate investment proposals and narrow them to shortlists based on statistical risk-return profiles; human experts then review shortlists, bringing qualitative considerations to select final investments. This augmented intelligence approach has been described as moving from AI as tool to AI as teammate (Johnson et al., 2022).

FRAMEWORKS AND TYPOLOGIES OF HYBRID DECISION SYSTEMS

The organisation of human-AI collaborative decision systems reflects varying allocations of autonomy, control, and oversight between human and artificial agents, with systematic frameworks distinguishing configurations based on timing, nature, and extent of human involvement in algorithmic processes. As organisations embrace human-AI partnerships, need arises to develop frameworks and typologies describing how these hybrid systems operate. Not all human-AI decision systems are alike – they vary in autonomy granted to AI, sequence of human versus machine actions, predictability of AI behaviour, and nature of tasks involved.

The predominant typology identifies three archetypal arrangements: human-in-the-loop, human-on-the-loop, and human-out-of-the-loop systems, each with distinct implications for agency, accountability and performance, as illustrated in Figure 1 (Cummings, 2004; Sheridan & Parasuraman, 2005).

Figure 1. Human–AI Collaboration Configurations (Human-in/on/out-of-the-loop Framework)

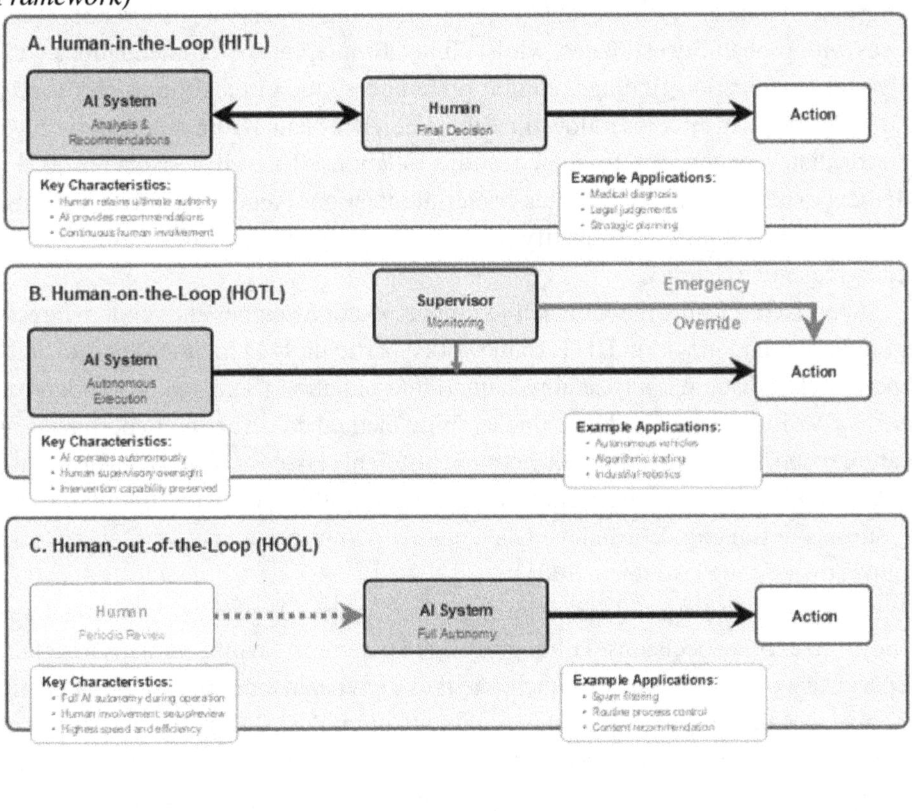

Human-in-the-loop (HITL) configurations position humans as active decision-makers receiving AI-generated recommendations but retaining ultimate authority. AI systems perform analytical functions including data processing, pattern recognition, hypothesis generation, and option evaluation, presenting findings to human operators who assess recommendations considering contextual factors, ethical implications, and domain expertise before making final decisions (Mosqueira-Rey et al., 2023; Shneiderman, 2020).

This arrangement proves appropriate when decisions carry significant consequences, operate in environments with high uncertainty or novelty, or require contextual understanding and ethical judgement exceeding current AI capabilities. HITL architectures align naturally with expert systems encoding domain knowledge through production rules, supporting human decision-makers with structured reasoning chains (Buchanan & Shortliffe, 1984). HITL systems require direct human involvement in decision execution, with AI providing recommendations or predictions that humans approve or override before action (Mosqueira-Rey et al., 2023).

Expert systems and supervised machine learning algorithms align with human-in-the-loop frameworks given their function as decision support tools rather than autonomous agents. In medical diagnosis contexts, AI systems analyse patient symptoms, medical histories, and diagnostic imaging to generate differential diagnoses with probability estimates, whilst clinicians integrate recommendations with physical examination findings, patient preferences, and clinical intuition (Topol, 2019). Rule-based systems follow traceable inference chains from premises through intermediate conclusions to final recommendations (Russell & Norvig, 2021). Medical expert systems diagnosing bacterial infections can enumerate symptoms, test results, and logical rules justifying particular diagnoses, enabling clinicians to verify reasoning validity.

Supervised learning models such as support vector machines and random forests commonly function within HITL frameworks, particularly in high-stakes decisions where probabilistic outputs require human interpretation (Safavian & Landgrebe, 1991). SVM is a supervised machine learning method constructing hyperplanes in feature space maximising distance between different classes of objects (Hearst et al., 1998). Random Forest is an ensemble model fitting multiple decision-tree classifiers on subsets of training data and using averaging to improve predictive accuracy and control over-fitting (Breiman, 2001).

Human-on-the-loop (HOTL) configurations grant AI systems operational autonomy to execute decisions within predefined parameters whilst humans maintain supervisory oversight, ready to intervene if system behaviour appears problematic or contexts shift beyond algorithmic training (Cummings, 2004). This arrangement balances efficiency gains from automated decision execution with safety assurance through human monitoring and exception handling. HOTL architectures reduce direct involvement, allowing AI to execute decisions autonomously whilst humans supervise and intervene when necessary (Parasuraman et al., 2000). HOTL frameworks place humans in monitoring roles with capacity to intervene upon anomalies, performance degradation, or out-of-scope situations (Tschiatschek et al., 2024; Mosqueira-Rey et al., 2023).

Reinforcement learning systems exemplify human-on-the-loop configurations, particularly in robotics and resource optimisation applications. In autonomous warehouse systems, reinforcement learning agents determine optimal routing, task allocation, and inventory management whilst human supervisors monitor aggregate performance metrics, respond to exceptional situations, and adjust operational parameters (Sutton & Barto, 2018). Exploitation versus exploration is a topic in reinforcement learning where autonomous agents balance contrasting needs in making decisions (Sutton & Barto, 1998). RL agents need to find best solutions quickly whilst avoiding committing to solutions too quickly without enough exploration (Keramati et al., 2011).

Deep learning systems including convolutional neural networks for computer vision and recurrent networks or transformers for sequential data processing operate within human-on-the-loop frameworks for real-time applications. Autonomous vehicle perception systems process sensory information and make continuous navigation decisions whilst human drivers maintain supervisory awareness ready to assume control (Grigorescu et al., 2020). CNNs excel at image and signal processing tasks, drawing on convolutional and pooling layers to extract local features and downsample inputs (Dosovitskiy et al., 2020).

Human-out-of-the-loop (HOOL) configurations grant AI systems full operational autonomy without ongoing human oversight or intervention capacity during execution. These systems operate independently based on predefined objectives and training, with human involvement limited to initial specification, periodic evaluation, and strategic adjustment (Cummings, 2004). HOOL systems remove direct human involvement: AI agents act autonomously based on predefined objectives or learned policies (Wooldridge, 2009). HOOL maximises speed but raises accountability and alignment challenges, especially in safety-critical domains (Tschiatschek et al., 2024; Rhim et al., 2021).

Certain supervised learning and expert system applications operate effectively in human-out-of-the-loop configurations when decision stakes remain low and environments stable. Automated spam filtering systems employ supervised learning algorithms to classify email without human review. Expert systems controlling routine industrial processes operate autonomously within narrowly defined parameters where clear rules govern appropriate responses (Russell & Norvig, 2021). Rule-based expert systems can operate in HOOL mode when stakes are low and conditions stable (Russell & Norvig, 2021).

Computer vision systems in retail contexts automatically counting inventory or tracking customer traffic operate without human supervision. Large language models serving in customer service chatbot capacities operate in human-out-of-the-loop configurations for routine inquiries, escalating to human agents when queries exceed capabilities or customer frustration rises (Bommasani et al., 2021). High-frequency trading systems in finance autonomously execute trades based on pre-set algorithms combining supervised ML and reinforcement learning.

Shrestha et al. (2019) identify three categories for managerial decision-making with AI: full AI delegation where agents have authority to make decisions without intervention; hybrid sequential decisions where AI provides recommendations and humans have final say; and hybrid collaborative decisions where humans and AIs interact iteratively to converge on decisions. The collaborative mode can be seen in applications such as planning or creative design, where generative AI systems iteratively propose ideas and humans refine or direct the AI's search.

Another dimension is determinism. Deterministic AI follows fixed rules yielding same output for identical inputs – desirable for auditability and accountability – whereas non-deterministic AI such as deep learning models with stochastic elements may yield different outputs under identical inputs. Deterministic AI aligns with classical rational choice models, whereas non-deterministic AI reflects behavioural perspectives in which decisions evolve with context (Rudin, 2019). Hybrid systems may combine both: deterministic guardrails atop probabilistic cores, ensuring reliability whilst preserving adaptability (Bryson, 2018).

Parasuraman et al. (2000) describe a continuum from fully manual to fully automatic control. Modern adaptive systems shift along this continuum, increasing human oversight during anomalies and granting AI greater autonomy in routine contexts (Abbass et al., 2018). Such adjustable autonomy prevents both over- and under-reliance.

Sectoral deployment patterns reflect systematic relationships between AI types, decision contexts, and appropriate collaboration frameworks. Healthcare applications predominantly employ human-in-the-loop configurations given high decision stakes, ethical complexity, and regulatory requirements mandating human oversight (Topol, 2019). Manufacturing contexts exhibit greater diversity, with high-consequence decisions such as quality control operating in human-in-the-loop frameworks whilst routine production processes employ human-on-the-loop or human-out-of-the-loop automation (Villani et al., 2018). Defence and security applications balance operational tempo requirements favouring autonomy against accountability demands necessitating human control, resulting in predominance of human-on-the-loop frameworks for weapons systems whilst intelligence analysis employs human-in-the-loop configurations (Cummings, 2004). Financial services employ varied configurations reflecting regulatory constraints and risk tolerances.

Logistics and supply chain management extensively employ human-on-the-loop frameworks given combination of continuous decision requirements, operational complexity, and need for strategic responsiveness (Sutton & Barto, 2018).

A fourth configuration has emerged in enterprise practice between 2024 and 2026 that does not fit neatly into the three archetypes above. Agentic AI systems combine large language models with tool-use, memory and planning scaffolds to operate autonomously over extended task trajectories, sometimes running thirty hours or more of continuous work without human intervention (Joshi et al., 2025). Industry surveys indicate that approximately half of enterprises using generative AI now deploy such agents in production, with expectations that AI systems will hold delegated decision-making authority growing substantially within three years (Joshi et al., 2025). The three-archetype framework still applies—agents operate under human-on-the-loop or human-out-of-the-loop regimes within the boundaries their principals set—but two new design problems surface. The first is bounded autonomy: each agent requires an explicit specification of what it may decide unilaterally, when it must escalate, and what it is forbidden to attempt, a problem that has revived interest in graduated-autonomy schemes moving from augmentation through automation to constrained autonomy. The second is ecosystem oversight: organisations deploying many agents in parallel require what BCG calls a virtual control tower, a shared registry that tracks every agent, names its human owner, and records its actions for audit (Joshi et al., 2025). Both problems are near-term operational challenges rather than speculative, and each translates directly into the governance and workforce implications examined in the sections that follow.

Case Vignettes: Real-World Examples Of Human–Ai Decision Systems

To illustrate practical consequences of hybrid decision systems, the following cases highlight successes and failures across sectors. Each vignette shows how algorithmic design, data quality and organisational context shape outcomes.

Amazon hiring tool bias

Amazon developed a machine-learning tool to rank job candidates, but an internal investigation revealed that the model penalised résumés containing words such as “women’s” and downgraded graduates of women’s colleges because the training data were dominated by male résumés. By 2015 the company abandoned the system when it became clear that it was not rating candidates in a gender-neutral way. The case underscores how biased historical data can propagate discriminatory patterns when used uncritically (Dastin, 2018).

IBM Watson for Oncology

IBM's Watson for Oncology aimed to provide evidence-based treatment recommendations, yet early deployments found that its advice often conflicted with local medical practices and even produced unsafe or incorrect recommendations. Persistent mismatches between the model's training data and real-world contexts led to declining revenues and eventual discontinuation of the programme. This vignette illustrates that even high-profile AI initiatives can fail when clinical complexity and localisation are not adequately addressed (Strickland, 2019).

Uber self-driving car accident

In 2018 an Uber automated vehicle struck and killed a pedestrian in Tempe, Arizona. The National Transportation Safety Board concluded that the probable cause was the operator's failure to monitor the driving environment due to distraction, coupled with Uber's inadequate safety risk assessments and the automated system's inability to correctly classify and track the pedestrian. Contributing factors included deactivation of the vehicle's factory collision-avoidance systems and lack of redundant safety layers. This tragedy shows how complacency and insufficient governance can lead to catastrophic outcomes (National Transportation Safety Board, 2019).

Risk Assessment Algorithm Bias

ProPublica's investigation of commercial risk assessment software used in US criminal sentencing found that the algorithm often rated black defendants as high risk while rating white defendants with similar or more serious histories as low risk. Statistical analysis revealed that black defendants were almost twice as likely to be falsely flagged as future criminals and that the algorithm's predictions were only marginally better than random chance. The case highlights structural bias and raises questions about the fairness of algorithmic decision tools in high-stakes domains (Angwin et al., 2016).

Quantitative Models and Practical Tools

Beyond narrative frameworks articulating theoretical relationships, practitioners require quantitative instruments enabling systematic evaluation and ongoing monitoring of hybrid decision system performance. The operationalisation of human-AI collaboration demands structured approaches encompassing evaluation matrices for assessing decision quality, workflow architectures guiding collaborative processes, and performance metrics tracking system effectiveness across multiple dimensions (Raji et al., 2020; Doshi-Velez & Kim, 2017).

Evaluation matrices for hybrid decision quality provide structured mechanisms for comparing outcomes across different collaboration configurations. Such matrices typically assess decisions against criteria including accuracy relative to ground truth where ascertainable, consistency across similar decision contexts, alignment with organisational objectives and values, stakeholder acceptance, and compliance with regulatory requirements. The matrix structure enables systematic comparison between human-only, AI-only, and various human-AI collaborative configurations, revealing where complementary strengths generate superior outcomes and where particular arrangements introduce vulnerabilities (Bansal et al., 2019; Steyvers & Kumar, 2023).

Decision workflows for human-AI collaboration establish structured sequences through which collaborative decisions proceed. The initial phase covers data collection and preprocessing, wherein relevant information is gathered and cleaned to ensure representativeness and quality sufficient for subsequent analysis. AI analysis follows, applying appropriate models to generate predictions, classifications, or recommendations based on the prepared data. Human interpretation then reviews AI outputs, contextualising results against factors the system may not capture and assessing alignment with organisational values and strategic priorities. Decision execution integrates human judgement with AI guidance in proportions appropriate to the specific context, stakes, and regulatory requirements. Post-hoc review completes the cycle, evaluating outcomes against expectations, auditing decision processes for procedural integrity, and updating models or procedures based on accumulated feedback (Amershi et al., 2019; Mosqueira-Rey et al., 2023).

Key performance indicators for human-AI collaboration extend beyond traditional decision quality metrics to capture dimensions unique to hybrid systems. Decision accuracy differential measures the gap between AI-only and human-AI collaborative performance, quantifying the value added through human involvement and identifying contexts where collaboration proves most beneficial. Training completion rates track the proportion of the workforce receiving preparation to work effectively with AI systems, reflecting organisational investment in building collaborative capabilities. Model drift incidence monitors the frequency with which algorithmic performance degrades due to evolving data distributions, signalling when recalibration or retraining becomes necessary. Stakeholder satisfaction captures feedback from users and affected parties regarding perceived fairness and utility of AI-supported decisions, ensuring that technical performance translates into legitimate organisational outcomes. Compliance adherence assesses the extent to which decisions conform to regulatory requirements and internal governance policies, maintaining accountability as decision authority becomes distributed across human and artificial agents (Raji et al., 2020; Morley et al., 2020). Together, these indicators provide a multidimensional view of hybrid system performance that informs continuous improvement efforts whilst ensuring alignment with broader organisational and societal expectations.

Worked Operationalisation: Applying the Framework to a Supervised-Learning Credit Decision

To show how the three-dimensional framework can be made operational rather than descriptive, consider a mid-sized European retail bank deploying a supervised-learning model (gradient-boosted decision trees) to score consumer credit applications. The framework locates this system by three observable indicators.

Architectural characteristics are measured as: determinism = deterministic at inference (identical feature vectors yield identical scores; training is stochastic but the production model is frozen); transparency = medium (individual predictions are explainable via SHAP values, Lundberg and Lee, 2017, though the learned pattern is not directly readable); adaptivity = low (monthly retraining with governance review, not continuous online learning).

Decision context is measured as: stakes = high (loan denial affects livelihoods and triggers anti-discrimination obligations under Article 14 of the EU AI Act when deployed for consumer credit); dynamism = moderate (macroeconomic conditions shift slowly but can spike); reversibility = partially reversible (appeals process exists); ethical complexity = high (fairness across protected attributes is legally mandated).

Collaboration mode is chosen by intersecting the above: high stakes combined with moderate dynamism and high ethical complexity rule out human-out-of-the-loop. The framework predicts a human-in-the-loop configuration for denials (a credit officer must review and ratify rejections) and a human-on-the-loop configuration for approvals (auto-approval within a tolerance band, with random and targeted audits). This matches Scenario 20 in the 36-scenario matrix. Testable propositions follow directly: if the bank deploys this system as human-out-of-the-loop, we predict an elevated adverse-action complaint rate within two quarters and a regulator-initiated audit within four quarters, relative to a matched peer running the in-/on-the-loop configuration. Complementary indicators—disparate-impact ratios across protected groups, SHAP-value drift across months, human override rates—provide continuous monitoring data against which the framework's predictions can be falsified.

GOVERNANCE, ETHICS, AND ACCOUNTABILITY

The integration of diverse artificial intelligence architectures into strategic decision-making necessitates governance frameworks addressing transparency, explainability, accountability, and alignment with human values across systems exhibiting fundamentally different operational characteristics. The infusion of AI into strategic decision-making introduces governance and ethical challenges as decision authority becomes shared between humans and machines. Traditional accountability structures are strained: who is responsible if an AI-informed decision leads to harm? The governance challenge extends beyond technical considerations to encompass legal liability allocation, ethical responsibility assignment, and procedural mechanisms ensuring AI systems operate within societal norms and organisational objectives (Dignum, 2019; Floridi et al., 2018).

Contemporary regulatory frameworks including the European Union AI Act, ISO/IEC 42001 artificial intelligence management systems standard, and OECD AI Principles establish baseline requirements whilst recognising that governance approaches must adapt to varied AI architectures and deployment contexts (European Commission, 2021; International Organization for Standardization, 2023; OECD, 2019). The European Union's Artificial Intelligence Act mandates human oversight for high-risk AI systems, meaning humans can intervene or override (European Commission, 2021). The Act classifies systems by risk, imposes conformity assessments, and bans unacceptable uses such as social scoring. High-risk AI systems deployed in employment, education, law enforcement, and critical infrastructure face requirements including human oversight, technical documentation, logging capabilities, and accuracy thresholds. The Act's implementation is now tangible rather than prospective: the Article 5 prohibitions (social scoring, untargeted facial-image scraping, real-time biometric identification in public spaces and workplace emotion recognition, among others) together with the AI literacy obligations of Article 4 entered into application on 2 February 2025; the General-Purpose AI Code of Practice was published by the European Commission's AI Office on 10 July 2025 and offers signatory providers a presumption of conformity with Article 53 GPAI obligations; and the Act becomes fully applicable on 2 August 2026 (European Commission, 2025; European Union, 2024).

Transparency and explainability requirements differ substantially across deterministic and non-deterministic AI systems, reflecting fundamental differences in how these architectures generate decisions. Expert systems built on explicit rule structures offer inherent transparency, as decision logic follows traceable inference chains (Russell & Norvig, 2021). Rule-based systems are highly transparent because rules governing decisions are explicit and understandable (Vale et al., 2022). This makes them ideal for tasks where transparency is critical, such as medical diagnosis or credit risk assessment (Ashley, 2017).

Supervised machine learning algorithms present more complex explainability challenges despite deterministic deployment behaviour. Models such as support vector machines and random forests learn implicit decision boundaries through training rather than incorporating explicitly programmed rules (Hastie et al., 2009). Post-hoc explainability techniques including LIME, which approximates complex models locally with interpretable linear models, and SHAP, which assigns importance scores to features based on game-theoretic principles, enable insights into which input variables influence predictions (Ribeiro et al., 2016; Lundberg & Lee, 2017).

Current methods used to explain AI are mainly technical ones, geared toward machine learning engineers for debugging purposes rather than toward end users affected by the system (Miller, 2019). The statistical nature of learned patterns means supervised learning systems may exploit spurious correlations appearing predictive within training distributions but lacking causal grounding. Although intrinsically interpretable models offer high transparency, they often lack complexity needed for highly nonlinear, high-dimensional problems (Vale et al., 2022).

Unsupervised learning systems present distinctive explainability challenges, as these algorithms discover latent structures without explicit labels or supervision. When clustering algorithms group customer segments or anomaly detection systems flag unusual transactions, explaining why particular instances belong to specific clusters requires articulating learned similarity metrics and distributional assumptions.

Deep learning systems including convolutional neural networks, recurrent networks, and transformers exhibit severe opacity given hierarchical feature learning across numerous layers with millions or billions of parameters (LeCun et al., 2015; Vaswani et al., 2017). Whilst techniques such as attention visualisation for transformers and gradient-based saliency mapping for CNNs provide partial insights, comprehensively explaining why particular hierarchical representations emerge and how they interact remains largely intractable (Arrieta et al., 2020).

Attention mechanisms in transformer architectures enable some transparency by revealing which input tokens influence output generation, supporting explainability particularly for large language models. However, attention constitutes only one component of complex architectures, and high attention scores do not necessarily indicate causal influence (Jain & Wallace, 2019). LLMs are highly susceptible to adversarial hallucination attacks, frequently generating false clinical details posing risks when used without safeguards (Zhang et al., 2023). Hallucination occurs when models generate outputs that neither align with factual information nor follow logical coherence (Huang et al., 2024).

Reinforcement learning systems introduce additional explainability dimensions through sequential decision processes and exploration-exploitation dynamics (Sutton & Barto, 2018). Explaining why reinforcement learning agents selected particular actions requires characterising learned value functions, policy structures, and how current decisions relate to anticipated future rewards. Generative AI systems producing synthetic images, videos, or text face explainability challenges regarding both individual outputs and broader distributional characteristics (Goodfellow et al.,

2016). GANs are deep learning architectures using two neural networks – generator and discriminator – to create new realistic data (Battaglia et al., 2018).

Table 2. Governance Requirements by AI Type

AI Type	Transparency Level	Explainability Method	Key Governance Challenges	Regulatory Considerations	Audit Approach
Expert Systems	High Explicit rule chains	Direct rule tracing IF-THEN logic paths	Rule completeness Edge case handling Maintenance burden	Minimal regulation due to transparency Focus on rule validation	Rule coverage testing Logic verification Exception case review
Supervised ML	Medium Learned patterns	LIME, SHAP, feature importance Post-hoc explanations	Spurious correlations Distribution shift Overfitting	Model documentation requirements Performance thresholds	Statistical validation Bias testing Holdout set evaluation
Unsupervised ML	Low Latent structures	Cluster characteristics dimensionality reduction visualisation	Ambiguous outputs Validation difficulty Parameter sensitivity	Challenging to regulate Focus on use cases	Stability analysis Parameter sensitivity Output consistency
Reinforcement Learning	Variable Policy transparency	Reward decomposition Policy visualisation Trajectory analysis	Reward hacking Exploration risks Non-stationary behaviour	Safety constraints Sandbox requirements	Behaviour boundary testing Reward alignment Safety violation checks
Deep Learning - CNNs	Very Low Black box	Saliency maps Attention visualisation Layer-wise relevance	Opacity Adversarial vulnerability Computational requirements	High-risk application restrictions Accuracy requirements	Adversarial testing Performance benchmarking Failure case analysis
Deep Learning - RNNs/Transformers	Very Low Complex interactions	Attention weights Gradient analysis Probe classifiers	Temporal dependencies Long-range interactions Gradient issues	Sequence-critical application oversight	Temporal consistency Attention pattern analysis Catastrophic forgetting tests
Large Language Models	Very Low Emergent behaviour	Chain-of-thought prompting Attention analysis Self-explanation*	Hallucination Inconsistency Prompt sensitivity Bias amplification	Content moderation Truthfulness requirements Bias mitigation	Factuality testing Consistency evaluation Harmful output detection

continued on following page

Table 2. Continued

AI Type	Transparency Level	Explainability Method	Key Governance Challenges	Regulatory Considerations	Audit Approach
Computer Vision Systems	Low Integrated pipelines	Component analysis Confidence scores Detection visualisation	Novel scenario failure Privacy concerns Demographic bias	Biometric data protection Surveillance restrictions	Performance across demographics Edge case testing Privacy compliance
Generative AI	Very Low Stochastic processes	Latent space analysis Generation process tracing	Unpredictable outputs Copyright issues Deepfake potential	Intellectual property Content authenticity Harmful content	Output quality metrics Copyright checking Authenticity verification
Robotics AI	Mixed Varies by component	Modular explanation Decision tree visualisation Trajectory planning	Physical safety Unpredictable environments Human interaction	Safety certification Liability frameworks Operational boundaries	Safety testing Human interaction protocols Failure recovery assessment

Note:

- **Transparency Levels:** Transparency varies significantly across AI architectures. High transparency characterises systems with directly observable and traceable decision logic. Medium transparency applies to systems where patterns are learnable but require interpretation. Low transparency occurs when internal representations are abstract or distributed. Very low transparency describes essentially opaque, black-box systems. Variable or mixed transparency depends on specific implementation or combines multiple approaches.
- **Explainability Methods:** Various methods enable understanding of AI decisions. Direct methods allow logic to be directly readable through rules or decision trees. Post-hoc explanations are generated after model training using techniques such as LIME and SHAP. Visualisation approaches provide graphical representations of internal states or attention mechanisms. Statistical methods characterise performance through metrics and testing. Self-explanation involves natural language explanations that may not accurately reflect actual processing.
  - * Self-explanation refers to AI systems (particularly LLMs) generating natural language descriptions of their reasoning; however, these explanations may be post-hoc rationalisations rather than accurate represen-

tations of internal processing, potentially providing plausible but misleading accounts of how decisions were reached

- **Governance Priorities by AI Type:** Deterministic systems (Expert Systems, trained Supervised ML) require focus on validation, completeness, and maintenance of decision logic, with regulatory emphasis on ensuring rules align with domain requirements and legal standards. Learning systems (RL, Deep Learning) emphasise behavioural boundaries, safety constraints, and continuous monitoring, with regulation focusing on preventing harmful emergent behaviours and ensuring robustness. Generative systems (LLMs, Generative AI) prioritise output quality, truthfulness, and harm prevention, with regulatory frameworks addressing content authenticity, intellectual property, and potential misuse. Embodied systems (Robotics AI) critically focus on physical safety, human-robot interaction protocols, and liability allocation, with regulations emphasising certification, operational boundaries, and failure recovery.
- **Risk-Based Regulatory Approach:** The EU AI Act and similar frameworks apply graduated requirements based on risk levels: minimal risk requires information provision only; limited risk entails transparency obligations; high risk demands comprehensive requirements including human oversight, documentation, and accuracy thresholds; whilst unacceptable risk applications are prohibited entirely.

Regulatory requirements continue evolving rapidly. Organisations must monitor jurisdiction-specific updates and sector-specific guidance. The table represents current typical governance approaches as of the knowledge cutoff date.

Accountability in hybrid decisions remains a practical concern. Organisations should not use AI as an "accountability shield" (Bryson, 2018). Leading approaches emphasise that accountability must remain with human actors and institutions, no matter how autonomous the AI. One strategy is the "human-in-command" approach – retaining human final decision-makers who remain accountable (Floridi et al., 2018). This is standard in medicine, where physicians sign diagnoses produced with AI assistance. Circuit-breakers and escalation protocols are part of governance logic: trading algorithms might halt if losses exceed thresholds and alert human supervisors (Hauer, 2023).

Beyond the European and ISO regimes, the United States National Institute of Standards and Technology published its AI Risk Management Framework 1.0 in January 2023 (National Institute of Standards and Technology, 2023), with a companion Generative AI Profile in July 2024 (National Institute of Standards and Technology, 2024). The framework organises AI risk work around four functions—govern, map, measure and manage—and is explicitly voluntary and sector-neutral, which positions it as a bridge between the prescriptive tiered structure of the EU AI Act and the management-system logic of ISO/IEC 42001:2023. Organisations operating across jurisdictions commonly combine the three in a layered stack: the EU Act for binding obligations on high-risk deployments, ISO/IEC 42001 for certifiable management-system practices, and the NIST framework for operational risk identification and mitigation guidance.

Audit trails reinforce accountability. Systems should log decision processes (input data, model output, human overrides) so post-hoc analysis can determine sources of error (Doshi-Velez & Kim, 2017). In regulated domains, reproducibility and traceability are non-negotiable – auditors require documentation for each AI-influenced decision (Leslie, 2019).

ISO/IEC 42001 provides international standards for AI management systems, establishing organisational processes for developing, deploying, and monitoring AI systems responsibly (International Organization for Standardization, 2023). The standard emphasises risk management, stakeholder engagement, transparency documentation, and continuous improvement. Implementation differs across AI architectures: deterministic systems benefit from clearly documented decision rules and validation procedures, whilst non-deterministic systems require statistical characterisation of typical behaviour.

OECD AI Principles articulate values including inclusive growth, sustainable development, human-centred values, transparency, robustness, security, and accountability (OECD, 2019) (revised in May 2024 to address generative-AI risks around safety and information integrity; OECD, 2024). These principles apply across AI types but require contextual interpretation reflecting system characteristics and deployment contexts. The OECD AI Principles, updated in 2025, articulate five value-based principles plus five policy recommendations promoting investment, enabling ecosystems, human capacity building, and international cooperation (OECD, 2019).

Algorithmic auditing emerges as an essential governance mechanism enabling independent assessment of AI system performance, bias, and alignment with intended objectives (Raji et al., 2020). Deterministic systems admit testing across enumerated input scenarios, whilst stochastic systems require statistical sampling and characterisation of distributional behaviour. AI can reproduce societal bias embedded in data (Noble, 2018; O'Neil, 2016). In workforce and strategic decisions, this manifests in hiring or credit algorithms disadvantaging protected groups (Angwin et al., 2016). Ethical governance entails fairness-aware design: bias audits, representative data, and algorithmic adjustments with fairness constraints (Morley et al., 2020).

Scenario Analysis: Regulatory Fragmentation and Cross-Border Coordination

Regulatory trajectories are not a settled variable. Two scenarios bound the plausible near-term space. Under mild fragmentation, the EU AI Act, the United States executive orders and sector guidance, the UK's pro-innovation framework and ISO/IEC 42001 converge around overlapping risk tiers, common documentation requirements (model cards, data sheets, incident logs) and reciprocal audit recognition. Multinationals deploy a single compliance stack with jurisdictional add-ons; mid-sized firms face manageable transaction costs; cross-border data flows for training and inference continue under adequacy arrangements.

Under severe fragmentation, regimes diverge on foundational questions—what constitutes a high-risk system, whether foundation-model developers or deployers bear primary liability, whether algorithmic explanations must be machine-readable or human-intelligible, and whether training-data lineage must be disclosed. Duplicative compliance, conflicting deletion and retention mandates, and blocked cross-border inference force regional model duplication. The strategic consequences differ sharply: dynamic capabilities in regulatory sensing and modular architecture become a first-order determinant of survival, while the 36-scenario matrix (read against mission-oriented policy logic; Mazzucato, 2018) (read against mission-oriented policy logic; Mazzucato, 2018) must be read jurisdiction by jurisdiction rather than as a single firm-level map. The analysis here applies cleanly to the mild case; under severe fragmentation, the collaboration-mode prescriptions remain, but the cost of operating the recommended configurations rises, and smaller firms may be forced into conservative configurations even where the framework would otherwise recommend more autonomous modes.

WORKFORCE TRANSFORMATION AND COGNITIVE ROLE EVOLUTION

The integration of diverse artificial intelligence architectures into organisational decision-making precipitates transformations in human workforce roles, required competencies, and cognitive identities as workers transition from direct task execution toward oversight, interpretation, and strategic direction of hybrid human-AI systems. These transformations extend beyond simple automation of routine tasks to encompass reconceptualisation of what constitutes valuable human contribution when computational systems increasingly perform analytical functions previously requiring specialised expertise. The rise of human-AI collaborative systems is altering the nature of work and roles humans play in organisations, with cognitive machines assuming routine thinking tasks, analysis, and even some creative work (Autor, 2015; Brynjolfsson & Mitchell, 2017).

The transition from operators executing prescribed procedures toward orchestrators coordinating human-AI ensembles constitutes the fundamental trajectory across sectors deploying AI systems, summarised across the 2010–2030 timeline in Figure 2 (Wilson & Daugherty, 2018). Traditional operator roles emphasise procedural adherence, domain-specific technical skills, and execution proficiency within established frameworks. As AI handles analytical tasks, humans shift toward higher-order cognition – judgement, creativity, and empathy (Davenport & Kirby, 2016).

Figure 2. Evolution of Human Workforce Roles from Operators to Orchestrators as AI integration advances. The timeline illustrates the transformation of required competencies, shifting from direct task execution toward meta-cognitive oversight of human-AI systems. Shaded areas represent the increasing proportion of tasks suitable for AI automation, whilst human roles evolve toward uniquely human capabilities including creativity, ethical reasoning, and complex systems management. Timeline based on observed patterns across manufacturing, healthcare, financial services, and technology sectors.

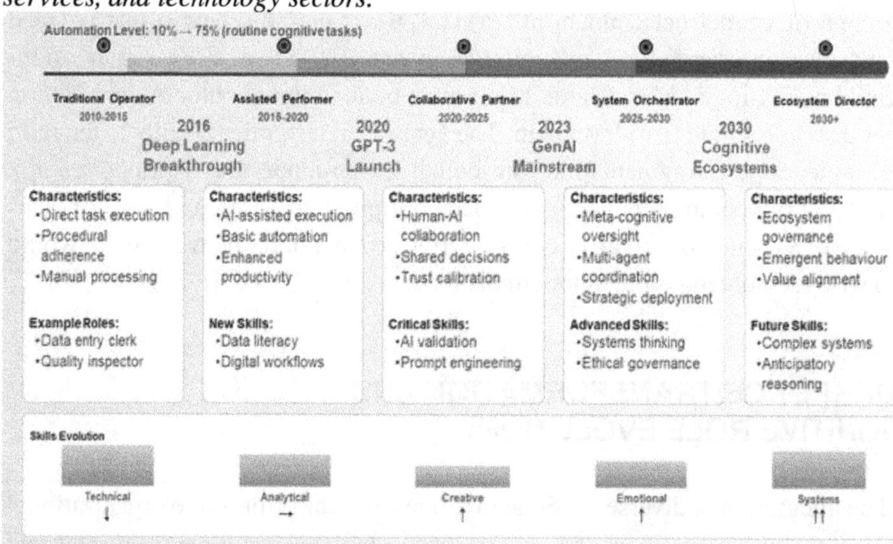

Expert systems and supervised machine learning algorithms automate portions of procedural tasks, particularly those characterised by clear rules or predictable patterns (Russell & Norvig, 2021; Hastie et al., 2009). Medical expert systems encode diagnostic protocols for routine conditions, enabling less specialised providers to deliver care previously requiring physician expertise. Supervised learning models assess credit risks using statistical patterns obviating detailed manual financial analysis. Computer vision systems inspect manufactured components with accuracy and speed exceeding human capabilities for repetitive visual tasks (Villani et al., 2018).

The orchestrator role emerging emphasises coordination across human and AI agents, as well as coordination across a workforce in which up to five generations now work side-by-side, each bringing different technological fluency, career expectations and tacit knowledge to the human side of the hybrid system (Roos, 2025), strategic prioritisation determining which tasks warrant human attention versus algorithmic handling, and meta-cognitive monitoring ensuring system performance remains aligned with objectives (Wilson & Daugherty, 2018). Orchestrators must understand AI system capabilities and limitations to delegate appropriately, recognise contexts where algorithmic recommendations prove unreliable, and intervene when human judgement proves necessary.

This interconnected strategy allows utilisation of cognitive resources, enabling humans to focus on tasks requiring creativity, intuition, and complex problem-solving (Akata et al., 2020). In manufacturing contexts, orchestrators monitor automated production systems, respond to anomalies flagged by unsupervised learning algorithms, and make strategic decisions regarding capacity allocation when demand patterns shift. Supply chain managers shift from manual logistics coordination toward oversight of AI-driven optimisation systems automating purchasing, inventory, and routing decisions whilst humans define service levels, ethical constraints, and strategic priorities (World Economic Forum, 2025).

Working effectively with different AI types demands distinctive skill sets reflecting varied operational characteristics and collaboration requirements. Collaborating with expert systems necessitates understanding explicitly programmed rules and decision logic. Workers must develop proficiency evaluating rule applicability, recognising edge cases, and exercising judgement when formal procedures prove inadequate.

Supervised machine learning systems demand competencies centred on interpreting statistical patterns, recognising distributional shifts, and calibrating trust toward probabilistic predictions (Hastie et al., 2009). Workers must understand that these systems learn correlations from historical data without necessarily capturing causal relationships, that performance degrades when operational contexts differ from training distributions, and that predictions carry uncertainty requiring contextual evaluation.

Large language models and generative AI systems necessitate competencies centred on prompt engineering, output evaluation, fact-checking, and creative direction (Bommasani et al., 2021). Effectively eliciting desired outputs from LLMs requires understanding how these systems respond to differently structured queries. Workers must develop critical evaluation skills recognising hallucinations, assessing factual accuracy, and identifying biased or inappropriate content.

Reinforcement learning systems operating in dynamic environments require human supervisors to understand exploration-exploitation dynamics, recognise when learned policies exploit unintended reward loopholes, and adjust objective specifications (Sutton & Barto, 2018). Deep learning systems pose explainability challenges requiring workers to calibrate trust based on aggregate performance metrics rather than comprehension of individual decisions (LeCun et al., 2015; Vaswani et al., 2017).

Robotics AI systems necessitate spatial reasoning regarding robot capabilities and motion patterns, communication skills enabling effective human-robot coordination, and safety awareness recognising hazards from shared workspace operations (Villani et al., 2018). Workers collaborating with industrial robots must anticipate robot movements, coordinate timing to avoid interference, and recognise contexts where human intervention proves necessary.

Beyond technical skills, successful human-AI collaboration demands cognitive flexibility adapting to evolving system capabilities, emotional intelligence managing AI-induced workplace changes, and ethical reasoning navigating value tensions from algorithmic decision support (Jarrahi, 2018). As AI systems improve, task allocations between human and artificial agents continuously shift, requiring workers to relinquish previously central responsibilities whilst developing new competencies.

Analysis by Indeed Hiring Lab (2025) found about a quarter of jobs could be highly transformed by generative AI, yet only about one percent of skills could be fully automated by AI alone. The majority of jobs are expected to be moderately transformed, meaning AI will change how tasks are performed without completely eliminating the human role. Those gains accrue unevenly across skill groups, occupations and regions, with coordinated government–industry–labour policy a decisive variable in how the surplus is shared (Roos & Shroff, 2017). Consider a marketing analyst: AI can automate data gathering and suggest campaign strategies, but the analyst's role shifts to validating AI insights, adding domain-specific angles, and focusing on interpersonal coordination. This phenomenon is described as a "hybrid workforce" where humans and AI agents work together (Wilson & Daugherty, 2018).

In manufacturing and logistics, cobots – collaborative robots – capture this idea: robots physically share tasks with workers, each doing what they excel at. In Amazon fulfilment centres, humans handle dexterity- or judgement-intensive tasks whilst robots perform heavy lifting and fetching. This has increased productivity without one-for-one worker replacement; job profiles have shifted toward robot operators and maintenance technicians (Choudhury et al., 2022).

Riso and Adăscăli ei (2024) finds that robots are not causing widespread job losses but changing job profiles. Many workers transition to monitoring, exception handling, and multi-machine supervision. Job quality considerations – autonomy and skill variety – should guide robot adoption to prevent stress and loss of control (Riso & Adăscăli ei, 2024).

The changing landscape spawns new occupations. Data scientists and ML engineers have long been in demand; now we see AI trainers, AI explainability specialists, and prompt engineers optimising interactions with generative AI (IBM Institute for Business Value, 2022). AI ethics officers and algorithm auditors are emerging to ensure compliance and ethical integrity.

Knowledge workers increasingly need digital and data literacy. The World Economic Forum (2023) identifies analytical thinking and technology literacy as top-growth skills, whilst routine information processing declines. Creativity, complex problem-solving, and emotional intelligence gain prominence (World Economic Forum, 2020). AI literacy covers understanding AI capabilities and limits, identifying suitable use cases, and critically evaluating outputs (Wilson & Daugherty, 2018).

Prompt engineering skills become essential for interacting with large language models – crafting effective prompts, refining outputs, and validating generated content for accuracy and relevance (Zhang et al., 2024). Algorithmic auditing competencies enable workers to monitor performance, detect drift or bias, and verify fairness (European Union AI Act, 2024; ISO, 2023).

Psychological adaptation required for effective collaboration with different AI types varies. Working with deterministic systems such as rule-based expert systems or trained neural networks with temperature 0 provides predictability many workers find reassuring (Vale et al., 2022). Non-deterministic systems such as reinforcement learning agents during exploration phases or large language models with higher temperature settings require workers to tolerate and manage uncertainty (Pursnani et al., 2023).

Organisations must grapple with ethical considerations such as ensuring AI systems do not perpetuate biases or infringe on privacy rights (Barocas & Selbst, 2016). Implementing hybrid intelligence is not simply a technology upgrade but a cultural transformation (Akata et al., 2020).

IMPLEMENTATION AND CHANGE MANAGEMENT

Successful implementation of hybrid decision systems requires more than technical deployment; it demands coordinated organisational change and capacity building. Resistance can arise from fear of job loss, uncertainty about new roles and distrust in algorithmic recommendations. Change management models help leaders work through these challenges.

The ADKAR model emphasises five building blocks—Awareness, Desire, Knowledge, Ability and Reinforcement—for guiding individuals through change. Awareness entails informing stakeholders about upcoming changes and their significance; Desire encourages active support; Knowledge provides necessary training; Ability ensures practical application; and Reinforcement sustains adoption (Hiatt, 2006).

The ADKAR sequence (Hiatt, 2006) provides a disciplined remedy at the individual level but must be coupled with organisational instruments that act on the whole system. Awareness-building for hybrid decision systems differs from generic change communication in two respects: it must demystify the probabilistic behaviour of the specific AI architecture in use, and it must explicitly locate accountability within the human-on-the-loop or human-in-the-loop configuration chosen. Desire is not generated by exhortation but by demonstration, which is why pilots that surface algorithmic errors in a low-stakes sandbox tend to produce stronger later adoption than pilots optimised for show performance (Bansal et al., 2019; Mosqueira-Rey et al., 2023). Knowledge acquisition now extends beyond tool operation to prompt engineering, algorithmic auditing and interpretation of uncertainty estimates, the three competency clusters that distinguish orchestrators from operators (Wilson & Daugherty, 2018; World Economic Forum, 2025). Ability is built through repeated task-embedded practice in realistic decision environments; Reinforcement is maintained through feedback loops in which outcomes of human–AI decisions are traced back to architectural or procedural causes.

Kotter's eight-step process for leading change begins with establishing a sense of urgency, creating a guiding coalition, developing and communicating a clear vision, empowering employees by removing obstacles, generating short-term wins, consolidating gains and producing more change, and anchoring new approaches in organisational culture. These steps help leaders build momentum, align stakeholders and embed change (Kotter, 1996).

Kotter's (1996) eight-step process operates at the level of the organisation and is complementary to ADKAR. Creating urgency for AI integration is rarely difficult; creating urgency around the quality of integration is considerably harder and usually requires linking hybrid decision performance to board-level strategic risks. A guiding coalition must span technical leadership, business-unit ownership, risk and compliance, data stewardship, and workforce representation; coalitions that omit any of these layers produce implementations that satisfy one constituency and fail others (Raisch & Krakowski, 2021). The communicated vision should name the collaboration configuration explicitly—HITL, HOTL, or HOOL—so that every employee understands what AI will and will not decide unilaterally, which limits over-reliance (Dzindolet et al., 2003; Lee & See, 2004) and under-utilisation (Logg et al., 2019). Short-term wins should favour decisions of moderate stakes with clear outcome measurement rather than high-profile applications whose performance is noisy.

Three barriers warrant specific treatment. The first is data-infrastructure debt. Most large organisations discover during AI deployment that their data systems, designed for operational reporting, are not fit for continuous model training or for the feature lineage required by ISO/IEC 42001:2023. Remediation usually requires eighteen to thirty-six months of platform investment before hybrid decision systems can operate reliably at scale (International Organization for Standardization, 2023). The second is accountability architecture. Responsibility for algorithmic decisions is often left to whichever team happens to own the model, leading to unclear escalation paths when outputs harm stakeholders. The Raji et al. (2020) internal algorithmic auditing framework, combined with the post-market monitoring duties imposed by the European Union (2024) AI Act, supplies the accountability scaffolding needed for high-risk applications. The third barrier is cultural: organisations that have historically rewarded individual expertise penalise orchestrators who succeed by recognising the limits of their own judgement and deferring to AI, and vice versa (Kellogg et al., 2020). Performance evaluation systems must therefore be redesigned to reward appropriate rather than frequent deference.

Sectoral specifics sharpen these general patterns. In healthcare, implementation failures such as IBM Watson for Oncology (Strickland, 2019) reflect not model performance alone but insufficient integration with local clinical guidelines and electronic health records, underscoring that hybrid decision systems succeed or fail at the boundary between architecture and context rather than within the model itself. In financial services, algorithmic credit and fraud decisions operate under pre-existing accountability regimes and consequently achieve implementation more easily than in sectors without comparable compliance infrastructure. In manufacturing and logistics, reinforcement-learning and computer-vision deployments have matured toward human-on-the-loop arrangements supported by safety standards that predate AI regulation (Villani et al., 2018; Choudhury et al., 2022). In public-sector deployments, legitimacy constraints frequently force the choice of interpretable models even at the cost of accuracy (Rudin, 2019; Leslie, 2019), a pattern consistent with the transparency-by-design principle (Felzmann et al., 2020) and with broader observations about the governance of AI in democratic contexts (Morley et al., 2020).

Implementing AI systems also encounters barriers such as organisational resistance, cultural inertia, data quality issues and technical debt. People may fear loss of control or job security; legacy systems may limit integration; and inadequate training can hinder effective adoption (Pressman & Wildavsky, 1984). Change strategies should therefore include investment in workforce upskilling, aligning skill sets with new cognitive roles, and establishing feedback loops to monitor impact and adjust processes.

Sector evidence indicates that implementation failures outnumber successes by a wide margin. A RAND Corporation study based on interviews with sixty-five experienced data scientists and engineers found that more than four in five AI projects fail to reach meaningful production deployment, roughly twice the failure rate of non-AI information-technology projects (Ryseff et al., 2024). A parallel meta-analysis of 106 experimental studies on human–AI decision-making found that, on average, human–AI combinations under-perform the better of the two when used alone, with gains concentrated in content-creation rather than decision tasks (Vaccaro et al., 2024). A parallel enterprise study by MIT Project NANDA (2025), drawing on fifty-two executive interviews, a survey of one hundred and fifty-three leaders and analysis of three hundred deployments, reports that around nineteen in twenty generative-AI pilots produced no measurable profit-and-loss impact, with tangible gains concentrated in the technology and media sectors and negligible gains reported in professional services, healthcare, finance, consumer and energy. The diagnosis across these sources points to four clusters of cause: misalignment between architectural choice and decision context; insufficient investment in human competencies and interface design; fragmented governance and ownership; and data infrastructures that cannot sustain continuous learning (Shrestha et al., 2019). Both findings warrant the framework advanced here: matching architecture to context is not a refinement but a precondition for AI projects that produce value.

STRATEGIC AND ORGANISATIONAL IMPLICATIONS

The deployment of diverse artificial intelligence architectures generates strategic and organisational implications extending beyond operational efficiency improvements to reshape competitive dynamics, organisational structures, capability requirements, and value creation mechanisms across industrial sectors. Human-AI cognitive collaboration can be a source of competitive advantage, but requires strategic realignment and organisational change. Understanding these implications necessitates examining how AI-enabled decision systems affect organisational agility responding to environmental changes, resilience maintaining performance amid disruptions, and dynamic capabilities continuously adapting to evolving contexts (Teece et al., 1997; Eisenhardt & Martin, 2000).

Organisational agility, defined as capacity to detect and respond rapidly to environmental changes, opportunities, and threats, becomes enhanced through AI systems enabling faster information processing, pattern recognition across vast datasets, and automated adaptation to shifting conditions (Teece et al., 1997). Supervised machine learning algorithms continuously monitoring market conditions, customer behaviours, and competitive actions enable organisations to identify trends and anomalies requiring strategic response more rapidly than manual analysis permits (Hastie et al., 2009).

However, AI-driven agility proves double-edged, as organisations may become over-responsive to algorithmic signals without adequate consideration of strategic coherence and long-term implications. Reinforcement learning systems optimising operational decisions may make locally efficient choices undermining strategic objectives when reward specifications inadequately capture broader organisational goals (Sutton & Barto, 2018). Financial trading algorithms responding to market movements at millisecond timescales may amplify volatility through herding behaviour when multiple systems react similarly to identical signals.

Strategic decision-making traditionally involves analysing competitive environments, forecasting trends, and creatively formulating plans – processes dependent on executives' experience and intuition (Mintzberg, 1994). AI changes this by providing analytic and simulation capabilities. Leaders draw on predictive analytics and scenario modelling to inform strategic choices. Csaszar et al. (2024) showed LLMs could generate business strategies attracting investor interest comparable to human-devised plans. This suggests AI can act as "strategy consultant" in ideation phases, expanding options considered. Evidence indicates hybrid human-AI approaches outperform either alone (Cortes et al., 2022).

The development of dynamic capabilities in hybrid organisations requires investment in both technical infrastructure and human capital effectively drawing on diverse AI architectures. To optimise decision-making within hybrid-system contexts, shifts involve use of hybrid intelligence in creation of integrated strategic decision-making approaches (Rodríguez-Espíndola et al., 2020). Organisations must cultivate capabilities in selecting appropriate AI architectures for specific decision contexts, integrating multiple AI types into coherent decision systems, and managing transitions between different levels of automation as circumstances change (von Krogh, 2018).

Competitive advantage increasingly derives from sophisticated orchestration of human and AI capabilities rather than from any single technology. Organisations excelling in combining pattern-recognition capabilities of deep learning systems, logical reasoning of expert systems, adaptability of reinforcement learning, and creative potential of generative AI with human intuition, ethics, and contextual understanding will outperform those relying on monolithic approaches (Anderson, 1983).

At macro level, "AI-first" firms design workflows around AI capabilities, pursuing data-driven value creation (Brynjolfsson & McAfee, 2017). Such firms exploit virtuous cycles – more data yields better models, improving service quality, attracting more users and data. Data-network effects underpin modern platform dominance. Proprietary algorithms and data become strategic assets (Barney, 1991).

Adopting hybrid decision systems calls for structural adjustment. Traditional silos hinder cross-functional AI integration. Agile "fusion teams" – combining domain experts and AI specialists – are increasingly common (Ransbotham et al., 2020). AI decision tools can flatten hierarchies by empowering frontline staff with real-time analytics. However, algorithmic centralisation may reassert control if AI is used for surveillance or micro-management (Kellogg et al., 2020). Leadership must balance empowerment with accountability.

Cultural transformation proves equally important. Effective adoption requires data-driven yet psychologically safe culture: employees must feel encouraged to challenge AI outputs and use them as learning tools, not unquestionable authorities. Training should celebrate both successful AI use and human overrides preventing errors. Raisch and Krakowski (2021) highlight an "automation-augmentation paradox": over-automation breeds inflexibility and social backlash, whilst neglecting automation forgoes efficiency gains.

Organisational resilience, capacity to maintain performance amid disruptions whilst adapting to fundamentally changed conditions, benefits from AI systems enabling robust operations through redundancy, rapid reconfiguration, and early warning of emerging threats (Linkov et al., 2014). Computer vision systems and unsupervised learning algorithms monitoring equipment condition enable predictive maintenance identifying potential failures before they occur (Hastie et al., 2009; Villani et al., 2018). Expert systems and supervised learning models encode critical organisational knowledge in explicit computational forms, reducing dependence on individual experts (Russell & Norvig, 2021). When senior engineers retire, expert systems capturing diagnostic reasoning preserve institutional knowledge. Trained ML models retain learned risk assessment patterns when experienced underwriters leave financial institutions.

Reinforcement learning systems show adaptive resilience by continuously learning from environmental feedback and adjusting strategies as conditions evolve (Sutton & Barto, 2018). Warehouse logistics systems employing RL adapt routing and resource allocation strategies as demand patterns shift, equipment availability changes, and new constraints emerge. Manufacturing scheduling systems adjust production sequences responding to supply disruptions, quality issues, and changing priorities.

AI introduces novel vulnerabilities. Model risk – systemic errors in algorithms – and dependency risk – reliance on third-party infrastructure – must be governed as enterprise risks. Resilience requires fallback procedures: maintaining human capacity to take over if AI fails. Aviation and healthcare already mandate such redundancy (Hauer, 2023). Firms should include AI incidents in enterprise risk registers and conduct scenario tests for model drift, bias amplification, and system outages.

Manufacturing sector implications reflect AI's impacts on production efficiency, quality assurance, supply chain coordination, and product innovation. Computer vision systems and deep learning models automate visual inspection identifying defects with accuracy exceeding human capabilities whilst operating continuously (Villani et al., 2018; LeCun et al., 2015). Predictive maintenance systems employing supervised and unsupervised learning analyse sensor data identifying equipment degradation patterns. Robotics AI systems transform manufacturing operations through automated assembly, material handling, and machine tending (Villani et al., 2018). Collaborative robots working alongside humans combine robotic consistency and strength with human adaptability and judgement. Reinforcement learning optimises production scheduling, resource allocation, and quality control parameters (Sutton & Barto, 2018).

Financial services witness AI transforming credit assessment, fraud detection, market risk evaluation, and regulatory compliance (Hastie et al., 2009). Supervised learning models assess credit risks using extensive applicant data and historical default patterns. These systems may perpetuate or amplify biases present in training data, necessitating careful fairness auditing (Mehrabi et al., 2021). Unsupervised learning and deep learning systems detect fraudulent transactions by identifying patterns deviating from typical behaviours.

Healthcare sector implications encompass diagnostic support, treatment optimisation, drug discovery, and administrative efficiency. Computer vision systems analysing medical imaging detect anomalies including tumours, fractures, and vascular abnormalities (Topol, 2019; LeCun et al., 2015). Supervised learning models predict patient risks for conditions including readmissions, complications, and disease progression. Expert systems encode clinical practice guidelines supporting standardised, evidence-based care particularly in resource-constrained settings (Russell & Norvig, 2021).

FUTURE SYNERGIES AND COGNITIVE ECOSYSTEMS

The evolution toward 2040 cognitive ecosystems represents transformation from isolated AI deployments to interconnected networks where multiple AI architectures, human agents, and organisational systems engage in continuous co-evolution. Looking towards the future, trajectory of human-AI collaboration suggests emergence of expansive cognitive ecosystems transcending individual organisations. These future ecosystems will transcend current limitations through emergent capabilities arising from synergistic interaction of diverse intelligence types, creating new forms of collective cognition reshaping industrial and societal decision-making (Akata et al., 2020).

Multi-agent systems incorporating all ten AI types will create opportunities for complex problem-solving through complementary processing. Integrated systems where computer vision CNNs identify patterns in visual data, unsupervised learning algorithms discover hidden structures, expert systems apply domain rules, reinforcement learning agents explore solution spaces, and large language models synthesise findings into actionable insights exemplify these possibilities (Battaglia et al., 2018). These systems will not merely aggregate individual capabilities but generate emergent intelligence through interactions, solving problems neither human nor individual AI systems could address alone (Surowiecki, 2004).

Swarm intelligence paradigms will extend beyond current applications to encompass heterogeneous AI swarms where different agent types specialise in distinct cognitive functions. Deterministic agents such as expert systems will provide stable anchoring points and rule enforcement, whilst non-deterministic agents explore novel solutions and adapt to changing conditions (Bonabeau et al., 1999). Transformers replaced traditional RNNs and CNNs with self-attention mechanisms, eliminating need for recurrent connections and convolutional layers (Vaswani et al., 2017). This design enables parallelisation, captures long-range dependencies, and efficiently handles sequential data. Human agents will participate as swarm members, contributing capabilities in ethical reasoning, creative insight, and contextual understanding complementing artificial agents (Dorigo & Stützle, 2004).

Industry 5.0 emphasises human-centricity, resilience, and sustainability, envisioning automation serving humanity rather than replacing it. In this paradigm, factories of future may consist of interconnected smart machines learning from one another, guided by human oversight focused on creativity and well-being (Nahavandi, 2019). Cognitive ecosystems in manufacturing might link suppliers, producers, distributors, and customers through AI-driven decision networks optimising value chains whilst humans supervise ethical and strategic trade-offs.

The interoperability challenges facing future cognitive ecosystems extend beyond technical integration to encompass semantic alignment, goal coordination, and trust establishment between diverse agent types. Standards for AI-to-AI communication must accommodate different representational formats and processing paradigms of various architectures (Blasch et al., 2019). Expert systems operating through symbolic logic must interface with neural networks processing continuous representations. Reinforcement learning agents optimising for specific reward functions must coordinate with large language models pursuing more abstract objectives.

Trust architectures for 2040 scenarios must address multi-level trust relationships: between humans and individual AI systems, among different AI systems, and at ecosystem level where collective behaviours emerge (Lee & See, 2004). Trust in deterministic systems such as expert systems and trained supervised models can build on predictable behaviour and transparent logic. Trust in non-deterministic systems requires more sophisticated approaches, including reputation systems tracking performance across multiple interactions. Blockchain-based architectures recording AI system credentials, performance histories, and audit trails provide potential mechanisms for establishing verifiable trust across organisational boundaries (Casino et al., 2019).

Ethical scalability in cognitive ecosystems presents challenges as collective behaviour of multiple AI systems may diverge from ethical constraints placed on individual components. When reinforcement learning agents interact with generative AI systems, unexpected strategies may emerge violating ethical principles even when each system operates within designed parameters (Bolander, 2019). Future ecosystems will require ethical governance mechanisms operating at multiple scales, constraining individual agent behaviours and shaping emergent ecosystem dynamics (Felzmann et al., 2020).

Human roles in these future ecosystems likely emphasise strategic direction establishing objectives and constraints for AI agent collectives, exception handling intervening when automated coordination proves inadequate, and ethical oversight ensuring collective behaviours align with human values (Wilson & Daugherty, 2018). Rather than directly executing tasks or supervising individual AI systems, humans increasingly operate at meta-levels establishing parameters for AI agent societies and monitoring emergent collective properties.

These elevated roles demand sophisticated competencies including systems thinking comprehending complex interactions among interdependent agents, anticipatory reasoning predicting potential emergent behaviours, and ethical judgement navigating value conflicts and distributional consequences (Jarrahi, 2018). Educational systems must evolve preparing workers through interdisciplinary training integrating technical AI literacy, complex systems understanding, ethical reasoning, and adaptive learning capabilities.

The capabilities needed to run hybrid decision systems well are themselves dynamic. Teece's (2007) microfoundations account separates sensing (detecting signals that current AI configurations are degrading or that a new architecture offers an advantage), seizing (reallocating data, talent and oversight capacity to exploit the signal) and transforming (reorganising workflows and retraining staff so the new capability sticks). Organisations with strong sensing detect model drift months before it surfaces in complaints; those with strong seizing move people and budget across silos without friction; those with strong transformation absorb new AI types without collapsing trust in the systems they already operate. Absorptive capacity—the ability to identify, assimilate and apply external knowledge (Cohen & Levinthal, 1990)—conditions all three stages: firms that invest in data literacy, model-auditing expertise and cross-functional AI governance enjoy compounding returns, while firms that treat each AI deployment as a one-off project pay rising integration costs.

Global cognitive infrastructure might emerge as AI infrastructure becomes global public utility – network of interoperable, ethical AI services accessible to all (Etzioni, 2018). This would democratise decision support so small firms or individuals can access intelligence once reserved for large corporations. Decentralised and open-source AI frameworks would be vital to maintaining fairness and innovation.

Future cognitive ecosystems could exhibit antifragility – systems improving through shocks (Taleb, 2012). Supply-chain AI networks might learn from disruptions, strengthening coordination after crises. This requires diversity in algorithmic approaches and human oversight preventing systemic convergence. Governments could employ large-scale AI simulations for proactive crisis management.

Successfully working through these challenges promises cognitive ecosystems exhibiting capabilities exceeding current human-AI configurations, accelerating innovation, enhancing quality of life, and addressing complex challenges including climate change, disease, and global coordination. Realising these benefits whilst avoiding scenarios of uncontrolled AI proliferation, technological unemployment, or erosion of human agency requires sustained attention to technical robustness, ethical alignment, and inclusive governance ensuring these technologies serve broad human flourishing.

KEY FINDINGS

This chapter has systematically addressed seven fundamental questions regarding human-AI strategic decision-making frameworks. The following synthesis presents the key findings emerging from this analysis.

Transformation of Cognitive and Procedural Decision Frameworks

The first question examined how different AI architectures transform cognitive and procedural decision frameworks within organisations. The analysis reveals that each AI type introduces distinctive modifications to decision processes. Deterministic systems including expert systems and supervised machine learning models transform decision-making by providing consistent, traceable analytical support that reduces cognitive load for routine decisions whilst maintaining transparency (Russell & Norvig, 2021; Hastie et al., 2009). These systems enable organisations to encode and scale expertise, ensuring consistent application of complex rule sets and statistical patterns across distributed decision points.

Non-deterministic systems including reinforcement learning agents, large language models, and generative AI reshape decision environments by introducing exploration, creativity, and adaptability at the cost of predictability (Sutton & Barto, 2018; Bommasani et al., 2021). These systems enable organisations to manage novel situations and generate innovative solutions but require sophisticated trust calibration and oversight mechanisms. The transformation extends beyond automation to reconceptualisation of cognitive labour distribution, with AI systems handling data processing, pattern recognition, and option generation whilst humans provide contextual understanding, ethical reasoning, and strategic direction (Jarrahi, 2018; Gill, 2020).

Frameworks for Human-AI Collaboration

The second question addressed appropriate frameworks for characterising human-AI collaboration configurations. The research establishes that the human-in-the-loop, human-on-the-loop, and human-out-of-the-loop typology provides essential structure for understanding collaboration modes (Cummings, 2004; Sheridan & Parasuraman, 2005). Human-in-the-loop configurations prove necessary for high-stakes decisions requiring contextual judgement and ethical consideration, regardless of AI sophistication. Healthcare diagnostics, strategic planning, and legal decisions exemplify contexts where humans must retain ultimate authority despite AI analytical support (Topol, 2019; Mosqueira-Rey et al., 2023).

Human-on-the-loop frameworks balance efficiency with safety for dynamic operational contexts where AI systems execute within parameters whilst humans monitor performance and intervene when necessary. Autonomous vehicle systems, warehouse robotics, and algorithmic trading demonstrate successful applications where operational tempo favours automation under human supervision (Grigorescu et al., 2020; Sutton & Barto, 2018). Human-out-of-the-loop configurations prove appropriate only for low-stakes, stable environments where decision logic remains consistent and consequences of errors prove minimal. The analysis reveals that optimal configuration depends on intersection of AI architectural characteristics, decision stakes, environmental dynamism, and ethical complexity.

Table 3. AI Architecture × Decision Context Mapping

AI Architecture	Low Stakes, Stable Context	Low Stakes, Dynamic Context	High Stakes, Stable Context	High Stakes, Dynamic Context	Ethical/Legal Complexity
Expert Systems	HOOL Automated compliance checking routine diagnostics	HOTL Process control with exceptions	HITL Medical diagnosis support legal analysis	HITL Crisis decision support	HITL Legal compliance ethical review
Supervised ML	HOOL Spam filtering product categorisation	HOTL Fraud detection quality control	HITL Credit risk assessment medical screening	HITL Financial trading decisions	HITL Hiring decisions loan approvals
Unsupervised ML	HOTL Customer segmentation data compression	HOTL Anomaly detection pattern discovery	HITL Security threat detection	HITL Market analysis risk identification	HITL Bias detection fairness auditing
Reinforcement Learning	HOTL Game playing simulation optimisation	HOTL/HOOL Resource scheduling routing	HOTL Industrial process optimisation	HITL Strategic planning portfolio management	HITL Autonomous weapons systems
Deep Learning - CNNs	HOOL Image tagging content moderation	HOTL Manufacturing inspection	HITL Medical imaging diagnosis	HITL Security screening threat detection	HITL Facial recognition surveillance
Deep Learning - RNNs/Transformers	HOOL Time series forecasting	HOTL Demand prediction trend analysis	HITL Clinical event prediction	HITL Financial market prediction	HITL Predictive policing
Large Language Models	HOOL Content generation chatbots	HOTL Customer service documentation	HITL Medical report generation legal drafting	HITL Strategic analysis policy writing	HITL Legal advice medical consultation
Computer Vision Systems	HOOL Inventory counting traffic monitoring	HOTL Quality inspection object tracking	HITL Medical diagnosis security screening	HITL Autonomous driving decisions	HITL Surveillance privacy-sensitive applications

continued on following page

Table 3. Continued

AI Architecture	Low Stakes, Stable Context	Low Stakes, Dynamic Context	High Stakes, Stable Context	High Stakes, Dynamic Context	Ethical/Legal Complexity
Generative AI	HOOL Marketing content stock images	HOTL Design prototyping creative assistance	HITL Architectural design drug discovery	HITL Strategic scenario generation	HITL Deepfake detection content authenticity
Robotics AI	HOOL Warehouse sorting routine assembly	HOTL Collaborative manufacturing	HOTL Surgical assistance hazardous operations	HITL Emergency response military operations	HITL Autonomous weapons care robotics

Note:

- **Key:** HITL (Human-in-the-Loop) indicates configurations where humans retain decision authority whilst AI provides recommendations. HOTL (Human-on-the-Loop) represents arrangements where AI operates autonomously with human supervision and intervention capability. HOOL (Human-out-of-the-Loop) denotes full AI autonomy within defined parameters.
- **Decision Context Dimensions:** Stakes range from low (reversible decisions with minimal consequences and limited liability) to high (irreversible outcomes with significant consequences and substantial liability). Environmental stability varies from stable contexts (predictable patterns, consistent rules, minimal variation) to dynamic environments (changing conditions, evolving patterns, high uncertainty). Ethical/legal complexity covers decisions involving human rights, privacy, fairness, regulatory compliance, or moral judgements.
- **Configuration Selection Principles:** Escalation to human control increases with higher decision stakes, greater environmental dynamism, increased ethical complexity, and reduced AI transparency. Conversely, automation feasibility increases with lower decision stakes, environmental stability, clear objective functions, and high AI explainability. Hybrid configurations (HOTL) prove optimal when operational efficiency matters, human oversight provides safety, context allows intervention time, and AI reliability is proven but not absolute.

Actual deployment configurations may vary based on regulatory requirements, organisational risk tolerance, and technological maturity. This matrix represents typical patterns observed across industrial deployments.

Workforce Transformation and Capability Evolution

The third question explored workforce transformation as AI integration shifts human roles from operators to orchestrators. The findings show that rather than wholesale replacement, jobs undergo reconfiguration with AI assuming routine analytical tasks whilst humans focus on creativity, empathy, and complex reasoning (Wilson & Daugherty, 2018; Indeed Hiring Lab, 2025). Approximately one quarter of jobs face substantial transformation, yet only one percent of skills prove fully automatable, indicating evolution rather than elimination of human work.

Required competencies vary systematically across AI types. Expert systems demand understanding of rule logic and exception handling. Supervised learning requires statistical reasoning and trust calibration toward probabilistic outputs. Large language models necessitate prompt engineering and critical evaluation skills. Reinforcement learning systems require understanding of exploration-exploitation dynamics. Deep learning poses challenges in trust calibration absent explainability (Hastie et al., 2009; Bommasani et al., 2021; Sutton & Barto, 2018). Successful adaptation demands cognitive flexibility, emotional intelligence, and continuous learning capabilities transcending technical skills (World Economic Forum, 2023; Wilson & Daugherty, 2018).

Governance Structures and Ethical Frameworks

The fourth question examined necessary governance structures and ethical frameworks for hybrid systems. The analysis reveals that transparency and explainability requirements must adapt to AI architectural characteristics, with deterministic systems offering inherent interpretability whilst non-deterministic systems require statistical characterisation and post-hoc explanation techniques (Vale et al., 2022; Ribeiro et al., 2016; Lundberg & Lee, 2017). Contemporary regulatory frameworks including the EU AI Act, ISO/IEC 42001, and OECD AI Principles establish risk-based governance mandating human oversight for high-risk applications whilst enabling innovation in lower-risk contexts (European Commission, 2021; International Organization for Standardization, 2023; OECD, 2019).

Accountability must remain with human actors and institutions regardless of AI autonomy, implemented through human-in-command approaches, audit trails, and circuit-breaker protocols (Bryson, 2018; Floridi et al., 2018). Algorithmic auditing emerges as essential governance mechanism for assessing performance, detecting bias, and ensuring alignment with objectives (Raji et al., 2020). The analysis shows that ethical governance requires embedding fairness-aware design from dataset curation through deployment, with continuous monitoring for bias amplification and distributional impacts (Morley et al., 2020; Angwin et al., 2016).

Strategic and Organisational Implications

The fifth question addressed strategic and organisational implications of AI deployment. The findings reveal that organisations gain enhanced agility through faster information processing and pattern recognition, though risk over-responsiveness to algorithmic signals without strategic coherence (Teece et al., 1997; Sutton & Barto, 2018). AI transforms strategic decision-making by providing analytical and simulation capabilities that expand option generation and evaluation, with evidence that AI-generated strategies can match human expert quality (Csaszar et al., 2024; Mintzberg, 1994).

Competitive advantage increasingly derives from sophisticated orchestration of diverse AI capabilities rather than any single technology, requiring dynamic capabilities in selecting appropriate architectures, integrating multiple systems, and managing automation transitions (von Krogh, 2018; Anderson, 1983). Organisational structures must evolve from traditional silos toward cross-functional fusion teams whilst cultures balance data-driven decision-making with psychological safety encouraging AI output challenge (Ransbotham et al., 2020; Kellogg et al., 2020). The automation-augmentation paradox reveals that over-automation breeds inflexibility whilst under-automation forgoes efficiency gains, necessitating adaptive strategies (Raisch & Krakowski, 2021).

Evolution Toward Cognitive Ecosystems

The sixth question explored evolution toward future cognitive ecosystems. The analysis projects transformation from isolated deployments toward interconnected networks where multiple AI architectures and human agents engage in continuous co-evolution (Akata et al., 2020; Surowiecki, 2004). Multi-agent systems incorporating diverse AI types will generate emergent intelligence through complementary processing, with deterministic agents providing stability whilst non-deterministic agents enable adaptation (Battaglia et al., 2018; Bonabeau et al., 1999).

Interoperability challenges extend beyond technical integration to encompass semantic alignment and trust establishment between heterogeneous agents (Blasch et al., 2019). Trust architectures must address multi-level relationships through reputation systems and blockchain-based verification mechanisms (Casino et al., 2019; Jøsang et al., 2007). Human roles will emphasise meta-level activities including strategic direction, exception handling, and ethical oversight of AI agent collectives (Wilson & Daugherty, 2018). Industry 5.0 paradigms envision human-centric automation where interconnected machines serve human creativity and well-being rather than replacing human agency (Nahavandi, 2019).

Integrated Conceptual Framework

The seventh question sought an integrated framework characterising human-AI decision systems. The analysis yields a three-dimensional model wherein decision-making emerges from interaction between AI architectural characteristics (determinism, transparency, adaptivity), decision contexts (stakes, uncertainty, dynamism), and human cognitive roles (execution, verification, orchestration). This framework reveals systematic patterns: deterministic transparent systems align with high-stakes human-in-the-loop configurations, whilst adaptable non-deterministic systems suit dynamic human-on-the-loop frameworks.

The framework maps interaction patterns including sequential processing chains, parallel configurations, and hierarchical structures enabling synergistic capabilities (Zhang & Patel, 2006; Caruana & Niculescu-Mizil, 2006; Vaswani et al., 2017).

Risk propagation patterns reveal cascade potential requiring resilient designs with validation checkpoints. Economic value creation operates at multiple levels from direct efficiency gains through cognitive liberation to systemic learning and network effects (Anderson, 1983; Jarrahi, 2018; March & Simon, 1958; Surowiecki, 2004). Success factors span technical transparency, organisational accountability and socio-cultural adaptation enabling productive human-AI complementarity whilst mitigating risk. Figure 3 depicts the resulting three-dimensional solution space, the thirty-six cells of which are enumerated in Table 4.

Figure 3. The Three-Dimensional Integrated Framework

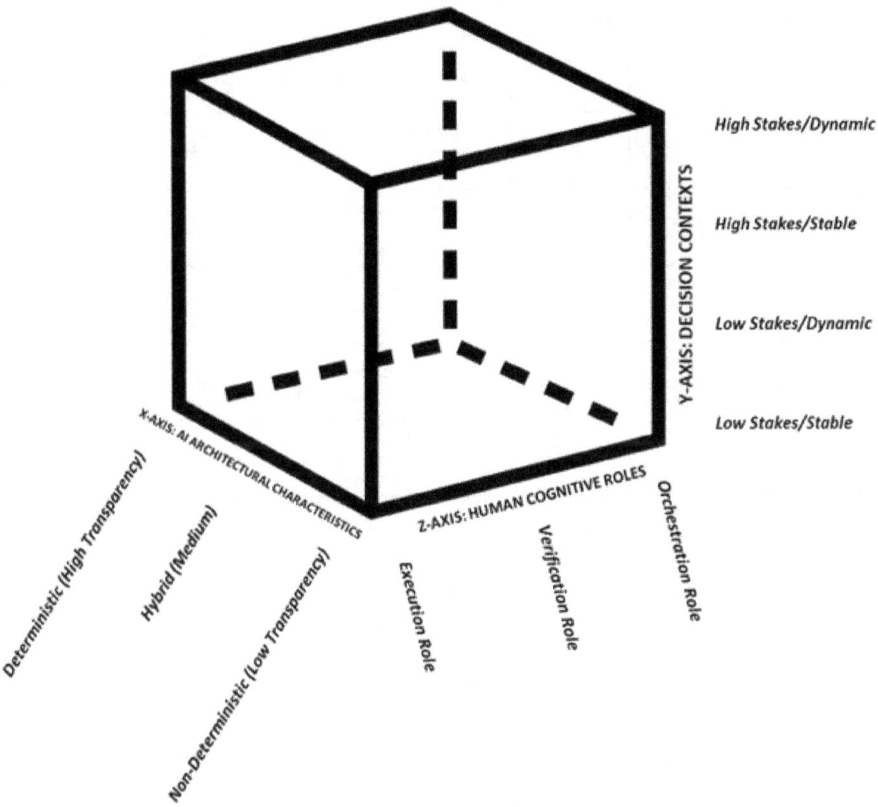

From Figure 3. 36 different combinations can be created, and they are shown in Table 4

Table 4. The Thirty-Six Combinations Arising Out of the Framework

#	AI Architecture	Decision Context	Human Role	Description	Complexity	Example
1	Deterministic	Low Stakes, Stable	Execution	Basic automation: Rule-based systems handling routine operations with minimal human oversight	Low	Data entry automation
2	Deterministic	Low Stakes, Stable	Verification	Quality control systems: Humans verify automated processes in predictable environments	Low	Invoice checking
3	Deterministic	Low Stakes, Stable	Orchestration	Process management: Coordinating multiple rule-based systems for efficiency	Low	Workflow automation

continued on following page

Table 4. Continued

#	AI Architecture	Decision Context	Human Role	Description	Complexity	Example
4	Deterministic	Low Stakes, Dynamic	Execution	Adaptive workflows: Rule-based systems in changing but low-risk environments	Medium	Seasonal inventory
5	Deterministic	Low Stakes, Dynamic	Verification	Dynamic monitoring: Checking transparent AI outputs in evolving contexts	Medium	Content filtering
6	Deterministic	Low Stakes, Dynamic	Orchestration	Agile coordination: Managing deterministic systems in changing environments	Medium	Campaign management
7	Deterministic	High Stakes, Stable	Execution	Critical operations: Transparent AI in high-risk but predictable scenarios	Medium	Medical dosage calculations
8	Deterministic	High Stakes, Stable	Verification	Safety validation: Human oversight of critical automated processes	Medium	Aircraft pre-flight checks
9	Deterministic	High Stakes, Stable	Orchestration	Strategic oversight: Managing critical systems with full transparency	High	Air traffic control
10	Deterministic	High Stakes, Dynamic	Execution	Crisis operations: Transparent AI in rapidly changing high-risk scenarios	High	Emergency response
11	Deterministic	High Stakes, Dynamic	Verification	Real-time safety: Continuous validation in critical changing conditions	High	Disaster monitoring
12	Deterministic	High Stakes, Dynamic	Orchestration	Crisis management: Coordinating transparent AI in emergency situations	Critical	Pandemic response
13	Hybrid	Low Stakes, Stable	Execution	Semi-automated tasks: Mix of AI and human actions in routine work	Low	AI-assisted customer service
14	Hybrid	Low Stakes, Stable	Verification	Balanced oversight: Checking partially transparent AI decisions	Medium	Recommendation review
15	Hybrid	Low Stakes, Stable	Orchestration	Mixed coordination: Managing hybrid systems in stable environments	Medium	Content curation
16	Hybrid	Low Stakes, Dynamic	Execution	Adaptive collaboration: Human-AI partnership in changing contexts	Medium	Real-time translation
17	Hybrid	Low Stakes, Dynamic	Verification	Dynamic validation: Checking hybrid AI in evolving scenarios	Medium	Fraud detection
18	Hybrid	Low Stakes, Dynamic	Orchestration	Flexible management: Coordinating mixed systems dynamically	Medium	Supply chain optimization
19	Hybrid	High Stakes, Stable	Execution	Critical collaboration: Human-AI teams in high-risk operations	High	Surgical robotics
20	Hybrid	High Stakes, Stable	Verification	Risk assessment: Validating hybrid AI in critical applications	High	Medical diagnosis verification
21	Hybrid	High Stakes, Stable	Orchestration	Strategic integration: Managing hybrid systems in critical contexts	High	Hospital system management
22	Hybrid	High Stakes, Dynamic	Execution	Crisis collaboration: Human-AI teams in rapidly evolving emergencies	High	Combat medicine

continued on following page

Table 4. Continued

#	AI Architecture	Decision Context	Human Role	Description	Complexity	Example
23	Hybrid	High Stakes, Dynamic	Verification	Dynamic risk control: Real-time validation of hybrid AI in crises	Critical	Financial crisis management
24	Hybrid	High Stakes, Dynamic	Orchestration	Adaptive command: Strategic management of hybrid systems in crisis	Critical	Military operations
25	Non-Deterministic	Low Stakes, Stable	Execution	Black-box automation: Opaque AI handling routine low-risk tasks	Medium	Image recognition
26	Non-Deterministic	Low Stakes, Stable	Verification	Output validation: Checking opaque AI results in stable contexts	Medium	ML model review
27	Non-Deterministic	Low Stakes, Stable	Orchestration	Black-box management: Coordinating opaque systems safely	Medium	Ad targeting systems
28	Non-Deterministic	Low Stakes, Dynamic	Execution	Adaptive black-box: Opaque AI in changing low-risk environments	High	Content generation
29	Non-Deterministic	Low Stakes, Dynamic	Verification	Dynamic auditing: Monitoring opaque AI in evolving contexts	High	Bias detection
30	Non-Deterministic	Low Stakes, Dynamic	Orchestration	Complex coordination: Managing opaque systems dynamically	High	Multi-agent systems
31	Non-Deterministic	High Stakes, Stable	Execution	Critical black-box: Opaque AI in high-risk stable operations (extreme caution)	High	Medical imaging AI
32	Non-Deterministic	High Stakes, Stable	Verification	Critical validation: Intensive oversight of opaque AI in high-risk contexts	High	AI safety monitoring
33	Non-Deterministic	High Stakes, Stable	Orchestration	Maximum complexity: Strategic management of opaque AI in critical contexts	High	Critical infrastructure AI
34	Non-Deterministic	High Stakes, Dynamic	Execution	Ultra-critical black-box: Opaque AI in crisis situations (maximum risk)	Critical	Autonomous weapons
35	Non-Deterministic	High Stakes, Dynamic	Verification	Crisis auditing: Emergency oversight of opaque AI in dynamic crises	Critical	Cybersecurity defense AI
36	Non-Deterministic	High Stakes, Dynamic	Orchestration	EXTREME COMPLEXITY: Managing opaque AI in life-critical dynamic crises	Critical	AGI governance

Discussion of the 36 Scenarios

The Foundation Layer: Low Stakes, Stable Environments (Scenarios #1-3, #13-15, #25-27)

The nine scenarios operating in low stakes, stable environments form the foundation of AI deployment, where organisations typically begin their automation journey. Scenario #1 represents the simplest possible configuration—deterministic AI in execution mode—where rule-based systems handle routine data entry with predictable outcomes and minimal risk. This progresses naturally to scenario #2, where humans verify these automated processes, introducing a quality control layer that builds organisational confidence. Scenario #3 elevates human involvement to orchestration, coordinating multiple deterministic systems to achieve broader operational efficiency.

The hybrid variants (scenarios #13-15) introduce partial opacity whilst maintaining environmental stability. Scenario #13's semi-automated customer service represents a widespread current application where AI suggestions augment human decision-making without fully replacing it. The verification role in scenario #14 becomes more nuanced than its deterministic counterpart (#2), as humans must now validate decisions from partially transparent systems, requiring different skills and potentially introducing new biases. Scenario #15's orchestration of hybrid systems in stable environments often manifests in content platforms where algorithmic and editorial decisions interweave.

The non-deterministic cluster (scenarios #25-27) presents an interesting paradox: whilst the AI is opaque, the stable, low-stakes environment provides a relatively safe space for deployment. Scenario #25's image recognition systems can operate with minimal human oversight precisely because errors have limited consequences. However, scenario #26's verification role becomes crucial for building trust in black-box systems, whilst scenario #27's orchestration challenges involve managing multiple opaque systems that may interact in unexpected ways, even within stable environments.

The Adaptive Layer: Low Stakes, Dynamic Environments (Scenarios #4-6, #16-18, #28-30)

When dynamism enters the equation, even low-stakes scenarios gain complexity. Scenario #4's seasonal inventory management must adapt deterministic rules to changing patterns, requiring more sophisticated programming than static automation. The verification demands of scenario #5 shift from checking static outputs to monitoring evolving patterns, whilst scenario #6's orchestration role must continuously rebalance system priorities as conditions change.

The hybrid configurations (scenarios #16-18) often represent optimal solutions for dynamic but low-risk environments. Scenario #16's real-time translation exemplifies successful human-AI collaboration where neither party alone could match their combined performance. Fraud detection in scenario #17 requires human verifiers who can adapt to evolving criminal tactics whilst drawing on AI's pattern recognition capabilities. Scenario #18's supply chain optimisation shows how hybrid orchestration can balance algorithmic efficiency with human intuition about market dynamics.

The non-deterministic variants (scenarios #28-30) push into genuinely complex territory despite low stakes. Scenario #28's content generation systems operate in constantly shifting cultural and linguistic contexts, making their opacity particularly challenging. Bias detection in scenario #29 presents a meta-challenge: using opaque systems to verify other opaque systems in dynamic environments. Scenario #30's multi-agent coordination resembles managing a team where no member's decision-making is fully understood, yet they must work together effectively.

The Critical Operations Layer: High Stakes, Stable Environments (Scenarios #7-9, #19-21, #31-33)

High stakes reshape even stable scenarios. Scenario #7's medical dosage calculations, whilst deterministic, carry life-or-death implications that demand exceptional reliability and validation protocols. The aircraft pre-flight checks of scenario #8 evolved from decades of aviation incidents, embodying hard-won knowledge about human-system verification in critical contexts. Scenario #9's air traffic control represents one of humanity's most successful examples of orchestrating transparent, critical systems, yet it requires years of training and sophisticated cognitive abilities.

The hybrid approaches (scenarios #19-21) often represent current best practice in high-stakes domains. Surgical robotics in scenario #19 exemplifies the delicate dance between human expertise and machine precision, where neither alone would achieve optimal outcomes. Medical diagnosis verification in scenario #20 increasingly involves AI systems that outperform humans in specific tasks whilst still requiring human oversight for contextual understanding and ethical decisions. Hospital system management in scenario #21 orchestrates a complex mix of automated and human-driven processes where failures cascade through interconnected dependencies.

The non-deterministic cluster (scenarios #31-33) ventures into controversial territory. Scenario #31's medical imaging AI may detect patterns invisible to human radiologists but cannot explain its reasoning, creating ethical dilemmas about accountability. The AI safety monitoring of scenario #32 presents recursive challenges: how do we verify systems we don't fully understand when the stakes are critical? Scenario #33's critical infrastructure AI management raises questions about societal dependency on opaque systems for essential services.

The Crisis Zone: High Stakes, Dynamic Environments (Scenarios #10-12, #22-24, #34-36)

The final nine scenarios represent the frontier of AI deployment challenges. Even with deterministic systems, scenario #10's emergency response operations must balance pre-programmed responses with unpredictable crisis evolution. Disaster monitoring in scenario #11 requires continuous recalibration of what constitutes "normal" as situations deteriorate. Scenario #12's pandemic response coordination showed real-world limitations when COVID-19 overwhelmed even transparent, well-understood systems.

The hybrid configurations (scenarios #22-24) may offer the best hope for managing certain crises. Combat medicine in scenario #22 combines AI's rapid analysis with human medical expertise under extreme pressure, where both speed and accuracy have immediate life consequences. Financial crisis management in scenario #23 must validate hybrid AI decisions whilst markets move in milliseconds, creating temporal paradoxes between necessary oversight and required response speed. Military operations in scenario #24 represent perhaps the most consequential orchestration challenge, where hybrid systems must be coordinated across multiple domains whilst adversaries actively attempt disruption.

The non-deterministic apex (scenarios #34-36) pushes into speculative but increasingly relevant territory. Autonomous weapons in scenario #34 crystallise ethical debates about machines making life-death decisions without human intervention or understanding. Cybersecurity defence in scenario #35 increasingly requires AI systems that can respond faster than human comprehension, yet oversight remains essential when national infrastructure is at stake. Scenario #36 represents the theoretical maximum: orchestrating artificial general intelligence during existential crises, where the AI may be more capable than its human overseers, yet accountability and control remain paramount.

Cross-Cutting Insights

The progression across these scenarios reveals several critical patterns. First, the transition between adjacent scenarios often requires entirely different skillsets—moving from execution to verification to orchestration isn't simply climbing a ladder but rather shifting between fundamentally different cognitive and operational demands. Second, certain combinations create unique emergent properties: hybrid systems in dynamic environments (scenarios #16-18) often outperform both their fully automated and fully manual alternatives, suggesting that the optimal human-AI configuration depends heavily on context.

The framework also illuminates resource allocation challenges. Whilst scenarios #1-3 might operate with minimal governance overhead, scenarios #34-36 require resources that may exceed their operational benefits, raising questions about when AI deployment becomes economically or ethically unjustifiable. The framework suggests that organisational maturity doesn't necessarily progress linearly through scenario numbers—an organisation might simultaneously operate scenario #3 in some departments whilst struggling with scenario #16 in others.

Perhaps most significantly, the framework reveals that complexity isn't merely additive. The jump from scenario #33 to #34—adding dynamism to an already complex configuration—represents a qualitative shift that may require entirely new governance paradigms rather than incremental improvements to existing frameworks. This suggests that organisations approaching these boundary transitions must prepare for discontinuous change rather than smooth evolution.

Falsifiability, Scope Conditions and Counterfactual Cases

Two scope conditions limit these predictions. They apply where AI systems are used for bounded decisions with observable outcomes; they weaken where decisions are genuinely novel, outcomes are delayed by years, or counterfactuals are unobservable. They also assume an institutional environment that tolerates human override; in regimes where AI outputs are legally binding without appeal, the recommended configurations cannot be implemented even where the framework prescribes them.

Counterfactual cases strengthen the account. The Japanese and South Korean robotics manufacturing transitions of the 2000s and 2010s show organisations that successfully stretched capabilities from programmable machines to learning systems through co-investment in workforce retraining (Autor, 2015)—a pattern the framework predicts, and which distinguishes them from otherwise-similar firms that automated without investing in orchestration capability and subsequently lost quality leadership. Where the framework encounters empirical anomalies—for example, demonstrably successful human-out-of-the-loop deployments in high-stakes, dynamic settings—the anomaly either refutes the framework or isolates a missing variable (typically redundant safety layers or constrained action spaces). Neither move is costless, and both provide a route to revision rather than defence of the model in spite of the evidence.

IMPLICATIONS AND RECOMMENDATIONS

The preceding analysis yields a structured set of recommendations for organisations implementing hybrid decision systems across the three dimensions of the integrated framework. Each recommendation is derived from evidence presented earlier and is stated in a form that managers and policymakers can operationalise directly.

First, architectural selection should be driven by decision context rather than technological availability. Low-stakes, stable, well-structured decisions are appropriate for expert systems or supervised learning in human-out-of-the-loop configurations; high-stakes, novel, or socially consequential decisions require non-deterministic architectures only when their performance advantages clearly justify the additional governance burden, and should operate in human-in-the-loop configurations (Shrestha et al., 2019; Parasuraman et al., 2000). Organisations should map their current decisions onto the three-dimensional framework before selecting architectures, rather than retrofitting decisions to whichever system is available.

Second, trust calibration should be engineered into the system rather than assumed. This requires two interventions: exposing calibrated uncertainty estimates alongside every recommendation so that users can distinguish confident from marginal outputs (Okamura & Yamada, 2020; Bansal et al., 2019); and building feedback mechanisms that surface AI errors to users after the fact, so that mental models update in the direction of accurate reliance (Lee & See, 2004; Endsley, 2017). Organisations that omit either intervention consistently produce over-reliance in some roles and under-reliance in others, often simultaneously (Dzindolet et al., 2003).

Third, governance should be layered. A single governance regime cannot accommodate the different transparency and explainability needs of rule-based systems, supervised learning, deep learning, and large language models. A layered approach applies minimum standards uniformly (documentation, audit trails, incident reporting), tightens requirements by risk tier in line with the European Union (2024) Artificial Intelligence Act, and supplements regulatory compliance with sector-specific best practices derived from ISO/IEC 42001:2023 and the OECD AI Principles (Organisation for Economic Co-operation and Development, 2019). Internal algorithmic auditing (Raji et al., 2020) operationalises this layering in organisations of any size.

Fourth, workforce transformation should be treated as a managed transition rather than a market outcome. The shift from operator to orchestrator is gradual and sector-specific; organisations that invest early in prompt engineering, algorithmic auditing and meta-cognitive training develop internal talent pipelines difficult for competitors to replicate (Brynjolfsson et al., 2025; World Economic Forum, 2025). Investment decisions in this area are usefully framed as dynamic-capability building (Teece et al., 1997; Eisenhardt & Martin, 2000), because orchestration capability is precisely the capability to reconfigure human and artificial resources in response to changing contexts.

Finally, organisations should plan for architectural obsolescence. The AI stack that supports today's decisions will be replaced two or three times over the working life of a decision system, so procurement, contractual and data-pipeline choices should preserve the option to substitute architectures without wholesale re-implementation. Organisations that treat AI as a capability rather than a set of point solutions achieve this more reliably than organisations that procure individual products (Ransbotham et al., 2020; Wilson & Daugherty, 2018).

FUTURE RESEARCH DIRECTIONS

The analysis raises eight research questions that are directly testable and together constitute a programme for the next decade of scholarship on hybrid decision systems. Each is stated in a form that admits falsification through observational or experimental work.

Proposition 1: Architectural fit dominates model performance. In controlled settings, well-matched architecture–context pairs outperform poorly matched pairs by a larger margin than best-in-class models outperform average models of the same architecture. This can be tested by varying context and architecture orthogonally in decision-quality experiments following the design of Vaccaro et al. (2024), and predicts that moderate-performance models in well-matched contexts will exceed high-performance models in poorly matched contexts. The prediction weakens where architectures are close substitutes (competing supervised-learning variants on the same task) and where the decision context is generic rather than sector-specific, so the test requires sufficient variation in both dimensions to be informative.

Proposition 2: The orchestrator role is associated with distinct and measurable cognitive signatures. Relative to operators and verifiers, orchestrators show higher meta-cognitive accuracy (calibrated confidence in their own judgements), greater attentional switching between information sources, and a different pattern of reliance on AI outputs across task difficulty (Rastogi et al., 2022; Steyvers & Kumar, 2023). Longitudinal studies of workers moving into orchestration roles should reveal these signatures emerging together rather than independently. Selection effects may confound the comparison if orchestrators are drawn disproportionately from workers with pre-existing meta-cognitive strengths; designs that track the same workers across roles address this threat.

Proposition 3: Trust calibration transfers imperfectly across AI architectures. Users who achieve well-calibrated trust with one architecture (for example, expert systems) exhibit systematic miscalibration when transferred to another (for example, large language models) until they accumulate architecture-specific experience (Lee & See, 2004; de Visser et al., 2018). This predicts that organisations deploying multiple architectures simultaneously will experience higher error rates during transition periods than organisations that sequence deployments. The prediction assumes architectures differ materially in their error patterns; where two architectures behave similarly under the operator's task range, transfer costs will be small.

Proposition 4: Regulatory stringency produces measurable changes in architectural choice. The tiered-risk structure of the European Union (2024) Artificial Intelligence Act will shift deployment choices toward more interpretable architectures in high-risk domains, producing a detectable gap between European and non-European architectural profiles by 2028 (Hauer, 2023). This is testable via comparative deployment data and through analysis of model-card disclosures. Confounding from general firm-level preferences for interpretable models would attenuate this gap; controlling for sector and firm size in comparative deployment data sharpens the test.

Proposition 5: Cognitive-ecosystem effects compound over time. Networks of interacting AI systems and humans produce emergent decision patterns that cannot be predicted from pairwise interactions alone, and the divergence grows with the number of agents and the frequency of interaction (Akata et al., 2020; Jarrahi et al., 2022). Longitudinal observation of ecosystems with three, five, and ten interacting agents should reveal this non-linearity. Where interactions are governed by rigid protocols rather than adaptive coordination, emergent effects are suppressed and the prediction does not hold.

Proposition 6: Algorithmic appreciation is context-dependent, not trait-based. The tendency to prefer algorithmic to human judgement, first reported by Logg et al. (2019), varies systematically with domain, outcome feedback, and accountability structure, and declines when individuals are made responsible for the algorithm's errors. Experimental manipulations varying only accountability should produce quantifiable shifts in algorithmic appreciation even within individuals. Laboratory manipulations of accountability may overstate effects observed in naturalistic settings where accountability signals are noisier; replication in field contexts is therefore necessary.

Proposition 7: The explainability–accuracy trade-off is over-stated for structured decisions. For many structured, high-stakes decisions, interpretable architectures achieve accuracy equivalent to or better than black-box models (Rudin, 2019). Replicating this finding across additional domains (healthcare, finance, criminal justice, hiring) would establish the boundary conditions under which the often-assumed trade-off actually binds. Counter-evidence accumulates where problem structure is highly non-linear with weak signal; the proposition is bounded by structured decision domains and does not claim generality for all high-stakes prediction.

Proposition 8: Governance transfers unevenly across jurisdictions. The translation of risk-tiered governance frameworks across legal systems produces different enforcement patterns and therefore different deployment patterns. Comparative analysis of enforcement events in the first three years of the European Union (2024) Artificial Intelligence Act will reveal whether convergence or divergence is the dominant trend. Convergence could be driven not by regulatory substance but by common commercial pressures on multinational deployers; comparative analysis needs to isolate regulatory from market mechanisms.

CONCLUSION

The seven questions posed at the outset have each been addressed: the cognitive and procedural effects of ten distinct AI architectures; collaboration frameworks and their match to decision contexts; workforce transformation from operators to orchestrators; governance structures adapted to architectural diversity; the strategic and organisational implications of hybrid systems; the trajectory toward cognitive ecosystems by 2040; and the integrated three-dimensional framework connecting architecture, context, and collaboration. This examination of strategic decision-making frameworks incorporating human-AI systems reveals a fundamental re-conceptualisation of organisational cognition that extends beyond technological implementation to encompass workforce transformation, governance evolution, and the emergence of novel cognitive ecosystems. The systematic analysis across ten distinct AI architectures—from deterministic expert systems to non-deterministic large language models—shows that effective human-AI collaboration cannot follow monolithic approaches but must adapt to the heterogeneous characteristics of different AI types and decision contexts.

The theoretical implications of this work extend foundational decision theory into new domains. Where Simon's bounded rationality characterised individual cognitive limitations, hybrid intelligence systems exhibit bounded complementarity—synergistic capabilities constrained by the interface between human contextual reasoning and artificial computational processing (Simon, 1955; Dellermann et al., 2019). The three-dimensional framework developed herein—integrating AI architectural characteristics, decision contexts, and human cognitive roles—provides a structured approach for working through the 36 distinct configuration scenarios organisations face when deploying hybrid systems. This framework reveals that complexity emerges non-linearly: transitions between adjacent scenarios often require qualitative transformations in governance, skills, and organisational structures rather than incremental adjustments.

The practical implications for organisations are immediate and consequential. The evidence shows that approximately one quarter of current jobs face substantial transformation, yet wholesale automation remains limited to narrow, well-defined tasks (Indeed Hiring Lab, 2025). Success in this environment requires organisations to develop dynamic capabilities in three critical domains: architectural selection (matching AI types to decision contexts), integration orchestration (coordinating multiple AI systems with human oversight), and adaptive governance (evolving accountability frameworks as capabilities advance). The automation-augmentation paradox identified by Raisch and Krakowski (2021) necessitates continuous re-calibration—over-automation breeds brittleness whilst under-automation forgoes competitive advantages.

The governance challenges posed by hybrid systems demand renewed attention to accountability architectures that preserve human responsibility whilst enabling AI autonomy where appropriate. The risk-based frameworks emerging through the EU AI Act, ISO/IEC 42001, and OECD AI Principles provide scaffolding, yet implementation requires context-specific interpretation (European Commission, 2021; International Organization for Standardization, 2023; OECD, 2019). Deterministic systems permit traditional audit approaches, whilst non-deterministic systems require statistical characterisation and continuous monitoring. The principle that accountability must remain with human actors regardless of AI sophistication proves essential for maintaining societal trust and legal coherence (Bryson, 2018; Floridi et al., 2018).

Looking toward 2040, the trajectory from isolated AI deployments to interconnected cognitive ecosystems presents both transformative opportunities and existential challenges. Multi-agent systems incorporating diverse AI architectures promise emergent intelligence exceeding current capabilities, yet require governance mechanisms operating at ecosystem scales (Akata et al., 2020; Surowiecki, 2004). The Industry 5.0 paradigm's emphasis on human-centric automation offers a potential path where technological advancement serves human flourishing rather than displacing human agency (Nahavandi, 2019). Realising this vision demands proactive investment in interoperability standards, trust architectures, and educational systems preparing workers for meta-cognitive roles orchestrating AI collectives.

Three critical uncertainties will shape this trajectory. First, the pace of AI capability advancement relative to governance adaptation will determine whether deployment proceeds responsibly or chaotically. Second, the distribution of benefits from hybrid systems—within organisations, across sectors, and between nations—will influence social acceptance and regulatory responses. Third, the preservation of human agency and meaningful work amid increasing automation will affect psychological wellbeing and social cohesion.

This examination establishes that hybrid human-AI decision systems represent neither simple tool adoption nor inevitable human obsolescence, but rather a co-evolutionary process reshaping the fundamental nature of organisational cognition. The frameworks, typologies, and insights developed provide structured approaches for working through this transformation, yet their application requires continuous adaptation as capabilities evolve and understanding deepens. The path forward demands sustained interdisciplinary collaboration among technologists developing AI systems, organisational scholars understanding decision processes, policy makers establishing governance frameworks, and practitioners implementing hybrid systems.

The transformation of decision-making through human-AI collaboration constitutes one of the defining challenges of the twenty-first century. How organisations, sectors and societies handle this transformation will determine whether artificial intelligence extends human potential or diminishes human agency. The evidence presented suggests that intentionally designed hybrid systems, appropriately governed and thoughtfully implemented, can enhance both organisational performance and human flourishing. Achieving this outcome requires moving beyond technological determinism to embrace the complex socio-technical nature of human-AI collaboration, recognising that the most sophisticated algorithm remains merely a component within broader systems where human judgement, creativity, and values provide essential direction. The ongoing evolution of these systems will be shaped not by technology alone but by the choices organisations and societies make about how to integrate artificial and human intelligence in service of collective goals.

REFERENCES

Abbass, H. A., Tang, J., Amin, R., Ellejmi, M., & Kirby, S. (2018). *Trusted autonomy: Conceptual foundations and experimental studies*. Springer. DOI: 10.1007/978-3-319-64816-3

Adadi, A., & Berrada, M. (2018). Peeking inside the black box: A survey on explainable artificial intelligence (XAI). *IEEE Access : Practical Innovations, Open Solutions, 6*, 52138–52160. DOI: 10.1109/ACCESS.2018.2870052

Akata, Z., Balliet, D., de Rijke, M., Dignum, F., Dignum, V., Eiben, G., Fokkens, A., Grossi, D., Hindriks, K., Hoos, H., Hung, H., Jonker, C., Monz, C., Neerincx, M., Oliehoek, F., Prakken, H., Schlobach, S., van der Gaag, L., van Harmelen, F., & Welling, M. (2020). A research agenda for hybrid intelligence: Augmenting human intellect with collaborative, adaptive, responsible, and explainable artificial intelligence. *Computer, 53*(8), 18–28. DOI: 10.1109/MC.2020.2996587

Amershi, S., Weld, D., Vorvoreanu, M., Fourney, A., Nushi, B., Caruana, R., Kamar, E., & Horvitz, E. (2019). Guidelines for human–AI interaction. *Proceedings of the 2019 CHI Conference on Human Factors in Computing Systems,* 1–13. DOI: 10.1145/3290605.3300233

Anderson, J. R. (1983). *The architecture of cognition*. Harvard University Press.

Angwin, J., Larson, J., Mattu, S., & Kirchner, L. (2016, May 23). *Machine bias: There's software used across the country to predict future criminals*. And it's biased against blacks. ProPublica. https://www.propublica.org/article/machine-bias-risk-assessments-in-criminal-sentencing

Arrieta, A. B., Díaz-Rodríguez, N., Del Ser, J., Bennetot, A., Tabik, S., Barbado, A., García, S., Gil-López, S., Molina, D., Benjamins, R., Chatila, R., & Herrera, F. (2020). Explainable artificial intelligence (XAI): Concepts, taxonomies, opportunities and challenges toward responsible AI. *Information Fusion, 58*, 82–115. DOI: 10.1016/j.inffus.2019.12.012

Ashley, K. D. (2017). *Artificial intelligence and legal analytics: New tools for law practice in the digital age*. Cambridge University Press. DOI: 10.1017/9781316761380

Augenstein, I., Baldwin, T., Cha, M., Chakraborty, T., Ciampaglia, G. L., Corney, D., DiResta, R., Ferrara, E., Hale, S., Halevy, A., Hovy, E., Ji, H., Menczer, F., Miguez, R., Nakov, P., Scheufele, D., Sharma, S., & Zagni, G. (2024). Factuality challenges in the era of large language models and opportunities for fact-checking. *Nature Machine Intelligence, 6*(8), 852–863. DOI: 10.1038/s42256-024-00881-z

Augier, M., & March, J. G. (2001). Remembering Herbert A. Simon (1916-2001). *Public Administration Review, 61*(4), 396–402. DOI: 10.1111/0033-3352.00043

Autor, D. H. (2015). Why are there still so many jobs? The history and future of workplace automation. *The Journal of Economic Perspectives, 29*(3), 3–30. DOI: 10.1257/jep.29.3.3

Bansal, G., Nushi, B., Kamar, E., Lasecki, W. S., Weld, D. S., & Horvitz, E. (2019). Beyond accuracy: The role of mental models in human-AI team performance. *Proceedings of the AAAI Conference on Human Computation and Crowdsourcing, 7,* 2-11. DOI: 10.1609/hcomp.v7i1.5285

Bansal, G., Nushi, B., Kamar, E., Lasecki, W. S., Weld, D. S., & Horvitz, E. (2021). Beyond accuracy: The role of mental models in human-AI team performance. *Proceedings of the ACM on Human-Computer Interaction, 5*(CSCW1), 1-22.

Barney, J. (1991). Firm resources and sustained competitive advantage. *Journal of Management, 17*(1), 99–120. DOI: 10.1177/014920639101700108

Barocas, S., & Selbst, A. D. (2016). Big data's disparate impact. *California Law Review, 104*, 671–732.

Baron, J. (2008). *Thinking and deciding* (4th ed.). Cambridge University Press.

Battaglia, P. W., Hamrick, J. B., Bapst, V., Sanchez-Gonzalez, A., Zambaldi, V., Malinowski, M., Tacchetti, A., Raposo, D., Santoro, A., Faulkner, R., Gulcehre, C., Song, F., Ballard, A., Gilmer, J., Dahl, G., Vaswani, A., Allen, K., Nash, C., Langston, V., & Pascanu, R. (2018). *Relational inductive biases, deep learning, and graph networks.* arXiv. https://arxiv.org/abs/1806.01261

Blasch, E., Xu, R., Chen, Y., Chen, G., & Shen, D. (2019). Blockchain methods for trusted information fusion in contested environments. In Kadar, I., Blasch, E. P., & Grewe, L. L. (Eds.), *Signal processing, sensor/information fusion, and target recognition XXVIII* (Vol. 11018, pp. 318–331). SPIE., DOI: 10.1117/12.2518957

Bolander, T. (2019). What do we lose when machines take the decisions? *The Journal of Management and Governance, 23*(4), 849–867. DOI: 10.1007/s10997-019-09493-x

Bommasani, R., Hudson, D. A., Adeli, E., Altman, R., Arora, S., von Arx, S., . . . Liang, P. (2021). On the opportunities and risks of foundation models (arXiv:2108.07258). arXiv. https://arxiv.org/abs/2108.07258

Bonabeau, E., Dorigo, M., & Theraulaz, G. (1999). *Swarm intelligence: From natural to artificial systems.* Oxford University Press. DOI: 10.1093/oso/9780195131581.001.0001

Breiman, L. (2001). Random forests. *Machine Learning*, *45*(1), 5–32. DOI: 10.1023/A:1010933404324

Brynjolfsson, E., Li, D., & Raymond, L. R. (2025). Generative AI at work. *The Quarterly Journal of Economics*, *140*(2), 889–942. DOI: 10.1093/qje/qjae044

Brynjolfsson, E., & McAfee, A. (2017). *Machine, platform, crowd: Harnessing our digital future*. W. W. Norton.

Brynjolfsson, E., & Mitchell, T. (2017). What can machine learning do? Workforce implications. *Science*, *358*(6370), 1530–1534. DOI: 10.1126/science.aap8062

Bryson, J. J. (2018). Patiency is not a virtue: AI and the design of ethical systems. *Ethics and Information Technology*, *20*(1), 15–26. DOI: 10.1007/s10676-018-9448-6

Buchanan, B. G., & Shortliffe, E. H. (1984). *Rule-based expert systems: The MYCIN experiments of the Stanford Heuristic Programming Project*. Addison-Wesley.

Caruana, R., & Niculescu-Mizil, A. (2006). An empirical comparison of supervised learning algorithms. In *Proceedings of the 23rd International Conference on Machine Learning* (pp. 161-168). DOI: 10.1145/1143844.1143865

Casino, F., Dasaklis, T. K., & Patsakis, C. (2019). A systematic literature review of blockchain-based applications: Current status, classification and open issues. *Telematics and Informatics*, *36*, 55–81. DOI: 10.1016/j.tele.2018.11.006

Chen, J. Y., & Barnes, M. J. (2015). Human-robot interaction for multi-robot systems. *Proceedings of the Human Factors and Ergonomics Society Annual Meeting*, *59*(1), 664–668.

Choudhury, P., Allen, R., & Wagner, K. (2022). Robots, automation, and the evolution of logistics work. *Industrial Relations Journal*, *53*(2), 145–164.

Choung, H., Oh, H., & Park, E. (2023). The influence of anthropomorphism on trust in AI-based decision support systems. *Computers in Human Behavior*, *147*, 107836.

Cohen, W. M., & Levinthal, D. A. (1990). Absorptive capacity: A new perspective on learning and innovation. *Administrative Science Quarterly*, *35*(1), 128–152. DOI: 10.2307/2393553

Cortes, M., Guetta, L., & Reuber, A. (2022). Human-AI collaboration and decision quality: Evidence from entrepreneurial evaluation experiments. *Entrepreneurship Theory and Practice*, *46*(5), 1153–1177.

Csaszar, F. A., Ketkar, H., & Kim, H. (2024). Artificial intelligence and strategic decision-making: Evidence from entrepreneurs and investors. *Strategy Science*, *9*(4), 322–345. DOI: 10.1287/stsc.2024.0190

Cummings, M. L. (2004). *Automation bias in intelligent time critical decision support systems*. In AIAA 1st Intelligent Systems Technical Conference (Paper 2004-6313). DOI: 10.2514/6.2004-6313

Cyert, R. M., & March, J. G. (1963). *A behavioral theory of the firm*. Prentice Hall.

Dastin, J. (2018, October 10). *Insight—Amazon scraps secret AI recruiting tool that showed bias against women*. Reuters. https://www.reuters.com/article/us-amazon-com-jobs-automation-insight-idUSKCN1MK08G/

Davenport, T. H., & Kirby, J. (2016). *Only humans need apply: Winners and losers in the age of smart machines*. Harper Business.

de Visser, E. J., Pak, R., & Shaw, T. H. (2018). From 'automation' to 'autonomy': The importance of trust repair in human–machine interaction. *Ergonomics*, *61*(10), 1409–1427. DOI: 10.1080/00140139.2018.1457725

Dellermann, D., Ebel, P., Söllner, M., & Leimeister, J. M. (2019). Hybrid intelligence. *Business & Information Systems Engineering*, *61*(5), 637–643. DOI: 10.1007/s12599-019-00595-2

Dignum, V. (2019). *Responsible artificial intelligence: How to develop and use AI in a responsible way*. Springer. DOI: 10.1007/978-3-030-30371-6

Döppner, D. A., Derks, P., & Schumann, M. (2019). A systematic review of the concept of hybrid intelligence. In *Proceedings of the 27th European Conference on Information Systems* (ECIS). Stockholm & Uppsala, Sweden.

Dorigo, M., & Stützle, T. (2004). *Ant colony optimization*. MIT Press. DOI: 10.7551/mitpress/1290.001.0001

Doshi-Velez, F., & Kim, B. (2017). Towards a rigorous science of interpretable machine learning (arXiv:1702.08608). arXiv. https://doi.org//arXiv.1702.08608DOI: 10.48550

Dosovitskiy, A., Beyer, L., Kolesnikov, A., Weissenborn, D., Zhai, X., Unterthiner, T., Dehghani, M., Minderer, M., Heigold, G., Gelly, S., Uszkoreit, J., & Houlsby, N. (2020). An image is worth 16×16 words: Transformers for image recognition at scale (arXiv:2010.11929). arXiv. https://arxiv.org/abs/2010.11929

Dzindolet, M. T., Peterson, S. A., Pomranky, R. A., Pierce, L. G., & Beck, H. P. (2003). The role of trust in automation reliance. *International Journal of Human-Computer Studies, 58*(6), 697–718. DOI: 10.1016/S1071-5819(03)00038-7

Eisenhardt, K. M., & Martin, J. A. (2000). Dynamic capabilities: What are they? *Strategic Management Journal, 21*(10-11), 1105–1121. DOI: 10.1002/1097-0266(200010/11)21:10/11<1105::AID-SMJ133>3.0.CO;2-E

Endsley, M. R. (2017). From here to autonomy: Lessons learned from human-automation research. *Human Factors, 59*(1), 5–27. DOI: 10.1177/0018720816681350

Etzioni, O. (2018). How to regulate artificial intelligence. *Issues in Science and Technology, 34*(4), 32–36.

European Commission. (2021). *Proposal for a regulation of the European Parliament and of the Council laying down harmonised rules on artificial intelligence (Artificial Intelligence Act) and amending certain Union legislative acts (COM/2021/206 final).* https://eur-lex.europa.eu/legal-content/EN/TXT/?uri=CELEX%3A52021PC0206

European Commission. (2025). General-Purpose AI Code of Practice (endorsed 10 July 2025). European AI Office. https://digital-strategy.ec.europa.eu/en/policies/contents-code-gpai

European Union. (2024). *Regulation (EU) 2024/1689 of the European Parliament and of the Council of 13 June 2024 laying down harmonised rules on artificial intelligence (Artificial Intelligence Act).* Official Journal of the European Union, L 2024/1689. https://eur-lex.europa.eu/eli/reg/2024/1689/oj

Felzmann, H., Fosch-Villaronga, E., Lutz, C., & Tamò-Larrieux, A. (2020). Towards transparency by design for artificial intelligence. *Science and Engineering Ethics, 26*(6), 3333–3361. DOI: 10.1007/s11948-020-00276-4

Floridi, L., Cowls, J., Beltrametti, M., Chatila, R., Chazerand, P., Dignum, V., Luetge, C., Madelin, R., Pagallo, U., Rossi, F., Schafer, B., Valcke, P., & Vayena, E. (2018). AI4People—An ethical framework for a good AI society: Opportunities, risks, principles, and recommendations. *Minds and Machines, 28*(4), 689–707. DOI: 10.1007/s11023-018-9482-5

Geels, F. W. (2002). Technological transitions as evolutionary reconfiguration processes: A multi-level perspective and a case-study. *Research Policy, 31*(8–9), 1257–1274. DOI: 10.1016/S0048-7333(02)00062-8

Gershman, S. J., Horvitz, E. J., & Tenenbaum, J. B. (2015). Computational rationality: A converging paradigm for intelligence in brains, minds, and machines. *Science, 349*(6245), 273–278. DOI: 10.1126/science.aac6076

Gill, T. G. (2020). Artificial intelligence and contextual reasoning in managerial work. *Journal of Management Inquiry*, *29*(2), 206–219.

Goodfellow, I., Bengio, Y., & Courville, A. (2016). *Deep learning*. MIT Press.

Grigorescu, S., Trasnea, B., Cocias, T., & Macesanu, G. (2020). A survey of deep learning techniques for autonomous driving. *Journal of Field Robotics*, *37*(3), 362–386. DOI: 10.1002/rob.21918

Hastie, T., Tibshirani, R., & Friedman, J. (2009). *The elements of statistical learning: Data mining, inference, and prediction* (2nd ed.). Springer. DOI: 10.1007/978-0-387-84858-7

Hauer, T. (2023). Global AI regulation landscape: Comparative analysis of policy approaches. *Computer Law & Security Review*, *49*, 106747.

Hearst, M. A., Dumais, S. T., Osuna, E., Platt, J., & Scholkopf, B. (1998). Support vector machines. *IEEE Intelligent Systems & their Applications*, *13*(4), 18–28. DOI: 10.1109/5254.708428

Hiatt, J. M. (2006). *ADKAR: A model for change in business, government, and our community*. Prosci Learning Center Publications.

Hollan, J., Hutchins, E., & Kirsh, D. (2000). Distributed cognition: Toward a new foundation for human-computer interaction research. *ACM Transactions on Computer-Human Interaction*, *7*(2), 174–196. DOI: 10.1145/353485.353487

Huang, Y., Zhang, Y., & Liu, X. (2024). Hallucination in large language models: Patterns and mitigation strategies. *Nature Machine Intelligence*, *6*(3), 234–245.

Hutchins, E. (1995). *Cognition in the wild.* MIT Press. DOI: 10.7551/mitpress/1881.001.0001

Hutchins, E., & Palen, L. (1997). Constructing meaning from space, gesture, and speech. In Resnick, L. B., Säljö, R., Pontecorvo, C., & Burge, B. (Eds.), *Discourse, tools and reasoning* (pp. 23–40). Springer-Verlag. DOI: 10.1007/978-3-662-03362-3_2

IBM Institute for Business Value. (2022). *AI literacy: Closing the artificial intelligence skills gap*. IBM Corporation. https://www.ibm.com/thought-leadership/institute-business-value/en-us/report/ai-literacy

Indeed Hiring Lab. (2025, September 23). *AI at work report 2025: How GenAI is rewiring the DNA of jobs*. https://www.hiringlab.org/2025/09/23/ai-at-work-report-2025-how-genai-is-rewiring-the-dna-of-jobs/

International Organization for Standardization. (2023). *ISO/IEC 42001:2023 Information technology — Artificial intelligence — Management system*. https://www.iso.org/standard/42001

Jain, S., & Wallace, B. C. (2019). Attention is not explanation. In *Proceedings of the 2019 Conference of the North American Chapter of the Association for Computational Linguistics: Human Language Technologies* (pp. 3543-3556). Association for Computational Linguistics.

Jarrahi, M. H. (2018). Artificial intelligence and the future of work: Human-AI symbiosis in organizational decision making. *Business Horizons*, *61*(4), 577–586. DOI: 10.1016/j.bushor.2018.03.007

Jarrahi, M. H., Lutz, C., & Newlands, G. (2022). Artificial intelligence, human intelligence and hybrid intelligence based on mutual augmentation. *Big Data & Society*, *9*(2), 1–16. DOI: 10.1177/20539517221142824

Johnson, M., Vera, A., & Chen, L. (2022). AI as teammate: Designing human-centered AI systems for collaborative work. *ACM Transactions on Interactive Intelligent Systems*, *12*(3), 1–35.

Jøsang, A., Ismail, R., & Boyd, C. (2007). A survey of trust and reputation systems for online service provision. *Decision Support Systems*, *43*(2), 618–644. DOI: 10.1016/j.dss.2005.05.019

Joshi, A., Iansiti, M., & Lakhani, K. R. (2025). The emerging agentic enterprise: How leaders must navigate a new age of AI. *MIT Sloan Management Review*.

Kahneman, D. (2011). *Thinking, fast and slow*. Farrar, Straus and Giroux.

Kahneman, D., & Tversky, A. (1979). Prospect theory: An analysis of decision under risk. *Econometrica*, *47*(2), 263–291. DOI: 10.2307/1914185

Kellogg, K. C., Valentine, M. A., & Christin, A. (2020). Algorithms at work: The new contested terrain of control. *The Academy of Management Annals*, *14*(1), 366–410. DOI: 10.5465/annals.2018.0174

Keramati, M., Dezfouli, A., & Piray, P. (2011). Speed/accuracy trade-off between the habitual and the goal-directed processes. *PLoS Computational Biology*, *7*(5), e1002055. DOI: 10.1371/journal.pcbi.1002055

Klein, G., Moon, B., & Hoffman, R. R. (2006). Making sense of sensemaking 1: Alternative perspectives. *IEEE Intelligent Systems*, *21*(4), 70–73. DOI: 10.1109/MIS.2006.75

Kotter, J. P. (1996). *Leading change*. Harvard Business School Press.

Lample, G., & Chaplot, D. S. (2017). Playing FPS games with deep reinforcement learning. *Proceedings of the AAAI Conference on Artificial Intelligence*, *31*(1), •••. DOI: 10.1609/aaai.v31i1.10827

LeCun, Y., Bengio, Y., & Hinton, G. (2015). Deep learning. *Nature*, *521*(7553), 436–444. DOI: 10.1038/nature14539

Lee, J. D., & See, K. A. (2004). Trust in automation: Designing for appropriate reliance. *Human Factors*, *46*(1), 50–80. DOI: 10.1518/hfes.46.1.50.30392

Leslie, D. (2019). *Understanding artificial intelligence ethics and safety: A guide for the responsible design and implementation of AI systems in the public sector.* The Alan Turing Institute.

Liaw, A., & Wiener, M. (2002). Classification and regression by randomForest. *R News*, *2*(3), 18–22.

Linkov, I., Bridges, T., Creutzig, F., Decker, J., Fox-Lent, C., Kröger, W., Lambert, J. H., Levermann, A., Montreuil, B., Nathwani, J., Nyer, R., Renn, O., Scharte, B., Scheffler, A., Schreurs, M., & Thiel-Clemen, T. (2014). Changing the resilience paradigm. *Nature Climate Change*, *4*(6), 407–409. DOI: 10.1038/nclimate2227

Logg, J. M., Minson, J. A., & Moore, D. A. (2019). Algorithm appreciation: People prefer algorithmic to human judgment. *Organizational Behavior and Human Decision Processes*, *151*, 90–103. DOI: 10.1016/j.obhdp.2018.12.005

Lundberg, S. M., & Lee, S. I. (2017). A unified approach to interpreting model predictions. *Advances in Neural Information Processing Systems*, *30*, 4765–4774.

Lyons, J. B. (2013). *Being transparent about transparency: A model for human-robot interaction.* In *2013 AAAI Spring Symposium Series.*

March, J. G., & Simon, H. A. (1958). *Organizations.* Wiley.

Mazzucato, M. (2018). Mission-oriented innovation policies: Challenges and opportunities. *Industrial and Corporate Change*, *27*(5), 803–815. DOI: 10.1093/icc/dty034

McKnight, D. H., Carter, M., Thatcher, J. B., & Clay, P. F. (2011). Trust in a specific technology: An investigation of its components and measures. *ACM Transactions on Management Information Systems*, *2*(2), 1–25. DOI: 10.1145/1985347.1985353

Mehrabi, N., Morstatter, F., Saxena, N., Lerman, K., & Galstyan, A. (2021). A survey on bias and fairness in machine learning. *ACM Computing Surveys*, *54*(6), 115.

Miller, T. (2019). Explanation in artificial intelligence: Insights from the social sciences. *Artificial Intelligence*, *267*, 1–38. DOI: 10.1016/j.artint.2018.07.007

Mintzberg, H. (1994). *The rise and fall of strategic planning*. Free Press.

MIT Project NANDA. (2025). The GenAI divide: State of AI in business 2025. Massachusetts Institute of Technology. https://nanda.mit.edu/research/state-of-ai-in-business-2025

Moawad, A., Barnes, G., & Johnson, C. (2022). Deep learning for medical image analysis: Current state and future directions. *Medical Image Analysis*, *79*, 102444. DOI: 10.1016/j.media.2022.102444

Morley, J., Machado, C. C. V., Burr, C., Cowls, J., Joshi, I., Taddeo, M., & Floridi, L. (2020). Ethical guidelines for trustworthiness in artificial intelligence. *Nature Machine Intelligence*, *2*(8), 501–507.

Mosqueira-Rey, E., Hernández-Pereira, E., Alonso-Ríos, D., Bobes-Bascarán, J., & Fernández-Leal, Á. (2023). Human-in-the-loop machine learning: A state of the art. *Artificial Intelligence Review*, *56*(4), 3005–3054. DOI: 10.1007/s10462-022-10246-w

Mukherjee, D., Basak, S., & Sengupta, K. (2023). Bias interaction in human-AI decision-making: Taxonomy and mitigation strategies. *AI and Ethics*, *3*(4), 719–732.

Nahavandi, S. (2019). Industry 5.0—A human-centric solution. *Sustainability (Basel)*, *11*(16), 4371. DOI: 10.3390/su11164371

National Institute of Standards and Technology. (2023). *Artificial Intelligence Risk Management Framework (AI RMF 1.0) (NIST AI 100-1)*. U.S. Department of Commerce., DOI: 10.6028/NIST.AI.100-1

National Institute of Standards and Technology. (2024). *Artificial Intelligence Risk Management Framework: Generative Artificial Intelligence Profile (NIST AI 600-1)*. U.S. Department of Commerce., DOI: 10.6028/NIST.AI.600-1

National Transportation Safety Board. (2019). *Collision between vehicle controlled by developmental automated driving system and pedestrian, Tempe, Arizona, March 18, 2018 (Highway Accident Report NTSB/HAR-19/03)*. https://www.ntsb.gov/investigations/accidentreports/reports/har1903.pdf

Newell, A., & Simon, H. A. (1956). The logic theory machine: A complex information processing system. *IEEE Transactions on Information Theory*, *2*(3), 61–79. DOI: 10.1109/TIT.1956.1056797

Noble, S. U. (2018). *Algorithms of oppression: How search engines reinforce racism*. NYU Press. DOI: 10.18574/nyu/9781479833641.001.0001

O'Neil, C. (2016). *Weapons of math destruction: How big data increases inequality and threatens democracy*. Crown.

Okamura, K., & Yamada, S. (2020). Adaptive trust calibration for human-AI collaboration. *PLoS One*, *15*(2), e0229132. DOI: 10.1371/journal.pone.0229132

Open, A. I. (2023). GPT-4 technical report (arXiv:2303.08774). arXiv. https://doi.org//arXiv.2303.08774DOI: 10.48550

Organisation for Economic Co-operation and Development. (2019). *Recommendation of the Council on Artificial Intelligence (OECD/LEGAL/0449).* https://legalinstruments.oecd.org/en/instruments/oecd-legal-0449

Organisation for Economic Co-operation and Development. (2024). Recommendation of the Council on Artificial Intelligence (OECD/LEGAL/0449, as amended on 3 May 2024). https://legalinstruments.oecd.org/en/instruments/oecd-legal-0449

Parasuraman, R., Sheridan, T. B., & Wickens, C. D. (2000). A model for types and levels of human interaction with automation. *IEEE Transactions on Systems, Man, and Cybernetics. Part A, Systems and Humans*, *30*(3), 286–297. DOI: 10.1109/3468.844354

Pea, R. D. (1993). Practices of distributed intelligence and designs for education. In Salomon, G. (Ed.), *Distributed cognitions: Psychological and educational considerations* (pp. 47–87). Cambridge University Press.

Perry, M. J., & Macredie, R. D. (2000). *Distributed cognition: Investigating collaboration in large and highly complex organisational systems (Working paper).* Brunel University.

Polanyi, M. (1966). *The tacit dimension.* University of Chicago Press.

Popper, K. R. (2002). *The logic of scientific discovery.* Routledge. (Original work published 1959)

Pressman, J. L., & Wildavsky, A. (1984). *Implementation: How great expectations in Washington are dashed in Oakland* (3rd ed.). University of California Press.

Pursnani, V., Sermet, Y., Kurt, M. C., & Demir, I. (2023). Performance of ChatGPT on the US fundamentals of engineering exam: Comprehensive assessment of proficiency and potential implications for professional environmental engineering. Computers and Education. *Artificial Intelligence*, *5*, 100183.

Raisch, S., & Krakowski, S. (2021). Artificial intelligence and management: The automation-augmentation paradox. *Academy of Management Review*, *46*(1), 192–210. DOI: 10.5465/amr.2018.0072

Raji, I. D., Smart, A., White, R. N., Mitchell, M., Gebru, T., Hutchinson, B., Smith-Loud, J., Theron, D., & Barnes, P. (2020). Closing the AI accountability gap: Defining an end-to-end framework for internal algorithmic auditing. In *Proceedings of the 2020 Conference on Fairness, Accountability, and Transparency* (pp. 33-44). DOI: 10.1145/3351095.3372873

Ransbotham, S., Candelon, F., Kiron, D., LaFountain, B., & Khodabandeh, S. (2020). Expanding AI's impact with organizational learning. *MIT Sloan Management Review, 62*(1). https://sloanreview.mit.edu/projects/expanding-ais-impact-with-organizational-learning/

Rastogi, C., Zhang, Y., Wei, D., Varshney, K. R., Dhurandhar, A., & Tomsett, R. (2022). Deciding fast and slow: The role of cognitive biases in AI-assisted decision-making. *Proceedings of the ACM on Human-Computer Interaction, 6*(CSCW1), 83:1-83:22. DOI: 10.1145/3512930

Rhim, J., Lee, G., & Lee, J. H. (2021). Ethical considerations in AI-powered decision-making: A cross-industry analysis. *AI & Society, 36*(4), 1127–1145.

Ribeiro, M. T., Singh, S., & Guestrin, C. (2016). "Why should I trust you?" Explaining the predictions of any classifier. In *Proceedings of the 22nd ACM SIGKDD International Conference on Knowledge Discovery and Data Mining* (pp. 1135-1144).

Riso, D., & Adăscăli ei, I. M. (2024). *Human-robot interaction and workplace transformation: Evidence from European shops*. Eurofound.

Rodríguez-Espíndola, O., Chowdhury, S., Beltagui, A., & Albores, P. (2020). The potential of emergent disruptive technologies for humanitarian supply chains: The integration of blockchain, artificial intelligence and 3D printing. *International Journal of Production Research, 58*(15), 4610–4630. DOI: 10.1080/00207543.2020.1761565

Roos, G. (2025). Managing five-generational workforces amid labour shortages and technological change. In Achour, Z. (Ed.), *Advanced labor studies research and modern practice* (pp. 131–168). IGI Global., DOI: 10.4018/979-8-3373-1962-9.ch005

Roos, G., & Shroff, Z. (2017). What will happen to the jobs? Technology-enabled productivity improvement — Good for some, bad for others. *Labour & Industry, 27*(3), 165–192. DOI: 10.1080/10301763.2017.1359817

Rudin, C. (2019). Stop explaining black-box machine learning models for high-stakes decisions and use interpretable models instead. *Nature Machine Intelligence, 1*(5), 206–215. DOI: 10.1038/s42256-019-0048-x

Russell, S. J., & Norvig, P. (2021). *Artificial intelligence: A modern approach* (4th ed.). Pearson.

Ryseff, J., De Bruhl, B. F., & Newberry, S. J. (2024). The root causes of failure for artificial intelligence projects and how they can succeed: Avoiding the anti-patterns of AI (RAND Research Report RR-A2680-1). RAND Corporation. DOI: 10.7249/RRA2680-1

Safavian, S. R., & Landgrebe, D. (1991). A survey of decision tree classifier methodology. *IEEE Transactions on Systems, Man, and Cybernetics*, *21*(3), 660–674. DOI: 10.1109/21.97458

Schneider, J., & Leyer, M. (2019). Me or information technology? Adoption of artificial intelligence in the delegation of personal strategic decisions. *MDE. Managerial and Decision Economics*, *40*(3), 223–231. DOI: 10.1002/mde.2982

Shen, H., Liao, K., Liao, Z., Doornberg, J., Qiao, M., van den Hengel, A., & Verjans, J. W. (2021). Human-AI interactive and continuous sensemaking: A case study of image classification using scribble attention maps. *CHI '21 Extended Abstracts. Article*, *290*, 1–8.

Sheridan, T. B., & Parasuraman, R. (2005). Human-automation interaction. In Nickerson, R. S. (Ed.), *Reviews of human factors and ergonomics* (Vol. 1, pp. 89–129). Human Factors and Ergonomics Society.

Shively, R. J., Lachter, J., Koteskey, R., & Brandt, S. L. (2017). *Crew resource management for automated teammates (CRM-AT)*. In *International Conference on Engineering Psychology and Cognitive Ergonomics* (pp. 215-224). Springer.

Shneiderman, B. (2020). Human-centered artificial intelligence: Reliable, safe, and trustworthy. *International Journal of Human-Computer Interaction*, *36*(6), 495–504. DOI: 10.1080/10447318.2020.1741118

Shrestha, Y. R., Ben-Menahem, S. M., & von Krogh, G. (2019). Organizational decision-making structures in the age of artificial intelligence. *California Management Review*, *61*(4), 66–83. DOI: 10.1177/0008125619862257

Simon, H. A. (1955). A behavioral model of rational choice. *The Quarterly Journal of Economics*, *69*(1), 99–118. DOI: 10.2307/1884852

Simon, H. A. (1997). *Models of bounded rationality: Empirically grounded economic reason (Vol* (Vol. 3). MIT Press. DOI: 10.7551/mitpress/4711.001.0001

Steyvers, M., & Kumar, A. (2023). Three challenges for AI-assisted decision-making. *Perspectives on Psychological Science*, *18*(4), 772–784.

Strickland, E. (2019). IBM Watson, heal thyself: How IBM overpromised and underdelivered on AI health care. *IEEE Spectrum*, *56*(4), 24–31. DOI: 10.1109/MSPEC.2019.8678513

Surowiecki, J. (2004). *The wisdom of crowds*. Doubleday.

Sutton, R. S., & Barto, A. G. (1998). *Reinforcement learning: An introduction*. MIT Press.

Sutton, R. S., & Barto, A. G. (2018). *Reinforcement learning: An introduction* (2nd ed.). MIT Press.

Taleb, N. N. (2012). *Antifragile: Things that gain from disorder*. Random House.

Teece, D. J. (2007). Explicating dynamic capabilities: The nature and microfoundations of (sustainable) enterprise performance. *Strategic Management Journal*, *28*(13), 1319–1350. DOI: 10.1002/smj.640

Teece, D. J., Pisano, G., & Shuen, A. (1997). Dynamic capabilities and strategic management. *Strategic Management Journal*, *18*(7), 509–533. DOI: 10.1002/(SICI)1097-0266(199708)18:7<509::AID-SMJ882>3.0.CO;2-Z

Topol, E. J. (2019). High-performance medicine: The convergence of human and artificial intelligence. *Nature Medicine*, *25*(1), 44–56. DOI: 10.1038/s41591-018-0300-7

Tschiatschek, S., Ghosh, A., Haug, L., Devidze, R., & Singla, A. (2024). Learner-aware teaching: Inverse reinforcement learning with preferences and constraints. *Advances in Neural Information Processing Systems*, *32*, 4147–4157.

Tversky, A., & Kahneman, D. (1974). Judgment under uncertainty: Heuristics and biases. *Science*, *185*(4157), 1124–1131. DOI: 10.1126/science.185.4157.1124

United States Air Force Office of the Chief Scientist. (2015). *Autonomous horizons: System autonomy in the Air Force — A path to the future*. AF/ST TR 15-01.

Vaccaro, M., Almaatouq, A., & Malone, T. (2024). When combinations of humans and AI are useful: A systematic review and meta-analysis. *Nature Human Behaviour*, *8*(12), 2293–2303. DOI: 10.1038/s41562-024-02024-1

Vale, L., Pereira, J., & Santos, R. (2022). Deterministic vs. probabilistic AI systems: Trade-offs in transparency and performance. *AI & Society*, *37*(3), 891–906.

Vaswani, A., Shazeer, N., Parmar, N., Uszkoreit, J., Jones, L., Gomez, A. N., Kaiser, Ł., & Polosukhin, I. (2017). Attention is all you need. In *NeurIPS 2017 Proceedings* (pp. 5998-6008).

Villani, V., Pini, F., Leali, F., & Secchi, C. (2018). Survey on human-robot collaboration in industrial settings: Safety, intuitive interfaces and applications. *Mechatronics, 55*, 248–266. DOI: 10.1016/j.mechatronics.2018.02.009

Vincent, N. (2021). When AI is a teammate: Revisiting human-algorithm interaction and trust. *International Journal of Human-Computer Interaction, 37*(5), 432–448.

von Krogh, G. (2018). Artificial intelligence in organizations: New opportunities for phenomenon-based theorizing. *Academy of Management Discoveries, 4*(4), 404–409. DOI: 10.5465/amd.2018.0084

Wang, D., Churchill, E., Maes, P., Fan, X., Shneiderman, B., Shi, Y., & Wang, Q. (2021). From human-human collaboration to human-AI collaboration: Designing AI systems that can work together with people. In *Extended Abstracts of the 2021 CHI Conference on Human Factors in Computing Systems* (pp. 1-6).

Weick, K. E. (1995). *Sensemaking in organizations*. Sage Publications.

Wilson, H. J., & Daugherty, P. R. (2018). Collaborative intelligence: Humans and AI are joining forces. *Harvard Business Review, 96*(4), 114–123.

Wooldridge, M. (2009). *An introduction to multiagent systems* (2nd ed.). Wiley.

World Economic Forum. (2020). *The Future of Jobs Report 2020*. World Economic Forum.

World Economic Forum. (2023). *The Future of Jobs Report 2023*. World Economic Forum.

World Economic Forum. (2025, January 8). *The future of jobs report 2025*. https://www.weforum.org/publications/the-future-of-jobs-report-2025/

Xiong, W., Liu, H., & Wang, J. (2022). Human-AI collaboration in decision-making: Balancing computational power and cognitive capabilities. *Decision Support Systems, 153*, 113688.

Zhang, J., & Norman, D. A. (1994). Representations in distributed cognitive tasks. *Cognitive Science, 18*(1), 87–122. DOI: 10.1207/s15516709cog1801_3

Zhang, J., & Patel, V. L. (2006). Distributed cognition, representation, and affordance. *Pragmatics & Cognition, 14*(2), 333–341. DOI: 10.1075/pc.14.2.12zha

Zhang, Y., Gao, Y., & Gong, M. (2023). Large language models for automated text summarization: A systematic review. *Information Processing & Management, 60*(2), 103234.

Zhang, Y., Mao, S., Ge, T., Wang, X., de Wynter, A., Xia, Y., Wu, W., Song, T., Lan, M., & Wei, F. (2024). LLM as a mastermind: A survey of strategic reasoning with large language models (arXiv:2404.01230). arXiv. https://arxiv.org/abs/2404.01230

Zhu, K., Wang, J., & Zhou, J. (2023). The impact of sampling temperature on engineering problem-solving with large language models. *Engineering Applications of Artificial Intelligence, 119*, 105782.

ADDITIONAL READINGS

Afroogh, S., Akbari, A., Malone, E., Kargar, M., & Alambeigi, H. (2024). Trust in AI: Progress, challenges, and future directions. *Humanities & Social Sciences Communications, 11*(1), 1568. DOI: 10.1057/s41599-024-04044-8

Awad, M., & Martín-Rojas, R. (2025). Driving entrepreneurial success: Navigating AI-driven transformation through workforce agility and sustainability. *Journal of Innovation and Entrepreneurship, 14*(1), 554. DOI: 10.1186/s13731-025-00554-0

Balasubramaniam, N., Kauppinen, M., Rannisto, A., Hiekkanen, K., & Kujala, S. (2023). Transparency and explainability of AI systems: From ethical guidelines to requirements. *Information and Software Technology, 159*, 107197. DOI: 10.1016/j.infsof.2023.107197

Filippelli, S., Troise, C., Bigliardi, B., & Corvello, V. (2025). Artificial intelligence and dynamic capabilities: How startups can thrive in entrepreneurial ecosystems amid disruptive change. *The International Entrepreneurship and Management Journal, 21*(1), 97. Advance online publication. DOI: 10.1007/s11365-025-01146-w

Gomez, C., Cho, S. M., Ke, S., Huang, C.-M., & Unberath, M. (2024). Human–AI collaboration is not very collaborative yet: A taxonomy of interaction patterns in AI-assisted decision making from a systematic review. *Frontiers of Computer Science, 6*, 1521066. DOI: 10.3389/fcomp.2024.1521066

Kulichyova, A., Kazantsev, N., White, L., & Islam, N. (2025). Digital transformation in large established organisations: Four restructuring dilemmas based on dynamic capabilities. *International Journal of Management Reviews, 27*(3), 420–450. DOI: 10.1111/ijmr.12395

Mathew, D. E., Ebem, D. U., Ikegwu, A. C., Ukeoma, P. E., & Dibiaezue, N. F. (2025). Recent emerging techniques in explainable artificial intelligence to enhance the interpretable and understanding of AI models for human. *Neural Processing Letters, 57*(1), 16. DOI: 10.1007/s11063-025-11732-2

Mehrotra, S., Degachi, C., Vereschak, O., Jonker, C. M., & Tielman, M. L. (2024). A systematic review on fostering appropriate trust in human–AI interaction: Trends, opportunities and challenges. *ACM Journal on Responsible Computing*, *1*(4), 26. Advance online publication. DOI: 10.1145/3696449

Wen, Y., Wang, J., & Chen, X. (2025). Trust and AI weight: Human–AI collaboration in organizational management decision-making. *Frontiers in Organizational Psychology*, *3*, 1419403. DOI: 10.3389/forgp.2025.1419403

KEY TERMS AND DEFINITIONS

Absorptive Capacity: The organisational ability to identify, assimilate, and apply external knowledge — originally Cohen and Levinthal's (1990) concept — which in the hybrid decision context conditions how quickly new AI architectures and regulatory requirements are integrated into practice.

Algorithmic Auditing: Independent assessment mechanisms enabling evaluation of AI system performance, bias detection, and alignment verification through statistical sampling and distributional characterisation that ensure systems operate within intended parameters and ethical boundaries.

Bounded Rationality: Herbert Simon's theoretical framework describing how decision-makers operate within cognitive constraints necessitating satisficing strategies rather than optimisation, extended in hybrid systems to encompass complementary limitations of human and artificial agents.

Cognitive Ecosystems: Interconnected networks of multiple AI architectures and human agents engaging in continuous co-evolution that generate emergent intelligence through synergistic interaction, transcending capabilities of isolated deployments.

Computer Vision Systems: AI architectures employing convolutional neural networks and integrated processing pipelines for visual analysis that transform image data into actionable insights whilst requiring consideration of privacy concerns and demographic bias.

Deep Learning Architectures: Hierarchical neural networks including CNNs, RNNs, and transformers that achieve exceptional performance through feature learning across vast datasets whilst exhibiting severe opacity challenges for explainability and governance.

Deterministic AI Systems: Artificial intelligence architectures including expert systems and trained supervised models that produce consistent, predictable outputs for identical inputs, enabling transparent governance and straightforward trust calibration.

Distributed Cognition: Theoretical framework showing how cognitive processes extend beyond individual minds to encompass tools, representations, and social structures, applied to human-AI systems where decision capabilities emerge from interaction between agents and artefacts.

Dynamic Capabilities: Organisational competencies in selecting appropriate AI architectures, integrating multiple systems into coherent frameworks, and managing transitions between automation levels as circumstances evolve.

Expert Systems: Rule-based AI employing IF-THEN logic and inference engines for deterministic decision support that provides transparent reasoning chains whilst struggling with situations outside programmed parameters.

Falsifiability: Popper's (2002) criterion that a scientific claim must specify the observations that would show it to be wrong; applied here to the framework's propositions, which each state the conditions and measurements under which they could be refuted.

Generative AI: Non-deterministic systems employing probabilistic content synthesis for creative applications including image, video, and text generation that enable rapid iteration whilst raising concerns about output unpredictability and intellectual property.

Human-in-the-Loop (HITL): Collaboration configuration where humans retain ultimate decision authority whilst AI systems provide analytical support, recommendations, and option evaluation, particularly appropriate for high-stakes contexts requiring ethical judgement.

Human-on-the-Loop (HOTL): Arrangement granting AI operational autonomy within predefined parameters whilst humans maintain supervisory oversight and intervention capability, balancing efficiency gains with safety assurance.

Human-out-of-the-Loop (HOOL): Configuration enabling full AI autonomy without ongoing human oversight during execution, limited to low-stakes, stable environments with clear rules and minimal consequences for errors.

Hybrid Intelligence: The capability to achieve complex goals through combining human and artificial intelligence, reaching superior results through complementarity and synergy rather than substitution or competition.

IntelliFusion Strategies: Organisational approaches embedding artificial intelligence into strategic decision processes through varied architectures, expecting productivity gains as AI handles routine tasks whilst humans focus on creative and strategic work.

Large Language Models (LLMs): Transformer-based AI systems with inherent non-determinism that demonstrate human-like language understanding and generation whilst exhibiting challenges including hallucination risk, prompt sensitivity, and training bias amplification.

Mission-Oriented Innovation: A policy logic, developed by Mazzucato (2018), that directs public investment towards solving defined societal missions rather than leaving technological direction to market signals alone; relevant to governance of AI deployment at national and regional scale.

Multi-Level Perspective: Geels's (2002) framework for socio-technical transitions analysing change across micro (niche innovations), meso (regimes) and macro (landscape) levels; used in this chapter to connect individual human–AI interactions to organisational practices and regulatory regimes.

Non-Deterministic AI Systems: Artificial intelligence architectures including reinforcement learning agents and generative models that exhibit variable outputs under identical conditions, requiring sophisticated trust calibration and statistical governance approaches.

Orchestration: The coordination of heterogeneous human and AI agents toward a shared decision outcome, encompassing task allocation, oversight, exception handling and continuous adaptation across multiple AI architectures; distinguishes the orchestrator role from both operator (task execution) and verifier (output check) roles.

Orchestrator Role: Evolved human position emphasising coordination across human and AI agents, strategic prioritisation of task allocation, and meta-cognitive monitoring ensuring system performance alignment with objectives.

Reinforcement Learning: AI systems optimising decision policies through trial-and-error environmental interaction, exhibiting non-deterministic exploration during training whilst showing adaptive resilience through continuous learning from feedback.

Robotics AI: Embodied intelligence systems integrating sensorimotor capabilities with AI decision-making for physical task execution, combining deterministic perception with variable planning whilst requiring stringent safety protocols.

Socio-Technical Transitions: Long-run shifts in the joint configuration of technologies, practices and institutions; in the hybrid-decision context, the transition from human-centred decision-making to human–AI collaborative decision systems occurs across technical, organisational and regulatory dimensions simultaneously.

Supervised Machine Learning: Statistical pattern learning algorithms including decision trees, random forests, and support vector machines that discover predictive relationships from labelled data whilst remaining limited to trained scenarios.

Three-Dimensional Framework: Integrated conceptual model wherein decision-making emerges from interaction between AI architectural characteristics (determinism, transparency, adaptivity), decision contexts (stakes, uncertainty, dynamism), and human cognitive roles (execution, verification, orchestration).

Transparency Levels: Graduated scale characterising AI system interpretability from high (explicit rule chains in expert systems) through medium (learnable patterns requiring interpretation) to very low (essentially opaque black-box systems).

Trust Calibration: Process of aligning human confidence with AI system reliability, varying systematically across deterministic systems enabling straightforward assessment versus non-deterministic systems requiring nuanced statistical understanding.

Unsupervised Learning: Pattern discovery algorithms identifying latent structures within unlabelled data for exploratory analysis and anomaly detection, presenting unique explainability challenges through ambiguous outputs and parameter sensitivity.

Workforce Transformation: Evolution from operators executing prescribed procedures toward orchestrators coordinating human-AI ensembles, requiring new competencies in prompt engineering, algorithmic auditing, and meta-cognitive monitoring of hybrid systems.

APPENDIX: TABLE OF ABBREVIATIONS

Table 5.

Abbreviation	Full Term	Meaning
ADKAR	Awareness, Desire, Knowledge, Ability, Reinforcement	Sequential change management model developed by Prosci emphasising individual-level transitions through five building blocks enabling organisational change adoption
AGI	Artificial General Intelligence	Theoretical AI system with human-level cognitive abilities across all domains, representing the extreme complexity scenario for governance and orchestration
AI	Artificial Intelligence	Computational systems performing tasks typically requiring human intelligence, encompassing diverse architectures from rule-based to learning systems
AIeSA	AI-enabled Strategic Assistant	Hybrid AI designed as partner-in-thought with human users, promoting higher cognition through dynamic partnering with authoritative data access
AR/VR	Augmented Reality/Virtual Reality	Immersive technologies overlaying digital information on physical environments or creating entirely virtual spaces for interaction
BCG	Boston Consulting Group	Global management consulting firm producing research on AI workforce transformation and hybrid intelligence implementation
CNNs	Convolutional Neural Networks	Deep learning architectures employing hierarchical visual feature extraction for exceptional image processing and pattern recognition capabilities
EU	European Union	Supranational political and economic union establishing regulatory frameworks including the AI Act governing artificial intelligence deployment
GANs	Generative Adversarial Networks	Deep learning architecture using generator and discriminator networks competing to create realistic synthetic data
GPT	Generative Pre-trained Transformer	Transformer-based large language model architecture using unsupervised pre-training on text corpora followed by task-specific fine-tuning
HITL	Human-in-the-Loop	Collaboration configuration where humans retain ultimate decision authority whilst AI provides recommendations and analytical support
HOOL	Human-out-of-the-Loop	Full AI autonomy configuration operating without human oversight, limited to low-stakes stable environments with minimal error consequences
HOTL	Human-on-the-Loop	Arrangement granting AI operational autonomy within parameters whilst humans maintain supervisory oversight and intervention capability
IBM	International Business Machines	Multinational technology corporation whose Watson for Oncology project provides a case study in AI implementation challenges
ISO/IEC	International Organization for Standardization/ International Electrotechnical Commission	International standards bodies establishing frameworks including ISO/IEC 42001 for AI management systems and responsible deployment

continued on following page

Table 5. Continued

Abbreviation	Full Term	Meaning
LIME	Local Interpretable Model-agnostic Explanations	Post-hoc explainability technique approximating complex models locally with interpretable linear models to understand decision influences
LLMs	Large Language Models	Transformer-based AI systems demonstrating human-like language understanding and generation with inherent non-determinism and hallucination risks
ML	Machine Learning	AI approach enabling systems to learn patterns from data without explicit programming, encompassing supervised, unsupervised, and reinforcement paradigms
OECD	Organisation for Economic Co-operation and Development	International organisation establishing AI principles emphasising inclusive growth, human-centred values, transparency, and accountability
PCA	Principal Component Analysis	Unsupervised learning technique for dimensionality reduction identifying principal variations in high-dimensional data
RL	Reinforcement Learning	Machine learning paradigm optimising decision policies through trial-and-error environmental interaction with reward-based feedback
RMF	Risk Management Framework	Structured approach to identifying, assessing, and mitigating risks, specifically the NIST AI RMF 1.0 for artificial intelligence systems
RNNs	Recurrent Neural Networks	Deep learning architectures processing sequential data with temporal dependencies through recurrent connections maintaining memory
SHAP	SHapley Additive exPlanations	Game-theoretic explainability method assigning importance scores to features based on their contribution to model predictions
SVM	Support Vector Machine	Supervised learning algorithm constructing optimal hyperplanes in feature space for classification and regression tasks
USAF	United States Air Force	Military service branch whose Office of the Chief Scientist developed frameworks for autonomous systems and human-AI collaboration

Chapter 5
Workforce Transformation and AI-Driven Skill Development in Business and Management

Chandan Maheshkar
http://orcid.org/0000-0002-7188-0696
SR University, Warangal, India & East Nimar Society for Education, India

Durgesh Nandan
SR University, Warangal, India

Anil Chourasia
http://orcid.org/0009-0008-2392-0332
Power Grid Corporation of India Limited (POWERGRID), India

ABSTRACT

AI integration into businesses and managerial practices is reshaping workforce structure, competencies, and organizational design. This article reviews changes in the workforce, focuses on skills developed with AI assistance, and emphasizes that the mere use of technology is not sufficient for gaining a competitive advantage. Rather, organizations should focus on aligning their workforce development with technological advancements and the skills required. AI serves as a substitute for certain organizational activities, heightens the level of abstraction in managerial reasoning, and results in the integration of algorithms for control and command of the organizational functions. Consequently, tensions arise between the devel-

DOI: 10.4018/979-8-3373-7036-1.ch005

opment of efficiency and capability, the development of control and autonomy, the achievement of short-term goals, and the development of long-term robustness of the organization. The article seeks to develop an understanding of these tensions by reviewing current literature on digital transformation, algorithmic management, and human–AI collaboration.

INTRODUCTION

The systematic intervention of artificial intelligence (AI) is transforming workforce dynamics and compelling employees to develop AI skills. AI-driven practices in business and management exhibited potential for enhanced efficiency and innovation, but uncertainty about workforce readiness and skill adaptability cannot be undermined. As processes become automated, employees have to acquire role-specific competencies. This has forced organizations across sectors to rethink the traditional framework of work and workforce development towards the rapid growth of AI technologies.

As broadly understood, AI is a set of computational systems proficient in performing tasks that traditionally require human cognition, now implanted in core business practices, from operations and HR management to strategic decision-making and customer engagement. The rapid increase in AI adoption has fundamentally changed employment patterns, skill demands, and the essence of managerial roles, creating both disruptive challenges and new opportunities for workforce transformation (Wang & Lu, 2025; Babashahi et al., 2024; Morandini et al., 2023). As AI systems take over routine and repetitive tasks, companies are experiencing a shift in job roles that emphasizes higher-order cognitive skills, interdisciplinary expertise, and human-centric abilities like creativity, ethical decision-making, and teamwork (Zirar et al., 2023; Babashahi et al., 2024).

Academic deliberations reveal a dual effect of AI on employment. First, automation shrinks the need for specific task-based roles, particularly those with expected manual or clerical tasks, and second, AI also supports human efforts by creating new hybrid job categories that combine technical and socio-cognitive skills (Brandao, 2025; Liu & Li, 2025; Deranty & Corbin, 2024; Zirar et al., 2023; Malang et al., 2025). This complementarity is evident in the evolving world of skills, where adaptability, digital savvy, and critical thinking are becoming increasingly important for people to stay employable and competitive in the age of AI (Przegalinska et al., 2025). Successfully integrating AI into organizational workflows involves more than just investing in technology; it also requires strategic development of human capital, inclusive learning ecosystems, and adaptable workforce policies that focus on ongoing reskilling and upskilling (Babashahi et al., 2024; Ekuma, 2023; Zirar et al., 2023). Organizations that proactively develop learning pathways aligned with AI-driven changes show greater agility in managing disruption and promoting innovation. Conversely, those ignoring human capability development risk growing skill shortages, employee disengagement, and labor market mismatches (OECD, 2025; Nyberg et al., 2025).

In addition, the change goes beyond the particular technical skills involved and includes changes in the design and management of work. AI complements human judgment rather than replacing it, which in turn drives changes in leadership style, team roles, and organizational culture toward greater experimentation and continuous learning (OECD, 2023; Babashahi et al., 2024; Ali et al., 2025). In this article, the transformed workforce is viewed as the combination of technological transformation and human development. This transformed workforce changes the work people do and the ways in which organizations build and strengthen their work. The ability to build and strengthen their work is the source of sustained competitive advantage in changing markets in which organizations operate. AI is changing the global and national patterns of work, including the structure and distribution of employment within and across occupations. There is a need for evidence-based approaches to support ethically aligned AI, equity in skills, and adaptable workforce ecosystems in the field and in management (Naoum et al., 2026). Therefore, this chapter seeks to provide a review-based synthesis of this particular complexity, utilizing current empirical and theoretical work to identify important lines of inquiry and consequences for the work of researchers, practitioners, and policymakers concerned with the future of work in an AI age.

EVOLUTION OF WORKFORCE DISRUPTION

A technological shift may be disruptive or sustaining, where disruption means a shift in paradigms and practices that lead to new developments, while sustenance lead advancements in existing ones. During the last five decades, the world witnessed immense transformations in societies and economies. These technological disruptions brought paradigm shifts in business and society that are constantly transforming interactions among organizations and the workforce (Didier, 2024). The changing nature of the workforce can be attributed to many causes, with the current wave of AI being just one of them. The relationship between automation and workforce disruptions has happened in the past. The Industrial Revolution was one of the first major shifts in relationships between technology, work, and labor (Didier, 2024). The introduction of mechanized and steam-powered engines to agrarian societies would displace thousands of agricultural workers and create factory jobs (Chen, 2025). Technological disruption changes the characteristics of jobs. It never eliminates work, but rather alters the composition of work being performed (Pavaloaia & Necula, 2023; Feder, 2018). The many waves of innovations, ranging from electric power to the latest in digital technology and automation, have required constant changes in the workforce, the tasks being performed, and the overall composition and structure of the economy (Shen & Zhang, 2025).

The current wave of AI-driven disruption is once again expanding. However, unlike new technologies that focus on the automation of physical and administrative tasks, AI combines automation with cognitive functions that include the processing of information, generating, recognizing, and making decisions, analyzing data, and assisting with decision-making. This dynamic shift in focus is viewed as a transformation that changes the workforce dynamics from primarily task-oriented to knowledge-based (Brynjolfsson & McAfee, 2014; Dong & McIntyre, 2014). Most of the new technologies have been said to have been developed to serve the professional, as well as the managerial, functions of society that were thought to be beyond the reach of automation. For instance, Eloundou et al. (2023) show that although whole jobs may be secure from being eliminated, a significant number of jobs may be subject to be transformed by AI. The transformations to jobs identified above demonstrate that the real disruption is no longer about losing jobs, but rather the content of the job and skills that are required to perform it.

The integration of AI into various fields is changing the nature of work. Job structures are being modified to combine automated, augmented, and human-only functions, with highly intelligent systems to manage predictable routine tasks, and to reserve highly critical and high-consequence roles for people, including those that are judgment-based, creativity-oriented, emotionally intelligent, and ethically proactive (Oliveira et al., 2025; Islami & Mulolli, 2024). This has led to what is called labor market polarization, defined by strong demand for high-end, technologically intensive jobs and even greater demand for middle-end routine jobs (Breemersch et al., 2017; Usabiaga et al., 2022; Siena & Zago, 2024; Ganuthula & Balaraman, 2025). This change does not affect people equally. Skilled jobs are being created faster in advanced economies because they can adopt AI more quickly, and developing economies can create jobs at a much slower pace due to high vulnerability and skill gaps (Bone et al., 2025). This means that work disruptions are on the rise in structurally unequal regions, sectors, and demographics.

The current phase of workforce disruption is characterized by speed. Previous technological shifts occurred over several decades. This gave labor markets and education systems the time to adequately adjust and respond to shifts and changes. However, the current shifts driven by AI, particularly generative AI, are being incorporated into existing systems at breakneck speeds. This leaves far less time to allow for the workforce to adapt and for organizations to learn and adjust. These changes place additional strain on the existing workforce, organizations, and education systems and structures, which simply are not keeping pace with the other changes (Bughin et al., 2018). Studies have also shown that AI can be a positive force that creates new tasks and even new levels of productivity when used alongside, making the case for reskilling and redesigning jobs even stronger (Marguerit, 2025).

Therefore, the evolution of workforce disruption shows both continuity and change. Technological advances have always impacted employment and the skills in demand. AI and cognitive work are an inflection point because they can reshape tasks, adjust them, and create ambiguity between tasks performed by machines and those performed by people. Leveraging this understanding is critical to addressing workforce challenges and designing strategies that advance skills, promote equity, and preserve the role of people in AI. This is the foundation that informs analyses of AI-human collaboration, and the framework for reskilling initiatives.

AI INTERVENTIONS IN BUSINESS AND MANAGEMENT

AI technology is essential to almost all components of business and management processes (Raina et al., 2026). Considerable activity is underway today using AI technologies to assist in strategic business planning, operations management, marketing, finance, human resources, supply chain management, and customer relationship management. AI technologies are different from previous digital technologies in that they do more than assist in keeping records or performing rule-based automation (Parycek et al., 2024; Homann, 2025). AI technologies assist in learning from data, recognizing and creating patterns and predictions, and, more recently, recommending courses of action. This move from technology-based automation to technology-based augmentation of management functions means that there is now algorithm-guided decision-making in management. This has the potential to improve the pace, scope, and depth of analysis of management decisions (Raina et al., 2026; Paul et al, 2021; Le et al., 2025). There are significant operational efficiency, cost savings, and improved forecasting in organizations that use AI technology, but these value outcomes rely mostly on human and organizational readiness (Murugesan et al., 2023; Ekuma, 2023; Dima et al., 2024; Mwita & Kitole, 2025; Hong et al., 2026).

In operations and supply chain management, AI facilitates predictive maintenance, demand forecasting, logistics, and real-time inventory management. Maintaining flexibility and adaptive capacity is achieved through the identification and real-time analysis of possible disruptions and the intelligent allocation of resources to address them. AI-powered analytics have been shown to increase the accuracy of forecasting and the quality of decisions, especially when forecasting time periods are more uncertain and the environment is more turbulent (Choi et al, 2018). Marketing is another area where AI has introduced advanced tools for customer segmentation, trend analysis, dynamic pricing, and personalized recommendation systems. AI enables the identification and implementation of micro-targeting strategies and increased customer engagement. Firms integrating AI into customer analytics and personalization strategies experience significant increases in customer retention and enhanced competitiveness, provided that data governance and human oversight are adequately structured (Huang & Rust, 2021).

HRM is also an important area for AI implementation in business. Companies are using AI technologies for recruitment and candidate screening, talent analysis, performance reviews, and workforce analytics (Madanchian, 2024; Dadaboyev et al., 2025; Chen, 2023a). These technologies have the capability to analyze CVs, predict employee turnover, and assess the skills needed (or missing) in the company. Although such technologies can improve efficiency and reduce the burden of administrative tasks, they also transform the skills needed from HR professionals. It is no longer about the transaction or operational level, but rather strategic talent management, alignment, and ethics (Jarrahi, 2018). Research indicates that AI HR tools can be biased and discriminate against people in the absence of careful design to mitigate harm, particularly in the context of historical data that reflects discrimination (Davidson et al., 2026; Chen, 2023b; Belenguer, 2022). Thus, the application of AI in HRM is not purely technical, but also a managerial and ethical problem that necessitates restrictions and transparency.

In scenarios such as analyzing competing businesses, predicting risk, and assessing future risk, AI tools assist executives and strategic managers. Using sophisticated analytic platforms, managers can access and analyze both structured and unstructured data. Despite the aforementioned advantages, scholars claim that AI does not replace managerial judgment (Deliu & Olariu, 2024; Alves et al., 2024; Booyse & Scheepers, 2024; Kiškis, 2026). Instead, it shifts the center of AI-related decision-making to the combination of human, algorithmic, and contextual AI intuition (Raisch & Krakowski, 2021). To assist managers in minimizing the aforementioned shift, they must develop *algorithmic literacy*. That is, managers must develop the cognitive and technical skills to ensure AI output does not contradict the mission of the organization and/or long-term goals. It is this integration of cognitive and technical skills that is the key to AI-enhanced workforce transformation in enterprises.

Although AI has the potential to revolutionize how work is done in businesses and management, its adoption is inconsistent and depends on specific organizational characteristics. Lack of access to AI technologies is not the main problem facing the implementation of technologies in the work people do. Some of the issues companies face include employee culture, resistance to change, continuous technological evolution, inadequate staffing, poor strategic focus, and poor-quality data. Often, the successful implementation of AI technology requires employees to redesign their roles and also requires the organization to provide more training. AI technologies are most effective when employees have the capacity to think creatively and ethically and come up with unique solutions. These are AI capabilities that other technologies do not possess. Therefore, in business and management, AI technologies should be viewed as having the potential to transform work, rather than simply adding to the current technological offerings.

Table 1. Analytical Review of AI Interventions in Business and Management

Business Function	AI Intervention	Impact on Workforce	Emerging Skill Requirements	Key Scholarly Sources
Strategic Management	Predictive analytics, scenario modeling, and competitive intelligence systems	Enhances executive decision-making; shifts managers from data gathering to interpretation and judgment	Understanding algorithms, interpreting strategies, and critically assessing AI results.	Raisch & Krakowski (2021); Brynjolfsson & McAfee (2014)
Operations & Supply Chain	Demand forecasting, predictive maintenance, logistics optimization	Automation of routine planning; higher reliance on real-time analytics; improved resilience	Data analytics, systems thinking, AI-enabled process management	Choi et al. (2018)
Marketing Management	Customer segmentation, personalization engines, sentiment analysis, and dynamic pricing	Enhanced targeting; automation of campaign optimization; human focus on creative and brand strategy	Data interpretation, customer analytics, and ethical AI use in consumer data	Huang & Rust (2021)
Human Resource Management	AI-driven recruitment screening, talent analytics, and performance prediction	Automation of administrative HR tasks; shift toward strategic talent management	Ethical AI governance, bias auditing, and workforce analytics	Jarrahi (2018); Morandini et al. (2023)
Finance & Accounting	Fraud detection, risk modeling, automated reporting	Reduced manual auditing; improved predictive risk analysis	Financial data science, AI-supported risk evaluation	Brynjolfsson & McAfee (2014)
Customer Service	Chatbots, conversational AI, automated complaint resolution	Replacement of routine queries; human focus on complex problem-solving and empathy	Emotional intelligence, escalation management, human–AI coordination	Davenport & Ronanki (2018)
Innovation and R&D	AI-driven design, generative modeling, and simulation tools	Faster product development through hybrid human–machine creativity	Creative problem-solving, AI tool integration, interdisciplinary collaboration	Raisch & Krakowski (2021)

Table 1 shows that AI implementations in business and management are more focused on restructuring processes rather than eliminating jobs altogether. AI primarily emphasizes automating tasks that are repetitive, data-driven, and routine across all functions, such as strategy, operations, marketing, HR, finance, customer service, and augments even higher-order cognitive and relational tasks. This exemplifies the automation-augmentation paradox in management literature, where technology replaces some tasks, but also elevates and enables the performance of tasks that are uniquely human. There is a growing emphasis on interpretative judgement, ethical oversight, creative problem-solving, and cross-functional collaboration. Managers have to shift the focus of their role from operational execution to orchestrating human-AI systems, which requires a degree of algorithmic literacy and the ability to critique AI recommendations rather than rely on them. The implementation of AI entails far more than a set of technical decisions and considerations. It also includes the adoption and implementation of new organizational structures and governance processes for reskilling, as well as shifts in organizational culture. The analysis in Table 1 reveals that transforming skill sets for the age of AI requires a new hybrid competency that involves digital literacy and social-emotional, ethical, and strategic thinking. This further reinforces the notion that AI transformations should not focus solely on technological advancement, but also on the redefinition of professional identities, managerial functions, and operational capabilities of the workforce in the context of enduring business change.

REDISCOVERY OF HUMANITY AT WORK

The incorporation of AI technologies into management may have helped organizations automate certain processes, but it has also helped identify the drawbacks of a purely mechanistic approach to management (Brandao, 2025; Nenni et al., 2025; Ritala et al., 2024). With the emergence of predictive and prescriptive analytics, automated algorithmic decision-making, and intelligent automation, organizations have started to realize that the efficiency with which a task is completed is not the same as the effectiveness with which it achieves organizational goals (Smyth et al., 2024; Wang, 2025). AI technologies have a revolutionary and unprecedented ability to optimize processes through pattern identification, variance reduction, and increased speed. However, such intelligent systems cannot clarify ambiguity or help corporate decision-makers choose between competing strategic alternatives. These distinctions have been observed in organizations that have integrated AI-driven algorithmic decision-making into recruitment, credit scoring, customer profiling, and employee performance evaluation processes (Kassa & Worku, 2025; Madanchian, 2024). Once machines handle the calculative parts of work, the tasks that remain will be exclusively within the realm of human beings. These will include negotiating competing claims of stakeholders, exercising moral agency, making decisions in the face of unmanageable historical uncertainty, and building trust among organizational members when changes are implemented. The new focus on humanity in work is not a philosophical movement but a direct operational result of AI's penetration into all parts of organizational life.

The human–AI symbiosis literature illustrates how technological systems work through a logic of probability while the organizational sphere works through a logic of norms and context (Jarrahi, 2018). An AI model can predict the best potential hires as a result of data correlations from past performance, but no AI model can decipher culture or the varying human goals and aspirations unless a human teaches it. This also applies to operational predictive maintenance systems that can automate and decrease downtime. However, how a workforce is adjusted post-automation is a question of ethics rather than the automation of processes. This is what Raisch and Krakowski (2021) call the automation-augmentation paradox (or the automation-augmentation dynamic). However, the essence of the paradox is that managerial roles will become increasingly complex, rather than vanish. If machines are to carry out lower-order tasks, managers will become more dominant in ambiguity, interdependence, and accountability to the organization, in contrast to data. What is scarce is not data, but judgment.

A structural recalibration is going on in how organizations assess competencies. For a long time, digital transformation initiatives have focused on technical skills and data-driven mindsets, often viewing relational or interpretive abilities as afterthoughts. The integration of AI slightly changes this. With abundant computational processing, relational intelligence, ethical reasoning, and adaptive reasoning, these traits will become distinguishing rather than merely nice-to-haves. Many AI failures don't result from algorithmic issues but from organizational culture resistance, mistrust, or misalignment (Behl et al., 2026; Davenport & Ronanki, 2018). These failures illustrate that even in a digital world, coordination, communication, and a shared understanding remain central to high performance. The human ability to understand, critique, and align technological outputs with organizational values is critical.

AI deeply affects accountability, which needs to be rethought. Automated systems make **decisions** on personnel promotions, pricing, risk assessments, and resource allocation, so the responsibility cannot be packaged and dealt with (Alabdali et al., 2024). Organizations need people with the capacity to be put in place for ethical judgment and accountability. Thus, the rediscovery of humanity at work is to re-center moral agency, or ethical principal agency, in the practice of management. The incorporation of AI may improve the quality of an organization's analysis, but for a decision to be deemed legitimate, a person must still make that decision (Voiseux et al., 2025; Khosravi et al., 2024; Zhao & Fariñas, 2023). Systemic algorithms require a greater demand for leaders who ethically integrate technology and social accountability, and do not allow the ethical principles of the dignity and fairness of people be subsumed by the principle of efficiency (Ghasemaghaei & Kordzadeh, 2024). Therefore, it has not decreased human relevance, but is transforming it. The worker's role becomes more interpretive, quasi-judicial, relational, strategic, and contributory in imaginative and new ways. AI helps clarify this boundary by performing tasks that can be programmed in the area of work with increasing consistency (Breckner et al., 2025; Trenerry et al., 2021). What is programmable will predominate and be controllable. What is programmable will be difficult to value. The return of humanity at work is not a loss of technology, but an adjustment centred on intelligence, which has also combined with AI, technology, and human responsibility, and must be clearly delineated in practice from one and inevitably the other.

HUMAN-AI COLLABORATION

While the earlier phases of technological transformation revolved around substitution frameworks, the current phase concerns itself with configurations, that is, how humans and AI are structured within decision-making frameworks (Engström et al., 2025). The main managerial concern has shifted from whether AI will substitute employees to how the different roles of human and algorithmic agents will be apportioned in terms of authority, discretion, and accountability (Callari et al., 2024). Such a shift in the distribution of roles is not automatic, and it is not a function of technology. It is a function of design, negotiation, and institutional processes. The best performing AI systems are not fully automatic but rather hybrid, integrating machine output into a system of human judgment. Examples of such systems are found in fields such as radiology, finance, logistics, and marketing analytics, where AI is used for first-level screening and prediction, while the human actor is responsible for identifying and contextualizing the output of the machine and making judgment calls (Brandao, 2025; Varnosfaderani & Forouzanfar, 2024). Jarrahi (2018) refers to these types of human-machine systems as symbiotic decision-making, in which machines extend the analytical capabilities of the human actor, while the human is responsible for managing the contextual uncertainty. What becomes evident is that the real essence of collaboration is not simply the integration of technology; it is the design of the cognitive processes. The quality of collaboration will depend on how components such as timing, override rights, transparency, explainability, and escalation are structured so that the collaboration results in a system that augments output rather than creates friction.

Crossing the boundaries of business collaboration means dealing with the "black-box" effect. Numerous sophisticated AI systems are built on deep learning and can produce statistically strong results. The "black-box" model results in a range of opacity injected within the flow of business collaboration (Tzionis et al., 2026; Nastoska et al., 2025). The Manager will have to choose between predictive accuracy and interpretability when making decisions, such as credit, hiring, or even medical analyses. Predictive analytics augmentation requires transparency of oversight. Predictive analytics that aren't interpretable are most likely to lead to passive collaboration and will weaken professional skills instead of strengthening them. Therefore, successful collaboration with an AI system is about the practical and predictive analytics (Raina et al., 2026). Trust is a little more than a lack of fear. Another, more subtle effect is a role change that is caused by an AI system taking over some analytical roles. Knowing how to transition to new roles is a skill in itself. Especially in the long-term future. The relationships between humans and systems are better described as co-evolution. In business, change is perpetual, and the future always holds potential for new and greater things. When new roles and functions are recognized and accepted, effort can be decoupled from tasks and placed on supervision, interpretation, and refinement of the automated functions.

In relation to automation in the workplace, too much can lead to "*automation complacency*", where humans are disengaged and ignore potential mistakes, while too little automation can lead to less productivity overall, as the technological advances are not fully utilized. Finding the right balance between oversight and delegation is important to organizations. Davenport and Ronanki (2018) state that many AI projects fail because businesses overstate the automation's capabilities and neglect the need to make managerial changes. Efficient collaboration often requires changes to the entire process, including new goals and metrics to evaluate success, and incentives to review and reward oversight responsibility, instead of just considering the volume of work done. Human–AI collaboration also changes how expertise is distributed as it becomes less centralized. This new configuration of power can lead to more influence for the data-literate employees. In many ways, the new organizational hierarchy created by collaboration also defines where the information and decision-making authority should lie, rather than eliminating it. It is up to leaders to determine where AI's advice falls in these categories, and if it is only a suggestion, a decision, or can be overridden. This is a significant ethical and reputational concern. When it comes to long-term collaboration, the goal is to develop adaptable systems that go beyond mere human-machine pairing (Kunz et al., 2025; Saravanan et al., 2026; Berretta et al., 2023). Organizations that view AI as a tool face more barriers. In contrast, organizations that integrate AI through participatory governance have role clarity, transparency, and ongoing learning to define participation, and are more likely to achieve augmentation over substitution (Mišić et al., 2025). Human–AI collaboration is an ongoing process of design and engineering that requires an exacting equilibrium of capability, accountability, and trust.

Referring to *Table 2*, it can be idealized that unlike traditional technology adoption models that focus on efficiency, this synthesis underscores that the success of human–AI collaboration relies heavily on structural design choices. Outcomes are influenced more by how authority, transparency, accountability, and learning mechanisms are structured than by technological sophistication alone. When decision rights are unclear or oversight mechanisms are weak, collaboration can weaken expertise and ethical governance. On the other hand, embedding AI within interpretive, accountable, and continuously learning organizational settings makes augmentation sustainable instead of extractive.

Table 2. Human-AI Collaboration in Organizations

Collaboration Dimension	Organizational Design Choice	Human Role Configuration	Risk if Misaligned	Key Scholarly Sources
Decision Authority	Advisory AI vs. Autonomous AI	Interpreter, approver, or supervisor of AI output	Over-reliance leading to deskilling; under-reliance leading to inefficiency	Raisch & Krakowski (2021)
Task Allocation	Task decomposition (AI handles planned tasks; humans handle ambiguous tasks)	Contextual integrator and exception handler	Automation complacency or task fragmentation	Jarrahi (2018)
Transparency & Explainability	Interpretable models vs. Black-box systems	Critical evaluator of logics	Blind trust, liability gaps, ethical exposure	Davenport & Ronanki (2018)
Learning Integration	Static deployment vs. iterative co-learning systems	Data curator, feedback provider, model refiner	Stagnant performance, skill erosion	Jarrahi (2018)
Accountability Structure	Clear human errors vs. diffused responsibility	Ethical arbiter and risk manager	Liability ambiguity, reputational risk	Raisch & Krakowski (2021)
Skill Adaptation Mechanism	Reactive training vs. continuous reskilling ecosystems	AI-literate decision partner	Capability gap, resistance to adoption	Davenport & Ronanki (2018)
Trust Calibration	Blind automation vs. calibrated reliance	Judgment-based override authority	Either algorithm aversion or automation bias	Raisch & Krakowski (2021)

Based on the nature of human-AI collaboration and synthesis present in *Table 2*, the process model in *Table 3* shows the first of many iterations of how AI-enabled workforce transformation develops over time. Each stage (or iteration) of transformation requires different degrees of human competency and different levels of sophisticated governance. In the early stages, most challenges lie in the technical and cultural fields. However, as AI is more deeply integrated into workflow automation and the core of strategic decision-making, the nature of the disruption evolves from operational change to transformative change. Employees who were formerly operational doers become system supervisors, interpreters, and ethical guardians of the algorithmic systems. The model illustrates the advanced phases of integration and the concerns. The advanced phases demonstrate the greatest risks not present in initial automation levels. Strategic AI integration into decision-making roles leaves organizations vulnerable due to unclear accountability, diminished human expertise, and reliance on the algorithm (Raisch & Krakowski, 2021). Therefore, workforce transformation is co-evolution in closed phases (i.e., technological advancement of AI and the human forces workforce transformation). This process-oriented synthesis reinforces the argument made throughout the chapter and emphasizes that AI adoption is not a singular technological shift. It is a complex, multifaceted organizational transformation of processes that requires continual attention across several levels (i.e., reskilling, cultural change, and governance redesign).

Table 3. Process of AI-Driven Workforce Transformation

Transformation Stage	Organizations' AI Focus	Nature of Work Redesign	Emerging Human Competency Needs	Strategic Risk if Neglected
Research & Pilot Deployment	Isolated AI applications for efficiency (e.g., analytics, automation tools)	Task-level automation; minimal structural redesign	Digital literacy; openness to technological learning	Resistance to adoption; symbolic implementation
Functional Integration	Embedding AI within specific departments (HR analytics, predictive marketing, financial forecasting)	Partial workflow redesign; hybrid task allocation	Data interpretation; algorithmic oversight; cross-functional coordination	Fragmentation of expertise; over-reliance on algorithms
Process Reengineering	AI is integrated across value chains and decision systems	End-to-end process redesign; human roles shift toward exception management	Systems thinking; collaborative problem-solving, and adaptive decision-making	Deskilling or loss of tacit knowledge

continued on following page

Table 3. Continued

Transformation Stage	Organizations' AI Focus	Nature of Work Redesign	Emerging Human Competency Needs	Strategic Risk if Neglected
Strategic Embedding	AI informs strategic planning, forecasting, and competitive positioning	Human–AI co-decision structures; strategic augmentation	Critical judgment; ethical reasoning; AI governance capability	Strategic rigidity; accountability diffusion
Continuous Co-Evolution	Organization-wide AI ecosystem with feedback loops	Iterative human–AI learning cycles; dynamic job redefinition	Lifelong learning orientation; cognitive flexibility; socio-emotional intelligence	Capability obsolescence; cultural fatigue

RESKILLING AS HUMAN GROWTH IN AN AI-DRIVEN WORKPLACE

Discourse around reskilling has changed as organizations increasingly use AI, defining the goals of their human capacity utilization to ensure sustainable organizational success in the long run. With AI, the perception of displacement of human labor is incorrect. It is the restructuring of the work landscape with the elimination of repetitive work and the elevation of the importance of work that requires human intervention to use adaptive and situational skills (Bakhshi et al., 2017; Sharma et al., 2025). In such a context, reskilling must go beyond the acquisition of tool-related skills to encompass the further development of the cognitive capacity and the socioemotional skills of the individual, including the capacity for critical and ethical judgment, situational analysis, and collaborative problem solving, that underscore the unique human contribution in the area of decision making where technology is either unable to function or is inappropriate to deploy. Employees with these skills are better positioned to work with AI and support the innovation, flexibility, and learning of their organizations (Mastrogiacomo, 2025; Przegalinska et al., 2025). This view of reskilling focuses on human development rather than treating it as a remedial measure to address the impact of job loss.

It can be observed that AI-powered reskilling should be considered as an ongoing, systemic process instead of discrete training programs. Organizations that take this position see capability-building as an intricate part of work design, performance feedback, and teamwork. For instance, firms with cross-hierarchical AI literacy training integrated from entry-level to executive positions had higher employee engagement, innovation, and quick strategic decision-making than companies that offered just basic tech training. Studies shown that the best learning pathways are iterative and adaptive, and facilitate an environment where employees can use AI, evaluate their work, and draw and integrate new lessons from their job roles (Córdova-Esparza, 2025; Sposato, 2025; Yan et al., 2026). Such ecosystems foster both foundational and advanced digital skills along with the meta-skills that, in addition to adaptability, establish a mindset of continual learning and the ability to think in systems, which are skills that are always in demand and are essential for thriving in unpredictable job markets (Bouwmans et al., 2024; Zenk et al., 2024; ITF, 2023).

Integrating psychological and identity-related elements is crucial for the successful implementation of reskilling initiatives aimed at developing employees. People experiencing workforce disruption often feel uncertainty, job insecurity, and identify conflict as traditional roles shift. New studies show that if reskilling is geared to the development of employees by providing clarity on their purpose, developing and communicating actionable career goals, and enriching their job roles, they become more satisfied with their jobs, less likely to leave the organization, and more assured that they will cope with the technology that needs to be integrated (Li, 2022; Braun et al., 2025; Babashahi et al., 2024). Reskilling for human growth offers dual benefits for both employees and the organization. Clarity and autonomy in roles, and growth-oriented reskilling, promote psychological safety, enabling employees to use AI more confidently, learn from errors, and make constructive changes (Shen & Zhang, 2024; Golgeci et al., 2025; Soomro et al., 2026; Yoo, 2025).

Strategically, reskilling influences more than just individual competencies and affects the culture and governance of the organization. Without the necessary capabilities, the implementation of AI can create skill gaps and inequities among employees. Some employees may gain advanced knowledge and become proficient with AI, and may have the opportunity to gain higher positions in the company, while others may be left behind and remain without that knowledge. To avoid these issues, successful companies have implemented reskilling programs, mentorship, peer learning, cross-functional rotations, and leadership opportunities as core elements of workforce transformation (Li, 2022; Appiah, 2025). These strategies indicate that in an artificial intelligence workplace, having technical skills alone is insufficient, and employees also need the skills to navigate social relationships, ethical challenges, and the influence of unequal access to technology. Reskilling that prioritizes dignity and fairness is more likely to reduce cross-organizational vertical barriers that strengthen silos.

WORKFORCE TRANSFORMATION STRATEGIES

AI does not merely digitize existing workflows; it changes the underlying structure of work. Previous generations of technologies focused on optimizing efficiency; in contrast, AI not only automates tasks, but also literally enhances the cognitive capacity of and embeds the algorithmic guardian of an organization's structural processes. This kind of change tends to create structural tensions distinct from the temporary "*adjustment costs*" that are typically associated with incremental changes. Studies have documented these tensions from various perspectives, such as the automation–augmentation paradox (Raisch & Krakowski, 2021), algorithmic management (Kellogg et al., 2020), and the paradox of digital capabilities (Meijerink & Bondarouk, 2023; Theres & Strohmeier, 2023). However, most of these studies articulate the tensions but do not define the processes that organizations use to navigate them. Here, workforce transformation is defined as the proactive adjustment of an organization's task architectures, relational authority configurations, and learning systems to maintain the structural balance organizations require to function amid the structural contradictions introduced by AI. Instead of viewing reskilling, governance changes, and digital initiatives as independent actions, workforce transformation is characterized as an integrated process for managing tensions across several different streams at the organizational level.

Structural Tensions in AI-Driven Workplace

AI-supported working systems incorporate conflicting organizational logics in everyday practices. AI technologies create permanent, not temporary, solutions because they automate, augment, and provide algorithmic oversight simultaneously. These tensions arise not from unsuccessful implementations but from the multifaceted functionality of AI. Efficiency and capability logics frictionally coexist in augmentation and automation. Data-driven governance undermines the established norms of professional autonomy. Short-term productivity goals contradict long-term capability development investments. The need for contextual judgment conflicts with computational standardization. Platform-based centralization is at odds with distributed innovation. These tensions in AI-supported organizing are not resolvable by purely technical systems. Instead, they require purposeful organizational arrangements. The framework in *Table 4* summarizes these embedded contradictions that influence workforce transformation.

Table 4. Structural Tensions in AI-Driven Workforce Transformation

Transformation Tension	AI-Centric Logic	Human-Centric Logic	Risk of Overemphasis	Strategic Balancing Mechanism
Automation vs. Autonomy	Algorithmic optimization, standardization, efficiency	Human discretion, contextual judgment, creative agency	Over-automation → deskilling & disengagement; Over-autonomy → inconsistency & inefficiency	Augmentation design: automate repetitive tasks while maintaining human control in complex areas.
Speed vs. Deliberation	Real-time analytics, rapid iteration, predictive acceleration	Reflective reasoning, ethical scrutiny, collective consultation	Excess speed→ ethical blind spots; Excess deliberation→ lost competitive advantage	Layered decision-making system: quick execution phases complemented by human review checkpoints.
Data Fairness vs. Interpretive Context	Quantification, measurable indicators, predictive scoring	Tacit knowledge, social nuance, experiential insight	Data dominance → reductionism; Human dominance→ bias & inconsistency	Hybrid intelligence protocols combine algorithmic outputs with structured human interpretation.
Control vs. Trust	Algorithmic monitoring, performance tracking, behavioral analytics	Psychological safety, intrinsic motivation, relational commitment	Excess control → surveillance culture; Excess trust→ weak accountability	Transparent governance + participatory AI design to sustain legitimacy

These tensions are fundamental aspects of AI-enabled organizing, not mistakes. If not effectively managed, they can cause resistance, legitimacy concerns, reduced capabilities, and obstacles to innovation. Therefore, effective workforce transformation depends on organizational mechanisms that manage these tensions rather than attempting to eliminate them.

Strategic Levers as Tension-Mediating Mechanisms

AI work systems for now still contain unresolved structural tensions that are not the subjects of incremental change or pure technological advancement. These tensions result in differing organizational logics—efficiency and capability; control and autonomy; standardization and discretion. Such tensions require structural design interventions. Strategic levers are those design interventions that balance differing organizational logics. Such levers do not aim to eliminate tensions, but rather help manage the balance of tensions, or logics in the restructuring of control, balance, cognition, learning, and innovation within the different levels of the firm. Instead of focusing solely on advancing AI technologies, the extent to which the integrated design of mediating or balancing controls, levers, or systems aligns with workforce transformation determines greater success.

Table 5. Strategic Levers in AI-Driven Workforce Transformation

Core Structural Tension	Strategic Lever	Organizational Level	Risk if Misaligned	Mediation Mechanism
Automation vs. Augmentation	Human–AI Task Redesign	Job/Process Level	Skill redundancy and morale erosion	Complementarity and capability amplification
Algorithmic Control vs. Professional Autonomy	Transparent Algorithmic Governance	Governance/ Policy Level	Surveillance culture and resistance	Trust-based digital oversight
Short-Term Productivity vs. Long-Term Capability	Strategic Reskilling Ecosystems	Strategic/HR Level	Training fatigue and symbolic upskilling	Sustainable skill renewal
Efficiency Standardization vs. Contextual Judgment	Hybrid Decision Architectures	Operational Level	Over-reliance on AI outputs	Balanced augmentation and human discretion
Centralization vs. Distributed Innovation	Cross-Functional Digital Labs	Structural Level	Innovation silos	Organizational learning diffusion

The convergence of *Tables 4* and *5* presents a multilevel explanatory framework. Structural tensions indicate the contradictions embedded within AI-based organizing. Strategic levers act as mediating design mechanisms that redistribute thinking, power, and the ability to learn across various levels of the organization. In several ways, these framing advancements prior scholarship. First, the automation–augmentation paradox (Raisch & Krakowski, 2021) is extended by identifying task redesign as the causal explanation that determines if automation will erode or enrich a person's capability. Without intentional task reallocation, automation will always prioritize efficiency. Second, the research on algorithmic management (Kellogg et al., 2020) is repositioned from a control narrative to a governance-design narrative. There are institutional design variables that influence the construction and perception of legitimacy and trust: Transparency, explainability, and participatory oversight. Third, the transformation of the workforce is reconceptualized as a changing capability. Organizations will be able to capture adaptive stability in times of technological turbulence through continuous reskilling embedded within the redesign of workflows and cross-unit experimentation (Meijerink & Bondarouk, 2023). Thus, the loss of capability to transform becomes a more permanent attribute of the organization rather than a transitional initiative.

ETHICS IN AI INTEGRATION AND WORKFORCE CHANGE

The application of AI in business processes is not only a mechanism of technological change, but an alteration of the ethical context of work itself. The influence of AI has permeated the realms of recruitment, evaluations, job allocation, promotions, restructuring, and any other organizational practices (Chen, 2023a; Murire, 2024). AI is beginning to alter the allocation, distribution, visibility, and equity of power among employees in organizations. AI is different from previous digital tools in that it has the ability to make decisions autonomously, instead of merely assisting managers in making decisions (Khosravi et al., 2024). AI has predictive power and can shape processes. These changes influence how decisions are made and the way responsibility, equity, and accountability are structured. AI-related ethical problems are intertwined with issues of fairness (or the lack of it) and pose fundamental issues about justice, legitimacy, and accountability, which are often ignored in techno-managerial workplaces (De Cremer & Narayanan, 2023; Hart et al., 2025). The accountability and fairness of work structures that rely on automation are contingent on the algorithms that drive those decisions when employees are automated, scored, or otherwise directed by opaque systems.

In the attempt to merge data-driven objectivity with its inherent biases lies an ethical concern. AI is trained using historical data. These past data may include structural biases, workforce discrimination, and discriminatory practices (Chen, 2023b). Unchecked predictive models may treat such data as neutral, but they can reproduce and even worsen bias. In such cases, bias may arise from the design process (e.g., over-design) rather than from an explicit managerial decision. Also, bias and discrimination are not the only ethical problems with AI. There is also criticism of surveillance, in which AI technologies can track and record behavior, thereby expanding managerial control over areas of work that have traditionally been ruled by trust and professional judgment (Afroogh et al., 2024; van Zoonen et al., 2025). Although the visibility enabled by AI may improve coordination and productivity, too much surveillance can be damaging to people's dignity, psychological well-being, and intrinsic motivation. Therefore, ethical AI workforce transformation must also ensure a reasonable and respectful balance between the level of surveillance and its actual purpose.

AI integration raises ethical concerns for workforce displacement and the need for workforce repositioning after displacement (Shrama et al., 2025; Chen, 2023b; Davidson et al., 2026). Organizational restructuring driven by automation creates ethical concerns beyond efficiency. Displaced stakeholders feel abandoned by organizations. Ethically rescheduling change in the workforce requires constructive reskilling pathways, inclusive redesigns, and clear communication regarding the workforce's technological futures (Li, 2022; Chen, 2023b). AI adoption without these will erode trust and undermine the legitimacy of institutions, even when outcomes improve. For organizations to engage in ethical AI integration, they need to be ready to incorporate ethics and not treat it as an afterthought (Faulds et al., 2025). Organizations that are ready to integrate fairness, accountability, and open governance into AI will most likely gain stakeholders' trust and commitment. Organizations that do not recognize the ethical implications of AI will increase their risk potential. For the AI-driven change in the workforce to be sustainable, organizations must integrate technology and ethical organizational design.

CONCLUSION

Incorporating AI into the systems of an organization means a different way of doing, managing, and experiencing work. AI has been framed as a type of technology for operations, but also as a structural technology. It decentralizes decision rights, the way power is exercised, and influences the kinds of capabilities people develop. The AI challenges differ from earlier phases of digitalization, as AI will perform and augment the capabilities of the worker, while also combining the ability to process algorithms and oversee the core work of the organization. The fundamental and ongoing challenges of an AI organization are managing the trade-offs between efficiency and capability, control and autonomy, oversight and context-specific judgment, centralization and distributed authority for innovation. The key will be how consciously the organization designs to address each of these challenges and manage the trade-offs and controls.

In this context, the primary mechanism enabling mediation is workforce transformation. The strategic tools of task redesign for human–AI collaboration, transparent algorithmic governance, embedded reskilling ecosystems, hybrid decision-making frameworks, and distributed experimentation structures respond to structural contradictions. When aligned, these tools direct outcomes towards augmentation, adaptive stability, and dynamic capability growth. On the flip side, misalignment reinforces concerns of displacement, surveillance, and skill erosion. The ethical concerns associated with AI also underscore the need for responsible integration, particularly because algorithmic systems shape Opportunity, Fairness, and Accountability trade-offs. Therefore, the governance arrangements for AI integration must incorporate transparency, accountability, and participation to facilitate sustainable transformation. AI-based organizational change is not merely the result of technological determinism or skill upgrading; it requires a major redesign effort that addresses structural integrity, strategic cohesion, ethical legitimacy, and capability rejuvenation. Sustainability is rooted in the combination of responsible organizational architecture with technological progress to ensure continued focus on the human core of the enterprise.

Declaration: This work is the authors' original work, and they have not used Generative AI tools to write any part of this work. They have used Grammarly to improve the language, grammar, and coherence of the work.

REFERENCES

Afroogh, S., Akbari, A., Malone, E., Kargar, M., & Alambeigi, H. (2024). Trust in AI: Progress, challenges, and future directions. *Humanities & Social Sciences Communications, 11*(1), 1568. Advance online publication. DOI: 10.1057/s41599-024-04044-8

Alabdali, M. A., Khan, S. A., Yaqub, M. Z., & Alshahrani, M. A. (2024). Harnessing the Power of Algorithmic Human Resource Management and Human Resource Strategic Decision-Making for Achieving Organizational Success: An Empirical Analysis. *Sustainability (Basel), 16*(11), 4854. DOI: 10.3390/su16114854

Ali, I., Nguyen, K., Ali, A. M., & Cui, T. (2025). Human–AI collaboration in knowledge ecosystems: A multidisciplinary review, integrative framework and future directions. *Journal of Knowledge Management*, 1–22. Advance online publication. DOI: 10.1108/JKM-03-2025-0431

Alves, M., Seringa, J., Silvestre, T., & Magalhães, T. (2024). Use of Artificial Intelligence tools in supporting decision-making in hospital management. *BMC Health Services Research, 24*(1), 1282. DOI: 10.1186/s12913-024-11602-y

Appiah, R. I. (2025). Corporate Strategies for Successful Workforce Upskilling and Reskilling in Response to AI Adoption - What Works, What Does not, and Why. *International Journal of Emerging Research in Engineering and Technology, 6*(1), 91–99. DOI: 10.63282/3050-922X.IJERET-V6I1P112

Babashahi, L., Barbosa, C. E., Lima, Y., Lyra, A., Salazar, H., Argôlo, M., Almeida, M. A., & Souza, J. M. (2024). AI in the Workplace: A Systematic Review of Skill Transformation in the Industry. *Administrative Sciences, 14*(6), 127. DOI: 10.3390/admsci14060127

Bakhshi, H., Downing, J. M., Osborne, M. A., & Schneider, P. (2017). The Future of Skills: Employment in 2030. Retrieved on Feb. 05, 2026 from https://media.nesta.org.uk/documents/the_future_of_skills_employment_in_2030_0.pdf

Behl, A., Bhardwaj, S., Jayawardena, N., Pereira, V., & Roohanifar, M. (2026). Grass is always dark(er) on the other side: Exploring the dark side of artificial intelligence humanitarian supply chain operations. *Technological Forecasting and Social Change, 224*, 124484. DOI: 10.1016/j.techfore.2025.124484

Belenguer, L. (2022). AI bias: Exploring discriminatory algorithmic decision-making models and the application of possible machine-centric solutions adapted from the pharmaceutical industry. *AI and Ethics, 2*(4), 771–787. DOI: 10.1007/s43681-022-00138-8

Berretta, S., Tausch, A., Ontrup, G., Gilles, B., Peifer, C., & Kluge, A. (2023). Defining human-AI teaming the human-centered way: A scoping review and network analysis. *Frontiers in Artificial Intelligence*, *6*, 1250725. DOI: 10.3389/frai.2023.1250725

Bone, M., Ehlinger, E. G., & Stephany, F. (2025). Skills or degree? The rise of skill-based hiring for AI and green jobs. *Technological Forecasting and Social Change*, *214*, 124042. DOI: 10.1016/j.techfore.2025.124042

Booyse, D., & Scheepers, C. B. (2024). Barriers to adopting automated organisational decision-making through the use of artificial intelligence. *Management Research Review*, *47*(1), 64–85. DOI: 10.1108/MRR-09-2021-0701

Bouwmans, M., Lub, X., Orlowski, M., & Nguyen, T. V. (2024). Developing the digital transformation skills framework: A systematic literature review approach. *PLoS One*, *19*(7), e0304127. DOI: 10.1371/journal.pone.0304127

Brandao, P. R. (2025). The Impact of Artificial Intelligence on Modern Society. *AI*, *6*(8), 190. DOI: 10.3390/ai6080190

Braun, G., Raty, P., Bokinge, M., Rikala, P., Hamalainen, R., Syberfeldt, A., & Stahre, J. (2025). The skill bridge – A global qualitative analysis of skill gap management. *Journal of Environmental Management*, *395*, 127738. DOI: 10.1016/j.jenvman.2025.127738

Breckner, K., Neumayr, T., Mara, M., Streit, M., & Augstein, M. (2025). The Changing Nature of Human-AI Relations: A Scoping Review on Terminology and Evolvement in the Scientific Literature. *International Journal of Human-Computer Interaction*, *41*(22), 14248–14305. DOI: 10.1080/10447318.2025.2482742

Breemersch, K., Damijan, J., & Konings, J. (2017). Labour Market Polarization in Advanced Countries: Impact of Global Value Chains, Technology, Import Competition from China and Labour Market Institutions. *OECD Social, Employment and Migration Working Papers*, No. 197, OECD Publishing, Paris. https://doi.org/ DOI: 10.1787/1815199X

Brynjolfsson, E., & McAfee, A. (2014). *The second machine age: Work, progress, and prosperity in a time of brilliant technologies*. W. W. Norton & Company.

Bughin, J., Hazan, E., Lund, S., Dahlström, P., Wiesinger, A., & Subramaniam, A. (2018). Skill shift: Automation and the future of the workforce. McKinsey Global Institute. https://www.mckinsey.com/featured-insights/future-of-work/skill-shift-automation-and-the-future-of-the-workforce?utm_source=chatgpt.com

Callari, T. C., Segate, R. V., Hubbard, E.-M., Daly, A., & Lohse, N. (2024). An ethical framework for human-robot collaboration for the future people-centric manufacturing: A collaborative endeavour with European subject-matter experts in ethics. *Technology in Society*, *78*, 102680. DOI: 10.1016/j.techsoc.2024.102680

Chen, J. (2025). Industrial Revolution: Definition, History, Pros, and Cons. https://www.investopedia.com/terms/i/industrial-revolution.asp

Chen, Z. (2023a). Collaboration among recruiters and artificial intelligence: Removing human prejudices in employment. *Cognition Technology and Work*, *25*(1), 135–149. DOI: 10.1007/s10111-022-00716-0

Chen, Z. (2023b). Ethics and discrimination in artificial intelligence-enabled recruitment practices. *Humanities & Social Sciences Communications*, *10*(1), 567. DOI: 10.1057/s41599-023-02079-x

Choi, T., Wallace, S. W., & Wang, Y. (2018). Big Data Analytics in Operations Management. *Production and Operations Management*, *27*(10), 1868–1883. DOI: 10.1111/poms.12838

Córdova-Esparza, D.-M. (2025). AI-Powered Educational Agents: Opportunities, Innovations, and Ethical Challenges. *Information (Basel)*, *16*(6), 469. DOI: 10.3390/info16060469

Dadaboyev, S. M. U., Abdullayeva, J., Abbosova, N., Suleymenova, A., & Mamadjanova, K. (2025). Role of artificial intelligence in employee recruitment: Systematic review and future research directions. *Discover Global Society*, *3*(1), 99. DOI: 10.1007/s44282-025-00246-w

Davenport, T. H., & Ronanki, R. (2018). *Artificial Intelligence for the Real World. Harvard Business Review*. HBR.

Davidson, J., Schnepp, J., Buttrick, H., & Gupta, A. (2026). A framework for ethical AI-HRM development and use. *AI and Ethics*, *6*(1), 28. DOI: 10.1007/s43681-025-00858-7

De Cremer, D., & Narayanan, D. (2023). How AI tools can-and cannot-help organizations become more ethical. *Frontiers in Artificial Intelligence*, *6*, 1093712. DOI: 10.3389/frai.2023.1093712

Deliu, D., & Olariu, A. (2024). The Role of Artificial Intelligence and Big Data Analytics in Shaping the Future of Professions in Industry 6.0: Perspectives from an Emerging Market. *Electronics (Basel)*, *13*(24), 4983. DOI: 10.3390/electronics13244983

Deranty, J. P., & Corbin, T. (2024). Artificial intelligence and work: A critical review of recent research from the social sciences. *AI & Society*, *39*(2), 675–691. DOI: 10.1007/s00146-022-01496-x

Didier, N. (2024). Turning fragments into a lens: Technological change, industrial revolutions, and labor. *Technology in Society*, *77*, 102497. DOI: 10.1016/j.techsoc.2024.102497

Dong, X., & McIntyre, S. H. (2014). The Second Machine Age: Work, Progress, and Prosperity in a Time of Brilliant Technologies. *Quantitative Finance*, *14*(11), 1895–1896. DOI: 10.1080/14697688.2014.946440

Ekuma, K. (2023). Artificial Intelligence and Automation in Human Resource Development: A Systematic Review. *Human Resource Development Review, 23*(2), 199 229. https://doi.org/DOI: 10.1177/15344843231224009

Eloundou, T., Manning, S., Mishkin, P., & Rock, D. (2023). *GPTs are GPTs: An early look at the labor market impact potential of large language models*. arXiv. https://doi.org//arXiv.2303.10130DOI: 10.48550

Engström, A., Pittino, D., Mohlin, A., Edh, N., & Johansson, A. (2025). A paradox perspective on early AI adoption: Understanding temporal and relational tensions. *Journal of Organizational Change Management*, *38*(7), 1145–1171. DOI: 10.1108/JOCM-02-2025-0086

Faulds, D. J., Raju, P. S., & Davidson, C. (2025). Ethical considerations in the use of artificial intelligence: An interview with Professor Justin Biddle. *Business Horizons*. Advance online publication. DOI: 10.1016/j.bushor.2025.12.003

Feder, C. (2018). The effects of disruptive innovations on productivity. *Technological Forecasting and Social Change*, *126*, 186–193. DOI: 10.1016/j.techfore.2017.05.009

Ganuthula, V. R. R., & Balaraman, K. K. (2025). *Skill-based labor market polarization in the age of AI: A comparative analysis of India and the United States*. arXiv. https://doi.org//arXiv.2501.15809DOI: 10.48550

Ghasemaghaei, M., & Kordzadeh, N. (2024). Understanding how algorithmic injustice leads to making discriminatory decisions: An obedience to authority perspective. *Information & Management*, *61*(2), 103921. DOI: 10.1016/j.im.2024.103921

Golgeci, I., Ritala, P., Arslan, A., McKenna, B., & Ali, I. (2025). Confronting and alleviating AI resistance in the workplace: An integrative review and a process framework. *Human Resource Management Review*, *35*(2), 101075. DOI: 10.1016/j.hrmr.2024.101075

Hart, M., Bavin, K., & Lynes, A. (2025). Artificial Intelligence, Capitalism, and the Logic of Harm: Toward a Critical Criminology of AI. *Critical Criminology*, *33*(3), 513–532. DOI: 10.1007/s10612-025-09837-0

Homann, M. (2025). What is the Difference Between AI and Automation? Retrieved on Feb 8, 2026 from https://www.leapwork.com/blog/ai-and-automation-what-is-the-difference

Hong, S., Zhong, D., & Um, K. (2026). The impact of artificial intelligence (AI) adoption on operational performance in manufacturing. *Journal of Manufacturing Technology Management*, *37*(2), 295–317. DOI: 10.1108/JMTM-03-2025-0227

Huang, M. H., & Rust, R. T. (2021). A strategic framework for artificial intelligence in marketing. *Journal of the Academy of Marketing Science*, *49*(1), 30–50. DOI: 10.1007/s11747-020-00749-9

Islami, X., & Mulolli, E. (2024). Human-Artificial Intelligence in Management Functions: A Synergistic Symbiosis Relationship. *Applied Artificial Intelligence*, *38*(1), 2439615. Advance online publication. DOI: 10.1080/08839514.2024.2439615

ITF. (2023). Adapting (to) *Automation: Transport Workforce in Transition*. International Transport Forum Policy Papers, No. 122, OECD Publishing, Paris.

Jarrahi, M. H. (2018). Artificial intelligence and the future of work: Human-AI symbiosis in organizational decision making. *Business Horizons*, *61*(4), 577–586. DOI: 10.1016/j.bushor.2018.03.007

Kassa, B. Y., & Worku, E. K. (2025). The impact of artificial intelligence on organizational performance: The mediating role of employee productivity. *Journal of Open Innovation*, *11*(1), 100474. DOI: 10.1016/j.joitmc.2025.100474

Kellogg, K. C., Valentine, M. A., & Christin, A. (2020). Algorithms at work: The new contested terrain of control. *The Academy of Management Annals*, *14*(1), 366–410. DOI: 10.5465/annals.2018.0174

Khosravi, M., Zare, Z., Mojtabaeian, S. M., & Izadi, R. (2024). Artificial Intelligence and Decision-Making in Healthcare: A Thematic Analysis of a Systematic Review of Reviews. *Health Services Research and Managerial Epidemiology*, *11*, 23333928241234863. Advance online publication. DOI: 10.1177/23333928241234863

Kiškis, M. (2026). AI on the path to good decisions. *Sustainable Futures : An Applied Journal of Technology, Environment and Society*, *11*, 101644. DOI: 10.1016/j.sftr.2026.101644

Kunz, W. H., Sajtos, L., & Flavián, C. (2025). Beyond replacement: Human-machine collaboration in the age of AI. *Journal of Service Management*, *36*(4), 477–494. DOI: 10.1108/JOSM-04-2025-0194

Le, D., T., Vu, M.-C., & Tran, G. T. C. (2025). Artificial Intelligence in SMEs: Enhancing Business Functions Through Technologies and Applications. *Information, 16*(5), 415. https://doi.org/DOI: 10.3390/info16050415

Li L. (2022). Reskilling and Upskilling the Future-ready Workforce for Industry 4.0 and Beyond. *Information systems frontiers: a journal of research and innovation*, 1–16. Advance online publication. https://doi.org/DOI: 10.1007/s10796-022-10308-y

Liu, X., & Li, Y. (2025). Examining the Double-Edged Sword Effect of AI Usage on Work Engagement: The Moderating Role of Core Task Characteristics Substitution. *Behavioral Sciences (Basel, Switzerland)*, *15*(2), 206. DOI: 10.3390/bs15020206

Madanchian, M. (2024). From Recruitment to Retention: AI Tools for Human Resource Decision-Making. *Applied Sciences (Basel, Switzerland)*, *14*(24), 11750. DOI: 10.3390/app142411750

Malang, G., Gavhane, Y., Dahake, N., Dalvi, D., & Patil, A. (2025). The Impact of AI on Job Automation and the Workforce. http://dx.doi.org/DOI: 10.2139/ssrn.5343313

Marguerit, D. (2025). *Augmenting or automating labor? The effect of AI development on new work, employment, and wages*. arXiv. https://doi.org//arXiv.2503.19159DOI: 10.48550

Mastrogiacomo, R. (2025). Human-AI Collaboration and the Future of Work. In *AI Identities*. Apress., DOI: 10.1007/979-8-8688-2034-2_30

Meijerink, J., & Bondarouk, T. (2023). The duality of algorithmic management: Toward a research agenda on HRM algorithms, autonomy, and value creation. *Human Resource Management Review*, *33*(1), 100876. DOI: 10.1016/j.hrmr.2021.100876

Mišić, J., van Est, R., & Kool, L. (2025). Good governance of public sector AI: A combined value framework for good order and a good society. *AI and Ethics*, *5*(5), 4875–4889. DOI: 10.1007/s43681-025-00751-3

Morandini, S., Fraboni, F., Angelis, M. D., Puzzo, G., Giusino, D., & Pietrantoni, L. (2023). The Impact of Artificial Intelligence on Workers' Skills: Upskilling and Reskilling in Organisations. Informing Science. *Informing Science*, *26*, 39–68. DOI: 10.28945/5078

Murire, O. T. (2024). Artificial Intelligence and Its Role in Shaping Organizational Work Practices and Culture. *Administrative Sciences*, *14*(12), 316. DOI: 10.3390/admsci14120316

Murugesan, U., Subramanian, P., Srivastava, S., & Dwivedi, A. (2023). A study of Artificial Intelligence impacts on Human Resource Digitalization in Industry 4.0. *Decision Analytics Journal*, *7*, 100249. DOI: 10.1016/j.dajour.2023.100249

Mwita, K. M., & Kitole, F. A. (2025). Potential benefits and challenges of artificial intelligence in human resource management in public institutions. *Discover Global Society*, *3*(1), 35. DOI: 10.1007/s44282-025-00175-8

Naoum, R. F., Szakadáti, T., & Balogh, G. (2026). *Artificial Intelligence (AI) in human resource management (HRM): a systematic review of its dual impact on diversity, equity, and inclusion (DEI)*. Manag Rev Q., DOI: 10.1007/s11301-025-00580-y

Nastoska, A., Jancheska, B., Rizinski, M., & Trajanov, D. (2025). Evaluating Trustworthiness in AI: Risks, Metrics, and Applications Across Industries. *Electronics (Basel)*, *14*(13), 2717. DOI: 10.3390/electronics14132717

Nenni, M. E., De Felice, F., De Luca, C., & Forcina, A. (2025). How artificial intelligence will transform project management in the age of digitization: A systematic literature review. *Manag Rev Q*, *75*(2), 1669–1716. DOI: 10.1007/s11301-024-00418-z

Nyberg, A. J., Schleicher, D. J., Bell, B. S., Boon, C., Cappelli, P., Collings, D. G., Dalle Molle, J. E., Feuerriegel, S., Gerhart, B., Jeong, Y., Korsgaard, M. A., Minbaeva, D., Ployhart, R. E., Tambe, P., Weller, I., Wright, P. M., & Yakubovich, V. (2025). A Brave New World of Human Resources Research: Navigating Perils and Identifying Grand Challenges of the GenAI Revolution. *Journal of Management*, *51*(6), 2677–2718. DOI: 10.1177/01492063251325188

OECD. (2023). *OECD Employment Outlook 2023: Artificial Intelligence and the Labour Market*. OECD Publishing., DOI: 10.1787/08785bba-

OECD. (2025). *OECD Skills Outlook 2025: Building the Skills of the 21st Century for All*. OECD Publishing., DOI: 10.1787/26163cd3-

Oliveira, P., Carvalho, J. M. S., & Faria, S. (2025). AI Integration in Organisational Workflows: A Case Study on Job Reconfiguration, Efficiency, and Workforce Adaptation. *Information (Basel)*, *16*(9), 764. DOI: 10.3390/info16090764

Parycek, P., Schmid, V., & Novak, A. S. (2024). Artificial Intelligence (AI) and Automation in Administrative Procedures: Potentials, Limitations, and Framework Conditions. *Journal of the Knowledge Economy*, *15*(2), 8390–8415. DOI: 10.1007/s13132-023-01433-3

Paul, D., Sanap, G., Shenoy, S., Kalyane, D., Kalia, K., & Tekade, R. K. (2021). Artificial intelligence in drug discovery and development. *Drug Discovery Today*, *26*(1), 80–93. DOI: 10.1016/j.drudis.2020.10.010

Pavaloaia, V.-D., & Necula, S.-C. (2023). Artificial Intelligence as a Disruptive Technology—A Systematic Literature Review. *Electronics (Basel)*, *12*(5), 1102. DOI: 10.3390/electronics12051102

Przegalinska, A., Triantoro, T., Kovbasiuk, A., Ciechanowski, L., Freeman, R. B., & Sowa, K. (2025). Collaborative AI in the workplace: Enhancing organizational performance through resource-based and task-technology fit perspectives. *International Journal of Information Management*, *81*, 102853. DOI: 10.1016/j.ijinfomgt.2024.102853

Raina, K., Sharma, G. D., Taheri, B., Dev, D., & Chavriya, S. (2026). Artificial intelligence-driven management: Bridging innovation, knowledge creation, and sustainable business practices. *Journal of Innovation & Knowledge*, *11*, 100860. DOI: 10.1016/j.jik.2025.100860

Raisch, S., & Krakowski, S. (2021). Artificial intelligence and management: The automation–augmentation paradox. *Academy of Management Review*, *46*(1), 192–210. DOI: 10.5465/amr.2018.0072

Ritala, P., Aaltonen, P., Ruokonen, M., & Nemeh, A. (2024). Developing industrial AI capabilities: An organisational learning perspective. *Technovation*, *138*, 103120. DOI: 10.1016/j.technovation.2024.103120

Saravanan, T. R., Samuel, P., Kanimozhi, N., & Poongothai, E. (2026). Human–Machine Synergy in Learning-Driven Data Fabrics. In Samuel, P., Arockiam, D., Balusamy, B., Gernal, L. M., & Prabha, C. (Eds.), *Learning-Driven Data Fabrics for Sustainability. Sustainable Artificial Intelligence-Powered Applications*. Springer., DOI: 10.1007/978-3-032-09090-4_15

Sharma, V., Deb, S., Mahajan, Y., Ghosal, A., & Kapse, M. (2025). Psychological impacts of AI-induced job displacement among Indian IT professionals: A Delphi-validated thematic analysis. *International Journal of Qualitative Studies on Health and Well-being*, *20*(1), 2556445. DOI: 10.1080/17482631.2025.2556445

Shen, Y., & Zhang, X. (2024). The impact of artificial intelligence on employment: The role of virtual agglomeration. *Humanities & Social Sciences Communications*, *11*(1), 122. DOI: 10.1057/s41599-024-02647-9

Siena, D., & Zago, R. (2024). Job Polarisation, Labour Market Fluidity and the Flattening of the Phillips Curve. *Economic Journal (London)*, *134*(661), 2141–2174. DOI: 10.1093/ej/ueae006

Smyth, C., Dennehy, D., Fosso Wamba, S., Scott, M., & Harfouche, A. (2024). Artificial intelligence and prescriptive analytics for supply chain resilience: A systematic literature review and research agenda. *International Journal of Production Research*, *62*(23), 8537–8561. DOI: 10.1080/00207543.2024.2341415

Soomro, S., Fan, M., Sohu, J. M., Soomro, S., & Shaikh, S. N. (2026). AI adoption: A bridge or a barrier? The moderating role of organizational support in the path toward employee well-being. *Kybernetes*, *55*(2), 1059–1085. DOI: 10.1108/K-07-2024-1889

Sposato, M. (2025). Beyond the hype: Evidence-based approaches to responsible AI integration in workplace learning. *Education + Training*, *67*(9), 928–939. DOI: 10.1108/ET-01-2025-0034

Theres, C., & Strohmeier, S. (2023). Met the expectations? A meta-analysis of the performance consequences of digital HRM. *International Journal of Human Resource Management*, *34*(20), 3857–3892. DOI: 10.1080/09585192.2022.2161324

Trenerry, B., Chng, S., Wang, Y., Suhaila, Z. S., Lim, S. S., Lu, H. Y., & Oh, P. H. (2021). Preparing Workplaces for Digital Transformation: An Integrative Review and Framework of Multi-Level Factors. *Frontiers in Psychology*, *12*, 620766. DOI: 10.3389/fpsyg.2021.620766

Tzionis, G., Mouratidis, P., Kougka, G., Gialampoukidis, I., Vrochidis, S., Kompatsiaris, I., & Vlachopoulou, M. (2026). A review of explainable AI methods and their application in manufacturing systems. *Discover Applied Sciences*, *8*(1), 52. DOI: 10.1007/s42452-025-07908-z

Usabiaga, C., Núñez, F., Arendt, L., Gałecka-Burdziak, E., & Pater, R. (2022). Skill requirements and labour polarisation: An association analysis based on Polish online job offers. *Economic Modelling*, *115*, 105963. DOI: 10.1016/j.econmod.2022.105963

van Zoonen, W., von Bonsdorff, M. E., & van der Heijden, B. I. J. M. (2025). Algorithmic surveillance and workers' compliance: The role of trust, privacy concerns, and fairness in online crowdwork. *Human Relations*, *0*(0), 00187267251379698. Advance online publication. DOI: 10.1177/00187267251379698

Varnosfaderani, S. M., & Forouzanfar, M. (2024). The Role of AI in Hospitals and Clinics: Transforming Healthcare in the 21st Century. *Bioengineering (Basel, Switzerland), 11*(4), 337. DOI: 10.3390/bioengineering11040337

Voiseux, G., Zhou, R. T., & Huang, H.-C. (2025). Accepting the unacceptable in the AI era: When & how AI recommendations drive unethical decisions in organizations. *Behavioral Science & Policy, 11*(2), 20–30. DOI: 10.1177/23794607251384574

Wang, K.-H., & Lu, W.-C. (2025). AI-induced job impact: Complementary or substitution? Empirical insights and sustainable technology considerations. *Sustainable Technology and Entrepreneurship, 4*(1), 100085. DOI: 10.1016/j.stae.2024.100085

Wang, Y. (2025). *Algorithmic decision-making in organizations: a systematic review toward an integrated tension alignment framework.* Organization Management Journal., DOI: 10.1108/OMJ-11-2024-2342

Yan, J., Husted, K., & Fath, B. (2026). Organizational learning with artificial intelligence: Balancing new tensions between explorative and exploitative learning through hybridization. *International Journal of Information Management, 86*, 102997. DOI: 10.1016/j.ijinfomgt.2025.102997

Yoo, J. (2025). Employee intention and resistance to AI policies: Perceived injustice, moral outrage, and social influence from an activity theory perspective. *Behaviour & Information Technology*, •••, 1–27. DOI: 10.1080/0144929X.2025.2581857

Zenk, L., Pausits, A., Brenner, B., Campbell, D. F. J., Behrens, D. A., Stockler, M., Oppl, S., & Steiner, G. (2024). Meta-competences in complex environments: An interdisciplinary perspective. *Thinking Skills and Creativity, 53*, 101515. DOI: 10.1016/j.tsc.2024.101515

Zhao, J., & Fariñas, B. G. (2023). Artificial Intelligence and Sustainable Decisions. *European Business Organization Law Review, 24*(1), 1–39. DOI: 10.1007/s40804-022-00262-2

Zirar, A., Ali, S. I., & Islam, N. (2023). Worker and workplace Artificial Intelligence (AI) coexistence: Emerging themes and research agenda. *Technovation, 124*, 102747. DOI: 10.1016/j.technovation.2023.102747

Chapter 6

Future-Proofing the Workforce in the Age of Human-AI Synergy

Duggirala Aravind
http://orcid.org/0009-0005-4362-7684
ASET College of Science and Technology, India

N. V. Suresh
http://orcid.org/0000-0002-0393-6037
ASET College of Science and Technology, India

R. N. Ravikumar
http://orcid.org/0009-0009-3705-1681
Marwadi University, India

Jyoti Rana
http://orcid.org/0000-0001-7474-3702
Amity University, Noida, India

Veena Christy
http://orcid.org/0000-0001-9987-6253
SRM Institute of Science and Technology, India

ABSTRACT

The pace of AI usage among industrial systems is changing the nature of skills that are important, the organization of work, and the uniqueness of things that humans can bring to the table. This chapter presents ways of future-proofing the workforce through marrying human cognitive abilities with AI capabilities. It introduces the so-called concept of hybrid skills, a fusion of technical, interpersonal, ethical, and

DOI: 10.4018/979-8-3373-7036-1.ch006

adaptive abilities, and the way they will be crucial in Industry 4.0/5.0 conditions. Based on the current empirical research, the chapter elucidates the new demands, gaps, and opportunities of upskilling, reskilling, and lifelong learning. Sectoral illustrations are used to look at the ways human-AI synergy enhances productivity and does not disadvantage human agency, creativity, and wellbeing.The purpose of the chapter is to give practitioners, educators, and policymakers a roadmap towards creating resilient workforces, able to continuously disruption-proof and to capitalize on the opportunities of the human-AI collaboration.

1. INTRODUCTION

The emergence of Artificial Intelligence (AI) has brought into the world an age of profound industrial revolution, in which human abilities and machine wisdom are ever more tangling. Human functions are no longer limited to manual or routine processes; instead, they are becoming more about cognitive work, where intuition, moral judgments, creativity and emotional intelligence complete the analytical accuracy of AI. This redefinition is an indication that humans are no longer the operator of machines but rather they are co-creators of the intelligent ecosystems. In its AI class of applications (e.g. manufacturing, healthcare, logistics, finance), humans have become a decision architect, to interpret, guide and validate AI outputs. The future employee will not, therefore, be judged based on the skill on the job alone, but on how adaptable, empathetic, and judging issues in context. Uniquely human abilities are the only guarantee that organizations can have a moral and emotional compass, even in the context of technological acceleration. To redefine the human role, work processes, leadership models and learning systems need to be redesigned to encourage synergy and not substitution. The aspect that needs to be addressed is not to oppose AI but to use it to enhance the potential of humanity-to turn the personnel into a workforce of performers instead of engineers of thought.

The history of automation has been fear-filled over the decades, fear of unemployment, redundancy and obsolescence of human beings. But recent evidence is pointing towards the fact that AI does not need to be used to replace people, but rather to be used to augment them. The augmentation attitude considers AI as a colleague that enhances the human wisdom, effectiveness, and quality of decisions. Such a paradigm shift puts technological disruption in a new perspective as a source of empowerment. To put it in another example, in medical care, AI diagnostic systems enable the physician to put more effort on patient care and compassion instead of data input. In logistics, predictive algorithms allow the planners to optimise routes, whilst human knowledge allows the planner to be adaptable in uncertain situations.

This shift in the automation fear to the augmentation mindset is cognitive and institutional; it transforms the resource of success based on efficiency needs to collective intelligence, in which human creativity and machine ability coexist symbiotically. It will not be enough to equip the future generation of workers with readiness to meet the challenges of AI; the entire structure of work, education, and identity must be reviewed. The most important issue is to make sure that workers are not just employable but also empowered in the face of constant innovation. With the automation of industries in performing cognitive tasks, unique human capabilities will create value such as critical thinking, empathy, cross-domain synthesis, and ethical reasoning.

This discrepancy is endangering to increase socioeconomic divisions such that there remain populations incapacitated to the AI-based economies. Future-proofing hence entails the congruence between education, policy and organizational design to present dynamic ecosystems upon which learning is a lifelong activity and multidimensional. Governments have to support everybody to be digitally literate, industries need to jointly design training models and people have to learn to be curious and deft. It is not about saving the past jobs but in developing an everlasting renewal in the bond between human competency and clever technology-so that human beings can always be a significant source in the succeeding industrial revolution.

2. BACKGROUND STUDY

The conversation about Artificial Intelligence (AI) and automation has changed in the last ten years as a shift in technological fascination to socioeconomic critique of the socioeconomic consequences of this technology. The initial research was mainly concentrating on the displacement impact of automation, estimating the massive loss of jobs in manufacturing, logistic, and administrative fields (Frey and Osborne, 2017). Later empirical studies, however, found a more complex truth-AI is more likely to transform than to simply kill work and require new skill sets and blended human-machines (Brynjolfsson and McAfee, 2014). With the industries taking up AI, the jobs are being reorganized in a way that incorporates human invention, situational intelligence and ethical control into automation. This paradigm has created a novel concept referred to as human-AI synergy where co-evolution and not replacement is stressed.

The recent literature emphasizes that the key success of AI implementation is not as much a matter of technological maturity but rather a human ability and the learning capacity of the organization. Research by Epstein (2015) suggests that companies that equip themselves with AI as a collaborative intelligence system, that is, humans train, guide and refine AI tools, have a higher innovation performance and workforce satisfaction. Trakadas et al. (2020) discuss an AI-based collaborative approach in Industrial IoT manufacturing. In addition, according to the World Economic Forum (2023), more than 40% of core skills will become obsolete in five years, which can be regarded as the alarm signal to adopt reskilling and lifelong learning ecosystems.

Daugherty and Euchner (2020) discuss a new paradigm for collaborative work between people and intelligent systems. In spite of this accumulation of literature, the current literature is still usually disseminated between disciplines in technology, management and education, without an integrative model that may relate competencies acquisition, organizational adjustment and ethical governance. Little is known about how the hybrid skills, which integrate cognitive, technical, and emotional skills can be systematically developed to maintain workforce resilience. Moreover, research is more likely to ignore the issue of social equity in AI transitions where the inequality in access to education and digital infrastructure can contribute to the unequal labor force (Bandi et al., 2020; Imran, 2023).

The current chapter is therefore a continuation of the previous ones but it takes the discussion a step further as it suggests a cognitive collaboration framework that integrates these areas. It establishes one to identify viable solutions to the transformation of the workforce, the connection between human flexibility, integration of AI, and ethical thinking to make future work systems welcoming, as well as innovative.

3. CONCEPTUAL FRAMEWORK

The conceptual basis of this chapter is based on the Cognitive Collaboration Framework (CCF) - a model that combines human intelligence, artificial intelligence, and organizational learning to facilitate sustainable transformation of the workforce. The framework is based on three complementary theoretical perspectives socio-technical systems theory, cognitive augmentation theory as well as human capital evolution.

The socio-technical systems theory assumes that the performance of organizations occurs due to the fit between the social system (people, culture, and skills) and the technical one (AI, robotics, and digital infrastructure). In the CCF, such alignment is required to support, but not interfere with, human capabilities through the deployment of AI. The Cognitive augmentation theory delves on this further to add co-evolution: AI systems do not substitute human cognition but instead complement perception, memory, and decision-making. It is a joint effort that creates hybrid intelligence as human intuition and machine precision are dynamically intertwined to produce better results together.

Human capital evolution is the third pillar that aims at constant refreshment of skills and mindsets via learning ecosystems. Future-proofing is brought about in this model when individuals and organizations develop adaptive competence- learning, unlearning and relearning and changing with technologies. The CCF identifies adaptive competence as existing within three domains including cognitive competence (critical thinking, creativity and problem-solving), technological competence (AI literacy and data interpretation) and ethical competence (Responsibility, empathy and fairness). Collectively, these domains are what make up a strong and people-focused workforce.

Figure 1. Conceptual diagram for Cognitive Collaboration Framework

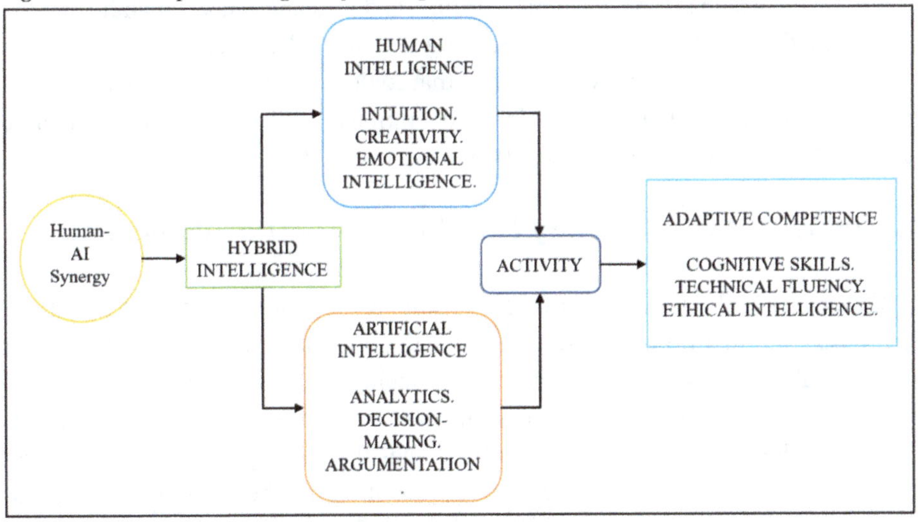

Evolution of a feedback loop is also incorporated in the framework, through which AI-based systems are constantly enhanced by human insight, and human learning is also steered by AI-based insights. This vicious cycle favors an augmentative equilibrium, and sustains a balance between the technological efficiency and human meaning. At the organizational level, the CCF proposes leaders with an inclusive leadership style, participative governance, and symbolic algorithmic framework to build on trust and accountability.

Summing up, the Cognitive Collaboration Framework theorizes that most sustainable kind of human-AI synergy can be achieved when technology can provide enhancement and not replacement to the human potential. It makes collaboration, ethics, and continuous learning the main drivers of industrial evolution to provide a multidimensional model to inform research, policy, and practice in the development of future-ready and human-centered workforce systems.

4. EVOLUTION OF HUMAN - AI SYNERGY

4.1 Mechanization to Cognition

The course of industrial development is indicative of the constant interaction of humanity with technology-mechanization during the First Industrial Revolution and digitalization during the fourth. The process of each phase redesigning human role in production systems introduced continuous transformation of labor, which turned more and more towards intellectual activity. The first machines were used to increase the human muscle, and the second, which were computers, were used to increase human precision and control. The present-day stage makes cognitive technologies, which are AI, machine learning, and data analytics, which further the human mind itself. This is a groundbreaking step since the machine becomes not a passive instrument but an active co-worker that could think, predict and learn. Hence, there is a shift in human work, which is operational implementation to cognitive orchestration (Hozdic & Makovec, 2023; Fang et al., 2025).

4.2 Industry 4.0 → Industry 5.0: Human-Centric Intelligence

Cyber-physical systems, automation, and digital interconnectivity are the features of Industry 4.0 that transformed the level of production efficiency and the use of data to make decisions. Nevertheless, its machine-oriented orientation provoked the fear of threatening workers displacement, moral transparency, and affectivity. In its turn, the new idea of Industry 5.0 changes the paradigm to human-centered intelligence, which is concerned more with collaboration between human ingenuity and tech accuracy. Instead of focusing on optimized systems to achieve pure productivity, Industry 5.0 is focused on sustainability, personalization, and social well-being. Man is not recast as a driver but as an architect, overseer of intelligent systems, which will have a role of steering ethical conduct, creating the ability to adapt to contexts, and adding compassion to automated actions (Enose Kamalabai et al., 2025).

This shift acknowledges the fact that humanistic values without technological progress bring about an unbalanced situation. Within the medical field, AI, when compared to a human clinician, is capable of processing medical data quicker than anyone, yet empathy, ethical care and situational intelligence are part of the human domain. Collaborative robots (cobots) are designed to help human workers, but not to substitute them, which results in safer, more versatile, and innovative production spaces. Therefore, Industry 5.0 is not only a technological development, but also a philosophical development the human well-being should lie in the heart of industrial smarts, and competitiveness must be redefined not as domination, but as cooperation (Toth et al., 2023).

4.3 Key Drivers of Workforce Transformation (AI, Robotics, IoT, Data Economy)

There are four key technological drivers, which include Artificial Intelligence (AI), Robotics, the Internet of Things (IoT), and the data economy that are accelerating the synergy between humans and machines. AI allows predictive decision-making and cognitive automation, changing the way organizations receive and distribute information and spend human resources. The element of robotics presents physical intelligence- artificial intelligence, it allows people to learn and adapt via the machine, as well as safely interact with it in the dynamic environment. IoT links devices, assets and processes and produces real-time roots of visibility extending into full value chains. Combined, these technologies create huge amounts of data that sustain the data economy that is known to maximize and innovate constantly (Dellermann et al., 2021).

To the employees, this convergence requires multidimensional literacy, technical fluency, analytical reasoning and ethical sensitivity. Employees should be able to decode AI insights, oversee robot activities, and make use of data in an ethical manner to be innovative. With the automation of routine tasks, the human contribution to the work process changes to creativity and the ability to solve problems across domains and design. Another commonality of these technologies is an erosion of sectoral boundaries: logistics is being infused with data science; healthcare is being combined with robotics and humanity with their empathy; and agriculture is adopting AI in creating precision-based sustainability. The overall effect is the redefinition of work itself, as an activity based on tasks, to a paradigm based on knowledge, in which human interpretation of intelligent products becomes the source of value (Langas et al., 2025; Borner et al., 2020).

4.4 Transition from Human vs. Machine to Human + Machine

The history of automation was hostile the last decades, with humans against machines, and technological unemployment as an unparalleled threat. Nevertheless, there is a different paradigm that can be seen today and in practice: Human + Machine. This symbiotic association appreciates the fact that machines are superior in speed, precision and data processing, whereas human beings get in touch with creativity, ethical judgment, and emotional intelligence. When properly integrated, these capabilities produce augmented intelligence a combination that is better than either of them (Jain et al., 2021; Jarrahi, 2018).

Companies like Deloitte, Siemens, and Microsoft have already adopted such philosophy and their cognitive teams consist of algorithms that perform pattern recognition and human beings that interpret the context. In logistics, exceptional management and ethical decision-making are left to the human supervisor as predictive AI takes care of demand forecasting. Generative AI can be used in a creative industry in collaboration with the designer, where it does not reduce the originality, but co-creates. But this change needs cultural and structural reengineering, the credence of AI systems, open communication and reformulated performance metrics. It is not aimed at automating humans, but humanizing automation so that collaboration increases the productivity and meaning of it. The Human + Machine model is simply a transition to a more intelligence-based evolution as opposed to efficiency-selective innovation - technology enhances humanity instead of overshadows it (Hosanagar & Ahn, 2024; Lepine et al., 2025; Chen et al., 2025).

5. MAPPING THE FUTURE WORKFORCE LANDSCAPE

5.1 Hybrid Work Ecosystems and Digital Collaboration Spaces

Hybrid ecosystems, a combination of physical, digital, and cognitive space, are redefining the future of work and enabling human beings and AI systems to work on the edges. These ecosystems do not limit themselves to remote work but to real-time work digitally, virtual workplaces that are more enjoyable to use, and intelligent systems that provide allocations dynamically and depending on human knowledge as well as machine potential. The AI-controlled coordination systems in such an ecosystem handle workflows, time schedule optimization, and can assist in global cooperation between time zones and industries.

The hybrid place of work focuses on fluidity and flexibility. Teams can consist of human workers, virtual characters, artificial intelligence, and robots that work in groups. Presence without proximity is made possible by cloud-based infrastructures, blockchain as a trust-checking tool, as well as augmented reality interfaces, which redefine what it means to work together. Nonetheless, these developments have necessitated reconsideration of the systems of governance, privacy, and equity in order to guarantee inclusion. The most thriving organizations in the future will not merely digitalize the existing processes, but they will develop digitally co-evolved ecosystems, in which technology sophistication boosts human creativity, health, and cooperation instead of undermining them.

5.2 Role of Emotional, Ethical, and Creative Intelligence

Although AI is better at the computation and prediction, human intelligence has not yet been able to replace it in the areas of emotion, ethics, and creativity, which are the core of the human thought. Emotional intelligence (EI) is another competency in AI-enabled work settings that is essential in the management of human-computer cooperation. With the help of EI, empathy-based leadership, reconciliation of algorithmically rather than human-driven results, and psychological safety in data-rich environments are reinforced. Ethical intelligence is an addition to EI that ensures that there is morality between human value and machine logic. With AI taking over decision-making in the fields of employment, healthcare diagnosis, distribution of resources, humans have to be the ethical guardians and audit biases, promote fairness and accountability. This ethical layer will be the basis of the reliable AI ecosystems with human control at the core.

Innovative thinking, on the other hand, is a force that leads to hybrid team innovation. The machines do not have the imaginative synthesis, narrative construction and contextual reasoning that is provided by humans. Combined with the generative potential of AI, creativity becomes augmented, that is, humans make AI their co-creator in order to seek new design space, solve more intricate problems, and achieve faster innovation. All these intelligences together keep technological advancement at a human moral and innovative level, a point of reference to which the evolution of the industry is socially and emotionally rooted.

5.3 AI as Co-worker, Mentor, and Decision Partner

The following industrial paradigm is that AI is not an instrument, it is a partner- a co-worker that learns, evolves, and is a co-creator with humans. Within this dynamic that is emerging, AI plays three interdependent roles, as co-worker, mentor, and decision partner. AI works as a collaborator doing repetitive and or data-intensive or precision based jobs, allowing human beings to do the higher-order jobs like problem-solving, strategy and relationship building. AI serves as an adaptive learning companion as a mentor. With the help of the real-time analytics system, feedback loops and performance insights, AI tailors the process of skills development and career paths, leading workers to a lifelong upskilling process. The application of AI in cultivating human potential and not eliminating it can be demonstrated by platforms such as intelligent learning assistants or simulation-based training environments.

Lastly, AI can also be a partner in the decision-making process as it assists in strategic foresight through the simulation and projection of future results, as well as modeling complicated systems. Human leaders with the help of AI-driven information use reflective judgment which balances accuracy of algorithms with moral and emotional judgment. This collaboration will produce a symbiosis driven decision system, in which human intuition and machine intelligence will help to enhance the advantages of both. This paradigm transforms productivity, purpose and performance in a manner that enhances the ability of man and organizational intelligence.

6. HYBRID SKILLS: THE NEW CURRENCY OF COMPETITIVENESS

6.1 Defining Hybrid Intelligence and Hybrid Skills

Competitiveness is not about technology but about the integration of human and artificial intelligence that Scholars are young adopting the term hybrid intelligence. This idea is a summary of the cognitive ecosystem of collaboration involving human intuition, creativity, and ethics with AI and its critical-analytical accuracy, scalability, and pattern recognition. The concept of hybrid intelligence goes beyond the man vs. machine dichotomy and looks to the combination of complementary advantages to bring forth adaptive problem solving and strategic innovation. This integration is operationalized by hybrid skills. They consist of human flexibility, digital literacy and interdisciplinary thinking to enable easy communication with AI systems. Not limited to coding or data-analysis-skills but are more about the understanding of how AI thinks, the interpretation of its output and the translation of insights into action that is moral and contextual. As an example, a logistics manager who is equipped with hybrid intelligence will be able to make sense of algorithmic forecasts taking into consideration the cultural, environmental, and human factors that cannot be conceived by a machine.

In that regard, it is the hybrid intelligence that is the new literacy of the AI economy, or a way of thinking about co-evolution, as opposed to competition. It gives workers the power not just to operate smart devices, but to also influence them in a purposeful way, so that technology can develop to work in unison with human judgment, empathy and creativity.

6.2 Taxonomy: Technical, Cognitive, Social, and Ethical skills

This taxonomy creates a multidimensional skill architecture that puts hybrid capability in both a technical and humanistic perspective. The development of the future workforce should hence not ignore integrative education systems in which data science and philosophy intersect and in which coding exists side-by-side with critical thinking to produce professionals that can spearhead ethically intelligent industries. Base- Technical Skills- they involve AI literacy and data analytics, human-machine interface design, and automation management. These capabilities enable the professionals to manage, analyze and optimize smart systems, which is a connection between engineering and strategic implementation. Cognitive Skills involve solving complex problems, analyzing and integrating systems thinking.

Human researchers in hybrid settings have to be able to synthesize vast amounts of AI-generated information and retain context and flexibility which machines cannot replicate. Communication, collaboration, and emotional intelligence are some of the Social Skills that can guarantee the effective operation of the hybrid teams that will consist of human beings and AI agents. These skills play a vital role in mediating between algorithmic logic and human requirements particularly in areas such as healthcare, education as well as creative sectors. The moral architecture of hybrid intelligence is comprised of Ethical Skills. These include the awareness of bias, online responsibility, and equity in judgment.

6.3 Adaptive learning and meta-cognition

Due to the increased rate of technological change, the competencies to become a lifelong learner and act strategically are turned to be the hallmark of competitive resilience. Adaptive learning is the tailoring of knowledge by the use of AI-based feedback tools that either detect strengths, weaknesses, and skills gaps in real-time. In contrast to the traditional approach in which the focus is on the fixed knowledge base, adaptive learning fosters dynamic intelligence, i.e. the ability to adapt to the new tools, processes and conditions with responsiveness. Meta-cognition- becoming aware of the thought processes of oneself is the core of adaptive learning. Meta-cognitive employees are vigilant in the way they think and challenge assumptions and they refreeze approaches, where new challenges arise. In this context, together with AI-driven feedback a potent feedback loop can work, where AI offers analytical feedback and humans reflective feedback, forming a process of iterative development.

This symbiosis is the nature of self-evolving systems of humans and AI. Companies utilizing adaptive learning designs harness the use of digital mentorship applications, simulators, and smart performance analytics in the development of lifelong learning cultures. What it produces is a strong workforce, one that is capable not only of surviving the technological disruption but also of adapting to it. Industries can make a fast technological change not out of anxiety, but a source of constant reinvention, by incorporating meta-cognitive awareness into the hybrid skill development.

7. RESKILLING, UPSKILLING, AND LIFELONG LEARNING FRAMEWORKS

7.1 Institutional Transformation for Skill Ecosystems

The shift in the global labor force to AI-enhanced sectors requires a paradigm shift in the institutional organization of education, corporate training and policy. Conventional organizations-universities, vocational centers, and corporations should transform their roles as being linear sources of knowledge to active points in the skill networks, where lifelong learning and digital flexibility become the core aspects. This is not merely a change of adding technology modules but yet incorporating social and economic systems with lifelong learning architectures.

Governments and the industrial outreach are also starting to understand the idea of competitiveness in the AI century is based on collaborative skill ecosystems, in which academia, enterprise, and technology platforms architecture the curriculum based on real-time labor market indicators. The institutions also need to incorporate anticipatory governance, which constantly predicts the future needs in skills and talent using AI-based labor analytics. Democratisation of learning, equal access to digital resources, mentoring are also aspects of such change that are required. Instead of having degrees awarded, future-ready institutions will be evaluated by their ability to create adaptive, AI-literate citizens that can re-skill continuously with the changing technology.

7.2 Micro-Credentialing and AI-Based Personalized Learning

Unlike conventional education models, which tend to run in terms of strict degree programs, micro-credentialing is a loose and modular form of skill acquisition. Students get specific, accumulative certifications to answer the demand of quickly evolving job specifications and technological transformation. Micro-credentials allow professionals to develop agile, just-in-time competencies without leaving the workforce, in the world of AI-driven industries, where new competencies are developed nearly every month. The AI technologies are transforming this space with the help of personalized learning ecosystems. Data analytics and machine learning algorithms are used by adaptive learning platforms to tailor their learning process, indicating what areas of learning require enhancement and recommending a special learning journey. This intimacy turns education into a process of one size fits all into the process of co-evolution where human students and AI trainers engage in a dynamic relationship.

In addition, credentialing, a blockchain-based system, guarantees interindustry, inter-border skills recognition transparency and portability. Employees have the ability to carry digital skill passports, which are authenticated by AI algorithms which evaluate competencies based on performance measures and not standardized tests. These innovations encourage solidarity, productivity and never-ending motivation. Finally, AI-based personalized learning transforms learning into a smart lifelong relationship between people and computers that will help invest career growth and cognitive development.

7.3 Human–AI Co-Training Systems and Adaptive Learning Pathways

In addition to personal learning, human-AI co-training systems, i.e., interactive learning systems in which humans instruct AI systems and in which they also learn, will define the future of the workforce development. These systems are based on the bidirectional feedback: AI models are trained on human knowledge, language representations, and contextual opinion, and humans receive information-driven insights and decision-making plans created by AI. The outcome is a positive feedback loop of learning which builds intelligence in both man and machine. Practically, such co-training environments utilize simulation based learning, in which AI agents simulate complex industrial domain, giving humans an opportunity to test decisions in risk free digital twins. Such experiences speed up experiential learning as well as allowing AI to develop a better sense of context. Based on this interaction, adaptive learning pathways are generated dynamically - in real time, always adapting to the developmental path of the learner, their emotional conditions, and their performance indicators.

In organizations that use such systems, there exists a culture of continual learning as both employees and intelligent systems develop together with the exchange of knowledge and ethical reflection. This paradigm transcends the workforce training to cognitive collaboration and integration of adaptability as a group competency. The future of AI and its expansion guarantees that its intellectual limits do not disregard the human mind and its capabilities, as the co-training of humans and AI will continue to provide a significant portion of intelligence to the future industries.

8. CORPORATE CASE STUDIES

8.1 Siemens: Cognitive Automation and the Human-in-the-Loop Model

Siemens has become a world leader in applying AI and digital twins and cognitive automation to industrial manufacturing. Siemens does not focus on full automation of operations, but instead, this is the philosophy of the human-in-the-loop, which means that human experience is one of the core components of machine decision-making. Siemens enhances engineers to monitor intricate industrial systems in real-time to predict maintenance, optimize the use of energy and reduce downtime through its MindSphere platform and AI-driven analytics. Importantly, Siemens transformed the tasks of a machine operator to an interpreter and orchestrator of the decisions of the AI in its engineers.

The company created Digital Industries Academies to help re-train thousands of employees in hybrid skills - such as data analytics, automation ethics, and cognitive systems management. Siemens inspired symbiotic relationship between human creativity and algorithmic efficiency by enabling employees to have interpretive authority over AI systems.

The result is not technological development alone but of a cultural change - of control to co-operation. Siemens shows that intelligent innovation in industry can be achieved by humans and AI co-designing intelligent processes together in a sustainable way. The success of the company highlights the fact that future-proofing does not imply that the human will be replaced, but rather the roles of humans are redesigned to ensure that human intuition and machine precision become the two equal members of the industrial ecosystem.

8.2 IBM: Cognitive Collaboration and the AI Workforce Ecosystem

The path of IBM to AI started several decades ago, though the transformation of a cognitive collaboration powerhouse is symbolised by its Watson ecosystem. The next wave of automation was Augmented Intelligence, or AIs that do not replace human mental capabilities, but augment them, as seen at IBM. This philosophy informed IBM workforce planning in the healthcare solutions, finance, and enterprise solutions. In IBM, workers not only use AI but also co-creators. Such tools like Watson Orchestrate can assist workers automatizing repetitive working and allowing them to concentrate on creativity, innovation, and contact with clients. In the meantime, the SkillsBuild and Open P-TECH programs of IBM allow the world to gain access to AI literacy and digital literacy as democratic.

These initiatives demonstrate the idea of lifelong employability rather than lifelong employment that IBM believes is the key to human-AI synergy since it is based on constant upskilling and flexibility. Another aspect that IBM is on the forefront with is the creation of ethical AI systems, which will make the automated decision systems transparent and accountable. The employees are educated about AI ethics and trust, which allows them to perceive machine recommendations in terms of morality and social aspects.

This convergence between technology and the ethics makes IBM a role model of responsible cognitive companies. Its business approach shows that the corporate resiliency in the age of AI hinges upon the facts and algorithms but also on human judgment, empathy, and integrity inherent in digital systems.

8.3 Tata Consultancy Services (TCS): AI for Scalable Human Development

Tata Consultancy Services (TCS) is a case in point of how extensive IT companies can use AI-based human capital transformation to remain competitive in the digital economy. Understanding that it has more than half a million employees worldwide, TCS introduced the so-called Machine First Delivery Model (MFDM) - the philosophy of operation incorporating AI as a digital partner helping humans at all stages of the project lifecycle. Instead of throwing human workers out of the job, TCS focuses on co-learning ecosystems.

Talent Development Platform of the company is based on AI analytics to recognize gaps in skills and suggest custom learning opportunities and anticipate new trends in the skills. With the help of micro-credentialing and digital academies, staff constantly upgrade their skills in hybrid skills like AI ethics, cognitive automation and data storytelling. The biggest difference of TCS is its Human-AI Co-Training model - in which the workers not only train AI systems using their domain knowledge but also learn about AI suggestions. This generates a feedback of learning which improves the intelligence of the machine and the human ability. Furthermore, the value-oriented leadership priorities of the company are such that the technological change will be aligned with human welfare and inclusivity, which is based on the long-term ethical tradition of Tata.

TCS shows that equitable growth can be engineered by AI in the emerging markets where it can be used to close the skill gap and egalitarianize the digital literacy gap. Its workforce transformation model is a model of how to incorporate industrial AI ecosystems with large-scale reskilling and adaptive learning as well as ethical governance.

Table 1. Summary table of case studies

Company	AI Approach	Human-AI Interaction	Outcomes
Siemens	Human-in-the-loop, Cognitive automation, Digital twins.	Humans act as interpreters and orchestrators of AI decisions. Human intuition + machine precision.	Real-time monitoring, Predictive maintenance, Energy optimization, Sustainable industrial innovation.
IBM	Augmented Intelligence, AI as a collaborator, Ethical and responsible AI.	Employees co-create with AI, AI assists in repetitive tasks to focus on creativity and client engagement.	Cognitive collaboration across healthcare, Finance, Enterprise, Transparent, Ethical AI, Empathy, and integrity integrated into digital systems.
TCS	Machine First Delivery Model (MFDM), AI as a digital partner in project lifecycle.	Human-AI co-training, Employees train AI and learn from AI feedback.	Skill gap reduction, Inclusive, Ethical workforce transformation, Improved AI and human capabilities simultaneously.

9. SECTORAL PERSPECTIVES ON HUMAN–AI COLLABORATION

9.1 Manufacturing: Cobots and Cognitive Automation

The manufacturing industry has moved out of mechanized production lines to cognitive automation as the emergence of collaborative robots (cobots) to work together in a safe and smart manner with humans. Cobots unlike the traditional industrial robots, which work in isolation, are fitted with sensors, vision systems and adaptive learning features, which allow them to work in real time with human operators. This paradigm transforms manufacturing into more of automation to augmentation where human dexterity and decision making is enhanced through machine accuracy and stamina.

In manufacturers like Siemens, ABB, and Fanuc, cobots can be used to do repetitive or dangerous jobs whereas human operators control the quality, interpolate anomalies, and provide instructions about production optimization. These hybrid teams show the usefulness of human-in-the-loop manufacturing, in which AI promotes human operational flexibility, but not labor replacement. In addition, AI-based digital twins are used to simulate production processes, and engineers can test designs in the virtual world and then implement them on the ground. The result is an efficient, creative enough, self-reactive manufacturing ecosystem that will ensure that human-created wisdom is central in the age of smart production.

9.2 Healthcare: Clinical Decision Support and Empathetic AI

The interaction between AI systems and clinicians in the field of healthcare has changed the way clinicians diagnose, manage their patients, and provide personal treatment. Clinical decision support systems (CDSS) based on AI analysis of large datasets, including medical records, genomic profiles, and scanning images are used to provide evidence-based recommendations. Nevertheless, the real innovation is empathetic AI - systems trained to increase the emotional and ethical aspects of care, and not only the analysis.

Jurist platforms like IBM Watson Health, Google DeepMind, and AI diagnostic tools developed by Philips help doctors by detecting hidden trends in radiology or them predicting patient degradations. Nonetheless, the human expertise is necessary to interpret the context, prioritize ethically, and communicate emotionally to patients. Medical workers are moralized, whose programmed output matches human values and compassion. The combination of intelligence driven by data and compassion of people promotes a new paradigm of cognitive care giving whereby the use of technology can improve the diagnostic accuracy without compromising the moral and emotional integrity of practice.

9.3 Logistics: Predictive Systems and Human-Centered Optimization

The logistics industry has provided a testing ground to predictive AI systems to transform global supply chain management. AI enhances efficiency and speed on multifaceted networks in robotics and warehouse, route optimization, and demand prediction. However, with an ever more autonomous operation it is essential to have human-centered control in place to guarantee resilience, flexibility, and ethical governance. The Amazon, DHL, and Maersk organizations apply models based on machine learning to predict disruptions, to manage fleets, and to dynamically manage stock.

Nonetheless, human operators are still important in the interpretation of predictive alerts, handling of exceptions, and judgment calls in unpredictable situations- e.g. weather-related delays or geopolitical interruptions. Such a co-decisional setup will make sure that AI does not override human thinking. Besides, explainable AI tool integration will enable human supervisors to be aware of the logic of algorithms, which enhances confidence, responsibility, and safety in autonomous logistics systems. The future of logistics will then rest on the cognitive orchestration, in which human foresight of the strategy and analytical foresight of AI will intertwine smoothly.

9.4 Agriculture: Precision Farming and Augmented Sensing

Precision farming, which is combined with augmented sensing systems, is radically changing one of the oldest industries in human history, agriculture. Artificial intelligence, including drone surveying, soil monitors, and crop forecasting AI, will allow farmers to make informed choices regarding the irrigation program, fertilizers, and pests. These instruments are more sustainable, resource-efficient and environmentally less harmful. To provide examples, John Deere, Watson Agriculture developed by IBM, and Microsoft FarmBeats are companies that use AI-based services that combine satellite imagery and sensor data to deliver real-time agronomic data. But human judgment can never be dispensed with in regards to understanding local situations, culture, and ethical issues that arise about the biodiversity and land stewardship.

The farmers and agronomists are now being transformed into interpreters of data and system partners, by combining the old ecological knowledge with the machine-generated knowledge. This merge is the first step to Augmented Agriculture-human-AI collaboration, a more productive and environmentally sound one. The AI in agriculture does not only reshape the production by making farmers cognitive partners, but also strengthens the human as the custodian of sustainable technology.

10. GOVERNANCE, ETHICS, AND WORKFORCE EQUITY

10.1 The ethics of algorithmic management and worker surveillance

The emergence of AI-based managerial systems has come with a range of efficiencies--yet with a set of thorny ethical challenges in regards to worker autonomy, privacy, and dignity. The application of machine learning to control employee behavior, forecast productivity, and distribute tasks is now widely used in the industrial field of logistics, ride-sharing, and e-commerce as algorithmic management. Although these systems increase efficiency and bias-reduction in the scheduling processes, they are capable of causing so-called digital Taylorism, as human workers are monitored, optimized, and quantified as data points, as opposed to being valuing as people. Such firms as Amazon and Uber have undergone criticism over their application of AI-driven systems that automatically judge and punish employees. These practices provide ethical concerns that are quite urgent. The principles of ethical governance require an attitude of transparency in algorithms, informed consent and participatory control to AI usage at the workplace. In order to be dignified and fair, organizations need to abandon surveillance and adopt augmentation ethics, whereby data is utilized to empower employees via feedback, competence development, and performance sustenance. The implementation of human rights principles in algorithmic design will lead to development of trust, autonomy and accountability, as it will turn the AI management process into the control mechanism, instead of the facilitator of fair cooperation.

10.2 Fairness, transparency, and accountability in AI decisions

The use of AI systems in the workforce decision-making is gaining influence; they are involved in hiring, performance evaluation, promotions, and training suggestion decisions. Otherwise, these systems will pose a threat of strengthening historical inequalities and introducing algorithmic bias, especially within gender, race, and socio-economic groups. Governance systems should hence be designed to be just, transparent and accountable throughout the entire AI implementation. Fairness involves making sure that data inputs are representative, bias free and of ethical origin. Transparency means that the algorithmic processes should be understandable to the users of these systems and to employees who are being impacted- what the European Union AI Act calls explainable AI. Accountability is also beyond compliance to moral responsibility whereby an organization should own what automated systems have created.

Ethics review boards, independent audits, and ethics bias bounties are some of the best practices. As an example, IBM and Microsoft have initiated internal AI ethics to measure algorithmic fairness prior to implementation. Moreover, by having human-in-the-loop validation, the final decisions are not made based on algorithms, but they are made based on understanding and social consciousness- qualities that cannot be simulated by algorithms. Essentially, fairness and transparency cannot be considered technical characteristics but governance principles. They make sure that AI does not replace human judgment but it is rather a collaborator. Operationalization of these principles allows organizations to develop ethical intelligence, which is one of the pillars of reliable and sustainable human-AI ecosystems.

10.3 Policy frameworks for inclusive AI adoption

The application of AI to the workforce ethically needs extensive systems and regulations that facilitate the idea of inclusion, rights, and equitable access to opportunity. Surgeons and policymakers on the national and international levels are starting to realize that unregulated automation may only worsen inequality unless it has been directed by human-based governance. Such frameworks as the OECD AI Principles (2019), EU AI Act (2024), and UNESCO Recommendation on the Ethics of AI (2021) focus on fairness, transparency, and human control as the main pillars of responsible AI. At the government level, governments need to invest in AI literacy, digital infrastructure and social safety nets to enact all workers- irrespective of their gender, region and socio-economic status- to be able to take part in the AI economy. The incentives to the policies should also be to encourage companies to create reskilling programs, ethical audits, and diversity-oriented AI teams.

The public-private partnerships may serve as an important source of filling the skill gaps, especially in the economies that are developing and in which digital divides continue to exist. Most importantly, the ethical pluralism should inform the adoption of AI inclusivity by taking into account cultural and contextual variations on fairness, accountability, and autonomy that define fairness in societies. The globally-coordinated but locally-adaptable policy ecosystem will help make sure that AI will bring about mutual prosperity, rather than stratification. Finally, governance will also need to be transformed into prospective stewardship, and thereby place ethics as a strategic contributor to human-AI co-evolution.

11. IMPLICATIONS

11.1 Managerial Implications

In the new landscape of AI-enhanced industries, managers need to transform by no longer focusing on automation productivity but rather on human-AI augmentation as a philosophy. The leaders should understand that the productivity and innovation are dependent on the effectiveness of the cooperation between the humans and the machines. Managers have to build organizations as learning eco-systems, where human understanding, creativity and moral thought augment machine reasoning.

In order to stay competitive, companies should include the continuous reskilling and hybrid intelligence in their workforce strategies. Human resource departments are able to use AI analytics to determine the changing skill requirements and offer micro-credentialed, adaptive education. Leadership programs are expected to foster cognitive flexibility, ethical literacy and emotional intelligence among technical proficiency.

Clear communication regarding the integration of AI is most appropriate in terms of trust and psychological safety of the staff. The managers should maintain the level of algorithmic transparency and fairness, by engaging the workers in decision-making processes associated with AI. Organizations can also leverage innovation, job satisfaction and resilience of the entire workforce by advocating AI as an exciting colleague, not as a substitute worker but gain a long-term strategic advantage during the cognitive collaboration age.

11.2 Practical Implications

In governance terms, the shift towards AI-driven economies requires farsighted and inclusive policy structures. The policy-makers ought to consider empowering fair access to digital infrastructure, AI education, and opportunities to acquire life experience, and all these in order to reduce polarization in the workforce.

The governments need to stimulate public-private relations that ensure better connection of the educational program to the new industrial requirements that produce skills, and that enable the ecosystems between academia, industry, and civic institutions. The design of policies should focus on the establishment of AI ethics and accountability principles, assure transparency, explainability and prevent biased occurrences in algorithms.

Furthermore, using fiscal and regulatory tools, it is possible to encourage innovation that follows a responsible approach and helps to boost the sectors employing human-centered AI systems and fair labour policies. Considerable focus should be made on women, rural labourers, and marginalized groups to make sure that they are engaged in the digital economy equally. Through combining the progress of workforce development with ethical governance, the policy frameworks will be able to make the development of AI a source of the improvement of the economic competitiveness as well as social well-being. Good governance is therefore the pillar to sustainable human-driven industrial future.

12. DISCUSSION

The changing interaction between humans and artificial intelligence is one of the far-reaching socio-technical shifts in the history of industry. The results and conclusions that are reached during this chapter keep culminating with a central assumption: the future of work is not seen in the efficiency and speed with which machines can substitute humans, but rather in how smart humans are in dealing with machines. The next phase of industrial evolution is the emergence of hybrid intelligence in which cognitive, emotional and ethical human capabilities are left to play with computational power.

In all industries, the implementation of AI has been taking a leap forward beyond automation to cognitive cooperation. Manufacturing is now using cobots which help in supplementing human dexterity; healthcare using clinical decision systems which improve diagnostic accuracy and logistics networks using predictive algorithms co-supervised by human operators. These illustrations address the fact that the potential of AI is not in autonomy, but in augmentation, it allows humans to do creative, strategic, and empathetic work that cannot be automated.

Nevertheless, the shift toward human-AI synergy does not occur without a problem. The issue of skill obsolescence, moral ambiguity, and digital inequality are still a menace to inclusive advancement. Employees in the developing areas have unequal access to reskilling opportunities, and algorithmic management poses the problem of surveillance and bias. The structure of the chapter, which is based on the socio-technical, cognitive, and human capital theories, hints at the fact that the importance of sustainable integration is that it will be necessary to redefine competence and trust. The concept of competence is to be broadened beyond technical competence, and it should include hybrid fluency (a mixture of analytical reasoning, empathy, and ethical judgment). Trust should develop based on transparency, explainability and participatory design where human beings are not seen as operators but as co-created.

Institutional adaptation is the other important dimension. Learning models and educational systems should be changed to focus not on the fixed qualification models, but on the dynamic learning ecosystems, facilitated by AI personalization and micro-credentialing. Governments and industries should also cooperate to introduce the ethical governance frameworks that would not threaten data privacy, eliminate algorithmic discrimination, or enhance the equitable access to AI.

Finally, the discussion indicates that the future-proofing of the workforce is a group exercise that should reflect a coordination between the technology designers, educators, policymakers, and workers. The way forward is not completely human or completely artificial but a cooperative and developing process based on the principles of empathy, creativity, and social responsibility. It is only with this comprehensive synergy that societies will be able to make certain that the AI-induced change will empower, rather than weaken, human intent, dignity, and competence.

13. FUTURE DIRECTIONS

13.1 Human–AI Symbiotic Intelligence

Intelligence will be redefined in the next decade as a co-evolutionary process between machines and people. Rather than creating AI to substitute cognitive labor, subsequent systems will focus on symbiotic intelligence where human intuition, empathy and morals are complimented by machine accuracy, size, and flexibility. This paradigm suggests both technological innovation and redesign of the organization, which allows human and AI agents to take shared decisions, make contextual decisions, and engage in adaptive learning.

Recent findings in neuro-symbolic AI, affective computing, and cognitive augmentation also suggest that the workplace of the future would be an adaptive ecosystem: constantly changing, in which human beings and AI teach one another. Human-AI symbiosis will also not be limited to productivity, but present a source of creativity, problem solving and social well-being. According to this description, AI would be seen as more than a digital helper because in this vision, it would be an intelligent partner of human cognitive abilities and perceiving of nuances and purpose. The difficulty is to establish frameworks that support this partnership in ethical terms making each accountable and transparent. With the current and further advances of automation into augmentation, symbiotic intelligence is the next horizon, with human values being the determinants of technological advancement, not vice versa.

13.2 Evolving Education and Learning Architecture

The future of transforming the workforce depends on the reconsideration of education systems in order to foster an ongoing, individualized, and dynamic learning process. The old system of education centered around earning degrees will be replaced by modular learning systems and lifelong learning ecosystems where people constantly reskill by earning micro-credentials, studying in digital academies and learning paths curated by AI. AI will become the teacher and learner- it will analyze the skills gaps, forecast career changes, and provide the contextual learning experience based on cognitive styles.

Higher education and business companies should work together to design the Human-AI Learning Architectures co-designed- linking university learning with immediate industry competencies. Those structures would combine machine analytics and human mentorship so that learners do not receive recommendations of AI but co-design their learning paths. The future education will appreciate meta-cognition, innovativeness, moral reasoning, and the ability to be fluent across disciplines, and it will equip the human being to work in dynamic positions that cannot be automated. By making AI literate in every field, the societies will be able to build an inclusive, flexible and robust talent pool that will survive in unforeseen futures.

13.3 Building Ethical and Sustainable AI Societies

As AI starts to be more and more entangled in the global economic and social frameworks, the future workforce will have to work in ethical, open, and sustainable AI ecosystems. This does not just stop at compliance; it requires AI stewardship, where individual people, organizations and government are all responsible in ensuring that technology is used in the best interest of humans collectively. The emphasis will be on AI that is sustainability-based, meaning that it is created not to provide financial efficiency but also to balance the environment and the social dimension. Green AI projects, carbon-conscious computing and circular digital economies, will transform the way that industries in their design, deployment and evaluation of technology. It is also vital to mention that there is the emergence of AI citizenship, where everyone has the right to know, challenge, and influence the algorithms that affect their lives.

This means that global collaboration by multilateral AI ethics councils, open-source governance forums and comprehensive policy development will be needed so that access is equal and shared responsibility established. Finally, sustainable AI societies need to be designed as futures in which technology enhances human flourishing that is, technology should make the future more innovation-focused, more empathetic, more diverse, and more committed to long-term well-being of the planet.

14. CONCLUSION

The era of industrial intelligence is not characterized by the automation anymore but the reciprocal development of the person and the machine. The chapter has touched upon how the human role in industries that are driven by AI is being redefined- as an element of displacement anxiety to one of collaborative empowerment. The automation to augmentation is a paradigm shift as the technology will enhance the human potential instead of reducing them. The future of work is however not the competitive version with AI, but the collaborative one: making a new social contract based on cognitive collaboration, ethics, and lifelong learning.

The theoretical background provided by the Cognitive Collaboration Framework has highlighted three interrelated pillars i.e., the socio-technical alignment, cognitive augmentation, and the growth of human capital. The combination of these dimensions helps to understand that the successful human-AI integration is not possible only due to the technological complexity, but also organizational empathy and cultural preparedness. Under the condition that people, processes, and intelligent systems are oriented on some common goals and ethical principles, AI is a necessary strategic partner in innovation, productivity and decision-making. Such alignment should also be backed up with open algorithmic governance in such a way that humans still have interpretive control and responsibility in automated ecosystems.

Just to name a few, manufacturing, healthcare, logistics, and agriculture, the evidence is convergent across industries. The most competitive organizations are those ones, which will not see AI as a tool of cost reduction, but as a force of multiplier. Cognitive automation and cobots are also facilitating safer and flexible production in the manufacturing industry. In health care, AI supports clinicians and still maintains the aspect of empathy and judgment. Predictive optimization is available in logistics to ensure resilience along with human control. The examples of these sectors also support the same point: the value of AI can be fully achieved only at the time when people could remain the part of its working cycle.

However, the chapter also identified the structural obstacles that should be cleared. The polarization of the workforce, the skill differences across the region, and bias in the algorithms remain the challenges of the equitable AI. In order to future-proof the workforce, therefore, three-dimensional approach is required: (1) creating adaptive, hybrid skillsets, which integrate technological fluency with creativity and ethics; (2) building learning ecosystems, which will support continuous reskilling; and (3) institutionalizing governance structures, which will ensure fairness and inclusion. In the absence of these pillars, AI will contribute to the widening of the gaps instead of sealing them.

One of the key facilitators of such transformation is education and training. Life-long learning should become the new standard in the professional field as the life of skills is getting shorter. The democratization of skill development will result in AI-enabled personalized learning spaces, micro-credentials, and competency-based learning, which will enable each worker to keep abreast of technology. In addition, the development of human-AI co-training, in which algorithms are trained by humans and at the same time trained by humans, will represent a huge step towards the development of self-sustaining learning systems.

The chapter requires a human-centric AI governance paradigm. The principles of fairness, transparency, and accountability should be the pillars of any AI implementation as algorithms are getting involved in employment, productivity, and policy. Sustainable innovation can be achieved through ethical AI, and not an add-on. It is time that governments, corporations, and academia work together to come up with regulatory standards and participation oversight mechanisms that will make technology be a servant of human progress rather than its master.

The vision of the chapter is a symbiotic intelligent ecosystem - a place where people and machines will co-evolve in a constant flow of knowledge, imagination and compassion. In this co-evolutionary future, the metrics of the success will not be the level of automation, but rather the richness of human-AI collaboration, the inclusiveness of new technologies, and the amount of innovation that adheres to the human dignity.

Finally, the future-proofing of the workforce is more of a human process rather than a technological one. It requires interest rather than apprehension, cooperation rather than antagonism, and flexibility rather than adherence. With industries shifting to cognitive collaboration, humanity does not have to be as fast as machines; instead, it is necessary to redefine what intelligence is like, with progress being based on compassion and creativity, as well as a common cause. The actual development of the industry will not be gauged at the capabilities of machines, but the extent to which people and AI can learn, create, and coexist in a harmonious manner.

Originality of the Chapter

The chapter is a novel, unpublished writing. It has not been sent or published in any form in part or otherwise. Everything discussed, analyzed, and thought of is the original work of the author, and it was created with the purpose of being included in Human-AI Cognitive Collaboration for Industrial Evolution.

Conflict of Interest

The author does not have any conflict of interest regarding the research, authorship or publication of this chapter. Any interpretation and conclusion is carried out on their own, without references to financial, institutional, or personal factors that influence the objectivity of the work.

ACKNOWLEDGEMENT

The author is very grateful to have been given a chance to contribute to the book Human-AI Cognitive Collaboration by the editors of the Industrial Evolution. It is also acknowledged to colleagues, researchers, and industry professionals who have contributed to this work by their insight into artificial intelligence, workforce transformation and ethical innovation. The enduring encouragement to learn more about human-centered technological development and education of the future is due to academic mentors and institutional support systems.

REFERENCES

Bandi, R. K., Klein, S., Madon, S., Monteiro, E., & Ranjini, C. R. (2020). *The future of digital work: The challenge of inequality.* In Bandi, R. K., C. R., R., Klein, S., Madon, S., & Monteiro, E. (Eds.), *The Future of Digital Work: The Challenge of Inequality* (IFIPJWC 2020, IFIP Advances in Information and Communication Technology, Vol. 601). Springer, Cham. https://doi.org/DOI: 10.1007/978-3-030-64697-4_1

Brynjolfsson, E., & McAfee, A. (2014). *The second machine age: Work, progress, and prosperity in a time of brilliant technologies.* W. W. Norton & Company. https://doi.org/DOI: 10.2139/ssrn.2724764

Chen, L., Song, Y., Guo, J., Sun, L., Childs, P., & Yin, Y. (2025). How generative AI supports human in conceptual design. *Design Science*, *11*, e9. DOI: 10.1017/dsj.2025.2

Daugherty, P., & Euchner, J. (2020). Human+ Machine: Collaboration in the age of AI: Paul Daugherty talks with Jim Euchner about a new paradigm for collaborative work between people and intelligent systems. *Research Technology Management*, *63*(2), 12–17. DOI: 10.1080/08956308.2020.1707001

Enose Kamalabai, N., Hannola, L., & Donoghue, I. (2025). Role of industrial artificial intelligence in advancing human-centric sustainable development of Industry 5.0. In *Contemporary Issues in Industry 5.0: Towards an AI Integrated Society* (pp. 325–371). Springer Nature Switzerland. https://doi.org/DOI: 10.1007/978-3-031-74779-3

Epstein, S. L. (2015). Wanted: Collaborative intelligence. *Artificial Intelligence*, *221*, 36–45. DOI: 10.1016/j.artint.2014.12.006

Fang, X., Tao, L., & Li, Z. (2025). Closer to language than steam: AI as the cognitive engine of a new productivity revolution. *arXiv preprint arXiv:2506.10281*. https://doi.org//arXiv.2506.10281DOI: 10.48550

Frey, C. B., & Osborne, M. A. (2017). The future of employment: How susceptible are jobs to computerisation? *Technological Forecasting and Social Change, 114*, 254–280. DOI: 10.1016/j.techfore.2016.08.019

Hosanagar, K., & Ahn, D. (2024). Designing human and generative AI collaboration. *arXiv preprint arXiv:2412.14199*. https://doi.org//arXiv.2412.14199DOI: 10.48550

Hozdić, E., & Makovec, I. (2023). Evolution of the human role in manufacturing systems: On the route from digitalization and cybernation to cognitization. *Applied System Innovation, 6*(2), 49. DOI: 10.3390/asi6020049

Imran, A. (2023). Why addressing digital inequality should be a priority. *The Electronic Journal on Information Systems in Developing Countries, 89*(3), e12255. DOI: 10.1002/isd2.12255

Jain, H., Padmanabhan, B., Pavlou, P. A., & Raghu, T. S. (2021). Editorial for the special section on humans, algorithms, and augmented intelligence: The future of work, organizations, and society. *Information Systems Research, 32*(3), 675–687. DOI: 10.1287/isre.2021.1046

Jarrahi, M. H. (2018). Artificial intelligence and the future of work: Human-AI symbiosis in organizational decision making. *Business Horizons, 61*(4), 577–586. DOI: 10.1016/j.bushor.2018.03.007

Lepine, B., Weerantunga, G., Kim, J., Mishkin, P., & Beane, M. (2025). Evaluations at work: Measuring the capabilities of GenAI in use. *arXiv preprint arXiv:2505.10742*. https://doi.org//arXiv.2505.10742DOI: 10.48550

Tóth, A., Nagy, L., Kennedy, R., Bohuš, B., Abonyi, J., & Ruppert, T. (2023). The human-centric Industry 5.0 collaboration architecture. *MethodsX, 11*, 102260. DOI: 10.1016/j.mex.2023.102260 PMID: 37388166

Trakadas, P., Simoens, P., Gkonis, P., Sarakis, L., Angelopoulos, A., Ramallo-González, A. P., Skarmeta, A., Trochoutsos, C., Calvo, D., Pariente, T., Chintamani, K., Fernandez, I., Irigaray, A. A., Parreira, J. X., Petrali, P., Leligou, N., & Karkazis, P. (2020). An artificial intelligence-based collaboration approach in industrial IoT manufacturing: Key concepts, architectural extensions and potential applications. *Sensors (Basel), 20*(19), 5480. https://www.mdpi.com/1424-8220/20/19/5480. DOI: 10.3390/s20195480 PMID: 32987911

World Economic Forum. (2023). *The future of jobs report 2023.* World Economic Forum. https://www.weforum.org/reports/the-future-of-jobs-report-2023

KEY TERMS AND DEFINITIONS

Algorithmic Governance: Application of AI and automated systems in organization or society management which involves decision making processes which demand transparency, fairness and accountability.

Cognitive Collaboration: Partnership model A model in which humans and AI systems collaborate in the cognitive tasks, i.e. human intuition and emotional intelligence with machine accuracy and data analysis.

Ethical AI: Artificial intelligence was created and implemented in accordance with the ethical guidelines of fairness, privacy, accountability, and inclusivity to serve the human in a responsible manner.

Human–AI Synergy: The interactive synergy of the human intelligence system with the artificial intelligence system so that the system can improve its performance, make decisions, and be more creative.

Human-in-the-Loop (HITL): A design rule that keeps human beings active observers of the AI decision-making processes that bring supervision on the process, show ethics and provide contextual insight.

Hybrid Intelligence: A synthesized type of intelligence that results as a result of constant collaboration between human beings and machines, and producing outcomes neither can do alone.

Lifelong Learning Ecosystem: An adaptable model that allows acquiring and reskilling new skills continuously throughout the professional lifetime of a person, with the help of digital learning tools and AI-based customization.

Socio-Technical Systems Theory: The organizational design that focuses on the interdependency between people and technology to attain the best efficiency and human well-being.

Chapter 7
Digitalisation in Industrial Design:
A Holistic Review of Wheelchair Design

Pınar Tabak
http://orcid.org/0009-0001-6216-1411
Istanbul Aydin University, Turkey

Sude Erdağ
http://orcid.org/0009-0008-2459-9405
Istanbul Aydın University, Turkey

Ayşe Sirel
http://orcid.org/0000-0002-5244-3001
Istanbul Aydın University, Turkey

ABSTRACT

Major transformation industrial design has undergone was not limited to tools but also brought about a shift in the design paradigm. This study brings wheelchairs into focus and investigates new perspectives introduced by digital technologies, utilising a literature review and case studies. A wheelchair is an exemplary product to represent universal design principles and allow full observation of concrete effects of digital transformation with its sensitivity towards individual needs. Study uses qualitative analysis techniques supported by the literature review and case studies. A comparative analysis was conducted using the criteria obtained from this literature review. With the help of this transformation, design processes have become a

DOI: 10.4018/979-8-3373-7036-1.ch007

more innovative and holistic structure by integrating technology. In this context, the digitalisation process in industrial design does not only refer to a technical phase but also a social transformation representing barrier-free and universal design.

INTRODUCTION

Industrial design is a problem-solving process that redefines the existing problems with creative solutions and increases the quality. American design historian Margolin indicates that architects follow a set of defined principles and rules while industrial designers invent their professional content over time (Riccini, 1998). The earliest known form of development in the industrial design field began to emerge during the 18th century, alongside the expansion of capitalist industry, while existing markets were inadequate to meet the new growing demands. Manufacturers have collaboratively worked with the artists in creating the products' design drawings to satisfy different tastes and fashion trends of the time. The role of industrial designers grew significantly in the 19th century, and industrial design began to blend with technological advancements, starting especially in certain European countries. New industrial technologies completely changed the design concept in the United States at the end of the 19th century and at the beginning of the 20th century. Large-scale enterprises used mass production techniques and advertising technologies by using innovative products to ensure widespread adoption of these. Thus, the diversity of the products increased because of competitiveness between similar companies within the same market and advertising activities took on a new dimension (Heskett, 2001). There is a strong link between industrial revolutions and digitalisation processes of industrial design (Figure 1). This relationship was set in the First Industrial Revolution, which started in the 1760s and extended through the 1830s, when energy production was based on steam power and fossil fuels that mechanised the total production (Güdek, 2023). The use of steam engines that were working as power sources instead of human labour increased productivity. These inventions, which had deep impacts on productivity, made it possible to transport the products overseas due to the improvements in the facilities to track over long distances (Sharma and Singh, 2020; Freeman and Soite, 2005).

Figure 1. Historical Development of Industrial Revolutions

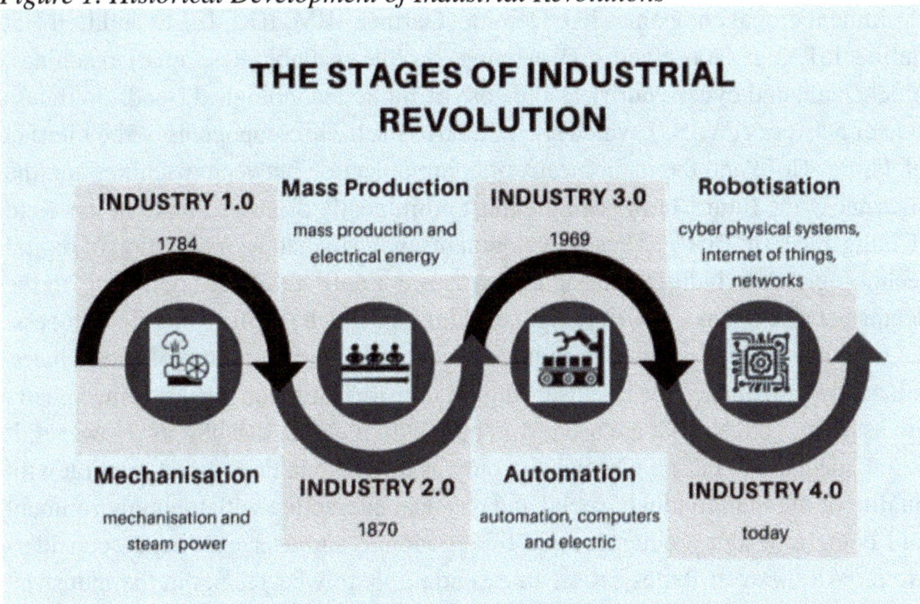

Created by the authors using Jean-Paul Rodrigue's (2020) studies

Second Industrial Revolution, which started in the 1870s, was different than the first one did not only depended on steam power but also on electricity, steel manufacturing, the chemistry industry, and the division of labour thanks to the assembly line systems that let mass production (Kılıç, 2023). In this period, chronologically, the shift to the capital-based production, productivity, and the growth of living standards, formation of large corporate hierarchies, emergence of excessive capacity, and consequently, shutdown of some companies due to this overcapacity can be distinguished (Jensen, 1993).

The Third Industrial Revolution started in the 1970s. Computer technologies, automation systems, 3D printers and artificial intelligence (AI) supported tools and robotics made contributions to product manufacturing efficiency and reduced the production costs. Computer-aided design (CAD) created some key solutions like rapid prototypes and product lifecycle management (Güdek, 2023).

The Fourth Industrial Revolution was a period where technological improvements created one of the pivotal points that transformed the industrial landscape deeply. While computer and automation systems were widely integrated in the previous periods, the use of intelligent and autonomous systems gave character to new alternatives and functionality in industrial applications with this new stage. At this point, a different dimension was created with the integration of advanced technologies that go beyond digitalisation, such as artificial intelligence, Internet of

Things (IoT), big data analytics, and cyber-physical systems (Apilioğulları, 2019). Prominent global companies like Deloitte, Gartner, IBM, IDC, IEEE, and Red Hat define IoT, big data, cloud, 3D printing, mobile technologies, smart machines, blockchain, and cybersecurity as the core of future technological trends. Wireless sensor networks (WSNs), which are defined as the basic components of the Internet of Things (IoT) and the main means of communication between machines and the Internet of the future, have been the most prominently discussed topic in this field (Chung and Kim, 2016). These developments indicate that the integration of digital technologies into industrial design processes not only introduces new tools to the sector but also opens a new paradigm of thinking, which is still an ongoing process.

Considering the favourable conditions and representative features of wheelchairs, this study investigates the effects of digital transformation on design. Wheelchairs are assistive gadgets that are used by people with physical disabilities. However, it is not adequate to define wheelchairs only as devices via their direct relation with quality of life in individuals' social and physical interaction with their environment and being a design product created by ergonomic and user experience considerations. All these attributes assign an essential role to wheelchairs in the industrial design context.

Conventional manufacturing techniques may be inefficient in terms of time and cost and may fail to satisfy the user's needs. This inefficiency causes the users to ask for assistance, makes them dependent on second bodies, and this may lead to feelings of less confidence, physical discomfort (Nguyen and Tung, 2024). The fact that health products like wheelchairs may be altered according to users' anatomic structures and personal preferences makes digital transformation-related features more inevitable. Thus, it is possible to present more flexible, user-focused solutions wherever the conventional methods may become insufficient. Revolutionary opportunities that were brought by digital technologies should be properly addressed to facilitate correct solution methods and to be applied more easily.

Along with numerous technological changes, digitalization has reshaped not only production tools but also the epistemological framework of industrial design. A gradual evolution has been observed from object-centred thinking to experience-centred systems focused on functionality. This change makes industrial design more inclusive, aligning it with global transformations.

Furthermore, the integration of digital design tools into the industrial design process helps designers predict performance, ergonomics, and user feedback without the need to create physical prototypes. This development significantly reduces waste and costs while offering a high degree of personalization.

These technologies contribute to the development of products that better reflect body proportions, behavioural patterns, and environmental conditions in the design of assistive devices for individuals with special needs, such as wheelchairs.

Thanks to these approaches, users actively participate in shaping the design process, ensuring the emergence of more inclusive and adaptable solutions and facilitating the creation of personalized products. Digital production technologies such as 3D printing and CNC machining enable these collaborative ideas to be directly applied to concrete prototypes, ensuring harmony between theoretical and real design stages.

Therefore, the convergence of digital transformation and industrial design not only accelerates production but also opens a new dimension within the design discipline. Wheelchair products with functional and aesthetic value demonstrate how technology can restore independence and strengthen social integration for individuals with physical limitations, thereby embodying this transformation.

METHODOLOGY

The study was conducted to reveal the production effects of the digitalisation processes of industrial design. Particularly wheelchairs were chosen for this purpose. The following research techniques were used:

- Textual Analysis (Based on document Analysis),
- Visual Analysis (Based on image-based Analysis),
- Historical Comparative Analysis,
    - *Pre-Digitalisation*
    - *Post-Digitalisation*

The general part of the study is divided into four main stages.

The first stage is composed of textual analysis with an extensive literature review to explain the conceptual framework. The general focus was on industrial design, industrial revolutions, digitalisation processes and the evolution of wheelchair design. This was achieved through databases like Google Scholar, Emerald, Routledge, Taylor & Francis, JSTOR, and Springer. Table 1 summarises the main sources reviewed during this stage. At this stage, where the conceptual framework is established, a better understanding of the integration of digital technologies into the field of industrial design and their contributions to wheelchair design has been achieved.

Table 1. Literature summary regarding the general outline of the study

Authors/ years	*General summary*
Jensen, 1993	Industrial Revolutions
Riccini, 1998	History of Industrial Design
Heskett, 2001	History, Current State and Future of Industrial Design
Freeman and Soite, 2005	Relationship Between Industrial Design, Inventions and the Economy
Chung and Kim, 2016	Industrial Revolutions, Industry 4.0
Apilioğulları, 2019	Components of Dijital Transformation, Industry 4.0
Sharma and Singh, 2020	Industrial Revolutions, Industry 1.0, Industry 2.0, Industry 3.0, Industry 4.0
Kumar et al., 2020	Smart Wheelchair Designs After Digitalisation
Gabrielski et al., 2021	Facilitating Transportation for People with Disabilities, Wheelchairs
Kılıç, 2023	Industrial Revolution Process, From Industry 1.0 to Industry 5.0
Güdek, 2023	Industrial Transformation
Fouad et al., 2025	History and Evolution of Wheelchairs
Rasoulivalajoozi and Farhoudi, 2025	Wheelchairs and General Characteristics

In the second stage, the synthesis was conducted according to the ideas that were outlined in the first stage, and eight production criteria was defined considering before and after the digitalisation transformation.

The third stage is mainly constructed on the comparison of wheelchairs belonging to two different major companies, resembling the before and post-digitalisation processes. Production criteria defined previously were analysed and evaluated by using a 3-Point Likert System. A 3-point Likert scale is a psychometric response scale commonly used in survey research to measure attitudes, perceptions, or opinions. It is a simplified version of the Likert-type scale in which respondents are asked to indicate their level of agreement or disagreement with a given statement across three ordinal categories (Joshi et al., 2015). The aim of this study was to identify the clear directional differences between designs representing two distinct periods (pre-digitalization and post-digitalization). Therefore, a simplified 3-point Likert scale (suitable, partially, not suitable) was deemed appropriate, as it reduces interpretation uncertainty and facilitates transparent comparison between multiple design dimensions within a holistic review structure. This approach supports consistency in evaluation while remaining consistent with the exploratory and qualitative nature of the analysis.

In the fourth stage, based on the findings of the third stage, the reasons behind the formative approaches to wheelchairs that have eliminated the user from being a passive recipient are explained.

DEVELOPMENT AND INTEGRATION OF DIGITAL TECHNOLOGIES IN INDUSTRIAL DESIGN WORLD

The historical development was examined to understand the integration of digital technologies into the industrial design world and the improvement of it in this chapter of the study. The state of how digitalisation is used in industrial design practices was analysed. The specific impact of wheelchair design before and after digital integration was examined and explained in detail.

Historical Development of Digital Technologies in Industrial Design

The use of digital technologies in industrial design processes can be considered as a natural extension of the tools and methods shaped by industrial development. These industrial revolutions resemble as milestones in shaping the production models (Mayda, 2019).

Industrial production experienced a significant increase as soon as steam-powered engines were introduced and utilized in factories, marking the transition from an agricultural society to an industrial version (Güdek, 2023).

The first Industrial Revolution lasted approximately a hundred years. During the Second Revolution, design became more systematic, achieving a balance between aesthetics and functionality. With the widespread use of electricity and the emergence of computers and information technology, society entered Industrial 3.0, enabling effective control of production and management processes (Kılıç, 2023; Güdek, 2023).

In the Fourth Industrial Revolution (Industry 4.0), technological integration has reached new peaks, wide range of domains like smart houses, cities, factories, health and agricultural applications, were introduced with the Internet of Things. Opportunities increased in all aspects of people's lives since these inventions (Madakam et al., 2015; Banger, 2016; Çetin, 2023). Internet of Things (IoT) endowed wheelchairs with sensors and web services valuating the user needs, analysing the environmental information and transferring these into the smart systems which can monitor from afar. The products in the industrial design framework are not only accessed through aesthetics, but also through prioritising the integration of digital technologies with user preferences. Virtual production processes aim at minimising production errors by testing the design in digital platforms (Fathy, et al., 2024; Tatlısu et al., 2025).

Anticipated Fifth Revolution is described as an approach that aims at the integration of human competences with technological improvements in industry, reevaluating the opportunities in a human-focused perspective from which was obtained by digital tools were obtained (Güdek, 2023; Kılıç, 2023). The main difference between

Industry 4.0 and Industry 5.0 is that the first one optimises performance through automation and system integration, but the second one restores the importance of human creativity, cognitive ability, and value-based decision-making in the design and production process (Yücebalka, 2020). In this stage, social responsibility, sustainability, and personalised production became indispensable components of the industrial design world and related systems of Industry 5.0 (Kılıç, 2023).

As summarized in the literature, the development process of digital technologies has become increasingly complex, and these processes redefined industrial design. Despite clear technological advances, theoretical clarity remains limited regarding the broader impacts of digital integration occurring alongside Industry 4.0 process. Consequently, it is necessary to better understand the relationship between digital production technologies, design processes, and social transformation.

Integration of Digital Technology into Today's Industrial Design Practices

Digital transformation in industrial design has been a pivotal turning point in reshaping the creative steps and changing the design paradigm from the ground up. Özdal (2024) explains digital transformation as the integration of digital tools, technologies, and processes into design workflows, tools that foster innovation, expand creative potential, and unlock new possibilities for problem-solving and product development. This new integration introduced new approaches, and a wide range of methodologies appeared, while the scope and impact of the industrial design context broadened.

The impact of digital technologies on industrial design is also reflected across one of the most basic stages of product creation: sketching. Sketching refers to the phase in which the rough drawings of the designer's ideas are visualised on paper, individual creativity is nurtured, and the basic ground for conceptual developments is prepared. Traditional sketching techniques improve hand-eye coordination and establish a direct, instinctive connection between thought and visual expression. On the other hand, digital sketching tools offer additional flexibility through easy error correction, quick revisions, layered compositions, and compatibility with multiple formats, thereby increasing efficiency and precision in the creative process (Şenler and Yıldırım, 2020).

Computer-aided design (CAD) technologies create a suitable ground for designers to model complex geometries and details practically (Eryayar, 2017). Augmented Reality (AR) and Virtual Reality (VR) technologies broaden the design toolkit by enabling user interaction, real-time simulation, and iterative testing of products (Bordegoni, 2011). All kinds of products can be visualised, tested with simulations, analysed, and modifications of designs in virtual environments are easily created

(Fathy et al., 2024). The design forms grew in online platforms by the impact of the internet, eliminating the need for traditional methods, which can be encountered among the most important improvements in this field (Özdal, 2024). Sharing the digital designs and creating global collaborations are among the other advantages of the process. Digital tools also strengthen interdisciplinary collaborations because they facilitate communication among various stakeholders involved in the design process. More integrated and comprehensive problem-solving approaches are promoted. Through interdisciplinary analysis and collaboration, industrial design can evolve to encompass new competencies and areas of knowledge, enriching the design ecosystem (Bodur and Yavuzcan, 2020).

One of the most groundbreaking inventions of these technologies is 3D printing. It is described as a process that uses CAD software in digital environments to convert models into three-dimensional physical objects by adding materials over layers one after another. Physical prototypes are obtained, and concrete feedback is received in early stages in this way (Qi et al., 2014). This technology is effective for producing personalised or small-batch products, aligning with sustainable design principles. Artificial intelligence (AI) plays an increasingly important role in industrial design as an analytical tool that supports and fosters innovation. AI systems can analyse historical design patterns, generate alternative solutions, and accelerate decision-making by presenting a variety of optimized design options (Tatlısu et al., 2025). A notable AI-based invention is called generative design, which enables designers to input special parameters (material, weight, etc.) into the system and the AI generates a set of design solutions according to these data. This process brings about complex geometries that cannot be achieved using traditional methods, making it one of the pioneering methods of digital transformation.

The Integration of digital technologies into industrial design re-evaluates user experience and interaction and brings a fresh perspective to the field (Bodur and Yavuzcan, 2020). Internet of Things (IoT) and smart devices offer a wide range of experiences thanks to the adaptability, productivity and personalisation. This fact reinforces the important role of users in the digital design era. The general summary of these digital technologies can be found on Table 2.

Table 2. Summary of Different Digital Technologies and Their Advantages

Digital Technology	***Image***	***Advantages***
Digital Sketch (Hamouda, 2025)		• Easy Error Correction • Rapid Revisions • Layered Drawings and Flexibility in Different Geometries
Computer-Aided Design (CAD) (Paknahad, 2024) **Rhino**	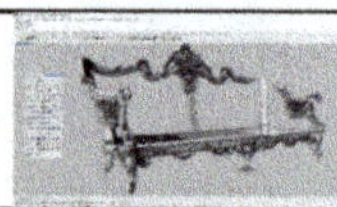	• Complex Geometry and Rapid Modelling of Details
Computer-Aided Design (CAD) (Therrien, 2021) **Solidwork**		• Complex Geometry and Rapid Modelling of Details
3D Printing (Hatch Duo, 2025)	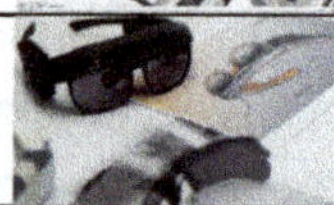	• Cost Advantages of Personalized Products • Minimum Waste Material and Source Efficiency
Artificial Intelligence Supported Design (Alderton, 2018)		• Evolution in Form Development

In summary, the integration of digital sketching, computer-aided design, and AI-powered tools signifies a digital-focused paradigm shift in industrial design. These technologies save designers' time and support them in efficiency and flexibility, with advantages such as complex form creation, rapid iteration, and mass customization. However, the long-term effects of increasing dependence on digital systems, due to the conveniences they offer, on design processes should not be overlooked (Borg et al., 2009; Boada et al., 2021; Saka & Das, 2025; Kesgin, 2025).

Implications and Limitations of Digital Technologies

The current literature mainly focuses on the advantages of digital technologies whereas global inequalities in relation to these technologies are also a major concern. In low-income countries, access to digital assistive devices is limited due to a lack of funding, inadequate infrastructure, and deficiencies in maintenance and service provision (Borg et al., 2009).

Boada et al. (2021) explains the critical issues concerning social assistive technologies (usually related with healthcare field). Following a bibliographical search, the ethical debates have been outlined as well-being (privacy / data control, deception, autonomy, loss of human contact, safety, dignity, emotional attachment, unauthentic intersubjectivity, freedom, objectification, human–human relationships, human moral skills, recognition, identity), care (loss of human contact, legitimacy, quality of practice, human moral practices, trust, impact on the concept of care, role disruption), justice (distributive justice, politics of technology, responsibility, recognition, social equality, robots' decision-making, ecological sustainability). In applying ethical principles, it is necessary to examine the relationships between all these concepts and reveal their connection to design. In this study, this situation is defined as a structural limitation.

Although Saka and Das (2025) focused on the usage of Internet of things technologies among older people, the risks associated with IoT prioritize common concerns like low technical ability, physical limitations affecting security practices, dependence on potentially insecure public networks, and increased vulnerability to social engineering tactics.

Kesgin (2025) summarises the fundamental challenges that may be encountered in AI-based support systems under two main headings: technical challenges alongside with ethical and legal challenges. The main potential technical challenges are listed as data quality and heterogeneity, explainability, real-time capability, and performance constraints. Ethical and legal challenges include bias and fairness, data privacy and sharing, and accountability. The author particularly draws attention to who would bear legal responsibility for erroneous decisions that AI-supported software could cause. The injustice that AI-supported software can bring about, regardless of the field in which it is used, is generally considered one of the most widely shared potential challenges in the literature. In particular, the right to equal use, centred on the concept of justice focused on global socio-economic class, is among the most important issues to be addressed in the future.

The increasing role of algorithms in maintenance, routing, and decision-support processes has brought various ethical issues, such as data privacy, algorithmic bias, and accountability, to the forefront in the context of assistive technologies. Literature emphasises that algorithms are not independent of values; they reflect the priorities of designers, developers, and institutions. Lack of transparency in decision-making processes makes it difficult for users to question and monitor these systems (Bozdag, 2013; Mittelstadt et al., 2016).

Wheelchairs in the Context of Accessible Design and Universal Design Principles

Accessible design consists of the applications that create the environments to ease people's lives. It encompasses the processes of designing environments that can accommodate a wide range of needs and making the necessary modifications to support this. According to Heiss et al., accessible design includes care-related facilities for three primary groups: individuals requiring patient care, elderly care, and general support for those in need. Accessible design forms the foundational concept for broader frameworks such as universal design. The ultimate objective is to foster equitable environments that empower all individuals in society to live independently and with dignity (2010).

Universal, as a term, is defined as a philosophical and metaphysical concept that encompasses all beings and everything that exists, expressing the common characteristics shared by things that have special and distinct features (Zimmerman, 2025). Universal design is defined as the creation of products and environments that are feasible on a wider scale and largely without any special adaptations and accessible to all humanity (Story et al., 1998). The Centre of Universal Design, which was established at North Carolina State University, conducted studies to make the "universal design" concept easy to understand.

The seven basic principles in this context are explained in the following.

1- Equitable Use – The design is useful and marketable to people with diverse abilities.

The first principle of Universal Design, Equal Access, states that products and environments must be accessible and usable for all users. Where possible, the same means of use should be provided; if this is not feasible, equivalent alternatives should be offered. Accessibility features should not discriminate against or stigmatize users, and privacy, safety, and security measures should be provided equally for everyone. Furthermore, the design should be aesthetically and functionally appealing, supporting the dignity and inclusion of all users.

2- Flexibility in Use – The design accommodates a wide range of individual preferences and abilities.

Design should accommodate a diverse range of individual preferences and abilities. It allows users to choose their method of use, adapts to different speeds or techniques, and accommodates both right-handed and left-handed users.

3- Simple and Intuitive Use – The design is easy to understand independent of the user's experience, knowledge, language skills, or concentration level.

Design should be simple to be understood regardless of the user's experience, knowledge, language skills, or concentration level. It should minimize complexity, provide clear instructions, and be consistent in its operation.

4- Perceptible Information – The design communicates necessary information effectively, independent of the environmental conditions or the user's sensory abilities.

The design effectively communicates necessary information to all users, regardless of their sensory abilities. It utilizes multiple modes (visual, auditory, tactile) and ensures clarity under a variety of conditions.

5- Tolerance for Error – The design minimises hazards and adverse consequences of accidental or unintended actions.

The design minimizes the dangers and negative consequences of accidental activities. To reduce this risk, it incorporates fail-safe systems, warnings, and the ability for unwanted actions to be easily reversed.

6- Low Physical Effort – The design can be used efficiently and comfortably with minimal fatigue.

The design allows for efficient and comfortable use with minimal fatigue. It accommodates a wide range of different body sizes and strengths, enabling working in various positions without excessive effort.

7- Size and Space for Approach and Use – Appropriate size and space are provided for approach, reach, manipulation, and use, regardless of the user's body size, posture, or mobility.

The design should provide all the things to provide approach, access, use, and manipulation regardless of body size, posture, or mobility, appropriate dimensions, and spaces. Open circulation and access areas are suitable for mobility aids and ensure accessibility (Centre for Excellence in Universal Design, 2025).

The improvements in universal design principles are linked with wheelchair design most clearly. The principle of creating barrier-free environments in industrial design has played a dual role. The first one is technical; the second one is a social transformation. In this context, wheelchair design is both triggered and nurtured by this process. In this study, wheelchairs were chosen as the case study product because of the consideration that the impacts of digital technologies in industrial design are well demonstrated.

Wheelchairs and Production Characteristics Before and After Digital Transformation

The evolution of the wheelchair has undergone significant transformation over more than two centuries, parallel to changes in functional and social needs. Initially developed to increase the comfort of individuals with disabilities, wheelchairs eventually became a tool used to provide mobility for war veterans. Today, they have evolved into high-performance, specially designed mobility devices for specific needs, such as those used by athletes. This process spans from the initial design of wheelchairs as furniture-like seating devices to their gradual evolution into increasingly mechanical and functional transportation systems.

Two main trends stand out in this evolutionary process. First, the emergence of wheelchairs focused on individual users represents the beginning of the wheelchair being designed as a personalized tool. In the 1930s, wheelchairs were mostly designed for shared use in hospital wards and were not assigned to individuals. This period prioritized shared use rather than personal use. During this time, wheelchairs were primarily a functional necessity in hospital environments and offered a user-specific ergonomic design.

The second fundamental trend is the customization of wheelchairs to meet individual needs and their design to adapt to the user. Over time, ergonomic and functional designs have been developed to suit the anatomical characteristics, physical abilities, and daily living requirements of users. Today, these developments have led to the emergence of customized wheelchairs that fit the user's body perfectly, like a glove or a special shoe, using complex seating systems and lightweight materials. These designs not only increase mobility but also significantly improve the comfort, safety, and quality of life of users (Sürmen, 2019; Wolfson, 2024).

Below is a chronological summary of the historical development and design transformations of the wheelchair designs. The following chapter explains the pre digital era of the transformation.

Pre-Digital Transformation Era: Mechanical and Function-Focused Solutions

The first wheelchair-like occurrence is a wheelbarrow that was thought to be created to assist people with disabilities between 6th and 4th BCE. In these early periods, the wheeled devices were projected related to people's mobility and rather than aesthetics, they prioritised functionality. This prepared foundation of the future developments (Fouad et al., 2025).

During the 12th century in Europe, the first wheeled furniture began to appear, but these devices were quite rudimentary and did not prioritize the comfort or autonomy (Wolfson, 2014). Although there is not a certain information about the first wheelchairs and the designer of it, it is known that the first specialized wheelchair emerged in 1595 by an unknown inventor and reported to be designed for King Philip II of Spain to operate as a "patient chair". The most important step towards developing a functional wheelchair was recorded in the 17th century. In 1655, German clockmaker Stephan Farfler designed a three-wheeled, self-propelled chair, and this innovation enabled users to move the chair independently using a rotating handle, marking a turning point. This paved the way for further innovations throughout the 18th century (Fouad et al., 2025; Wolfson, 2014).

Wheelchairs began to proliferate in public settings by the 18th century. Bath, a renowned English spa town, became a destination for individuals seeking treatments. Various types of wheelchair models have been designed to serve visitors with limited mobility to access healing waters and have been available for rent. That is when John Dawson's "Bath Chair" gained popularity during this period in 1783. It featured two large wheels beneath the seat and a smaller, steerable wheel at the front, allowing limited manoeuvrability with the aid of a caregiver (Science Museum Group, 2025).

The wheelchairs started to be commercialized and were found their places in health catalogues with their ornamental design features resembling chairs (Fouad et al., 2025). This shows that an increasing consciousness arose between the function and aesthetics in devices that assist mobility. The 19th century major developments emerged in terms of material and production processes. These significant developments helped lighter and more functional designs to be created. The wheelchairs became more accessible and available in a broader context thanks to the increased use of rubber tires, metal frames and lightweight wooden materials. Besides, Industrial Revolution contributed to the mass production of wheelchairs making them more affordable and widespread (Baş and Yapıcı, 2015 ; Fouad et al., 2025).

The 20th century introduced a major milestone with the invention of the first foldable wheelchair. This invention enhanced portability and usability that make wheelchairs more adaptable to daily life. Following World War II, the development of the first electric wheelchair marked another revolutionary step. Motorization allowed users to move independently without physical exertion, greatly improving accessibility and autonomy (Fouad et al., 2025; Wolfson, 2019).

Post-Digital Transformation Era: Smart, Personalized, and Autonomous Solutions

Along with digital transformation, the era has ushered in a paradigm shift in wheelchair design, incorporating artificial intelligence (AI), robotics, and brain-computer interface (BCI) technologies (Fouad et al., 2025). These innovations have transitioned wheelchairs from passive gadgets to user-focused, personalized and data-driven devices.

Today's wheelchairs offer innovative solutions to increase the life quality with the addition of sensor technologies, mobile application integrations, navigation assistance and autonomous driving capabilities. The evolution of wheelchair technologies is one of the most concrete examples that starts with its journey in a mechanical and physical level, extending through a transition where digital technologies are integrated. Ultimately, wheelchairs did not only become more functional but at the same time, they enhanced users' independence, comfort, and social participation.

Figure 2. Improvements of Wheelchairs Before and After the Digital Transformation

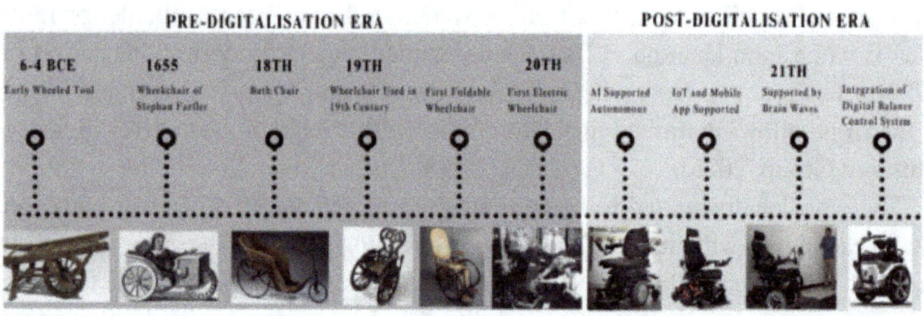

(Fouad, 2025; Tekerlekli sandalyeci, 2024; Science Museum Group, 2025; CSR, 2020; Aventus, 2025; Permobil, 2025; Accessible Madrid, 2025)

Contributions of Digital Technologies to Wheelchair Design

The contribution of digital technologies in industrial design is visible in personal health products. As a member of these products, wheelchairs have high needs in terms of both functional and user experience. User-focused design has to be programmed in products like wheelchairs that are designed for the use of individuals who have physical limitations due to a congenital condition or accident, have suffered damage to their skeletal, muscular, or nervous systems, and as a result who experience difficulties in continuing their daily routine activities (Tarımer and Akpınar, 2024). Digitalisation increases the opportunities of personalisation, accelerates the production processes, and strengthens a user-oriented design approach. This section of the study categorizes the contributions of digital technologies to wheelchair design and the criticisms of these technologies.

Role of Rapid Prototyping in Wheelchair Design in the Digitalisation Process

Rapid prototypes, together with digital technologies, which continue to grow today, appear in many fields like medicine, automotive, education, architecture, and industrial design (Çelik et al., 2013; Sürmen, 2019; Tarımer and Akpınar, 2024). While conventional methods rely on subtractive manufacturing techniques (cutting, carving, and chip removal), the 3D printing methods are known as additive manufacturing, and as opposed to, the materials are piled as layers and added on top of each other (Aydın and Kahraman, 2018). While conventional prototyping methods have time-consuming production processes, high-cost mould preparations, and limited revision opportunities, it is much easier to obtain physical samples in short durations with digital manufacturing technologies. This situation enables design processes to be faster, and errors to be noticed, and the final product to be reached in a shorter time (Çelik et al., 2013).

Rapid prototypes are effective in establishing feedback cycles and testing in the early stages of design, which helps to improve responsive solutions to the user's needs. Digital models that have been created in Computer-Aided Design (CAD) environments can be converted into physical samples in a short time with 3D printers and thus fundamental criteria like material, form, function, and ergonomics could be observed and evaluated. Thanks to this technology, designers find an opportunity to apply the trial-and-error method at a lower cost and with more variations (Junk and Matt, 2015; Qi et al., 2014; Seyhan et al., 2021; Sürmen, 2019). When wheelchair design is specifically a concern, the importance of rapid prototypes becomes more apparent. This is because in this product group, there are variables like anatomic structure, mobility, ergonomic factors, physical conditions, and individual comfort that directly affect the design (Baş and Yapıcı, 2015). Thanks to rapid prototypes, customized seating surfaces, back supports, and modular components can be produced and easily tested compared to conventional methods. They can be revised and optimised in a very short time based on user feedback. This situation increases the life quality of everyone, saves time, and resources during product development process.

In addition, rapid prototypes offer an ideal methodology for innovative designs with limited production quantities, or which have a special user target group. The flexibility and speed this technology offer have great advantages in personalised health products like wheelchairs, both for the designer and the user. Therefore, the production approach that digitalisation offers is not only a variable of a technical process, but it also resembles a humanistic value that touches the user. Advantages that rapid prototypes offer in industrial design create significant value, especially in user-focused and function-oriented products (Wang, 2024).

The development of personalized products that are sensitive to users' individual needs has become one of the most effective tools for designers. Personalization, supported by the research results presented, stands out as an important sector in rapid prototyping, but it is also important to consider the limitations of this new achievement.

Role of Digitalisation in Wheelchair Personalisation

Personalisation can be described as changing a system's aesthetics and functional features to increase the personal link of the product with its user (Ozan and Doğan, 2014). Digitalisation in industrial design created a radical transformation by offering new tools and approaches that enable special adaptation of the products for the benefit of the users (Bozdemir, 2014). In the last decades, thanks to digital technologies such as generative design and 3D printing, the products are tailored according to the user's needs specifically and have become more accessible and faster (Seyhan et al., 2021). Personalisation strengthens the product's life by extending the user-product bond and therefore, has the potential to make consumption more sustainable (Ozan and Doğan, 2014).

Wheelchairs show differences according to many variables such as sex, age, type of disability, physical structure, and personal preferences. This diversity highlights the importance of personalisation which is an era as the conventional production approaches are inefficient. The personalisation process starts with the data (body measurements, posture, and mobility) from the users, and this data is analysed by the digital tools, the creation of suitable models. CAD software and artificial intelligence-supported design tools are able to operate to find the best solution for the individual needs (Eryayar, 2017; Tatlısu et al, 2025). This creates advantages in terms of comfort and security (Bozdemir, 2024).

Personalisation is not limited to only physical dimensions. Alongside, the aesthetic preferences (colour, pattern, structure etc.) of the users can easily be integrated to the design in digital platforms. The applications that have an interaction with the user (as an example, engine systems that can be controlled by the mobile applications) are also part of the personalisation digital transformation created. In this context, the options of personalisation present after digitalisation increase the quality of individuals' lives and make them more independent.

Role of Digitalisation in the Use of Wheelchair Material

The integration of digital technologies of into industrial design processes transformed both design methods and the way materials are selected.

Material selection during design and manufacturing processes has a direct relationship with the product functionality, ergonomics, and aesthetic features. Material selection in personal and physical products like wheelchairs has become more strategic alongside the digitalisation. The conventional materials like steel and aluminium were replaced with lighter, more resistant, and environmentally friendly materials such as polymer-based ones. Especially together with 3D printing technologies, the use of sustainable materials was promoted. The environmental responsibility was increased by minimizing the waste rates and using the material as minimum. These materials do not only offer lightweight properties but also allow for multiple material combinations within the same product due to their varying degrees of hardness. For instance, the materials that have impact-absorbing features are used in the seating area to upgrade the user experience comfort while the high-resistance materials are preferred in wheels and load-bearing frames (Piselli et al., 2016; Shercliff and Lovatt, 2001).

The transformation of digitalisation makes it possible to take more conscious, functional, and sustainable decisions while it accelerates the design processes. All these features become important during the design and planning stage, throughout the usage period, and after the product's life span has ended in user-focused products, like wheelchairs. These improvements significantly demonstrate the interdisciplinary approach at the intersection of material science and digital design, shaping the future of accessible and innovative products.

Limitations of Digital Transformation in Wheelchair Design

Wheelchairs, defined in the assistive products category, are estimated to be needed by 80 million people worldwide. As stated by the World Health Organization (WHO), approximately 1% of the world's population requires these products. Again, as stated by the WHO, only % 5-35 of those who require wheelchairs worldwide have access to these devices (World Health Organisation, 2026). While this access rate is already low, the limitations on access to and use of digitally assisted wheelchairs should be addressed as a serious concern.

While digitalisation processes in wheelchair design bring various advantages in comfort and security for the users, accelerated productions processes and offer user-oriented solutions, there are ethical and social implications of artificial intelligence assisted wheelchair technologies to be considered. Yenugula et al. (2024); explain these challenges and limitations as the complexity of the technology and the need for specialized training for the operation. This situation poses a serious obstacle for low-income countries with limited resources. Another challenge is that when artificial intelligence algorithms are supported by data that reflects existing social biases, ethical and social issues are likely to arise in decision-making processes.

Sahoo and Choudhury (2023) consider privacy, autonomy, and equal access to be the main issues related to ethical and social impacts. Particularly in the concept of equal access, the fact that these technologies are only available to those who can afford them creates a conflict with universal design principles.

Comparative Analysis of Wheelchair Products Before and After Digital Transformation

In this chapter of the study, the selection criteria of the products that resemble the situation before and after the digitalisation of wheelchair axes are explained.

Subsequently, the features of the two products are detailed. Finally, the two selected wheelchairs are compared and analysed according to the criteria relevant to the purpose of the study.

Products Selection Criteria

Wheelchairs were identified based on Everest & Jennings (pre-digital transformation) and Whill (post-digital transformation); both companies show high representativeness before and after digital transformation. Both products were analysed in Table 3 according to the sales volume, technological innovation, and user impacts. Detailed explanations about each product are given after the product selection criteria.

Table 3. Everest & Jennings and Whill Electric Wheelchairs Selection Criteria Analysis

Criteria	*Everest & Jennings (Folded Wheelchair) (1932- USA)*	*Whill (C2) (Electric Wheelchair) (2022- Japan/USA)*
Product visual		
Sales Volume	Global sales leadership in the mid-20th century, the first mass production	Sales volume and development between companies (Whill Inc., 2017)
Representative strength	First standardised sample of a folded wheelchair from the 1930s	Smart wheelchair example from the modern digitalised era
Production method	Mass production, conventional methods	Technology-based digital production
User satisfaction	The satisfaction level is high due to the resistance, limited ergonomics (Wolfson, 2024)	User satisfaction, sense of independence, and driving control in feedback to WHILL (Whill Inc., 2017)

Everest & Jennings Standard Folded Wheelchair

This model was developed in the early 1930s by American engineers Herbert Everest and Harry Jennings and is accepted as the first folded manual wheelchair. Henry Jennings was a mining engineer who had become paraplegic following an accident. It was 1927 when he met Herbert Everest who was a mechanical engineer. The two collaborated to design a wheelchair using lightweight metal aircraft tubing and established a small factory in California in 1932. They started manufacturing wheelchairs of "Everest and Jennings" (Everest and Jennings, 1974).

Manual wheelchairs have historically and currently been characterised by overall higher usability, convenience of maintenance, and functionality compared to electric wheelchairs (Gabryelski et al., 2021).

Everest & Jennings, with its lighter steel tube chassis, foldable "X" shaped skeletal structure, and large rear wheels, has been preferred by various institutions and individuals, primarily hospitals, clinics, and long-term care centres. It has been a standard in the sector due to its design advantages during the 20th century. It has been manufactured along with its mechanical connections and convenient steel tube chassis, which makes it suitable for mass production. This manufacturing method is based on folding the tubes and connecting them by welding them together. The overall structure is resistant and robust, but its personalisation options are quite limited. It does not offer special dimensions and ergonomic form options. Standardisation of its parts is high, while the manufacturing cost is low. Thanks to these, mass production was obtained, and the product has been widely adopted as a necessity, mainly by the health systems.

This classical model is an iconic symbol of pre-digitalisation era. Sensors, engine, joysticks and mobile apps have been integrated. Its working principle is based on manual operation, dependent on the user's strength. The rotation radius is wide, while its capacity is limited. In the evolution of wheelchair design context, Everest& Jennings has a significant and historical importance since it is accepted as the founder of pre-digitalisation design paradigm.

Whill Model C2 Portable Electric Wheelchair

This model, which was developed in 2022 by Japan and US-based company Whill Inc., is a representative example of the post-digital transformation era with its user-focused design and technological features. Equipped with advanced features such as an electric drive system, joystick command, and mobile app integration, the C2 offers users not only mobility but also comfort, independence, and a personalised experience (Rasoulivalajoozi and Farhoudi, 2025). With its omnidirectional wheels, it achieves high manoeuvrability in narrow spaces; it can easily be dismantled and transported thanks to its modular structure, which is among the most important requirements of users (Yüksel and Börklü, 2021). With its powerful dual engines and innovative wheels, it can deal with challenging terrains that a standard wheelchair cannot handle (Yüksel and Börklü, 2021). It targets long-term comfort with its ergonomic seating systems and its battery that has a range of 12.4 miles (approx. 20 km), providing sufficient energy for daily usage (Whill Inc., 2017).

Compared to the conventional models, Whill Model C2 aligns with the principles of sustainable design and contributes to the minimisation of environmental effects thanks to its minimal carbon footprint. There is an under-seat storage basket addition with a capacity of 20 litres, which serves as a comfort-enhancement for daily life (Whill Inc., 2017). Whill Model C2 is manufactured with digital production techniques (CNC machining, IoT compatibility) and is supported by functions like speed control, navigation systems, battery monitoring, connected with mobile apps. It has been chosen as the winner of CES 2022 Innovation Awards and has gained a wide range of use in Europe and North America markets (Whill Inc., 2017).

In this context, Whill Model C2 is a solid example of the next generation in digitalised industrial design that has a direct contribution to the quality of life. The integration of digital components, such as mobile applications, sensor systems, and electronic control units, functions in sync with digital transformation processes. It offers special solutions to the users and can be personalised with ergonomic settings, control options, and colour preferences (Whill Inc., 2017).

Whill team has experienced some problems due to the complexity of the required parts of the wheelchair. The low manufacturing volume with conventional manufacturing methods forced the company to search for ways to decrease the production cost of the wheelchair. Flexible resin in parts that require tactile feedback, transparent resin in the opaque units around the wheelchair display, and rigid resin in components that will be subject to durability testing are chosen. The ability to print multiple units made of different materials provided a significant advantage in time and cost for a hardware startup (Formlabs, 2025). The team created a single-part body for the centre of the wheelchair. Since producing this kind of complex and monolithic part would cost high amounts and would not be feasible, they employed castable wax resin. As a result, they prototyped the product after modelling the main structure in Autodesk Fusion 360 software using generative design components. This maintained the chassis of the wheelchair to be 40% less in weight, making a significant structural advancement for the team (Formlabs, 2025).

The Comparative Analysis of Production Criteria of Everest& Jennings Folded Wheelchair and Whill Model C2 Electric Wheelchair

In this part of the study, the two brands have been analysed according to defined principles. The comparative analysis is composed of design criteria, material, manufacturing alternatives, mobility, ergonomics, user-centred state of the product, aesthetics and personalisation and accessibility. The findings of the two examples chosen for this study are represented in Table 4. The criteria used for the comparative analysis of the two examples representing pre-digitalisation and post-digitalisation era's general characteristics were created considering literature review, chronological evolution of wheelchairs, and fundamental features of these two wheelchairs.

To evaluate the products' general characteristics, various fundamental features were considered, and these comparisons, together with the visuals of the two brands, are presented in Table 5.

Table 4. Production Criteria Analysis of Everest& Jennings and Whill C2 Electric Wheelchairs

Criteria	***Everest & Jennings (Folded Wheelchair) (1932- USA)***	***Whill (C2)* (Electric Wheelchair) *(2022- Japan/USA)***
Design Criteria	Functional and robust	Personalised, robust, user-experience focused
Material	Steel skeleton	Light composite + ABS polymers
Manufacturing Alternatives	Craftsmanship, moulded manufacturing, standard mass production	3D Print, rapid prototype, CAD software
Mobility	Manual drive, the most convenient areas are indoor (hospital, rehabilitation centres) / outdoor	Electric, joysticks and mobile apps, indoor/ outdoor, and in urban cities
Ergonomics	Issues in long-term use, load capacity of 115 kg, a weight of 9 kg, foldable, and portable	Adjustable backrest, armrests, and footrest, suitable for long-term use, load capacity of 136 kg, a weight of 52,8 kg, can be dismantled into modular parts, portable
User-centred	The user is dependent on assistance	User is independent, direction with mobile apps
Aesthetics and personalisation	Personalisation via different colours	Personalisation via different colours, apps equipped with personalised information, driving control
Accessibility	150-250 USD	4000-5500 USD

Table 5. General Design Criteria Analysis of Everest& Jennings and Whill C2 Electric Wheelchairs

General Design Criteria	*Everest & Jennings (Folded Wheelchair) (1932- USA)*	*Whill (C2) (Electric Wheelchair) (2022- Japan/USA)*
Material		
Colour		
User-centred		 

DISCUSSION

This study conducted a comprehensive literature review covering a wide range of sources, including academic books, peer-reviewed journal articles, and symposium proceedings related to industrial design and digitalization. The collected materials were systematically analysed to identify recurring trends, conceptual frameworks, and methodological orientations in this field. This qualitative research revealed that the impact of digitalization in industrial design cannot be limited to technological innovation or material development alone. Rather, it represents a comprehensive paradigm shift that redefines the essence of design practice.

This transformation manifests itself not only in the tools and processes used by designers, but also in the ways design is conceptualized, produced, and experienced by users. Digital technologies have enabled adaptable, data-driven, and user-specific design outcomes by creating a new synthesis between creativity and computation. As a result, the nature of production has evolved from mass production to more decentralized and customizable forms of creation. Furthermore, digitalization has broadened the scope of design thinking, making user experience an integral and measurable component of product development.

A comparative analysis using the example of wheelchair design further validates these findings. As shown in Table 6, digital production technologies such as additive manufacturing, parametric modelling, and computer-aided ergonomics place great importance on user-centred design criteria, particularly personalization, ergonomic sensitivity, and increased mobility. These criteria demonstrate how digitalization has made industrial design a more inclusive, responsive, and human-centred discipline. Table 6 explains the analysis of wheelchair product before and after the digitalisation processes in industrial design. Table 7 analyses the universal design principles explained in earlier chapters of this study, focusing on wheelchair design and considering pre- and post-digitalization aspects.

Table 6. Analysis of Wheelchair Product Pre and Post Digitalisation

Product Criteria	***Pre-Digitalisation***	*Post-Digitalisation*
Design Criteria	Mechanics, function, mobility, and user-centred	**Smart, data-driven, personalized to individual needs**
Material	Steel, wooden, rubber, heavy material	**Carbon fibre, a lightweight composite, polymer, a sustainable and environmentally friendly material**
Manufacturing Alternatives	Handcrafting, moulded production, standard mass production	**3D printer, rapid prototyping, generative design, digital modelling**
Mobility	Manual, need for assistance	**Electric motor, AI-assisted autonomous driving, brain-controlled systems**
Ergonomics	Limited	**Advanced ergonomics, auto-balance, and comfort settings**
User-centred	Requiring physical aid	**Sensor-supported, IoT-based, mobile application-integrated driving assistants**
Aesthetics and personalisation	Standard forms for general users	**Personalisation based on colour, texture, form, and technology**
Accessibility	**Easy availability, modest cost**	**Unavailability of certain products, high cost, and quick revision capability**

Table 7. Analysis of Wheelchair Product in Pre and Post Digitalisation According to Universal Design Principles

Universal Design Principles	*Pre-Digitalisation*	*Post-Digitalisation*
1- Equitable use	✓	-
2-Flexibility in use	-	✓
3-Simple and intuitive use	✓	-
4-Perceptible information	-	✓
5-Tolerance for error	-	✓
6-Low physical effort	-	✓
7-Size and space for approach and use	-	✓

The findings according to Table 6 and Table 7 can be summarised in the following:

Manufacturing in traditional periods was mainly limited to mass, standard, and based on physical intervention-based solutions. As a result, ergonomics and individual compatibility concerns were given secondary importance. On the other hand, production processes were reshaped with high-tech infrastructures and user-centred, along with digital transformation.

Advanced digital transformation products such as the Whill Model C2 Electric Wheelchair have evolved beyond the traditional, pre-digital transformation products like the Everest & Jennings Standard Folded Chair. They are not only designed for transportation but also serve as a life support system that empowers users with a sense of independence and enables them to engage more actively with their environments (Rasoulivalajoozi and Farhoudi, 2025). In terms of sustainability, which is becoming increasingly important today, concerns such as minimising the carbon footprint, evolving the materials used, and strengthening the relationship between the product and the user to increase its lifespan have become more critical.

The first principle of universal design promotes equal access. This is explained as equality in the opportunities provided for privacy, security, and safety. Personalization processes introduced after digitalization may not parallel the increasingly complex design process that accompanies the development of digital processes. The positive impact of the digital design process on flexible use includes benefits such as providing perceptible information with extensive AI support, minimizing risks arising from user errors, and reducing physical effort. Another advantage is its adaptability to individuals with different physical characteristics through personalization options. The simplicity of use of pre-digital wheelchair designs is well-known. Post-digital products can have complex structures and may require specific experience and knowledge. The ability to cater to users of different languages with AI support can be seen as an additional benefit (Table 7).

Current research shows that the design of wheelchairs is not only related to physical aspects, but at the same time sensations and social experiences have also direct correlations with the design phase. Some user responses like "independence", "confidence", "appreciation", and "being stylish" concepts can be measured and transformed into product features. Users of advanced designs such as Whill Model C2 could feel these kinds of sensations strongly, and this makes them more active in their social interactions, enabling them to become more confident individuals (Rasoulivalajoozi and Farhoudi, 2025). The biggest barrier between individuals with disabilities and social life is access to the social environment (Mülayim and Azsöz, 2020). At this point, digital transformation made wheelchairs a smart solution tool that contributes social and physical interaction of the individuals rather than being only a functional device. The production processes gained flexibility thanks to the influence of digitalisation, rapid prototypes, and productive design techniques fastened production development cycles (Çelik et al., 2013; Seyhan and et al., 2021). Also, the feedback mechanisms are directly integrated with the user experience (Baş and Yapıcı, 2015). This triggered design solutions that are sensitive to individual needs, adaptive, and sustainable.

CONCLUSION

This study is about the digitalisation processes in industrial design. To evaluate the influence of digitalisation in industrial design processes, the wheelchair as a product was chosen. The research carried out aims to reveal that digital transformation in industrial design is not merely a technological advancement but also refers to a conceptual evolution. To achieve this goal, it was considered convenient to select a product with high representative potential to analyse the change in concrete terms. In this context, wheelchairs have been selected as the product within the scope of the study, as they effectively illustrate the parallel contributions of digitalisation and industrial revolutions to the field of industrial design.

Table 8 shows the analysis of two major brands (Everest and Jennings, Whill C2 Model) of wheelchairs that are representative of pre- and post-digitalisation eras. The data was analysed using a 3-point Likert scale. The evaluations of the scale are accepted as suitable, partially suitable, or not suitable.

According to the study, the Everest& Jennings Folded Wheelchair can be accepted as partially suitable in terms of design approach and material type, considering today's conditions. Manufacturing alternatives are not considered suitable because the users will still require manual labour and the assistance of a second person for mobility. The product is not suitable in terms of personalisation, one of the most positive gains of post-digitalisation products, as it is ergonomically limited. It stands out as an economically accessible product with high purchasing power and production capacity, as well as widespread availability. On the other hand, the Whill Model C2, one of the most preferred products among wheelchairs after digitalisation, has been partially deemed suitable due to its price-performance ratio and relatively high segment in its field. In terms of all other criteria, it is seen as more suitable than the Everest & Jennings Folded Chair.

Table 8. Analysis of the Impact of Digitalisation on Production Criteria of Wheelchairs

	Everest & Jennings	*3 Point Likert Scale*			*Whill Model C2*	*3 Point Likert Scale*		
Production Criteria		*Suitable*	*Partially*	*Not Suitable*		*Suitable*	*Partially*	*Not Suitable*
Design Criteria			✓			✓		
Material			✓			✓		
Manufacturing Alternatives				✓		✓		
Mobility				✓		✓		
Ergonomic			✓			✓		
User-centred				✓		✓		
Aesthetics and personalisation				✓		✓		
Accessibility							✓	

Production processes that are supported with digital technologies have made it possible to implement user-specific solutions, strengthening the basic quality of life criteria like accessibility, comfort, and independence for individuals with physical disadvantages (Rasoulivalajoozi and Farhoudi, 2025). During the process, starting from conventional to digital, the user has moved its role from passive into an active stakeholder.

However, it is not possible to say that digital solutions are open and accessible to all equally. The reason is that digital technologies are limited due to high cost, inefficient in terms of infrastructure, and lack of adequate distribution channels. These are among the realities that limit users in a wide range to benefit from digital technologies and create a social injustice. Digitalisation in industrial design not only changes the form or function but also influences the relationship with life. Wheelchair design shows the impacts of this transformation both in design processes and daily life. Digitalisation has enabled a holistic approach to multidimensional criteria such as aesthetics, ergonomics, personalisation, and environmental sensitivity.

Wheelchair design in the literature is mostly addressed in terms of accessibility, ergonomics, and functionality, and is generally presented positively thanks to the positive effects of digital technologies on personalization. However, it is necessary to mention some limitations brought about by digital production and advanced technologies. In particular, the high cost of CAD, AI-supported systems, and customized production solutions can weaken the principle of inclusivity in universal design and exclude low-income user groups. In addition, digitally supported designs have ethical sensitivities such as data privacy, algorithmic concerns, and reduced user control. The contradiction between academic discourse and practice arises from the fact that access to digital technologies conflicts with the principle of equality, one of the foremost tenets of universal design principles on a global scale, due to reasons such as lack of technical infrastructure and cost.

As a result, it is expected that future industrial design approaches will evolve into a structure that balances technology with human needs, prioritises sustainability, and positions the user not only as a target but also as an active partner in the process. One of these expectations is the importance of policies that facilitate access to products such as wheelchairs, which should be among the global priority policies. In this process, it will be important for the industrial design discipline to embrace the universal design visions.

REFERENCES

Accessible Madrid. (2025). *Genny 2.0 Urban*. https://www.accessiblemadrid.com/en/accesorios/genny-20-urban

Alderton, M. (2018). *Design & make with Autodesk*. Autodesk. https://www.autodesk.com/design-make/articles/automotive-design

Apilioğulları, L. (2019). Üretim endüstrisi dijital dönüşüm süreci: Kavramsal ilişki haritası. *Anemon Muş Alparslan Üniversitesi Sosyal Bilimler Dergisi*, *7*(3), 153–162. DOI: 10.18506/anemon.460721

Aydın, A., & Kahraman, F. (2018). 3B baskı teknolojisi kullanılarak hızlı prototipleme yapımı ve nesnelerin üretimi. *6. Uluslararası Yenilikçi Teknolojiler Sempozyumu (ISITES 2018) Bildiri Kitabı*, 1–15.

Banger, G. (2016). *Endüstri 4.0 ve akıllı işletme*. Dorlion.

Baş, H., & Yapıcı, F. (2015). Ergonomik tasarım ve üretimde hızlı prototipleme teknolojisi. *Suleyman Demirel University Journal of Engineering Sciences and Design*, *3*, 199–204.

Boada, J. P., Maestre, B. R., & Genís, C. T. (2021). The ethical issues of social assistive robotics: A critical literature review. *Technology in Society*, *67*, 101726. DOI: 10.1016/j.techsoc.2021.101726

Bodur, G., & Yavuzcan, H. G. (2020). İşbirlikli tasarım süreçlerinde tasarımcının rolü ve dijital adaptasyon. *4. Ulusal Tasarım Araştırmaları Konferansı (UTAK 2020) Bildiri Kitabı*, 89–103.

Bordegoni, M. (2011). Product virtualization: An effective method for the evaluation of concept design of new products. In *Innovation in product design: From CAD to virtual prototyping* (pp. 117–141). Springer London. DOI: 10.1007/978-0-85729-775-4_7

Borg, J., Lindström, A., & Larsson, S. (2009). Assistive technology in developing countries: National and international responsibilities to implement the Convention on the Rights of Persons with Disabilities. *Lancet*, *374*(9704), 1863–1865. DOI: 10.1016/S0140-6736(09)61872-9 PMID: 19944867

Bozdag, E. (2013). Bias in algorithmic filtering and personalization. *Ethics and Information Technology*, *15*(3), 209–227. DOI: 10.1007/s10676-013-9321-6

Bozdemir, M. (2014). *Engelli insanlar için elektrikli araç tasarımı. I. Uluslararası Endüstriyel Tasarım Mühendisliği Sempozyumu (ISIDE14)*. Bildiri Kitabı.

Çelik, İ., Karakoç, F., Çakır, M. C., & Duysak, A. (2013). Hızlı prototipleme teknolojileri ve uygulama alanları. *Journal of Science and Technology of Dumlupınar University, 31*, 53–70.

Centre for Excellence in Universal Design. (2025). *The 7 principles.* https://universaldesign.ie/about-universal-design/the-7-principles

Çetin, B. (2023). Dijital erişilebilirlik: Kapsam, kavramlar, standartlar, yasalar. *Tasarım Mimarlık ve Mühendislik Dergisi, 3*(1), 57–68. DOI: 10.59732/dae.1246531

Chung, M., & Kim, J. (2016). The internet information and technology research directions based on the fourth industrial revolution. *KSII Transactions on Internet and Information Systems, 10*(3), 1311–1320. DOI: 10.3837/tiis.2016.03.020

CSR. (2020). *History and evolution of wheelchairs.* https://www.karmanhealthcare.com/history-and-evolution-of-wheelchairs/

Eryayar, E. (2017). Endüstri ürünleri tasarımı eğitiminde hesaplamalı tasarım yaklaşımı. *Sanat-Tasarım Dergisi, 8*(8), 15–19. DOI: 10.17490/Sanat.2018.16

Everest, H. A., & Jennings, H. (1974). *Marketing brochure.*

Fathy, A. T. M., Ghareeb, S. E., & Shaban, Y. (2024). Endüstriyel tasarımda sürdürülebilirlik için dijital dönüşüm ve tasarım. *Digital Transformation and Design for Maintainability in Industrial Design, 3*, 111–120.

Formlabs. (2025). *Industrial parts at blazing speed.* https://formlabs.com/global/

Fouad, A. I., Ismail, A. K., Ali, A. N., Mansour, B. S., Salah Eldin, K., Ramadan, M. A., Muhammad, M. H., & Elsisy, M. (2025). Evolution of wheelchair technology: A comprehensive overview of history, disabilities, types and control mechanisms. *Advanced Sciences and Technology Journal, 2*(2), 1–20. DOI: 10.21608/astj.2025.347587.1046

Freeman, C., & Soete, L. (2005). 14. A digital society for us all: "Old" and "new" policy reflections. In *The economics of the digital society* (p. 330).

Gabryelski, J., Kurczewski, P., Sydor, M., Szperling, A., Torzyński, D., & Zabłocki, M. (2021). Development of transport for disabled people on the example of wheelchair propulsion with cam-thread drive. *Energies, 14*(23), 8137. DOI: 10.3390/en14238137

Güdek, B. (2023). Endüstriyel dönüşüm ve endüstri 5.0. *Ömer Halisdemir Üniversitesi İktisadi ve İdari Bilimler Fakültesi Dergisi, 16*(4), 1129–1142.

Hamouda, W. O. (2024). *Automotive engineering, a new specialty in higher education.* https://www.horizons.dz/?p=130661

Hatch Duo. (2025). *Thirdeye Gen.* https://www.hatchduo.com/works/thirdeye-gen

Heiss, O., Degenhart, C., & Ebe, J. (2010). *Barrier-free design* (G. Soffker & P. Thrift, Trans.). Edição Detail. https://www.detail.de/de_en/barrier-free-design

Heskett, J. (2001). Past, present, and future in design for industry. *Design Issues, 17*(1), 18–26. DOI: 10.1162/07479360152103804

Jensen, M. C. (1993). The modern industrial revolution, exit, and the failure of internal control systems. *The Journal of Finance, 48*(3), 831–880. DOI: 10.1111/j.1540-6261.1993.tb04022.x

Junk, S., & Matt, R. (2015). New approach to introduction of 3D digital technologies in design education. *Procedia CIRP, 36*, 35–40. DOI: 10.1016/j.procir.2015.01.045

Kesgin, K. (2025). Yapay zekâ tabanlı karar destek sistemlerinde teknik, etik ve yönetişimsel yaklaşımlar: Çok sektörlü bir derleme çalışması. *International Journal of Management Information Systems and Computer Science, 9*(1), 29–48. DOI: 10.33461/uybisbbd.1691218

Kılıç, R. (2023). Sanayi devrimlerinin serüveni: Endüstri 1.0'dan Endüstri 5.0'a. *Takvim-i Vekayi, 11*(2), 276–291.

Kumar, D., Malhotra, R., & Sharma, S. R. (2020). Design and construction of a smart wheelchair. *Procedia Computer Science, 172*, 302–307. DOI: 10.1016/j.procs.2020.05.048

Madakam, S., Ramaswamy, R., & Tripathi, S. (2015). Internet of things (IoT): A literature review. *Journal of Computer and Communications, 3*(5), 164–173. DOI: 10.4236/jcc.2015.35021

Mayda, M. (2019). Endüstri 4.0 sürecinde tasarımda dönüşüm. *RESS Journal, 6*(11), 1–21.

Mittelstadt, B. D., Allo, P., Taddeo, M., Wachter, S., & Floridi, L. (2016). The ethics of algorithms: Mapping the debate. *Big Data & Society, 3*(2), 2053951716679679. DOI: 10.1177/2053951716679679

Mülayim, A., & Azsöz, G. P. (2020). Tekerlekli sandalye kullanan engellilerin sosyal hayata katılımında en büyük engel: Yapılı çevrede yaşanan problemler ve çözüm önerileri. *Sosyal Politika Çalışmaları Dergisi*, 357–382.

Nguyen, V. T., & Tung, T. T. (2024). Design and development of a wheelchair prototype. *Engineering, Technology &. Applied Scientific Research, 14*(2), 13376–13379.

Ozan, E., & Doğan, Ç. (2014). Kişiselleştirme yoluyla kullanıcıları tasarım sürecinde etkin kılan yöntem ve yaklaşımların sürdürülebilirlik için ürün tasarımı açısından değerlendirilmesi. *UTAK 2014 Bildiri Kitabı* (10–12 Eylül 2014).

Özdal, M. A. (2024). Teknolojik yeniliklerin tasarım alanları üzerine etkisi. *Baçini Sanat Dergisi, 2*(4), 82–100.

Paknahad, A. (2024). *Classic furniture modeling with Rhino software for production in the furniture industry.* https://discourse.mcneel.com/t/classic-furniture-3d-modeling-with-rhino-5/196996/1

Permobil. (2025). *Permobil M3 Corpus.* https://www.permobil.com/en-us/products/power-wheelchairs/permobil-m3-corpus

Piselli, A., Simonato, M., & Del Curto, B. (2016). Holistic approach to materials selection in professional appliances industry. In *DS 84: Proceedings of the Design 2016 14th International Design Conference* (pp. 865–874).

Qi, N., Zhang, X., & Yin, G. (2014). Opportunities and challenges of industrial design brought by 3D printing technology. In *Moving Integrated Product Development to Service Clouds in the Global Economy* (pp. 369–376).

Rasoulivalajoozi, M., & Farhoudi, M. (2025). Integrating user perceptions of socio-emotional aspects in wheelchair design: A pilot study using Kansei Engineering. *Journal of Transport & Health, 42*, 42. DOI: 10.1016/j.jth.2025.102002

Riccini, R. (1998). History from things: Notes on the history of industrial design. *Design Issues, 14*(3), 43–64. DOI: 10.2307/1511893

Rodrigue, J. P. (2020). *The geography of transport systems.* Routledge. DOI: 10.4324/9780429346323

Sahoo, S. K., & Choudhury, B. B. (2023). AI advances in wheelchair navigation and control: A comprehensive review. *Journal of Process Management and New Technologies, 11*(3–4), 115–132. DOI: 10.5937/jpmnt11-45181

Saka, S., & Das, S. (2025). SoK: Reviewing two decades of security, privacy, accessibility, and usability studies on Internet of Things for older adults. *arXiv preprint* arXiv:2512.16394.

Science Museum Group. (2025). *Collection of fifteen wheelchairs and other mobility aids.* https://collection.sciencemuseumgroup.org.uk/objects/co127491/collection-of-fifteen-wheelchairs-and-other-mobility-aids

Şenler, F., & Yıldırım, K. (2020). Tasarımcıların yaratıcılık, bütünleştirme, kavramsallaştırma ve uygulama becerileri üzerine çizim tekniklerinin etkileri. *Gazi Üniversitesi Fen Bilimleri Dergisi Part C: Tasarım ve Teknoloji, 8*(3), 486–499.

Seyhan, M., Bayram, G., & Togay, A. (2021). Hızlı prototipleme ve endüstriyel tasarım ilişkisinin sektörel üretim süreçlerinde mevcut durum tartışması. *International Journal of 3D Printing Technologies and Digital Industry, 5*(2), 326–338.

Sharma, A., & Singh, B. J. (2020). Evolution of industrial revolutions: A review. *International Journal of Innovative Technology and Exploring Engineering, 9*(11), 66–73. DOI: 10.35940/ijitee.I7144.0991120

Shercliff, H. R., & Lovatt, A. M. (2001). Selection of manufacturing processes in design and the role of process modelling. *Progress in Materials Science, 46*(3–4), 429–459. DOI: 10.1016/S0079-6425(00)00013-X

Story, M. F., Mueller, J. L., & Mace, R. L. (1998). *The universal design file: Designing for people of all ages and abilities.*

Sürmen, H. K. (2019). Eklemeli imalat (3B baskı): Teknolojiler ve uygulamalar. *Uludağ Üniversitesi Mühendislik Fakültesi Dergisi, 24*(2), 373–392.

Tarımer, İ., & Akpunar, A. (2024). Dişlisiz ve direkt sürmeli elektrik motorlarının yerleştirildiği ergonomik bir elektrikli tekerlekli sandalye tasarımı.

Tatlısu, E., Bağlı, H. H., Turan, A. Z., Hamurcu, A., & Eser, A. (2025). Yapay zekâ çağında endüstriyel tasarım: Uygulamalar ve eğitim üzerine etkileri. *Öneri Dergisi, 20*, 154–188.

Tekerlekli Sandalyeci. (2024). *Tekerlekli sandalyenin icadı.* https://www.tekerlekli sandalyeci.com/tekerlekli-sandalyenin-icadi

Therrien, C. (2021). *SOLIDWORKS 3D CAD 2022 is here: Discover the top enhancements now.* https://blogs.solidworks.com/solidworksblog/2021/09/top-enhancements -in-solidworks-3d-cad-2022.html

Wang, J. (2024). *Prototyping in smart product development: Exploring new approaches for low-fidelity prototyping to enhance communication across disciplines* (Doctoral dissertation, Swinburne University of Technology).

Whill Inc. (2017). *The new era of personal mobility has finally arrived.* https://whill .inc/product-literature/ca/model-ci-brochure.pdf

Wolfson, P. (2018). *Enwheeled: History, design, and the wheelchair.* Cooper Hewitt, Smithsonian Design Museum.

World Health Organization. (2025). *Assistive technology* (Fact sheet). https://www.who.int/news-room/fact-sheets/detail/assistive-technology

Yenugula, M., Sahoo, S., & Goswami, S. (2024). Cloud computing for sustainable development: An analysis of environmental, economic, and social benefits. *Journal of Future Sustainability, 4*(1), 59–66. DOI: 10.5267/j.jfs.2024.1.005

Yücebalkan, B. (2020). Endüstri 4.0'dan endüstri 5.0'a geçiş sürecine genel bakış. *PEARSON Journal, 5*(9), 241–250. DOI: 10.46872/pj.181

Yüksel, N., & Börklü, H. R. (2021). Yapay zekâ destekli kavramsal tasarım: Tekerlekli sandalye tasarım seçenekleri değerlendirmede bulanık mantık kullanımı. *Gazi Mühendislik Bilimleri Dergisi, 7*(3), 309–319.

Zimmerman, D. W. (2025). *Universal*. https://www.britannica.com/topic/universal/Possibilia

ADDITIONAL READING

Cross, N. (2011). *Design thinking: Understanding how designers think and work*. Berg. DOI: 10.5040/9781474293884

Desmet, P. M. A., & Hekkert, P. (Eds.). (2007). *Framework of product experience*. Delft University of Technology / International Journal of Design.

Hitch, D., & Larkin, H. (2021). Designing mobility: An inclusive design framework for wheelchairs. *Disability and Rehabilitation. Assistive Technology, 16*(3), 221–232. DOI: 10.1080/17483107.2019.1664780

Kalay, Y. E. (2004). *Architecture's new media: Principles, theories, and methods of computer-aided design*. MIT Press.

Kırmızı, M., & Kocaoğlu, B. (2024). Design features of digital transformation maturity models: A systematic literature analysis and future research directions. *Journal of Modelling in Management, 19*(2), 313–341. DOI: 10.1108/JM2-11-2022-0271

Ming, X., Lu, W., Zhang, X., & Zheng, M. (2017). Industry 4.0 and intelligent manufacturing: A review. *Engineering, 3*(5), 616–630. DOI: 10.1016/J.ENG.2017.05.015

Sirel, A., & Sirel, O. Ü. (2018). Universal design approach for the participation of the disabled in urban life. *Journal of Civil Engineering and Architecture, 12*, 11–21.

KEY TERMS AND DEFINITIONS

Digital Transformation: Digital transformation (DT) is the process of adopting and implementing digital technology that will replace existing manual tools by creating new products, services and operations that convert business processes into digital format.

Industrial Design: a discipline that focuses on product design, acting with the aim of providing equal benefit to both manufacturers and users.

Industrial Revolution: The transition from an agriculture and handicrafts-based economy to an industry and machine dominated one.

Industry 4.0: the last industrial revolution that enabled industrial design more productive thanks to some technological inventions like internet of things (Iot), cloud, big data etc.

Industry 5.0: the latest new industrial revolution in which designers collaborate with robotic technologies and manufacturing and product designs favour personalised processes.

Universal Design: without any special organization or adaptation, usability of all the design products or environments by all kinds of people.

Wheelchairs: Equipment that helps people who have difficulties or can't walk due to an illness, injury or an age-related reason and assists them to easy mobilization. It resembles the design of a chair with wheels.

Chapter 8
Human and Machine Synergy in Agriculture and Food Systems

Kannan M.
http://orcid.org/0009-0003-9810-3496
Amrita Vishwa Vidyapeetham, India

Parthasarathy Seethapathy
http://orcid.org/0000-0003-0660-9311
Amrita Vishwa Vidyapeetham, India

ABSTRACT

The increasing complexity of agricultural and food systems necessitates human intelligence to engage in a successful collaboration with the artificial intelligence (AI) to transform agricultural systems into more productive, sustainable, and accurate in their judgments. The given chapter examines the idea of human-AI cognitive collaboration in agriculture with a specific focus on the complementary nature of human expertise and machine intelligence. It discusses the ways in which the state-of-the-art AI methods, such as machine learning, deep learning, and computer vision, contribute to the agricultural processes, such as crop monitoring, disease detection, yield prediction, and food quality evaluation. The chapter also talks about the human-in-the-loop models that combine the farmer knowledge and feedback with the AI-driven insights to enhance the trust and usability. . The chapter ends by giving future directions of research and industrial implications of human-machine synergy in sustainable agriculture and food systems.

DOI: 10.4018/979-8-3373-7036-1.ch008

1. INTRODUCTION TO HUMAN-AI COGNITIVE COLLABORATIONS IN AGRICULTURE

The agricultural and food systems are changing at a fast rate due to the rising population pressure, climate change, resource depletion and sustainable production methods (Wijerathna-Yapa & Pathirana, 2022). The scale and complexity of the modern agricultural operation continue to challenge the traditional farming methods that are highly dependent on human experience and manual decision-making. In this regard, the combination of artificial intelligence (AI) and human thinking power has become an offer to be productive, resilient, and better in decisions throughout the agricultural value chain (Sivasamy, 2026). Human-AI Cognition collaboration is a synergistic method where human knowledge and machine intelligence are used to solve complex agricultural issues in a better manner than either one could do on its own (Bao et al., 2023). The development of AI technologies (such as machine learning (ML), deep learning (DL), computer vision, sensor-driven analytics, etc.) has facilitated monitoring, prediction, and optimization of agricultural and food systems with the help of automation (Sivasamy, 2026). The technologies are applicable in a large variety of tasks, including the detection of crop diseases, crop yield prediction, irrigation precision, soil health, and food quality inspection.

1.1 Background and Motivation for Human-Machine Synergy

Human-machine synergy is based on the notion that the intelligent systems are supposed to complement the human abilities and not substitute them. Farmers, agronomists and agricultural professionals in the field of agriculture have extensive tacit knowledge in agriculture that has been gained through years of observation and practice. This is informed by knowledge of crop behaviour, environmental interpretation and making situation-specific decisions in the face of uncertainty. Although AI systems are good at handling large amounts of data and detecting complicated patterns, they are frequently black-box models, and their results are challenging to interpret and rely on (Hassija et al., 2024). The rationale of human-AI partnership is the necessity to merge the advantages of both sides to develop strong, clear, and practical decision-support frameworks. The increasing access to agricultural data through satellite, drones, Internet of Things (IoT), and mobile devices have contributed to accelerating the use of AI in agriculture further (Qazi et al., 2022). Such data sources produce heterogeneous and high-dimensional datasets that cannot be analyzed manually. Such data can be processed effectively by AI models to make appropriate insights available in time, but the interpretation and use of these insights is often human-dependent. The Figure 1 illustrate the overview of human–AI cognitive collaboration in agriculture.

Figure 1. Overview of Human-AI Cognitive Collaboration in Agriculture

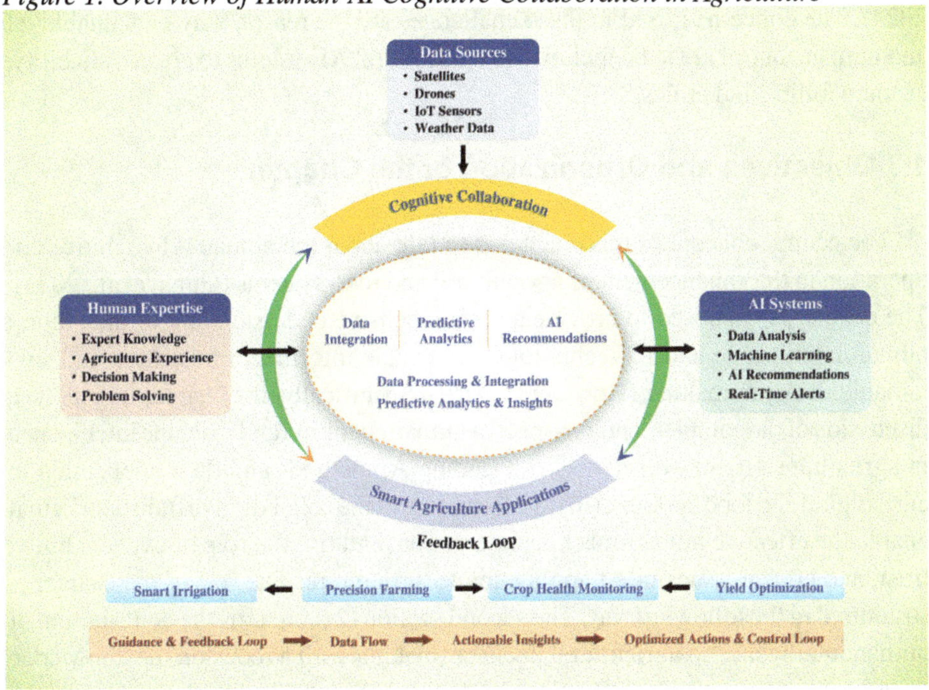

1.2 Challenges in Modern Agriculture and Food Systems

The modern-day agriculture is characterized by a wide range of related issues which need to be solved smartly and flexibly. Climate change has been characterized by unpredictable weather, pest and diseases outbreak and crop variability (Nega, 2025). The scarcity of resources, including water scarcity, soil erosion, and rise in inputs cost also complicates the agricultural decision-making. Additionally, the food systems ought to address post-harvest losses, quality assurance, and inefficiencies in the supply chains and be in a position to meet high safety and sustainability standards (Jariwala, 2025). Traditional agricultural systems are usually reactive whereby decisions are made after the harm or loss of produce has been realized. Although AI-based tools possess predictive and proactive capabilities, they possess few predictive and proactive opportunities because of data quality issues, model generalization and contextual blindness. In addition, the digital divide in the rural population, a low level of technical literacy, and the fear of information safety and privacy and ethical use of AI are the key barriers to ubiquitous application. Another problem of concern is the interpretability of AI models. In agriculture, where economic, environmental and social impacts of a choice are concerned, lack of explainability can reduce trust and slow the process of adoption (Mallinger et al., 2024). The concerns raised in these challenges are the reason why explainable AI and human control are to be included in agricultural AI systems to create reliability, accountability, and ethics.

1.3 Objectives and Organization of the Chapter

The central concept of this chapter is to talk about the human-AI cognitive co-operation in the enhancement of agricultural and food systems (Gupta et al., 2025). The chapter is supposed to provide a comprehensive understanding of integrating human intelligence and AI technologies to enable informed, transparent and sustainable decision-making (How et al., 2020). Specifically, the chapter targets (i) a discussion of the complementary aspect of human cognition and machine intelligence in agriculture, (ii) an overview of the leading AI methods and their application in agricultural and food systems, (iii) the discussion of the collaborative frameworks that enable the effective human-machine interaction, and (iv) the role of explainability, trust, and ethical issues in AI application to agriculture. The rest of the chapter is structured in the following way. The second section (2) concerns the contribution of human intelligence to agricultural decision-making with a focus on the knowledge of farmers and cognitive mechanisms. The third section (3) discusses some of the major AI methods in smart agriculture and food systems. Part 4 presents the human-machine synergy models and cognitive teamwork models. The practical uses and case studies in the agricultural and food systems are discussed in Section 5. Lastly, Section 7 provides the directions for future research and closes the chapter.

2. ROLE OF HUMAN INTELLIGENCE IN AGRICULTURAL DECISION-MAKING

Traditionally, agricultural systems have been based on human intelligence, and the decisions were based on the choice of crops, the management of the soil, the timing of irrigation, pest control, and the time of harvest (Shams et al., 2024). The human cognitive capabilities still play a central role in the successful agricultural decision-making despite the recent progress in automation and AI because of the ability to make decisions based on context, learn through experience, and make ethical decisions.

2.1 Farmer Knowledge, Experience, and Cognitive Skills

They are also able to evaluate the options between immediate benefit of yield and long-term soil health to strike a balance between productivity and sustainability ambitions (Shahane & Shivay, 2021). Moreover, human intelligence enables one to learn during unusual or infrequent occurrences, e.g. extreme weather conditions or sudden outbreaks of pests. Although AI models heavily depend on past data, and might not perform well in unobservable cases, humans are capable of making extrapolation and analogical reasoning based on the previous experience and apply them to new circumstances. Such capability is especially significant in the agriculture sector, where climate change is bringing new variability that has never been experienced before.

2.2 Human-Centered Decision Processes in Farming Systems

The process of making decisions in agriculture is multi-dimensional in nature and touches on technical, economic, social and ethical concerns. The decision processes in human orientation consider these various aspects and therefore incorporate quantitative data with qualitative decision making. The choice of the type of crop to grow, the amount of inputs to use, whether to use new technologies or not, etc. are made based not just on the yield projections, but also on the cost, labour supply, access to the market, and risk-taking. Most agricultural systems have been characterized by collective decision making between farmers, extension officers, agronomists and policymakers (Liao et al., 2024). This collective intelligence allows sharing of learning and transfer of knowledge that cannot be easily obtained by automated systems alone. Human based approaches also lay emphasis on participatory decision making that is based on stakeholders being active in the information and tools instead of having them passively receive a recommendation. Ethical responsibility is also associated with making decisions that are human-centered. The use of pesticides, the allocation of water and land management decisions all have direct consequences on the environmental sustainability and the health of people (Zhou et al., 2025). The human being is the only creature that is endowed with the ability to take these larger implications into consideration and harmonize the action with the societal values and current regulatory systems. Consequently, the responsible and sustainable innovation requires human control in the decision-making processes in agriculture. The table 1 list the comparison of human intelligence and AI in agriculture

2.3 Limitations of Manual and Conventional Approaches

Although human intelligence cannot be neglected, inherent limitations are also present in the traditional manual and experience-based approaches to large-scale and data-intensive agricultural systems. Cognitive biases, limited memory, and inability to process huge volumes of heterogeneous data are limitations to human decision-making processes. Since agricultural processes become more large-scale and intricate, manual observation and intuition begin to be more and more inefficient and prone to errors (Vinod Chandra et al., 2024). Traditional agricultural methods are usually reactive in nature, in which responses are made when the issues become apparent. This procrastination may lead to huge loss in yields, high cost of inputs and degradation of the environment. Also, the process of manual tracking of crops and resources can be very labour-intensive and cannot be applicable to large farms or areas where there is a limited availability of workforce. These issues underscore the importance of analytical tools which can give timely and predictive information. The other weakness of entirely human-based methods is the inconsistency and scalability. Relying solely on personal experience in decision-making can also differ a great deal among the practitioners, thus resulting in inconsistencies. Moreover, when farmers retire or move, the valuable experience they have gained is frequently lost and this leads to institutional lapses (Conway et al., 2022). When well developed, AI systems may contribute to maintaining and sharing knowledge within regions and generations. Nevertheless, it is not feasible and desirable to substitute human intelligence with fully automated systems. Rather, the shortcomings of the manual systems highlight the necessity of group models that combine human judgement with machine intelligence. An integration of human capabilities including contextual insights and moral decision-making with the AI abilities in data interpretation and forecasting will help agricultural systems become more accurate, efficient, and resilient.

Table 1. Comparison of Human Intelligence and Artificial Intelligence in Agriculture

S.No.	Aspect	Human intelligence in agriculture	Artificial intelligence in agriculture	Reference
1	Perception & context	Uses experiential, tacit knowledge and local context; good with unusual situations, but limited scale	Uses sensors, satellites, and models to monitor large areas frequently; struggles with novel cases	(Holzinger et al., 2022)
2	Decision-making	Holistic, value-laden decisions; can integrate social and ethical considerations, but may be inconsistent	Data-driven, consistent recommendations for irrigation, fertilization, disease control, yield prediction	(Elbasi et al., 2022)
3	Labor & scalability	Flexible manual skills; constrained by fatigue, labor shortages, and time	Automation and robotics perform repetitive tasks (weeding, spraying, harvesting) at scale and	(Vinod Chandra et al., 2024)
4	Learning & knowledge	Relies on experience, local traditions; often undocumented and hard to transfer widely	Learns from big datasets; models can be copied and updated for many farms and regions	(Nath et al., 2024)
5	Sustainability & risk	Can balance yield with long-term soil, community and ethical concerns, but may resist innovation	Optimizes inputs and reduces waste; raises issues of cost, dependence on tech, and data risks	(Taneja et al., 2023)

3. ARTIFICIAL INTELLIGENCE TECHNIQUES FOR SMART AGRICULTURE AND FOOD SYSTEMS

The use of AI in agriculture and food systems has allowed shifting the traditional experience-oriented farming towards data-oriented and precision-focused farming (Camarena, 2020). AI methods help to analyze difficult agricultural data automatically, which allows identifying issues in time, using the resources optimally, and making decisions. This part of the review examines the most popular AI methods used in smart agriculture, such as ML, DL, computer vision, and sensor-based data processing, and discusses how these methods can be used to improve agricultural productivity and sustainability.

3.1 Machine Learning and Deep Learning Models in Agriculture

The convolutional neural networks (CNNs) are especially useful in image-oriented agricultural tasks, including plant and animal disease detection and weed identification, as well as the quality of fruits (Li et al., 2022; Ayyaswamy et al., 2025). The CNNs automatically create discriminative features of raw images without manual creation of features and enhance the classification accuracy. Time-series analysis Time-series RNNs and long short-term memory (LSTM) networks are typically applied to agriculture, such as weather prediction, crop development, and irrigation control (Kurumatani, 2020). These models represent temporal correlations in the sequential data, which makes them predictive of the future more accurately. Transformer-based architectures are also recent developments and are better at large agricultural datasets due to their increased scalability and performance. ML and DL models are effective even though they need significant amounts of large and quality data to train and validate. Class imbalance, noise, and domain variability are some of the problems that may impact model generalization.

3.2 Computer Vision and Sensor-Based Monitoring Systems

Computer vision is a critical element in automated visual analysis in the agricultural industry. Computer vision systems can be used to monitor the health of crops and animals, detect diseases, estimate biomass, and determine the level of maturity using images obtained with smartphones, drones, ground-based cameras, and satellites (Ayyaswamy et al., 2025; Cao et al., 2025). Such systems save time and effort and allow massive field inspection since they do not need manual field inspection. Under precision agriculture, multispectral and hyperspectral cameras mounted on unmanned aerial vehicles (UAVs) enable high-resolution images which are useful in detecting stress early and variable-rate input application. Such imagery vegetation indices, together with AI models, provide information about the state of plants and their activity. Computer vision applications are also common in post-harvest processes, such as in grading, sorting and defect detection of fruits and vegetables. Monitoring systems that are sensor-based complement computer vision by giving real-time measurements of parameters in the environment and the soil. The sensors of the IoT measure temperature, humidity, soil moisture, pH, and nutrient content. Table 2 list the different AI techniques used in smart agriculture with applications.

Table 2. AI Techniques and Their Applications in Smart Agriculture

S. No.	AI technique	Main smart-agriculture applications	Reference
1.	Machine learning (ML)	Crop yield prediction, crop/soil suitability, disease and weed detection, irrigation optimization	(Mirani et al., 2021)
2.	Deep learning (DL / CNNs)	Plant disease and pest detection, weed/crop discrimination, fruit counting, yield estimation from images	(Saleem et al., 2021)
3.	Computer vision	Seed and soil quality analysis, plant health and stress monitoring, weed management, livestock monitoring	(Dhanya et al., 2022)

continued on following page

Table 2. Continued

S. No.	AI technique	Main smart-agriculture applications	Reference
4.	IoT-enabled AI systems	Sensor-driven monitoring of soil, water, microclimate; smart irrigation; greenhouse and machinery automation	(Subeesh & Mehta, 2021)
5.	UAV/remote-sensing + AI	Large-scale crop and land-cover mapping, disease and stress detection, variable-rate input application	(Wang et al., 2022)

4. HUMAN–MACHINE SYNERGY AND COGNITIVE COLLABORATION FRAMEWORKS

4.1 Human-in-the-Loop and Collaborative Intelligence Models

HITL techniques help farmers and experts to do data labelling, train models, and validation and final decisions within the agriculture sector. This interaction increases the precision of the models and does not inhibit the human control and responsibility. Collaborative intelligence is another evolution of the HITL concept: this one enables human and AI systems to engage with each other in a continuous manner throughout the decision lifecycle (Elia et al., 2025). The forecast or suggestions generated by AI systems in this kind of model are evaluated, refined or disregarded by human beings, based on the knowledge and experience of the situation. The human input feedback is then used to improve and optimize the AI models to create a dynamic learning process (Song et al., 2024). This iterative process is particularly applicable in agriculture where conditions differ by season, geographically and also between varieties of crops. As an illustration, in crop disease detection, the AI model can be trained to detect possible infections on the basis of image processing, whereas the agricultural specialists verify or refute such infections. This partnership minimizes the number of false positives and increases the level of confidence in the system. Moreover, HITL frameworks can be used to mitigate data scarcity and bias by including an expert judgment in a situation where data have few or few labels. Consequently, human-machine partnership results in stronger and more flexible solutions of AI in agriculture. The Figure 2 shows the human–AI cognitive collaboration framework for Smart Agriculture.

4.2 Decision Support Systems for Agricultural Applications

Human-machine synergy in agriculture has been applied in decision support systems (DSS). DSS are made to support the user, not to substitute them, as opposed to fully automated systems, and provide recommendations with the surrounding contextual information (Jin et al., 2025). Irrigation scheduling, nutrient management, pest control and harvest planning are some applications of DSS in smart agriculture. AI models take sensor data, weather predictions, and historical data to come up with optimal suggestions and human operators decide on the suggestions, depending on the conditions on the ground, economic factors, and risk factors. Effective human-AI interaction is impossible without the involvement of visualization tools, dashboards, and alerts that help to make the complex data and model outputs clear to non-technical users. Usability and transparency are the keys to the effectiveness of agricultural DSS (Htun et al., 2022). The systems that explain the recommendations properly and give the opportunity to see other situations have a higher chance of being adopted. Also, DSS may be used to support the knowledge sharing process by recording the expert decisions and make them available to less skilled farmers. This is an ability that facilitates capacity building and standard decision making in the various farming situations. In the industrial context, DSS allow the human control of AI technologies that can be scaled. Integrating AI into the user-based decision-making systems, agricultural organizations will be able to become more efficient and resilient without becoming less accountable.

4.3 Integration of Human Feedback with AI Predictions

Human feedback is a vital element to the successful human-machine synergy. Feedback in agriculture may be explicit, e.g., by hand-correction of model predictions, or implicit, e.g., by modifying a farmer based on AI advice. The adaptability of the model in such a way is enabled by feedback-based learning and mitigates the decline in performance due to alterations in environmental conditions (Song et al., 2024). An example of this is when an AI system fails to identify a specific crop disease in a specific condition of light, feedback by experts can be utilized to retrain the model or alter preprocessing. This is an ever-improving process that makes AI systems useful and precise in the actual agricultural environment. Personalization of AI tools is also achieved by the use of feedback. Automation is necessary to balance the user efforts to avoid cognitive overload. The issues can be addressed with the help of well-designed human-machine synergy frameworks that focus on simplicity, transparency, and meaningful interaction.

Figure 2. Human–AI Cognitive Collaboration Framework for Smart Agriculture

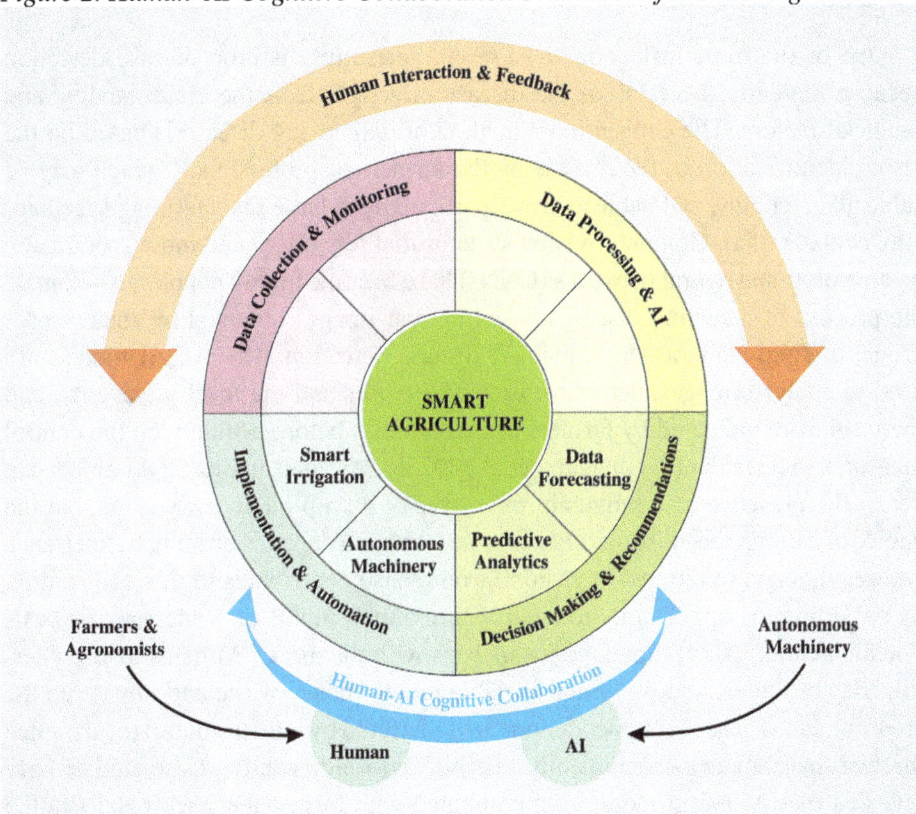

5. APPLICATIONS AND CASE STUDIES IN AGRICULTURE AND FOOD SYSTEMS

The practical use of human-AI cognitive cooperation is realized in the sphere of the real world of agricultural and food systems (Jariwala, 2025). Collaborative systems can be used to intervene in a timely manner, manage resources better, and increase the reliability of the decisions that are made by combining human knowledge with AI-driven analytics (Holzinger et al., 2022). This section outlines some of the main areas of application and some of the typical case scenarios that have seen human-machine synergy being of substantial benefit, especially in crop disease detection, precision farming, yield forecasting, and food quality estimation.

5.1 Crop Disease Detection and Precision Farming

One of the most influential AI uses in agriculture is crop disease detection because the early detection of the disease directly affects the yield, quality, and financial factors. The conventional method of diagnosing disease is based on the visual identification of the disease by the farmers or professionals which may be subjective, lengthy, and liable to errors particularly in large scale farming (Seethapathy et al., 2025a). Computer vision systems that rely on AI and more specifically on convolutional neural networks (CNNs) have become highly popular to automate the process of detecting diseases based on leaf images obtained by smartphone, drone, or field cameras. In human-AI disease detection systems, AI models are used to analyze images, detect patterns of diseases and the level of severity, and the results are validated by farmers or agronomists before deciding on the control measures to be taken (Seethapathy et al., 2025b). As an example, an AI model can detect the presence of a fungus in the leaves of a crop on an early stage, but the choice of a fungicide, dose, or other organic treatment depends on human experience and regulatory limitations. Precision farming also contributes to this cooperation by providing the opportunity to control agricultural inputs on a site-specific basis (Getahun et al., 2024). Spatial data analysis with the use of AI helps to apply fertilizers, pesticides, and water at a variable rate to reduce waste and impact on the environment. These recommendations are monitored by human control to guarantee the fact that they are economically feasible and field-specific. Case studies have revealed that AI-based monitoring combined with farmer feedback results in the enhanced efficiency of disease management and crop productivity in addition to the reduction of chemicals. The flow diagram of AI-driven crop disease detection with explainable given in Figure 3.

5.2 Yield Prediction and Resource Optimization

Farm planning, market prediction and supply chain management are dependent on the accurate prediction of yields. AI models use previous yield data, weather conditions, and soil type and crop management strategies to make predictions of anticipated output. Regression models and recurrent neural networks are ML and DL methods that have proven to be highly predictive with regard to yield estimation (Rashid et al., 2021). Nonetheless, predicting yield in the model is usually a problem that is associated with variability and uncertainty of data. Human-AI partnership is a solution to these issues because it uses the experience of farmers and local information to integrate into prediction models. Farmers are able to give contextual data on dates of planting, pest pressure or unforeseen weather conditions that may not be well represented in datasets. Such data is useful in calibration of models and also in predicting reliability. Yield prediction is directly associated with resource optimization because the effective utilization of water, fertilizers and energy directly influences productivity and sustainability (Xing & Wang, 2024). The decision support systems based on AI process sensor data and provide the best irrigation schedules and nutrient application rates. Human decision-makers consider these recommendations against the economic limitations, labor supply, and soil health in the long-term. Research has indicated that those farms implementing collaborative AI-based strategies of optimizing resources realize substantial water savings and input cost reduction without affecting the output. These applications underscore the importance of the human-machine synergy in the balancing between the productivity and the sustainability objectives. Predictive analytics combined with human judgment will help the agricultural systems to adjust to the changing conditions and use the resources more efficiently.

5.3 Food Quality Assessment and Supply Chain Monitoring

In addition to the main production, the use of human-AI cognitive cooperation in post-harvest processing, food quality evaluation, and monitoring supply chain is crucial. Food quality and food safety is a complicated process that entails inspection, grading, storage, transportation, and distribution (Wu & Zhong, 2025). Manual inspection processes are labour intensive and might not be consistent especially in a large-scale operation. Computer vision systems based on AI are becoming common in automated detection of defects and grading of fruits, vegetables, and processed food products. These systems measure size, color, texture and surface defects with a high level of accuracy. Nevertheless, human inspectors are still needed to check AI outputs, deal with unclear situations and adherence to quality standards. This is a cooperative method which is more efficient and quality assuring. In the context of supply chain monitoring, AI models process sensor, logistics platform, and market system information to monitor the movement of products, the conditions of their storage, and demand trends. Human control can guarantee that AI-based decisions are taken into consideration of regulatory needs, ethical sourcing, and consumer preferences (Du & Xie, 2021). As an example, the temperature monitoring systems can alert operators that the cold-chain might be interrupted and take corrective measures on a timely basis depending on the judgment of human operators. Transparency and accountability are also facilitated by human-AI collaboration and are becoming significant in the contemporary food systems.

Figure 3. AI-Driven Crop Disease Detection with Explainable Outputs

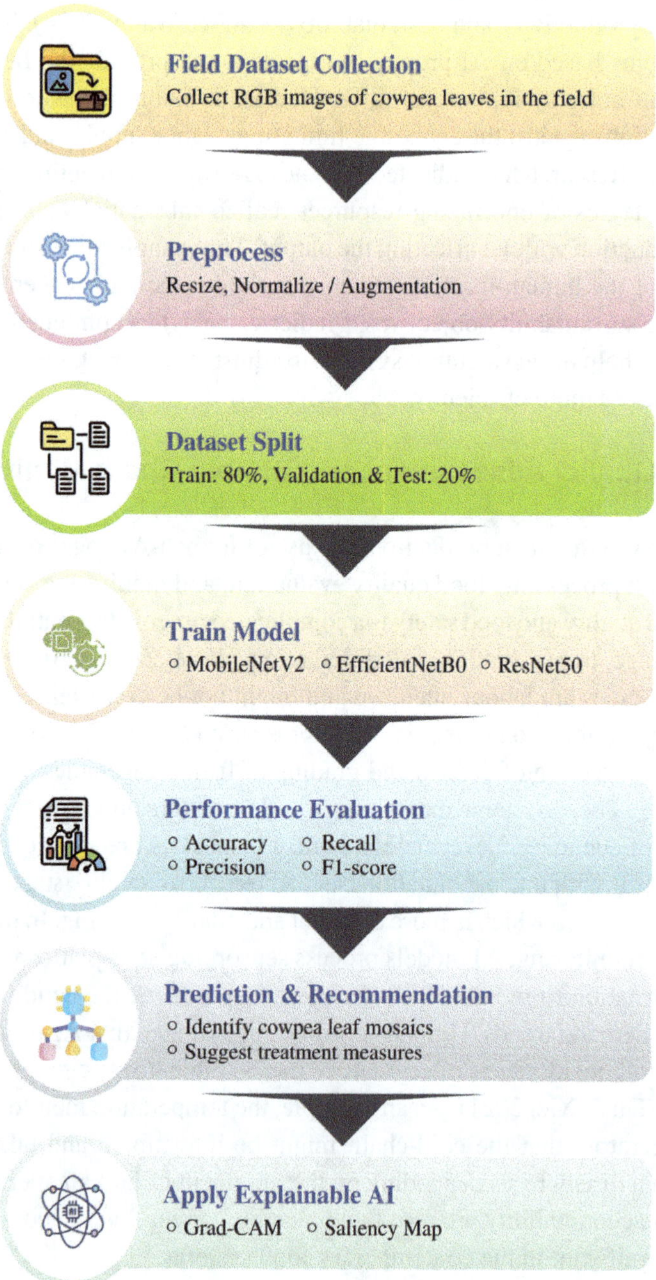

6. EXPLAINABLE AI, TRUST, AND ETHICAL CONSIDERATIONS

6.1 Explainable AI Techniques for Agricultural Models

Explainability is especially relevant in the agricultural sector since farmers, agronomists, and policymakers, as stakeholders, need to clarify the reasons behind AI-generated recommendations before acting on them. Even very precise models can be discarded without further clarification or can be poorly used. Some of the common XAI methods in agriculture are feature importance analysis, saliency maps, gradient-based visualization, and model-agnostic explanation methods (Kaler & Kaur, 2025). In the case of image-based tasks, like crop disease detection, visualization techniques can be used to identify which parts of the plant images are the most important to prediction by a model. These graphical explanations enable the farmers to confirm whether the AI models concentrate on biologically relevant symptoms as opposed to useless background facts. Equally, feature attribution algorithms in sensor-based and tabular data applications can be used to determine the important environmental or management variables that determine predictions. Through the XAI in the human-AI collaborative systems, users are able to test model behavior, identify the possible errors and trust the AI results. Consequently, XAI can be used to bridge the gap between the difficult ML models and human cognitive comprehension, enhancing cooperation and responsibility.

6.2 Trust, Transparency, and User Acceptance

Another basic factor that determines technology adoption in agriculture is trust. More importantly, farmers and practitioners are more inclined to use AI-driven tools when they feel that they are reliable, transparent, and do not contradict their values and practices. Trust cannot be only established based on the performance metrics but is formed as a result of the reliability of the systems (Mallinger & Baeza-Yates, 2024). The AI systems should be transparent enough so that users can know how the data are gathered, processed and utilized to come up with recommendations. The transparency of human-in-the-loop frameworks is increased by the ability to engage with AI models and make feedback and see the impact of input on the result. This interaction changes AI systems into black-box decision-makers to that of a partner. System usability and relevance also affect user acceptance. AI tools that are designed to work well with the current workflows and are displayed in the form of user-friendly presentations will be better accepted. Training and capacity-building programs are also vital to trust building through enhancing digital literacy and empowering users to work with AI technologies in a confident manner. In agriculture, trust goes further than an individual user to an institution and community

(Gruber, 2020). Open governance structures, articulated benefits and restrictions, as well as engagement of the local stakeholders in the design of the system, promote a communal trust. The human-AI cognitive collaboration models, which are based on trust and transparency, eventually enable the sustainability and inclusion of AI in the agricultural sector. The figure 4 shows the graphical relationship of trust, explainability, and ethics in human–AI agricultural systems.

6.3 Ethical, Social, and Sustainability Implications

The ethical application of AI in agriculture must pay special attention to the data ownership, privacy, fairness, and environmental consideration. Agricultural data are usually sensitive in terms of land use, productivity and economic conditions (Uddin et al., 2024). Privacy of data and ownership of the farmers are necessary to avoid abuse and exploitation. Prejudice and impartiality are other ethical issues because AI trained using a small or unrepresentative amount of data can offer unfair results. Indeed, as an illustration, models that are constructed based on big commercial farms are not likely to be well generalized to smallholder or resource-limited farming systems. Such biases can only be detected and countered through human supervision and involvement of various stakeholders. Ethical AI adoption in agriculture revolves around sustainability issues. Although AI-based optimization can help to consume less and produce less harmful effects on the environment, the improper design of the system can promote over-intensification or overuse of chemicals. The table 3 list the benefits and challenges of human–AI collaboration in agriculture.

Figure 4. Trust, Explainability, and Ethics in Human–AI Agricultural Systems

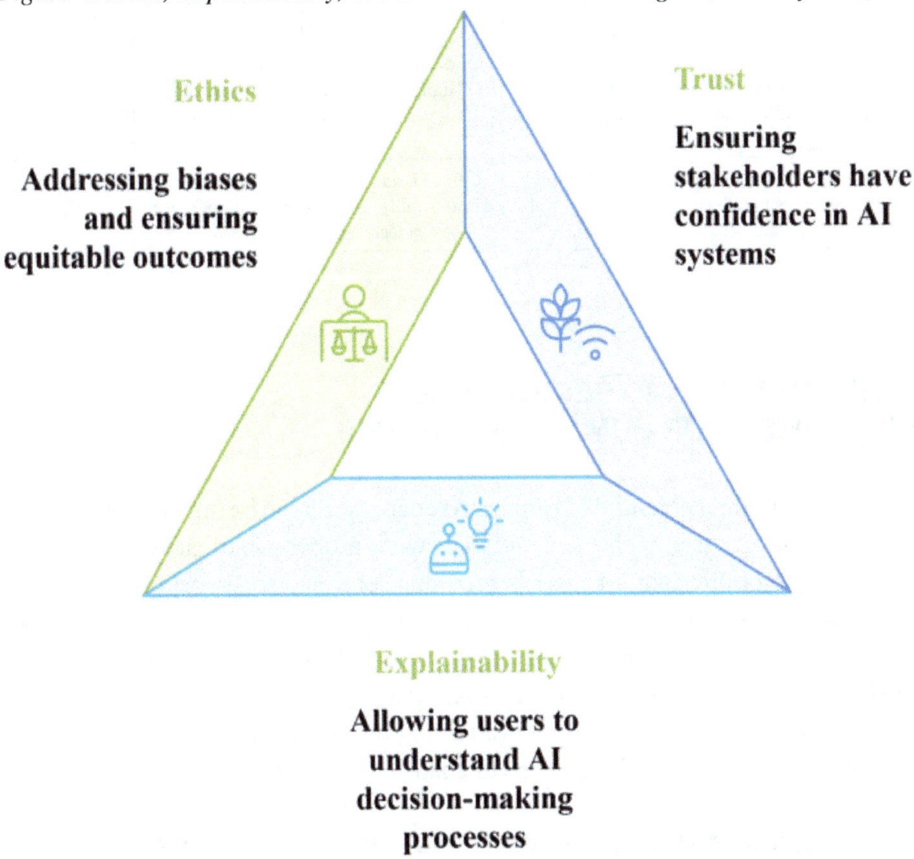

Table 3. Benefits and Challenges of Human–AI Collaboration in Agriculture

S. No.	Benefit	Challenge	Reference
1.	Higher productivity and farmer field level support system	Unequal access to AI can widen gaps between large and small farms	(Elbasi et al., 2022)
2.	Reduced physical workload and improved safety through human–robot collaboration in repetitive or hazardous tasks	Safety, ergonomics, and reliability of collaboration in unstructured field environments remain open issues	(Benos et al., 2023)
3.	Human-centered, explainable AI increases trust and usability of decision support for farmers	Low digital literacy and lack of interdisciplinary design reduce adoption and effective use	(Ryan et al., 2023)

continued on following page

Table 3. Continued

S. No.	Benefit	Challenge	Reference
4.	AI can support sustainability goals: optimized inputs, less waste, climate-smart management	High costs, infrastructure gaps, and "digital divide" limit benefits in low-resource contexts	(Maraveas, 2022)
5.	New AI-enabled business models and services for advisory, monitoring, and value-chain optimization	Ethical issues: data ownership, fairness, responsibility, and concentration of power in tech/agribusiness firms	(Ryan et al., 2023)

7. EVALUATION AND PERFORMANCE ASSESSMENT OF HUMAN–AI SYSTEMS

However, the real potential of Human-AI cooperation can be fulfilled only in case these systems are strictly checked and their work properly estimated. Assessment also serves to ensure that AI models provide useful and trustworthy information as well as to determine gaps, constraints, and aspects of improvement in system design (Fischer et al., 2020). In this chapter section, the methodologies, metrics and considerations of evaluating the Human-AI systems are discussed with three major aspects of metrics and benchmarks, validation in real-world agricultural context, and the importance of feedback loops to enable continuous improvement.

7.1 Metrics and Benchmarks for Human–AI Collaboration

The Human-AI systems analysis in the agricultural sector needs a complex of metrics that will examine both the quality of AI models and the efficiency of human-AI interaction (Holzinger et al., 2024). Classical AI performance measures, including accuracy, precision, recall, F1-score, and area under the curve (AUC) are still important to determine the predictive performance of ML and DL models. To illustrate, in crop disease detection, these measures give information about the capacity of the model to identify infected and healthy plants correctly. On the same note, regression measures such as mean absolute error (MAE) and root mean squared error (RMSE) are popular in tasks of predicting yields and optimizing resources. Nonetheless, Human-AI systems are not only to be evaluated within the traditional AI metrics (Holzinger et al., 2024). It is also vital to have performance indicators which are human-focused and include decision-making efficiency, cognitive load, user satisfaction, and trust in AI recommendations. An example is given of a highly accurate model which can in itself be ineffective when the outputs cannot be easily interpreted or translated by farmers. Thus, time to decision, decision errors in human-verified decisions, and perceived usefulness can be used to offer a good insight into the usability and practical efficiency of collaborative systems. Standard evaluation systems and benchmark datasets are essential in the evaluation of performance. Agro image datasets, soil moisture data and weather data are publicly available and can be used to perform a uniform assessment and comparison of AI models across studies. However, benchmarks also need to take into account diversity of farming practices, climatic conditions, and local types of crops so that to make sure that models are generalized. Also, simulation environments and testbeds may offer controlled environments in which system responses are studied in different scenarios, which would aid in quantifying robustness and resilience.

7.2 Validation of AI Models in Real Agricultural Scenarios

Although metrics and benchmarks are necessary, the real-life validation is important to evaluate the practicality of the Human-AI systems. Agricultural environments are too variable, uncertain and complex, to be fully represented in controlled laboratory or simulation-based tests (Cetegen & Stuber, 2021). Field-based validation enables the researcher to validate AI models in the real farming environment, where researchers can manipulate dependent variables, including different soil types, weather conditions, pest infestation, and random human actions. Real world validation requires the use of AI models with human operators in the operational environment. As an example, a farm field can be used to test a precision irrigation system, which is an AI-based prediction, where the actual water application, crop growth, and yield are measured. In the same manner, the model of disease detection can be used in crops that are subjected to stressors in the natural environment to determine their strength. Watching the farmers engage with the AI suggestions, change the system outputs, and incorporate the insights into their decision-making process will be important information regarding usability, acceptance, and effectiveness (Cetegen & Stuber, 2021). AI is likely to give false positives and negatives, and it is critical to know how such mistakes are applied in real-world settings. In this case, a false diagnosis will result in unjustified use of pesticides or loss of interventions. Assessment of the effects of such mistakes on human judgments as well as the productivity of the farms in general, aids in determining the risk reduction methods. In addition, practical validation gives the AI models feedback on how to undergo iterative refinement to make them adaptive and contextual.

7.3 Feedback Loops and Continuous Improvement

Human-AI systems are dynamic and their performance may change as time passes with increased availability of data and user interactions influence system behavior.

Continuous learning, improvement and adaptation require feedback loops to work in collaborative systems. These loops can be characterized by the gathering of information on the outputs of the AI system and human reactions, performance trend analysis, and the revision of models or interfaces. Farmer annotations, retraining of the models based on new environmental information, or sensor recalibration are some of the mechanisms that can be applied in agriculture to apply feedback loops (Wang et al., 2022). In the case of a farmer correcting a disease detection warning that an AI model has made, that data can be included in the training data to allow subsequent predictions to be more accurate. Similarly, system changes can be made by use of performance logs, user interaction metrics as well as error tracking to enhance its usability and reliability. Continuous improvement is also present and it involves the use of explainable AI (XAI) that fosters transparency and trust. The visualizations, the scores of confidences and the reasoning applied to AI suggestions can empower farmers to be more aware of how models behave and assure the results with their own expertise (Wang et al., 2022). Further, it is the long-term learning that can be attributed to feedback loops and AI systems can adapt to dynamic agricultural conditions, seasonal changes and evolving pest or disease trends.

8. CHALLENGES AND LIMITATIONS OF HUMAN–AI COLLABORATION IN AGRICULTURE

The adoption of AI into the agriculture processes is challenged and limited despite its massive potential. These challenges are of technical, human and socio-economic levels, which influence the adoption, performance and stability of Human-AI collaborative systems. These issues are important to understand so as to create more robust, equitable, and viable solutions to precision agriculture. The central limitations and challenges are described in this chapter section in three categories, namely, technical and infrastructural, human factors, and socio-economic and regulatory barriers.

8.1 Technical and Infrastructural Challenges

Agricultural Human-AI systems are dependent on the presence of strong technical infrastructure to be successfully implemented. The modern AI methodology, especially ML models and DL models need huge amounts of quality data to be trained and tested. Nevertheless, agricultural statistics tend to be non-homogeneous, incomplete or inconsistent because soil types, weather patterns, crop types and farming methods vary (Ahady et al., 2025). In most instances, the gathering of real-time field information by use of IoT sensors, drones, and remote sensing technologies is limited by the aspect of connectivity, equipment prices, and the nature of the environment. The infrastructural issues are especially high in rural and developing areas, where the internet connection is not so high or is not stable. AI services based on clouds that are commonly employed to analyze massive data and provide predictive analytics are reliant on consistent internet access (Ahady et al., 2025). The actual functionality of AI-generated suggestions is limited in places with intermittent connectivity. These constraints could be reduced with edge computing, although it makes system design and maintenance more complicated. The other major technical constraint is associated with AI model generalization and flexibility. Numerous AI models are trained on certain datasets and cannot work well in a variety of agricultural conditions.

8.2 Human Factors and Cognitive Constraints

Human factors are important in the adoption of AI in agriculture. The factors that will influence the acceptance of AI recommendations among farmers are the trust, perceived usefulness and ease of use (Holzinger et al., 2024). Both cognitive limitations and biases and variations in technical literacy may hinder human-AI collaboration. This can reduce trust and reduce adoption as seen in cases where the output of the complex AI or visualizations, which are difficult to understand, is provided. Also, automated recommendations may be doubted by farmers particularly in cases where the outcome is against their experience or experience. Human-in-the-loop approach where human input is applied to AI decision-making is significant to give context-specific recommendations, which are precise. However, it is possible to overload human operators with the AI system when they generate excessive alerts or require humans to intervene with the system too frequently (Holzinger et al., 2024). The other weakness is based on the fact that AI use will require constant human monitoring in some applications. Although AI can be used to automate certain processes like irrigation scheduling or pest detection, the contextual decision-making process like the assessment of unusual weather conditions or the interpretation of local market needs cannot be automated (Fertier et al., 2020). The issue of automation and human control is hard to strike the right balance because insufficient supervision can lead to errors, whereas excessive supervision can diminish AI effectiveness advantages. Human decision-making is susceptible to cognitive biases: confirmation bias, excessive dependence on conventional heuristics, and so on can intervene in the process of effective collaboration and restrain the potential of AI-assisted farming.

8.3 Socio-Economic, Regulatory, and Adoption Barriers

In addition to technical and human factors, socio-economic and regulatory issues can be very influential on the implementation of Human-AI collaborative systems in agriculture. Smallholder farmers or cooperatives would not be able to afford the high cost of AI technologies, such as IoT sensors, drones, computational hardware, and cloud services (Dhillon & Moncur, 2023). Unequal access is typically caused by financial constraints, which result in differences between the large-scale commercial farms and small subsistence farmers in terms of technological adoption. Such disparities can only widen the socio-economic disparities and reduce the large-scale adoption of AI-based agriculture. There are also regulatory issues that curtail the use of Human-AI systems. Data privacy, security and ownership are some of the primary concerns particularly when data in the farm is sent to cloud-based systems. The farmers might not be willing to provide sensitive data on land utilization, yield, or production techniques, thereby reducing the access to all-inclusive datasets to train AI models (Dhillon & Moncur, 2023). Also, not having uniform policies that would regulate the use of AI in the agricultural sector in terms of ethical, safety, and liability may present a legal gray area. The issue of responsibility appears when AI-based decisions lead to crop failures, resources gone to waste or unforeseen environmental effects. There can be no technical assistance, incentives, or training programs available to farmers. The extension services that have long been the source of technology transfer in the agricultural industry might not be ready to offer guidance to AI-based systems (Usigbe et al., 2024). Automation may be slowed down by resistance to change and perceived risk of automation. To conclude, the difficulties and constraints of Human-AI collaboration in the agricultural industry are complex, and they include technical infrastructure, human mental, and socio-economic and regulatory aspects. To overcome these issues, strategic approaches must be multifaceted involving technological innovation, user-focus, policy backing and capacity building programs.

9. FUTURE RESEARCH DIRECTIONS AND CONCLUSION

Human-AI cognitive cooperation is a revolutionary paradigm of developing agriculture and food systems. Although a lot has been done in the application of AI to decision-making in agriculture, there are still various research opportunities to explore more to improve collaboration, trust, and sustainability. This part summarizes the most important research directions of the future and summarizes the contribution of the chapter.

9.1 Emerging Trends in Human–AI Collaboration for Agriculture

Future studies on the collaboration of humans and AI in agriculture are likely to be directed to adaptive, context-aware and user-centric systems. The creation of customized AI models, which are adjusted to the circumstances of a particular farm, management, and preferences of a farmer, is one of the emerging trends (Holzinger et al., 2024). These systems have the capacity to dynamically revise their suggestions and become more relevant with time due to continuous human feedback. The other potential trend is the multimodal combination of data, such as imagery, sensor data, textual evidence, and farmer notes. The improved methods of fusion can be used to provide a better situational awareness and better decisions, especially in the situations which are uncertain and changing at a fast pace. Explainable and interpretable AI models will remain an important factor in subsequent studies. The trust and adoption can be further supported by efforts to create inherently interpretable models and domain-specific methods of explanation. Furthermore, cross-cultural studies of agricultural science, cognitive science, and human-computer interaction will be needed to develop easy and practical collaboration models.

9.2 Industrial Impact and Policy Implications

The agricultural value chain, which includes production, processing, distribution, and consumption, is affected by the industrial effect of the human-AI cognitive co-operation. The AI-based decision support systems can help to enhance the efficiency of operations, decrease the number of resources wasted, and promote the quality of products at a large scale (Khosravy et al., 2023). In the case of agribusiness and technology suppliers, collaborative models provide chances to provide scalable solutions that can be flexible to local situations. Policy implications are also very important. Regulatory agencies and governments can be instrumental in ensuring the provision of enabling environments to responsible AI in agriculture. The adoption can be accelerated by means of the policies that favour the development of the digital infrastructure, interoperability of data, and training farmers to be inclusive. Transparency, accountability, and data sovereignty should be regarded as ethical principles and standards of agricultural AI to secure the interests of stakeholders (Khosravy et al., 2023). The role of human-AI collaboration can be enhanced further because of the ability of the partners to make the technological advancement more significant and relevant to the needs of the society through the application of the public-private partnerships and participatory innovation models. The AI solutions can be more adapted to the real-world challenges and sustainability objectives through the involvement of farmers, researchers, industry stakeholders, and policymakers in the co-design processes.

10. CONCLUSION

The chapter has discussed how human-AI cognitive collaboration can be used to improve agriculture and food systems. The chapter reveals the significance of the synergy between human expertise and machine intelligence by analyzing the human intelligence, AI methods, collaborative systems, practical uses, and the ethical concerns. Instead of substituting human decision-makers, AI technologies are potent weapons that can provide a significant addition to human abilities to make informed, transparent, and sustainable agricultural activities. Explainable, trustworthy and ethically controlled AI systems should be integrated to achieve long-term adoption and impact. With agriculture ever changing to meet the global demands, human-AI cognitive collaboration provides a resilience and flexible means of improving productivity, sustainability and food security. The future of this paradigm in developing the future of agriculture and food systems will be dependent on continued research, interdisciplinary collaboration and responsible governance.

ACKNOWLEDGEMENT

The authors are grateful to the ANRF, New Delhi, India, for funding the research on Light Weight Deep Learning Based Mobile Application for the Early Detection, Identification, and Spatiotemporal Monitoring of Plant Diseases (Grant No. SUR/2022/004268).

REFERENCES

Ahady, A. B., Klopries, E. M., Schuttrumpf, H., & Wolf, S. (2025). Drought analysis methods: A multidisciplinary review with insights on key decision-making factors in method selection. *Water (Basel)*, *17*(15), 2248. DOI: 10.3390/w17152248

Ayyaswamy, K., Gobinath, V. M., Kathirvel, N., & Maheswaran, C. P. (2025). AI-Based Automated Disease Detection Based on Symptoms Healthcare for Pets and Birds. In AI and the Revival of Big Data (pp. 311-346). IGI Global Scientific Publishing. https://www.igi-global.com/chapter/ai-based-automated-disease-detection-based-on-symptoms-healthcare-for-pets-and-birds/369506

Bao, Y., Gong, W., & Yang, K. (2023). A literature review of human–AI synergy in decision making: From the perspective of affordance actualization theory. Systems, 11(9), 442. https://www.igi-global.com/chapter/ai-innovations-in-sustainable-food-production-and-agriculture/385485

Benos, L., Moysiadis, V., Kateris, D., Tagarakis, A. C., Busato, P., Pearson, S., & Bochtis, D. (2023). Human–robot interaction in agriculture: A systematic review. *Sensors (Basel)*, *23*(15), 6776. DOI: 10.3390/s23156776 PMID: 37571559

Camarena, S. (2020). Artificial intelligence in the design of the transitions to sustainable food systems. *Journal of Cleaner Production*, *271*, 122574. DOI: 10.1016/j.jclepro.2020.122574

Cao, Z., Sun, S., & Bao, X. (2025). A review of computer vision and deep learning applications in crop growth management. *Applied Sciences (Basel, Switzerland)*, *15*(15), 8438. DOI: 10.3390/app15158438

Cetegen, S. A., & Stuber, M. D. (2021). Optimal design of controlled environment agricultural systems under market uncertainty. *Computers & Chemical Engineering*, *149*, 107285. DOI: 10.1016/j.compchemeng.2021.107285

Conway, S. F., Farrell, M., McDonagh, J., & Kinsella, A. (2022). 'Farmers don't retire': Re-evaluating how we engage with and understand the 'older' farmer's perspective. *Sustainability (Basel), 14*(5), 2533. DOI: 10.3390/su14052533

Dhanya, V. G., Subeesh, A., Kushwaha, N. L., Vishwakarma, D. K., Kumar, T. N., Ritika, G., & Singh, A. N. (2022). Deep learning-based computer vision approaches for smart agricultural applications. *Artificial Intelligence in Agriculture, 6*, 211–229. DOI: 10.1016/j.aiia.2022.09.007

Dhillon, R., & Moncur, Q. (2023). Small-scale farming: A review of challenges and potential opportunities offered by technological advancements. *Sustainability (Basel), 15*(21), 15478. DOI: 10.3390/su152115478

Du, S., & Xie, C. (2021). Paradoxes of artificial intelligence in consumer markets: Ethical challenges and opportunities. *Journal of Business Research, 129*, 961–974. DOI: 10.1016/j.jbusres.2020.08.024

Elbasi, E., Mostafa, N., AlArnaout, Z., Zreikat, A. I., Cina, E., Varghese, G., Shdefat, A., Topcu, A. E., Abdelbaki, W., Mathew, S., & Zaki, C. (2022). Artificial intelligence technology in the agricultural sector: A systematic literature review. *IEEE Access : Practical Innovations, Open Solutions, 11*, 171–202. https://ieeexplore.ieee.org/abstract/document/9999442. DOI: 10.1109/ACCESS.2022.3232485

Elia, M., Ziethmann, P., Krumme, J., Schlogl-Flierl, K., & Bauer, B. (2025). Responsible AI, ethics, and the AI lifecycle: How to consider the human influence? *AI and Ethics, 5*(4), 1–18. https://link.springer.com/article/10.1007/s43681-025-00666-z. DOI: 10.1007/s43681-025-00666-z

Fertier, A., Barthe-Delanoe, A. M., Montarnal, A., Truptil, S., & Benaben, F. (2020). A new emergency decision support system: The automatic interpretation and contextualisation of events to model a crisis situation in real-time. *Decision Support Systems, 133*, 113260. DOI: 10.1016/j.dss.2020.113260

Fischer, L., Ehrlinger, L., Geist, V., Ramler, R., Sobiezky, F., Zellinger, W., Brunner, D., Kumar, M., & Moser, B. (2020). AI system engineering—Key challenges and lessons learned. *Machine Learning and Knowledge Extraction, 3*(1), 56–83. DOI: 10.3390/make3010004

Getahun, S., Kefale, H., & Gelaye, Y. (2024). Application of precision agriculture technologies for sustainable crop production and environmental sustainability: A systematic review. *TheScientificWorldJournal, 2024*(1), 2126734. https://onlinelibrary.wiley.com/doi/full/10.1155/2024/2126734. DOI: 10.1155/2024/2126734 PMID: 39421732

Gruber, S. (2020). Personal trust and system trust in the sharing economy: A comparison of community-and platform-based models. *Frontiers in Psychology*, *11*, 581299. DOI: 10.3389/fpsyg.2020.581299 PMID: 33362644

Gupta, P., Nguyen, T. N., Gonzalez, C., & Woolley, A. W. (2025). Fostering collective intelligence in human–AI collaboration: Laying the groundwork for COHUMAIN. *Topics in Cognitive Science*, *17*(2), 189–216. https://onlinelibrary.wiley.com/doi/full/10.1111/tops.12679. DOI: 10.1111/tops.12679 PMID: 37384870

Hassija, V., Chamola, V., Mahapatra, A., Singal, A., Goel, D., Huang, K., Scardapane, S., Spinelli, I., Mahmud, M., & Hussain, A. (2024). Interpreting black-box models: A review on explainable artificial intelligence. *Cognitive Computation*, *16*(1), 45–74. https://link.springer.com/article/10.1007/s12559-023-10179-8. DOI: 10.1007/s12559-023-10179-8

Holzinger, A., Fister, I., Kaul, H. P., & Asseng, S. (2024). Human-centered AI in smart farming: Toward agriculture 5.0. *IEEE Access : Practical Innovations, Open Solutions*, *12*, 62199–62214. https://ieeexplore.ieee.org/document/10510441/. DOI: 10.1109/ACCESS.2024.3395532

Holzinger, A., Saranti, A., Angerschmid, A., Retzlaff, C. O., Gronauer, A., Pejakovic, V., Medel-Jimenez, F., Krexner, T., Gollob, C., & Stampfer, K. (2022). Digital transformation in smart farm and forest operations needs human-centered AI: Challenges and future directions. *Sensors (Basel)*, *22*(8), 3043. DOI: 10.3390/s22083043 PMID: 35459028

How, M. L., Cheah, S. M., Chan, Y. J., Khor, A. C., & Say, E. M. P. (2020). Artificial intelligence-enhanced decision support for informing global sustainable development: A human-centric AI-thinking approach. *Information (Basel)*, *11*(1), 39. DOI: 10.3390/info11010039

Htun, N. N., Rojo, D., Ooge, J., De Croon, R., Kasimati, A., & Verbert, K. (2022). Developing visual-assisted decision support systems across diverse agricultural use cases. *Agriculture*, *12*(7), 1027. DOI: 10.3390/agriculture12071027

Jariwala, M. (2025). AI and Data Science in Sustainable Agriculture and Food Production. In Artificial Intelligence and Data Science for Sustainability: Applications and Methods (pp. 123-158). IGI Global Scientific Publishing. https://www.igi-global.com/chapter/ai-and-data-science-in-sustainable-agriculture-and-food-production/369749

Jin, C., Xiao, Y., Wu, H., Ji, X., Li, G., Shuai, J., Yang, P., & Xiong, L. (2025). Human-model interaction-based decision support system for optimizing food safety assessment. *Food Research International*, *208*, 116156. DOI: 10.1016/j.foodres.2025.116156 PMID: 40263826

Kaler, B., & Kaur, A. (2025). A systematic survey on explainable artificial intelligence (XAI) for plant health monitoring: Challenges and opportunities. *Applied Intelligence*, *55*(12), 889. https://link.springer.com/article/10.1007/s10489-025-06790-x. DOI: 10.1007/s10489-025-06790-x

Khosravy, M., Gupta, N., Pasquali, A., Dey, N., Crespo, R. G., & Witkowski, O. (2023). Human-collaborative artificial intelligence along with social values in industry 5.0: A survey of the state-of-the-art. *IEEE Transactions on Cognitive and Developmental Systems*, *16*(1), 165–176. https://ieeexplore.ieee.org/document/10288567. DOI: 10.1109/TCDS.2023.3326192

Kurumatani, K. (2020). Time series forecasting of agricultural product prices based on recurrent neural networks and its evaluation method. *SN Applied Sciences*, *2*(8), 1434. https://link.springer.com/article/10.1007/s42452-020-03225-9. DOI: 10.1007/s42452-020-03225-9

Li, Y., Wen, W., Miao, T., Wu, S., Yu, Z., Wang, X., Guo, X., & Zhao, C. (2022). Automatic organ-level point cloud segmentation of maize shoots by integrating high-throughput data acquisition and deep learning. *Computers and Electronics in Agriculture*, *193*, 106702. DOI: 10.1016/j.compag.2022.106702

Liao, C., Ross, H., Jones, N., & Palaniappan, G. (2024). System lacks systems thinking: Top-down organization and actor agency in China's agricultural extension system. *Systems Research and Behavioral Science*, *41*(1), 82–99. https://onlinelibrary.wiley.com/doi/full/10.1002/sres.2943. DOI: 10.1002/sres.2943

Mallinger, K., & Baeza-Yates, R. (2024). Responsible AI in Farming: A Multi-Criteria Framework for Sustainable Technology Design. *Applied Sciences (Basel, Switzerland)*, *14*(1), 437. DOI: 10.3390/app14010437

Mallinger, K., Corpaci, L., Neubauer, T., Tikasz, I. E., Goldenits, G., & Banhazi, T. (2024). Breaking the barriers of technology adoption: Explainable AI for requirement analysis and technology design in smart farming. *Smart Agricultural Technology*, *9*, 100658. https://www.sciencedirect.com/science/article/pii/S2772375524002636. DOI: 10.1016/j.atech.2024.100658

Maraveas, C. (2022). Incorporating artificial intelligence technology in smart greenhouses: Current State of the Art. *Applied Sciences (Basel, Switzerland)*, *13*(1), 14. DOI: 10.3390/app13010014

Mirani, A., Memon, M. S., Chohan, R., Wagan, A. A., & Qabulio, M. (2021). Machine learning in agriculture: A review. *LUME, 10*, 5.

Nath, P. C., Mishra, A. K., Sharma, R., Bhunia, B., Mishra, B., Tiwari, A., Nayak, P. K., Sharma, M., Bhuyan, T., Kaushal, S., Mohanta, Y. K., & Sridhar, K. (2024). Recent advances in artificial intelligence towards the sustainable future of agri-food industry. *Food Chemistry, 447*, 138945. DOI: 10.1016/j.foodchem.2024.138945 PMID: 38461725

Nega, A. (2025). Climate change impacts on agriculture: A review of plant diseases and insect pests in Ethiopia and East Africa, with adaptation and mitigation strategies. *Advances in Agriculture, 2025*(1), 5606701. https://onlinelibrary.wiley.com/doi/full/10.1155/aia/5606701. DOI: 10.1155/aia/5606701

Qazi, S., Khawaja, B. A., & Farooq, Q. U. (2022). IoT-equipped and AI-enabled next generation smart agriculture: A critical review, current challenges and future trends. *IEEE Access : Practical Innovations, Open Solutions, 10*, 21219–21235. https://ieeexplore.ieee.org/abstract/document/9716089. DOI: 10.1109/ACCESS.2022.3152544

Rashid, M., Bari, B. S., Yusup, Y., Kamaruddin, M. A., & Khan, N. (2021). A comprehensive review of crop yield prediction using machine learning approaches with special emphasis on palm oil yield prediction. *IEEE Access : Practical Innovations, Open Solutions, 9*, 63406–63439. https://ieeexplore.ieee.org/document/9410627. DOI: 10.1109/ACCESS.2021.3075159

Ryan, M., Isakhanyan, G., & Tekinerdogan, B. (2023). An interdisciplinary approach to artificial intelligence in agriculture. *NJAS: Impact in Agricultural and Life Sciences, 95*(1), 2168568. DOI: 10.1080/27685241.2023.2168568

Saleem, M. H., Potgieter, J., & Arif, K. M. (2021). Automation in agriculture by machine and deep learning techniques: A review of recent developments. *Precision Agriculture, 22*(6), 2053–2091. https://link.springer.com/article/10.1007/s11119-021-09806-x. DOI: 10.1007/s11119-021-09806-x

Seethapathy, P., Kannan, M., Gopakumar, G., Manivasagam, V. S., & Manalil, S. (2025a). Comparative analysis of pretrained CNN architectures for mobile-based early detection of brinjal little leaf disease. In *AI and Sustainable Transformations* (pp. 108–114). CRC Press., DOI: 10.1201/9781003642886-15

Seethapathy, P., Kannan, M., Gopakumar, G., Manivasagam, V. S., & Manalil, S. (2025b). *A Mobile Application for the Early Detection of Dolichos Yellow Mosaic Virus in Lablab-Bean Using Pre-Trained CNN Models and Deep Learning. In 2025 IEEE International Conference on Emerging Technologies and Applications (MPSec ICETA)*. IEEE., https://ieeexplore.ieee.org/document/11118776/

Shahane, A. A., & Shivay, Y. S. (2021). Soil health and its improvement through novel agronomic and innovative approaches. *Frontiers in Agronomy*, *3*, 680456. https://www.frontiersin.org/journals/agronomy/articles/10.3389/fagro.2021.680456/full. DOI: 10.3389/fagro.2021.680456

Shams, M. Y., Gamel, S. A., & Talaat, F. M. (2024). Enhancing crop recommendation systems with explainable artificial intelligence: A study on agricultural decision-making. *Neural Computing & Applications*, *36*(11), 5695–5714. DOI: 10.1007/s00521-023-09391-2

Sivasamy, S. (2026). AI innovations in sustainable food production and agriculture. In AI Innovations for Transforming Food Production (pp. 1-24). IGI Global Scientific Publishing.

Song, C., Shin, S. Y., & Shin, K. S. (2024). Implementing the dynamic feedback-driven learning optimization framework: A machine learning approach to personalize educational pathways. *Applied Sciences (Basel, Switzerland)*, *14*(2), 916. DOI: 10.3390/app14020916

Subeesh, A., & Mehta, C. R. (2021). Automation and digitization of agriculture using artificial intelligence and internet of things. *Artificial Intelligence in Agriculture*, *5*, 278–291. DOI: 10.1016/j.aiia.2021.11.004

Taneja, A., Nair, G., Joshi, M., Sharma, S., Sharma, S., Jambrak, A. R., Roselló-Soto, E., Barba, F. J., Castagnini, J. M., Leksawasdi, N., & Phimolsiripol, Y. (2023). Artificial intelligence: Implications for the agri-food sector. *Agronomy (Basel)*, *13*(5), 1397. DOI: 10.3390/agronomy13051397

Uddin, M., Chowdhury, A., & Kabir, M. A. (2024). Legal and ethical aspects of deploying artificial intelligence in climate-smart agriculture. *AI & Society*, *39*(1), 221–234. https://link.springer.com/article/10.1007/s00146-022-01421-2. DOI: 10.1007/s00146-022-01421-2

Usigbe, M. J., Asem-Hiablie, S., Uyeh, D. D., Iyiola, O., Park, T., & Mallipeddi, R. (2024). Enhancing resilience in agricultural production systems with AI-based technologies. *Environment, Development and Sustainability*, *26*(9), 21955–21983. https://link.springer.com/article/10.1007/s10668-023-03588-0. DOI: 10.1007/s10668-023-03588-0

Vinod Chandra, S. S., Hareendran, A., & Albaaji, G. F. (2024). Precision farming for sustainability: An agricultural intelligence model. *Computers and Electronics in Agriculture*, *226*, 109386. DOI: 10.1016/j.compag.2024.109386

Wang, D., Cao, W., Zhang, F., Li, Z., Xu, S., & Wu, X. (2022). A review of deep learning in multiscale agricultural sensing. *Remote Sensing (Basel)*, *14*(3), 559. DOI: 10.3390/rs14030559

Wijerathna-Yapa, A., & Pathirana, R. (2022). Sustainable agro-food systems for addressing climate change and food security. *Agriculture*, *12*(10), 1554. DOI: 10.3390/agriculture12101554

Wu, P., & Zhong, Y. (2025). Artificial intelligence in sustainable agriculture: Towards a socio-technical roadmap. *Smart Agricultural Technology*, *101578*, 101578. Advance online publication. DOI: 10.1016/j.atech.2025.101578

Xing, Y., & Wang, X. (2024). Precise application of water and fertilizer to crops: Challenges and opportunities. *Frontiers in Plant Science*, *15*, 1444560. DOI: 10.3389/fpls.2024.1444560 PMID: 39711591

Zhou, W., Li, M., & Achal, V. (2025). A comprehensive review on environmental and human health impacts of chemical pesticide usage. *Emerging Contaminants*, *11*(1), 100410. https://www.sciencedirect.com/science/article/pii/S2405665024001112. DOI: 10.1016/j.emcon.2024.100410

Chapter 9
Collaborative AI for Smart and Adaptive Educational Environments

N. Raghu
http://orcid.org/0000-0002-2091-8922
Jain University, India

Ergashev Nuriddin
http://orcid.org/0000-0002-8274-6193
Karshi State Technical University, Uzbekistan

Gulomjon Sapaev
Mamun University, Uzbekistan

Ismailov Azizbek
Faculty of Economics, Alfraganus University, Uzbekistan

Bobojonov Azizjon
Tashkent State University of Economics, Uzbekistan

ABSTRACT

Education is entering a new era where Artificial Intelligence (AI) is no longer seen merely as a tool for automating tasks but as a partner in shaping meaningful learning experiences. This chapter moves beyond the traditional view of AI as a replacement for routine processes, to focus on how humans and machines can co-create knowledge, foster critical thinking, and build collaborative intelligence in education. The chapter begins by tracing the shift from automation-driven applications such as grading, scheduling, and basic tutoring to more dynamic, interactive forms of human AI collaboration. It highlights how AI can act as a learning companion

DOI: 10.4018/979-8-3373-7036-1.ch009

that adapts to students' needs, supports teachers in identifying hidden learning patterns, and creates personalized pathways for skill development. Importantly, it also underscores the irreplaceable role of human educators in nurturing empathy, ethical reasoning, and creativity areas where machines remain limited.

1. INTRODUCTION

The Changing Face of Technology in Education Throughout history technological progress has continuously transformed educational methods which include knowledge delivery and learning processes. The invention of writing allowed ideas to be preserved beyond memory. The printing press placed books in the hands of ordinary people, transforming access to knowledge. Radio and television expanded the reach of classrooms far beyond their walls, while the arrival of computers opened up new methods of instruction and assessment (Roschelle, 2021). The internet then created a vast, interconnected world of resources, allowing learners to engage with information and with one another on a global scale as shown in figure 1. Artificial Intelligence now represents the latest and perhaps most transformative step in this long evolution. AI systems operate differently from previous tools which focused on information storage and display and transmission because they can adapt and predict and interact. The first steps toward this vision appeared during the mid-20th century when researchers explored computer-assisted instruction and intelligent tutoring systems through limited experiments (Andersen et al., 2022). The educational applications of artificial intelligence have evolved into systems which perform automated grading and develop customized tests and deliver continuous virtual support throughout the day. The new systems in place have helped reduce administrative work and students now get their results faster.

These systems focus primarily on performing automated tasks with increased speed and expanded operational reach. The system provides valuable functions yet it fails to activate the social and emotional and creative educational aspects of learning according to (Rensfeldt & Rahm, 2023). The educational challenge of preparing students for complex future challenges demands we move past automated systems to create partnerships between human intelligence and machine intelligence.

Figure 1. The Changing Face of Technology in Education across.

This chapter illustrates how collaborative intelligence is transforming classrooms, vocational training, and lifelong learning. The text shows methods to create learning spaces which use AI to boost human control instead of taking it away while promoting exploration and collaborative problem-solving and open communication. The chapter examines ethical and educational aspects of AI implementation while supporting open disclosure and inclusive practices and trust-building methods. The educational path leads away from basic automation toward a learning future that links students with intelligent systems through the combination of technological accuracy and human creativity to build learning environments which adapt and respond to multiple dimensions.

1.1 Automation to Collaboration: A New Mindset

Automation in education has often been praised for its efficiency, but it also tends to reduce human roles to supervision. In this view, the machine handles the work, while the teacher or student passively receives the results. The educational system faces a risk of transforming into a purely mechanical operation which disregards its fundamental nature as a human-based process that includes interactive learning and creative exploration and collective understanding (Padovano & Cardamone, 2024). The emerging mindset of collaboration offers a more balanced relationship. Intelligent systems function through their ability to support both students and educators in their learning activities. The automated system would grade the essay but the collaborative system evaluates student writing through its ability to identify strengths and weaknesses which helps students develop their critical thinking abilities. The application functions as a tool which enables users to analyze their data instead of delivering automated results.

This shift acknowledges the complementary strengths of humans and machines. AI systems handle large datasets through their ability to detect patterns which leads to exact resource matching. Humans provide essential educational qualities through their ability to empathize and reason ethically and understand culture and generate creative ideas. The combination of these strengths results in learning that becomes more enriching and adaptable (Yadav & Shrawankar, 2025). The broader world of work already illustrates this trend. Employers now use human-AI partnerships to achieve innovation through collaborative work between people and machines for problem-solving. The foundation of education needs to adapt to current times because it exists to prepare people for work and civic participation (Correia et al., 2024). The classroom serves as an educational environment which demonstrates how human intelligence and artificial intelligence systems can work together successfully.

1.2 Defining Collaborative Intelligence

Collaborative intelligence functions as the joint operation of human beings with machines to generate results which both entities by themselves would not be able to achieve. Educational settings need systems which unite machine accuracy with human creative thinking for successful collaboration rather than giving up control to technology (Bulut et al., 2024). The fundamental characteristics of collaborative intelligence in learning emerge through multiple dimensions:

1. Co-creation of knowledge – Intelligent systems deliver both knowledge and adaptive guidance but their outputs undergo evaluation and development through student and teacher feedback processes.
2. Personalization with human guidance – AI systems generate personal educational plans through their recommendation process yet teachers transform these plans to match the actual social and emotional aspects of their students and their cultural environments.
3. Interactive partnership – AI uses dynamic feedback instead of static responses which enables students to develop their critical thinking abilities and their natural curiosity.
4. Augmentation rather than substitution – The focus remains on expanding human capacity, not replacing it.

Consider a classroom scenario. An AI system recognizes that several students struggle with scientific reasoning. The system functions beyond alert generation because it partners with educators to create inquiry-based educational activities according to (Zou et al., 2025). Students receive data-based support from the system but the teacher maintains their role to connect activities with what students find interesting and what they know from experience. The educational approach achieves success through its combination of data-based methods with human-centered educational approaches.

1.3 The Historical Trajectory of AI in Education

1.3.1 Early Automation in Education: Grading, Scheduling, Tutoring Systems

The first encounters between education and artificial intelligence were modest but ambitious. The mid-twentieth century brought a fresh vision about how machines could perform repetitive tasks which teachers needed to handle (Zhou, 2024). The first educational tools built for process optimization instead of educational transformation (Raman et al., 2025). The grading system at the time operated with Automated grading systems as its first visible method. The 1960s brought multiple-choice scanners and computerized test scoring systems which showed that assessment standardization could happen through machine processing instead of human labor (Smith, 1998). The systems provided two separate benefits which included efficient handling of student answers and unbiased automated grading systems that removed human interference (Smith, 2005). The systems delivered two distinct promises: efficiency in processing large volumes of student responses and objectivity in removing potential human bias in grading. The systems provided two distinct advantages which included fast processing of student answers and automated grading systems that eliminated human bias (Smith, 2005). The systems delivered two distinct promises: efficiency in processing large volumes of student responses and objectivity in removing potential human bias in grading. The systems provided two separate advantages which included quick processing of student responses and unbiased assessment through automated grading systems (Smith, 2005).

The initial development of "intelligent tutoring systems" (ITS) started with ambitious goals during their early stages. The development of ITS systems came from the combination of cognitive science with computer engineering to replicate personal tutoring through specialized domain feedback and guidance for students in fields like mathematics and computer programming (Patil, 2024). The first systems SCHOLAR and GUIDON emerged as early examples of machine learning systems which used student interaction to change their educational approach. However, limited computing power, narrow subject focus, and rigid programming constrained their effectiveness. The research work produced an essential base for future studies by showing how artificial intelligence can handle logistics operations and directly participate in educational processes.

1.3.2 The Rise of Adaptive Learning Platforms

The late 1990s and early 2000s, advances in computing and internet connectivity gave rise to more sophisticated educational technologies. Adaptive learning platforms developed into the following stage of educational technology evolution. The systems operated beyond basic automation because they used real-time data to customize learning materials and delivery speeds and performance responses. The Cognitive Tutor from Carnegie Learning represented an early example of educational technology which used cognitive science principles to teach math according to (Zehle et al., 2021). The system provided immediate feedback to students who solved problems while it used their performance to adjust the difficulty level of upcoming questions. The educational field witnessed the rise of comparable models which provided teachers with tools to create customized learning paths for large groups of students. Adaptive learning platforms became popular because of multiple factors which (Babu et al., 2025) explains in his research. Educational psychology research shows that students need different learning approaches because they learn at their own speed. The system developers intended their solution for educational organizations which needed to improve student outcomes and retention rates at the university level. Digital platforms provided student progress tracking through data analysis which allowed educational leaders to choose suitable academic programs and support strategies.

Commercial learning management systems (LMS) incorporated adaptive features, blurring the line between administrative infrastructure and pedagogical innovation. The educational platforms Blackboard and Moodle and Canvas (which came later) developed tracking systems that enabled instructors to monitor student resource usage and their academic progress throughout the course (Nguyen et al., 2023). The educational sector started using adaptive systems as mainstream tools during the past ten years after their initial development as experimental projects.

Figure 2. sophisticated educational technologies.

1.3.3 The Current State: Personalization and Predictive Analytics

AI technology in education has expanded from its initial use of adaptive feedback systems to include personalized learning approaches and predictive analytic tools. The process of personalization requires changing both the order of lessons and the selection of content materials and the way information is presented. Students who want to study environmental science will get examples about climate problems but students who want to study engineering will see physics problems related to building design. The educational method of personalization seeks to boost student motivation through the delivery of content that fits their individual learning identities and interests and existing knowledge base (Martinez-Maldonado, 2023). Predictive analytics serves as an advanced stage of progress. AI systems analyze student interaction data from large datasets to identify patterns which predict academic success and areas where students may face difficulties. Universities implement predictive modeling to spot students who might quit their studies so advisors can help them before their situation gets worse (Baer & Norris, 2023). The platforms employ prediction methods to determine which abilities will become essential in upcoming job markets so they can develop appropriate educational programs and career guidance services (Chen et al., 2024).

Virtual assistants that operate through natural language processing technology serve as an extension of AI capabilities to reach further into various applications. Students can access basic course information and deadline details (Narayanan et al., 2022).

1.3.4 Lessons from Automation-Driven Models

The educational sector has learned multiple lessons from the historical development of automation in teaching methods. First, efficiency gains are undeniable. The ability of machines to perform repetitive rule-based tasks at scale became evident through automated grading and scheduling systems which allowed teachers to dedicate themselves to their more important duties (Lameras & Arnab, 2021). Adaptive learning platforms demonstrated that data-based methods achieve better results when detecting student learning gaps and creating customized educational solutions. The implementation of automated systems has generated multiple discussions about proper allocation methods and their ability to maintain neutrality. Early advocates of computerized grading highlighted the reduction of human bias, but critics pointed out that algorithms embed their own limitations, often reinforcing inequities (Kim et al., 2022). AI systems currently face extensive examination because experts need to determine whether their data sets and design structures include any discriminatory elements.

Third, automation systems have proved that different systems need to work together to achieve optimal results. The educational systems which replaced only partial teacher duties won student acceptance but total teacher replacement systems faced student opposition. Students showed better results when technology worked as an addition to human teaching methods instead of trying to replace them (Raghu et al., 2025). The design of this pattern shows that people remain vital within educational settings. Finally, automation taught institutions the value of data. The initial computer systems started producing data which showed student behavior patterns and academic results. Organizations have developed data management into their core business approach because this data now shapes their academic program choices and accreditation evaluation methods. Student data collection for educational purposes has created privacy dangers and ethical issues about how student information should be handled.

1.3.5 Limitations of "Automation" Approaches

Automation-based applications perform essential tasks yet their operational range remains restricted to specific domains. Organizations focus primarily on test scores and completion rates and attendance yet they fail to evaluate students' creative abilities and ethical judgment and emotional understanding (Varsha et al., 2024). Machines perform manufacturing operations at peak efficiency when following standard procedures yet they lose performance power when tasks need human interpretation of ambiguous situations. The definition of learning tends to shrink because automated systems create a rigid interpretation of what learning actually means. The switch to machine-graded assessments through multiple-choice and short-answer questions could lead to the disappearance of essential skills like critical thinking and teamwork. Adaptive platforms deliver strong performance but they operate through predetermined paths which restrict students to specific option choices for their learning activities. Another limitation lies in the risk of over-reliance (Veena et al., 2025). Teachers must select between following professional judgment or algorithmic recommendations which create conflicts in their decision-making process. Students will become passive recipients of machine guidance because the system no longer requires their direct involvement in the learning process. The two situations show how automation technology creates situations which reduce human control instead of enhancing it. The quest for operational excellence tends to overshadow the fundamental human nature which defines educational institutions. Educational institutions serve as knowledge delivery systems which simultaneously shape students' self-perception and their social bonds and ethical development

(Balamurugan et al., 2024). The human growth components essential for preparing students to handle modern life remain beyond the reach of automated systems.

The historical development of AI in education demonstrates a clear sequence which started with basic administrative automation before reaching adaptive learning systems and finally achieving personalized learning path prediction systems. Each stage has delivered benefits, yet each has also exposed limitations when technology is treated as a substitute rather than a partner. The historical record demonstrates that automated systems need human oversight to maintain operations in contemporary educational settings. The process of developing curiosity and empathy and creativity requires educational approaches that concentrate on teamwork as their main focus. AI education requires a shift from automated processes to collaborative intelligence systems which unite human and machine capabilities for developing versatile and moral educational spaces.

2. LITERATURE REVIEW

Yadav et al., illustrates Artificial Intelligence (AI) is reshaping education in exciting ways. The technology of Artificial Intelligence provides students with individualized learning experiences through customized feedback and adaptive educational content which adjusts to their learning speed and individual learning methods. But it's not just about machines doing the work; it's about using AI to help teachers focus more on what they do best connecting with and inspiring students. The educational system needs to handle issues of fairness and privacy protection and student inclusion to achieve its goals. The development of AI tools needs to focus on creating systems that serve all users while maintaining human involvement as the core element in educational processes.

Correia et al achieved their research goal by showing that AI can make education more interactive and tailor learning approaches to individual students. Students can receive continuous assistance through chatbots and virtual assistants which deliver personalized guidance and immediate answers to their questions. The research community has discovered that AI systems perform basic operational functions yet these systems present two primary concerns about data privacy and student dependence on automated educational tools. To make the most of AI, schools need clear rules and a focus on ethics so that technology works alongside people, respecting privacy and encouraging critical thinking.

Patil explained the educational system receives twofold benefits and challenges from artificial intelligence technology according to Patil. AI technology enables students to obtain individualized learning experiences but educational institutions must change their methods for teacher education and curriculum development. AI systems face two primary challenges which involve maintaining privacy and fairness because they must operate transparently while preventing the worsening of existing social inequalities.

Bulut et al. advocate for AI in education to extend past efficiency gains since it should create inclusive learning environments that support educational equality. AI technology enables educational access for students from underrepresented backgrounds while providing modern learning approaches. We need to move forward carefully by creating rules which protect students while honoring their cultural backgrounds. Education leaders and policymakers should unite their efforts to direct AI capabilities toward educational equality so that all students gain equal access to learning opportunities.

Andersen et al., described on AI technologies which examine team interactions enable educators to give better assistance through collaborative learning enhancement. Research shows that student collaboration improves when AI supports their work which results in better academic results. The systems need precise development to protect privacy while maintaining human social interaction patterns. AI technology functions as a collaborative tool which improves team operations while supporting teachers and students to build their critical thinking and social abilities.

3. THE CONCEPT OF COLLABORATIVE INTELLIGENCE

3.1 Definition and Foundations: Human + AI Synergy

Collaborative intelligence describes the process through which human intelligence and artificial intelligence combine to achieve results neither could accomplish alone. The objective requires establishing a collaborative system which enables each participant to apply their unique abilities for mutual achievements. Artificial intelligence provides rapid data processing and exact results from enormous data analysis but people offer essential advantages through their emotional connection and cultural knowledge and ethical judgment and creative thinking. According to Achar (2025) educational collaborative intelligence stems from the combination of these abilities. The theory develops from the understanding that learning requires multiple components which include cognitive processes and emotional elements and social interactions. Machines process data through pattern recognition to create adaptive solutions and provide instant feedback but they fail to duplicate the complex human bonds and cultural ways of constructing meaning. Education systems can boost human learning and teaching abilities through intentional design of machine-system interactions that enhance human capabilities instead of replacing them.

3.2 Distinction Between Automation, Augmentation, and Collaboration

The distinction between automation, augmentation, and collaboration clarifies how technology functions within education.

- Automation: refers to machines taking over routine tasks. Automated grading systems for multiple-choice tests and software tools for attendance tracking and scheduling fall under this category. The functions operate at peak performance levels while requiring minimal human involvement.
- Augmentation: The technology enhances human capacity without fundamentally altering the relationship. The educational tools which include spell checkers and plagiarism detectors and teacher data analytics dashboards fall under the category of augmentation. The core operations depend on human involvement yet technology enables better decision-making processes.
- Collaboration: The humans and machines operate as partners in co-creation. An AI system employs two main functions to help students with scientific reasoning difficulties: it identifies their challenges and recommends appropriate solutions which teachers need to analyze through their cultural and emotional background. Collaboration functions through human-machine dialogue which produces results through interactive processes instead of single-directional operations.

This distinction matters because education is often tempted to stop at automation or augmentation, where technology provides efficiency without fundamentally reimagining the learning experience. The educational system faces the problem of AI tools creating cheating opportunities because students use these resources to bypass their learning responsibilities.

3.3 Cognitive Science and AI in Co-Creation of Knowledge

The educational system encounters a major problem because students use AI tools to cheat which stops them from learning. The main problem with AI tools in education is students use them to cheat which prevents them from developing their own knowledge. The educational system faces a major challenge because students use AI tools to cheat which blocks their learning process. The main problem with AI tools in education is students use them to cheat which prevents them from developing their own knowledge. The educational system faces a major challenge because students use AI (Raghu et al., 2025).

Teachers, too, benefit from this co-creative dynamic. The recognition of effective teaching through misconception diagnosis and strategy adaptation has been established by cognitive science for many years. AI systems discover hidden patterns which stay hidden from human detection because they identify both repeated assignment errors and distinct learning patterns that develop through behavioral patterns over extended periods. Teachers develop targeted interventions by combining this knowledge with their understanding of specific educational contexts. The machine provides the "what," but humans determine the "why" and "how." The collaborative approach to learning works in harmony with constructivist educational principles because it enhances the learning process based on constructivist methods. The system achieves knowledge generation by uniting machine accuracy with human interpretation through an interactive learning process between students, educators and the system.

Figure 3. cognitive science supports the idea of human–AI co-creation.

3.4 Case Examples of Collaborative Intelligence in Education

Examples of collaborative intelligence in education are already visible across diverse contexts.

- Intelligent Writing Assistants in Classrooms: Tools that go beyond automated grammar correction illustrate collaboration. Some systems engage students in reflective dialogue about tone, style, and clarity. These systems offer correction suggestions which students need to assess before they can apply them. The educational process enables students to develop their critical thinking abilities about writing decisions while teachers direct their attention toward teaching students advanced skills in argument construction and creative thinking.
- STEM Learning with Adaptive Tutors: The educational platform Carnegie Learning's Cognitive Tutor demonstrates how educational partnerships prove their value through their mathematical and scientific knowledge delivery. The system follows each step of problem-solving while it detects student errors and generates customized practice activities. The insights which teachers gain from this data help them develop group discussions and projects which strengthen student understanding through automated accuracy and human-based discovery.
- Collaborative Project-Based Learning: AI-powered systems in certain universities help student teams who work on projects by managing their resources and providing research suggestions and identifying areas where students lack knowledge. Students evaluate these suggestions through discussions which let them share their personal perspectives with their classmates. The system joins the learning team as a silent member who shapes the team's direction through its presence.
- Vocational Training Simulations: AI-driven simulations enable students to practice their skills effectively through simulated learning which takes place in controlled learning environments. Nursing students practice patient care through AI-guided scenarios which they use for their learning activities. The system generates instant responses for every action that users perform while instructors lead students through ethical and emotional aspects during their debriefing sessions. The two systems work together to create a continuous process which combines computer-assisted training with human-led evaluation of learning experiences.

These cases illustrate that collaborative intelligence is not confined to theory. It is unfolding in classrooms, labs, and training centers, reshaping the dynamics of teaching and learning.

3.5 Benefits for Learners, Teachers, and Institutions

The promise of collaborative intelligence lies in the benefits it delivers across multiple stakeholders.

- For Learners: Students can access individualized direction through collaborative systems which help them maintain their social connections with others. The system provides users with immediate feedback which actively involves them in the learning process instead of making them passive recipients of information. Students who examine AI prompts develop their metacognitive abilities which helps them become self-directed learners while they acquire their academic material. Students who watch human-AI team work will become ready for their future jobs which will demand them to work with smart computer systems.
- For Teachers: The system reduces teachers' daily work load so they can devote their time to teaching creativity and building relationships and ethical understanding with students. The system enables teachers to track student learning progress at different stages which helps them provide students with the right help at the right time. Teachers have taken on new responsibilities which require them to guide students through their critical thinking development.
- For Institutions: The system enables collaborative intelligence to achieve scale because it keeps quality standards intact. Institutions can identify learning outcome patterns through predictive analytics while they distribute funding effectively and develop educational programs which match current market demands. The collaboration models provide students with customized learning support which helps students from different backgrounds receive equal educational chances.

4. THE ROLE OF HUMAN EDUCATORS IN THE AGE OF AI

4.1 Teachers Remain Central: Empathy, Ethics, Creativity

Artificial intelligence systems now operate in school environments which allow them to provide immediate responses to student work while tracking their learning development and creating customized learning paths. The current advanced technological environment continues to require human teachers because no machine system exists which can replace them. Teachers need to build emotional and ethical bonds with their students because this connection stands as the fundamental educational element which enables learning to occur. Machines maintain their computational abilities but they lack the ability to understand human emotions because they do not possess human-like empathy. The complete understanding of student struggles and their emotional reactions to success or failure remains beyond the capability of computer algorithms. Teachers need to understand these subtle signals because they use their knowledge to provide students with support through motivational backing and emotional comfort and ethical direction.

Teachers develop their ability to practice ethical decision-making in their educational work which goes past their natural ability to feel empathy. Artificial intelligence systems provide suggested learning paths and corrective actions yet humans must decide which methods prove fair and which ones promote inclusion while evaluating how learning processes affect people. Teachers who understand predictive analytics data about student performance use their knowledge to create helpful guidance which enables students to reach their full potential. The core function of teachers becomes more essential because creativity demands their essential role. Human beings possess the ability to inspire curiosity while creating fresh questions and original thinking approaches which no computer algorithm can duplicate. AI systems work with existing knowledge to create new examples but teachers need to trigger student creativity while showing them alternative ways to analyze data and motivate them to explore new ideas. The educator needs to maintain their position because AI technology has turned their role into an essential vital element of teaching.

4.2 Teachers as Facilitators, Mentors, and Co-Learners

The evolution of artificial intelligence in education shifts the teacher's role from a transmitter of knowledge to a facilitator of learning. The combination of artificial intelligence with classroom learning requires educators to build educational resources which they should use to develop interactive learning activities that help students interpret digital assessment results. The focus now centers on building environments which support open communication and self-assessment and team-based work. Teachers function as mentors who provide students with advice which reaches past academic topics to include essential life abilities and methods for building strength and community involvement. The teacher uses AI to find student knowledge gaps which she turns into learning opportunities that help students develop their perseverance instead of feeling worried. The mentorship system which education provides helps students develop their character while they master their academic subjects.

Teachers now increasingly adopt the role of learning partners who study alongside their students. AI systems generate new insights which help teachers discover patterns that educators must continue to learn about and develop their skills. The open learning environment serves as a model for students to develop their lifelong learning skills because it shows how human beings maintain their ability to adapt and explore new things even though smart machines exist. The collaborative approach of co-learning minimizes the mental gap which students view between their teachers and themselves because it shows that everyone involved in learning experiences shared exploratory processes.

Figure 4. evolution of artificial intelligence in education

4.3 AI as Support, Not Substitute

Teachers face the risk of being replaced by AI because this technology performs repetitive tasks with remarkable speed. Automated grading systems perform assessment processing at high speed while adaptive platforms independently generate customized exercises which teachers do not need to monitor. The idea that AI can replace human teachers completely fails to grasp the intricate nature which education demands. AI systems perform well with big datasets and fast processing but they cannot understand subtle details. Students who perform poorly in school might have personal issues or face language difficulties or cultural obstacles which affect their academic work. The algorithms struggle to understand these signals because educators must interpret them by referring to human experiences which exist beyond the data points.

Positioning AI as support rather than substitute recognizes the complementary nature of the relationship. The system enables teachers to spend more time with their students because it handles all administrative management functions automatically. The personalized dashboard system delivers pattern alerts to teachers but they must interpret the data and modify their teaching approach to make the information useful. The best classrooms function with AI technology working in the background because teachers use their human abilities to create trust and help students feel belonging and develop their moral thinking skills.

4.4 New Teacher Competencies for AI-Augmented Education

The educational field demands teachers to develop fresh abilities because collaborative intelligence has become more common. Digital tool knowledge has become a fundamental requirement which educators need to succeed in their teaching work. Teachers need to understand how algorithms work, what kinds of data are collected, and how outputs should be interpreted responsibly. Users who understand AI systems will gain more than basic technical skills because they will recognize all the system's operational limits together with its built-in prejudices. The educational system needs to revive its focus on pedagogical design because of its growing importance. Teachers must develop learning activities which use AI-generated feedback to support student learning while keeping human control in their teaching practice. The educational platform generates personalized practice exercises which teachers must combine with their existing group work and discussion activities and creative assignments. The system needs to perform well when it connects automated operations with human-based search processes which require physical movement.

Another emerging competency is ethical stewardship. Teachers need to protect their students' personal information while they assess the fairness of algorithms and create AI environments which welcome all students. Teachers can protect student equity from machine learning systems which create discrimination through their decisions by learning data ethics and cultural responsiveness and bias detection techniques. Teachers need to develop their students' soft skills through teaching them collaboration and adaptability and communication abilities. People who want to work with others in AI-rich environments need to learn how to join forces with technologists and policymakers and their educational colleagues. Teachers who want to build connections between data-based tools and human-centered education need to learn technological skills and human understanding at the same time.

4.5 Teacher–AI Co-Design in Curriculum and Instruction

The most promising applications of AI in education emerge when teachers and intelligent systems co-design learning experiences. Teachers work with artificial intelligence systems to build learning environments which they use for designing educational content. Teachers use AI systems to build educational content which they implement for teaching activities. The educational process benefits from AI technology because teachers can work with it to create custom learning environments for their students. The educational process benefits from AI technology because teachers can work with it to create custom learning environments for their students. The educational process benefits from AI technology because teachers can work with it to create custom learning environments for their students. AI systems enable teachers to build virtual classrooms which help them develop personalized educational content. AI systems enable teachers to build virtual classrooms which help them develop personalized educational content. AI systems enable teachers to build virtual classrooms which help them develop personalized educational content. The system needs admins to handle maintenance tasks and users need to be trained so they can use the system correctly. AI systems enable teachers to build virtual classrooms which help them develop personalized educational content. The system needs admins to handle maintenance tasks and users need to be trained so they can use the system correctly. AI systems enable teachers to build virtual classrooms which help them develop personalized educational content. The system needs admins to handle maintenance tasks and users need to be trained so they can use the system correctly. AI systems enable teachers to build virtual classrooms which help them develop personalized educational content. The system needs admins to handle maintenance tasks and users need to be trained so they can use the system correctly.

In the age of AI, teaching by humans is indispensable. Empathy, ethical judgment, and creativity anchor teaching in ways no algorithm can replicate. The teacher becomes a facilitator of learning, a mentor, and a co-learner, guiding the young through not only items of the curriculum but also through more complex, not always rational, elements of personal growth and social responsibility. Artificial intelligence forges an indispensable support in the service of teachers, making the routine burdens lighter and providing the sources of data-driven insights with no potential substitution for the relational, ethical, and imaginative aspects of teaching. The future of education depends on how teachers are reskilled with new competencies to respond to technological fluency, ethical stewardship, and adaptive pedagogy and make use of AI without losing human dignity and agency. More than anything, when teachers and AI co-design curricula and instruction, classrooms become a shared laboratory of collective intelligence. Here, technology amplifies human capacity rather than diminishing it, generating educational experiences that are both severe and humane. In this vision, teachers do not disappear into the background of technological progress but stand at the center, reading the data, designing learning, and safeguarding human life.

4.6 Adaptive Learning Environments

Artificial intelligence is thereby transforming classrooms to make them responsive spaces that can adjust to the requirements of learners. In traditional learning methodologies, the pace at which topics are covered is standard. Every learner is covered at the same pace irrespective of whether they are ready to move on to the next topic. With the integration of artificial intelligence in the student-learning environment, it is possible to attain responsiveness that would previously never have been possible.

Such malleability goes beyond issues of efficiency. Adaptive platforms take into consideration the learning style differences of students and offer multiple means of engaging with content. A mathematical notion, for example, can be represented in terms of pictorial representations, applications in the real world, or reasoning logic based on the student's background. The dynamic environment changes in real-time scenarios, thus offering an enabling context for students of all abilities to succeed.

4.7 Encouraging Curiosity, Dialogue and Exploration

While data and personalization are central, the real value of AI as a learning companion lies in how it fosters curiosity and exploration. Education is not just about mastering the content but about creating a disposition toward inquiry. Intelligent systems can engage this, for example, by asking open questions, offering options to the way a problem has been approached, or relating learners to further resources.

AI can foster deeper reflection, too. More than providing answers, well-designed systems proactively ask learners to engage in critical thinking, reason across opinions, and justify perspectives. A student learning about environmental science, for instance, may be challenged to consider the economic, social, and ecological aspects of climate change rather than memorize isolated facts. In this encouragement of dialogue, AI aligns the broader goals of education-developing independent thinkers through exposure to complexity. As curiosity becomes the driver, learning shifts from a task to be done into an adventure to be pursued.

4.8 AI in Vocational Training and Lifelong Learning

As a learning companion, the application of AI goes beyond classroom settings to vocational learning. In professional capacities, simulations by AI allow students to practice their skills in simulated settings. Students undertaking nursing programs can engage with virtual patients who will react to their actions, while mechanics can practice operating virtual machinery before getting to real machinery. Such training tools are crucial in instilling confidence in practitioners without posing any danger of mistakes in real situations.

For adult learners involved in lifelong education, AI presents flexible learning routes, which adjust as per evolving requirements and timetables. AI helps optimize personalized learning platforms, encouraging and helping employees enhance their skills according to new technological developments, acquire new skills to adopt new occupations, or engage their minds through lifelong learning programs when they grow older. Most importantly, vocational education and lifelong learning illustrate the collaborative aspect of AI. In such learning, the role of AI is merely to enhance scaffolding and simulation, whereas human teachers frame learning inputs in terms of their social, ethical, and professional contexts.

AI as a learning companion signals a shift in thinking about education. Adaptive learning remains inclusive as it reaches out to students wherever they are, and learning pathways offer distinct routes to capabilities and dreams. Big data insights offer insights into unnoticed difficulties, thus facilitating early intervention and optimization. However, beyond optimization, optimization, and precision, the potential of AI as a learning companion lies in its potential to stimulate curiosity, conversations, and exploration. In schools, learning institutions, or life-long learning, intelligent systems act as companions rather than substitutes. When coupled together, AI enhances education as it becomes flexible, customized, and dynamic, and human teachers ensure the ethical, relational, and imaginative part remains.

Figure 5. AI as a companion extends beyond formal classrooms.

5. DESIGNING COLLABORATIVE LEARNING ENVIRONMENTS, ETHICAL AND PEDAGOGICAL CONSIDERATIONS

5.1 Frameworks for Human–AI Classroom Collaboration

Across history, classrooms have often been imagined as places where knowledge flows from teacher to student in a linear fashion. The integration of artificial intelligence reshapes that image, encouraging more fluid and participatory forms of learning. Collaborative frameworks position AI not as a distant machine working in isolation, but as a partner woven into the very fabric of classroom interaction.

At the heart of these frameworks lies the idea of complementarity. Machines excel in analyzing vast amounts of data, recognizing patterns, and generating recommendations with speed and precision. Humans, by contrast, bring creativity, ethical judgment, and empathy—qualities that cannot be reduced to code. When both dimensions converge, learning environments transform into spaces of distributed intelligence, where responsibility for knowledge-building is shared among students, teachers, and intelligent systems. Another guiding principle comes from socio-constructivist traditions. Knowledge does not grow in isolation; it develops through dialogue, debate, and shared exploration. Within this view, AI can serve as a facilitator, linking students with complementary perspectives, curating questions that spark deeper inquiry, or offering insights that enrich collective discussion. Rather than replacing the social aspects of learning, AI has the potential to amplify them.

5.2 Pedagogical Strategies: Problem-Based, Inquiry-Based, and Project-Based Support

The frameworks become most concrete when linked to classroom strategies. Many traditional pedagogical methods are given new life when underpinned by AI.

- Problem-Based Learning: PBL involves presenting students with complex problems similar to those found in the real world. AI enhances this process by generating simulations, revealing different scenarios, or directing learners toward sources of information that are underexploited. Take a team of medical students studying a patient's case using an AI diagnostic assistant. The system flags possible conditions, but the actual learning occurs as students investigate, debate, and come to a decision.
- Inquiry-Based Learning: At its heart is questioning. AI serves as a prompt for curiosity: providing leads, pointing out strange correlations, or suggesting new avenues for investigation. An environmental science course might use an AI system to uncover associations between climate data and human

migration, which would push students toward an integrated exploration. The questions the system raises are not answers themselves, but rather firestarters of human thought and discussion.

- Project-Based Learning: Extended projects often require coordination, creativity, and applied knowledge. AI becomes a scaffold in this context, supporting design, organization, and analysis. A student team working on sustainable architecture might employ AI to test energy efficiency models while focusing human effort on the cultural and ethical dimensions of the project. The technology strengthens the technical side, yet the meaning-making remains a human responsibility.

5.3 Digital Platforms and Intelligent Tutoring Systems

Collaboration requires a supportive infrastructure. The digital platforms, especially ITSs, form the core of this change. Such systems integrate customized tutoring with collaboration opportunities. An ITS may change exercises based on individual progress but simultaneously suggest collaborative challenges where learners, in teams, apply knowledge. Individual session data informs group composition such that different strengths complement each other. Indeed, several platforms today embed dialogue systems that foster the participation of quieter voices to keep discussions in balance.

Beyond the classroom, digital platforms extend learning communities across institutions and even across continents. AI can connect learners with peers interested in a subject but bring different cultural or disciplinary perspectives to bear. These global collaborations nurture not only intellectual growth but also empathy and cross-cultural understanding-qualities indispensable in an interconnected world.

5.4 Transparency and Explainability in AI

In all the classrooms where intelligent systems have been involved, questions have arisen about the decisions of such systems. An algorithm for recommending learning materials or predicting a student's struggle is not outside the world that hid from learners and educators the data, the design choices, and the mathematical models behind it. Without transparency, these processes risk being perceived as black boxes-powerful, yet opaque. Explainability addresses this issue. When recommending that a student review a particular concept, the system must also specify that recommendation's grounds. Was the learner having trouble with that specific problem type? Did their performance-patterns mirror earlier difficulties seen in other contexts, thereby making the reasoning visible? AI nurtures understanding by not cultivating suspicion. Transparency demystifies the technology and allows it to collaborate toward educational integration rather than unquestioned authority.

In this respect, explainability empowers the educator to make informed judgments. The teacher who understands the grounds of a recommendation issued by a system will have far more opportunity to validate, modify, or even question it. Thus, AI continues to be a source of intelligence rather than an unquestionable textbook.

5.5 Building Trust in AI Systems

Trust is everything in education, Students need to trust their teachers to guide them, and teachers need to trust that the tools they bring into the classroom will actually help learning, not get in the way. When it comes to AI, trust doesn't come from flashy features or technical wizardry alone. It comes from being reliable, fair, and aligned with what really matters to people. Consistency is key here. When students see that an AI system gives them accurate, helpful feedback time after time, they start to believe in it. But if the AI gives confusing or inconsistent answers, that trust disappears fast. Trust also depends on feeling like the AI is genuinely trying to help you learn. If its suggestions seem random, or worse, if they feel manipulative, that connection between human and machine breaks down. Teachers play a huge role in building this trust too. When educators understand what AI can and can't do, they can present it as a helpful partner rather than a magic solution. They can set realistic expectations, So trust isn't about blindly believing in machines. It's about having confidence that both human and artificial intelligence are working together toward the same goal: helping students learn and grow.

5.6 Student Privacy and Data Ethics

AI bias is a serious ethical issue that deserves our attention. The problem is that algorithms learn from the data they're given. If that data contains historical inequalities or cultural imbalances, the AI systems can end up repeating those same problems. In education, this matters a lot. A biased system might misunderstand students who speak different dialects. It could undervalue contributions from underrepresented groups. Or it might overestimate the abilities of students who already have advantages from their background. Making things fair takes real work. Designers need to test their systems for bias across all kinds of people. Schools and institutions must make sure their training data and evaluation processes include everyone. But fairness is just the beginning. True inclusivity asks a bigger question: does AI just treat everyone equally, or does it actually support different needs? The good news is that adaptive platforms can be designed to help all students. They can offer voice-based navigation for students with disabilities. They can provide visual adjustments or different ways to take tests. Multilingual features let students learn in their preferred language, which removes barriers to participation. Inclusivity means

AI doesn't just avoid causing harm. It actively creates conditions where every kind of learner can succeed.

5.7 Pedagogical Responsibility: Ensuring Human Agency

Amidst these ethical considerations, one principle remains clear: education should protect human agency. AI can suggest, analyze, and simulate, but decisions regarding learning objectives, teaching methods, and personal development belong to humans. Educators are at the center of this agency. Interpreting AI suggestions through professional judgment, teachers prevent automation from reducing options. No learner should be made to feel defined by an algorithmic prediction; instead, this should be an impetus to improve upon, question, or redefine such predictions. Agency extends to the students themselves. AI may guide, yet learners should have the chance to make choices about their paths, question system outputs, and be owners of their development. But encouraging critical engagement with AI nurtures not only knowledge acquisition but also digital literacy-skills one would presumably need in a world where intelligent systems increasingly influence so many aspects of life.

Ethical and pedagogical considerations form the backbone of responsible AI integration in education. Transparency and explainability shed light on the logic behind machine decisions. Trust grows from reliability, fairness, and mediation through educators. Confronting bias, actively cultivating inclusivity, and protecting privacy are possible with transparent ethical frameworks. Most importantly, human agency needs to remain central in order for technology to act as a guide, not as a governor. Education is ultimately a human enterprise, replete with creativity, empathy, and ethical complexity. AI can enrich this process, but it cannot replace educator and learner responsibilities in the shaping of meaningful experiences. When grounded in ethical and pedagogic reflection, classrooms will transform into innovative, yet also more humane, equal, and trustworthy settings.

6. APPLICATIONS ACROSS EDUCATIONAL CONTEXTS AND CHALLENGES, BARRIERS

6.1 Classrooms: Primary, Secondary and Higher Education

Most AI applications within the primary classroom take forms that can be best described as entertaining companions that guide the early learner through foundational skills. Adaptive storybooks modulate language complexity to match reading levels, and games use intelligence to provide feedback that is immediate yet encouraging. The intention is not to hurry children through content but to foster curiosity at their own comfortable pace. To this end, forming learning into a responsive journey, AI supports young learners to develop confidence in literacy, numeracy, and problem-solving skills. In secondary education, different possibilities are brought into focus. Here, intelligent platforms fit coursework around individual strengths and struggles, ensuring no learner is left behind, while others are challenged to move ahead. In science and mathematics, AI-driven simulations allow students to experiment with variables that would be too complex or costly to replicate in physical laboratories. These experiences extend traditional teaching methods by marrying rigor with imagination.

AI in higher education becomes a partner in exploration and investigation. Intelligent systems dig through boundless academic databases, highlighting patterns across disciplines and suggesting new directions for inquiry. Students at universities often juggle both professional specialization and general intellectual development; AI can customize those paths while also supporting interdisciplinary collaborations. The university classroom thus evolves into a space in which human insight and machine precision meet.

6.2 Corporate Learning Ecosystems

In corporate environments, the pace of change is relentless, demanding agile approaches to learning. AI reshapes training programs into continuous ecosystems where employees upskill and reskill. Intelligent systems curate personalized learning plans in tune with both organizational goals and individual aspirations. A financial analyst may be advised on modules related to emerging regulatory frameworks, while a manager may be recommended for leadership and communication training. AI also encourages collaboration in corporate learning. Digital platforms bring together employees across regions and functions to solve problems collaboratively, supported by insights enabled by AI. Real-time translation tools make sure teams from different linguistic backgrounds can work together seamlessly. Integrating learning with daily work, AI helps corporations nurture cultures of continuous improvement rather than episodic training events.

6.3 Lifelong Learning and Reskilling for Future Jobs

These fast-paced changes within and from industry to industry have made learning a lifelong endeavor. AI, therefore, serves as an all-time companion for individuals at all stages of life. These workers transition into other occupations, and thus intelligent systems would be able to enlighten any user on the transferable skills and suggest reskilling pathways. For former employees, who retired, but now thirst for intellectual gratification, they can also join AI-powered platforms, which will take to their pace and interests. Especially because of the developing technologies and increased automation, reskilling will play an important role. Besides identifying in-demand skills, AI also assists in acquiring those skills through adaptive modules, simulations, and feedback. AI offers the structure and personalization that make lifelong learning possible, accessible, relevant, and sustainable.

6.4 Technological Limitations

While AI holds a promise for transformation, it has a number of limiting technological constraints. Many systems require large data, high computer capacity, and constant connectivity. This is real life; not all educational facilities can afford to maintain such a setup. Errors in speech recognition, limitation of natural language understanding, or lack of capturing the subtlety of human expression often reduces the effectiveness of AI in classrooms. Moreover, AI systems are only as strong as the data behind them. If data sets are not diverse or relevant, then outputs have the potential of being inaccurate or misleading. This fragility serves to remind educators that while AI may enhance learning, it cannot operate perfectly in every context. The promise of collaborative intelligence is challenged by the realization of imperfect systems.

7. AI + HUMAN CREATIVITY: SHAPING NEW KNOWLEDGE LANDSCAPES

he fusion of artificial intelligence with human creativity is just beginning to reshape how knowledge is built and shared. Rather than being separate domains, human intuition and technological precision increasingly interlink. While AI can generate insights from data on scales impossible for individuals, human creativity brings interpretation, context, and imagination. It is this partnership that opens pathways toward knowledge landscapes where learners will not be passive recipients but active designers and innovators. Such synergies could nurture learning environments in classrooms and outside them, where discovery will become a collaborative affair. For example, while an AI system identifies areas of misunderstanding, it is the learners who develop creative solutions that outstrip statistical patterns. This interplay essentially highlights a vision for the future of education, where it is less about fixed-content transmission and more about cultivating dynamic ecosystems through which knowledge is continuously evolving.

7.1 Resilient and Adaptive Education Systems

The global disruptions of the last few years have made very real the fragility of conventional education systems. Holistic ecosystems must, therefore, be resilient and adaptive-capable of withstanding uncertainty while continuing to nurture growth. AI contributes adaptability through flexible learning pathways, while human guidance ensures that learning remains inspired by empathy and purpose. Resilience requires inclusivity. Systems that adapt only to privileged contexts risk entrenching inequality. Holistic ecosystems must consider diverse learners, cultural settings, and resource realities so that AI tools remain accessible and meaningful across a range of settings. The strength of such systems does not lie in uniformity but in the capacity to adapt intelligently to change.

7.2 Cross-Disciplinary Collaborations

The effective integration of AI requires an interdisciplinary effort-psychology, neuroscience, pedagogy, computer science, and ethics. Understanding from psychology and cognitive science helps reveal how learners engage emotionally and cognitively with technology. Pedagogical theories underpin the creation of meaningful curricula. Technologists provide models and platforms, while ethicists ensure that values related to fairness, inclusion, and transparency are observed. Such cross-disciplinary collaborations nurture systems that are human-centered, rather than merely state-of-the-art. Indeed, a truly holistic ecosystem incorporates multiple knowledge domains in its design to assure AI-enhanced learning fosters respect for diversity and promotes human flourishing, rather than mere efficiency.

8. CONCLUSION AND FUTURE SCOPE

Artificial Intelligence in education has come a long way from its humble beginning of automating tasks. Today, it stands tall as a partner in shaping learning experiences that are more adaptive, inclusive, and deeply personal. Intelligent systems can adapt to the needs of the individual, make visible learning patterns not seen before, and create new pathways for growth. The center of education remains machines cannot replace: the human qualities of empathy, imagination, and ethical judgment. The future lies in combining these strengths-educators leading with wisdom and care, and AI with precision and adaptability-leading together to spaces that nurture curiosity, dialogue, and resilience. Challenges exist-bias, inequality, and governance do need attention-but the vision is clear. By cultivating collaborative intelligence, education moves toward ecosystems that prepare learners not just for changing careers but for thriving in a complex, interconnected, and human-centered world.

The future of AI in education can be seen as building holistic learning ecosystems in which technology and human creativity go hand-in-hand. Advancements will go beyond personalization to include the development of collaboration, problem-solving, and lifelong adaptability. Cross-disciplinary research linking pedagogy, psychology, and intelligent systems will shape tools that are both ethical and inclusive. Since global challenges require resilience, education should be preparing learners for jobs that have not yet been invented, focusing on critical thinking, empathy, and innovation. By embracing collaborative intelligence, societies can create fair pathways where AI enhances learning while preserving the irreplaceable human core of education.

REFERENCES

Achar, H. B., & Raghu, N. (2025, February). The Impact and Future of Artificial Intelligence in Learning. In *2025 3rd International Conference on Integrated Circuits and Communication Systems (ICICACS)* (pp. 1-6). IEEE.

Andersen, R., Mørch, A. I., & Litherland, K. T. (2022). Collaborative learning with block-based programming: Investigating human-centered artificial intelligence in education. *Behaviour & Information Technology, 41*(9), 1830–1847. DOI: 10.1080/0144929X.2022.2083981

Babu, C. S., William, M. P., & Kowsika, S. (2025). Generative AI Tools and Education: Enabling Collaborative Intelligence Through ChatGPT, Canva, InVideo AI, and InShot. In *Humans and Generative AI Tools for Collaborative Intelligence* (pp. 441-460). IGI Global Scientific Publishing.

Baer, L. B., & Norris, D. M. (2023). AI+ HI: Collaborative Intelligence Can Significantly Improve Student Success. *Planning for Higher Education, 52*(1).

Balamurugan, M., Raghu, N., Dayananda, L. N., Deepthi, M., & Vinod Kumar, M. (2024). AI-Enhanced Leadership Decision Support System Using ChatGPT for Ethical Learning. In *Impact of New Technology on Next-Generation Leadership* (pp. 137–161). IGI Global Scientific Publishing.

Bulut, O., Beiting-Parrish, M., Casabianca, J. M., Slater, S. C., Jiao, H., Song, D., . . . Morilova, P. (2024). The rise of artificial intelligence in educational measurement: Opportunities and ethical challenges. *arXiv preprint arXiv:2406.18900.*

Chen, Y. C., Liu, H., & Wang, Y. F. (2024, June). Governance design of collaborative intelligence for public policy and services. In *Proceedings of the 25th Annual International Conference on Digital Government Research* (pp. 146-155). DOI: 10.1145/3657054.3657075

Correia, A. P., Hickey, S., & Xu, F. (2024). Beyond the virtual classroom: Integrating artificial intelligence in online learning. *Distance Education, 45*(3), 481–491. DOI: 10.1080/01587919.2024.2338706

Kim, J., Lee, H., & Cho, Y. H. (2022). Learning design to support student-AI collaboration: Perspectives of leading teachers for AI in education. *Education and Information Technologies, 27*(5), 6069–6104. DOI: 10.1007/s10639-021-10831-6

Lameras, P., & Arnab, S. (2021). Power to the teachers: An exploratory review on artificial intelligence in education. *Information (Basel), 13*(1), 14. DOI: 10.3390/info13010014

Martinez-Maldonado, R., van Leeuwen, A., & Swiecki, Z. (2023). Artificial intelligence techniques for supporting face-to-face and online collaborative learning. In *Handbook of Artificial Intelligence in Education* (pp. 422–439). Edward Elgar Publishing. DOI: 10.4337/9781800375413.00031

Narayanan, A., Korium, M. S., Melgarejo, D. C., Hussain, H. M., De Sena, A. S., Silva, P. E. G., & Nardelli, P. H. (2022). Collective intelligence using 5G: Concepts, applications, and challenges in sociotechnical environments. *IEEE Access : Practical Innovations, Open Solutions, 10,* 70394–70417. DOI: 10.1109/ACCESS.2022.3184035

Nguyen, L. T., Nguyen, H. N., & Nguyen, S. H. (2023, September). From fragmented data to collective intelligence: a data fabric approach for university knowledge management. In *International Conference on Computational Collective Intelligence* (pp. 16-28). Cham: Springer Nature Switzerland. DOI: 10.1007/978-3-031-41456-5_2

Padovano, A., & Cardamone, M. (2024). Towards human-AI collaboration in the competency-based curriculum development process: The case of industrial engineering and management education. *Computers and Education: Artificial Intelligence, 7,* 100256. DOI: 10.1016/j.caeai.2024.100256

Patil, D. (2024). Impact of artificial intelligence on employment and workforce development: Risks, opportunities, and socioeconomic implications. *Opportunities, And Socioeconomic Implications (November 12, 2024).*

Raghu, N., Mahesh, T. R., Vivek, V., Yogesh Kumaran, S., Kannanugo, N., & Vishwanatha, S. (2025). IoT-Enabled Safety and Secure Smart Homes for Elderly People. In Ertuğrul, D., & Elçi, A. (Eds.), *Future of Digital Technology and AI in Social Sectors* (pp. 297–328). IGI Global Scientific Publishing., DOI: 10.4018/979-8-3693-5533-6.ch011

Raghu, N., Trupti, V. N., Dayananda, L. N., Kumar, H., & Shankaregowda, S. A. (2025). Enhancing Critical Thinking Through Problem-Based Learning: A Comprehensive Approach to Modern Education. In *Fostering Teacher Skills and Critical Thinking in Modern Education* (pp. 259-282). IGI Global Scientific Publishing.

Raman, R., Kowalski, R., Achuthan, K., Iyer, A., & Nedungadi, P. (2025). Navigating artificial general intelligence development: Societal, technological, ethical, and brain-inspired pathways. *Scientific Reports, 15*(1), 1–22. DOI: 10.1038/s41598-025-92190-7 PMID: 40069265

Rensfeldt, A. B., & Rahm, L. (2023). Automating teacher work? A history of the politics of automation and artificial intelligence in education. *Postdigital Science and Education, 5*(1), 25–43. DOI: 10.1007/s42438-022-00344-x

Roschelle, J. (2021, July). Intelligence augmentation for collaborative learning. In *International Conference on Human-Computer Interaction* (pp. 254-264). Cham: Springer International Publishing.

Sumeet Kaur, , Raghu N, , Shivangi Gupta, , & Jagmeet Sohal, . (2024). Adoption of Advanced Systems for Digital Resource Management in Academic Institutions. *Evolutionary Studies in Imaginative Culture*, •••, 913–923. DOI: 10.70082/esiculture.vi.1089

Veena, V., Manjunath, R., Asha, V., Gangaraju, V., Swetha, G., & Raghu, N. (2025). Smart Homes With IoT Technology Provide Safety and Security for Elderly Individuals. In *Real-World Applications of AI Innovation* (pp. 425–450). IGI Global Scientific Publishing.

Yadav, U., & Shrawankar, U. (2025). Artificial intelligence across industries: A comprehensive review with a focus on education. *AI applications and strategies in teacher education*, 275-320.

Zehle, S., Kollegala, R., & Crombie, D. (2021, April). Co-creating value with the cooperative turn: Exploring human-machinic agencies through a collective intelligence design canvas. In *International Conference on Human Interaction and Emerging Technologies* (pp. 155-161). Cham: Springer International Publishing. DOI: 10.1007/978-3-030-74009-2_20

Zhong, N., Wang, Y., Xiong, R., Zheng, Y., Li, Y., Ouyang, M., Shen, D., & Zhu, X. (2024). Casit: Collective intelligent agent system for internet of things. *IEEE Internet of Things Journal, 11*(11), 19646–19656. DOI: 10.1109/JIOT.2024.3366906

Zou, H., Zhao, Q., Lasaulce, S., Bariah, L., Bennis, M., & Debbah, M. (2025). GenAINet: Enabling wireless collective intelligence via knowledge transfer and reasoning. *IEEE Access : Practical Innovations, Open Solutions, 13*, 77764–77777. DOI: 10.1109/ACCESS.2025.3565859

Chapter 10
Human-AI Collaborative Intelligence and Pedagogies of Hybrid Cognition in Learning Ecosystems:
Infrastructure in Post-Industrial Education Design

Ashraf Alam
http://orcid.org/0000-0001-6178-1187
Department of Education, National Institute of Technology Calicut, India

ABSTRACT

This chapter develops a conceptual architecture for Human-AI Collaborative Intelligence and Hybrid Cognition in post-industrial learning ecosystems. Drawing on bounded rationality, distributed cognition, joint cognitive systems, and socio-technical alignment, it shows how collaboration reliability depends on interaction loops, interface regimes, role ecologies, and audit-ready accountability. It maps sectoral task ecologies to modality stacks that build uncertainty articulation, provenance discipline, escalation competence, and recovery-from-error performance. A multilevel capability model aligns individual, team, organizational, and ecosystem strata with curriculum design, assessment validity, and integrity-by-design institutions, concluding with polycentric governance and metric families grounded in procedural justice, capability expansion, ethics of care, privacy, and contestability for globally portable praxis.

DOI: 10.4018/979-8-3373-7036-1.ch010

1. INTRODUCTION

1.1 Industrial Evolution as a Cognitive-Infrastructural Transition

Post-industrial transformation can be read less as a sequence of machines and more as a reconfiguration of cognition, coordination, and control across socio-economic regimes, where the locus of value shifts from muscular effort to epistemic labor, interpretive judgment, and auditable decision-making. The classical coupling of labor with mechanization, standardization, and procedural compliance has matured into cyber-physical production, platform-mediated service delivery, and algorithmically orchestrated workflows, where *Hybrid Cognition* becomes an infrastructural condition rather than a personal talent. In this regime, *Human-AI Collaborative Intelligence* is not synonymous with automation-as-substitution, it is a design problem concerning interaction loops, role ecologies, failure-boundary management, and governance of epistemic authority. Education therefore functions as the cognitive interface of industrial evolution, a societal subsystem that enculturates learners into calibrated trust, anomaly-sensitive oversight, and value-consistent refusal of algorithmic overreach (Padovano & Cardamone, 2024; Chiu & Rospigliosi, 2025; Nguyen et al., 2024). The consequential question is not whether AI is used, but how responsibility, explainability, and contestability are engineered into learning-to-work transitions.

1.2 The Central Design Mismatch and Its Consequential Stakes

The core mismatch is structural, educational systems still privilege individualistic cognition, static syllabi, and credentialing logics optimized for human-only performance, while contemporary work is increasingly organized as distributed, tool-saturated, continuously updating socio-technical production. In AI-mediated environments, competence is expressed through oversight quality, uncertainty articulation, cross-tool triangulation, and documented justification, yet many assessment regimes reward polished outputs and penalize iterative verification (Edwards et al., 2025; Brusilovsky, 2024; Atchley et al., 2024). This misalignment produces predictable pathologies, including automation bias, skill atrophy through passive delegation, brittle reliance under distribution shift, accountability laundering through opaque pipelines, and surveillance creep via analytics infrastructures that normalize extractive monitoring. It also amplifies inequality, because differential access to compute, proprietary platforms, mentorship, and authentic workplace placements becomes a hidden curriculum that stratifies opportunity (Kim, 2024a; Luckin, 2025). The stakes are therefore simultaneously safety-critical, economically material, and civically normative, because AI-mediated judgment is increasingly embedded in healthcare triage, supply-chain allocation, public welfare targeting, and industrial quality assurance. A post-industrial curriculum must therefore be redesigned around reliability, auditability, and rights-consistent decision governance.

1.3 Conceptual Questions and Theoretical Commitments Guiding the Chapter

This chapter is a conceptual-theoretical narrative review, its purpose is to consolidate actionable constructs that can be translated into curriculum architectures, institutional governance, and ecosystem-level policy design, without reliance on empirical case narration. The guiding questions are deliberately interactional and multi-level, asking what constitutes *Cognitive Collaboration* when one agent is a statistical inference system with updateable behavior, how *Meaningful Human Control* can be specified as teachable competence rather than regulatory rhetoric, and which learning ecosystem conditions foster contestability, error recovery, and dignified agency under algorithmic mediation. The chapter commits to *Socio-Technical Systems Theory* as a baseline, where outcomes are emergent properties of coupled subsystems, and to *Distributed Cognition* and *Joint Cognitive Systems* lenses, where cognition is enacted across people, artifacts, representations, and governance routines. It also commits to a normative spine grounded in *Capability-Oriented Governance*, *Procedural Justice*, and *Ethics of Care*, treating equity, privacy, and accountability as design constraints rather than downstream add-ons. These commitments motivate the comparative theoretical synthesis that will be consolidated in Table 1 in Section 2.

1.4 Core Definitions, Scope Delimitations, and Analytic Units

For conceptual precision, *Human-Tool Use* is distinguished from *Human-AI Joint Activity*, where the AI system exhibits partial autonomy in generating recommendations, classifications, simulations, or plans that can materially shape consequential action. *Collaborative Intelligence* refers to coordinated complementarity under explicit accountability, requiring calibrated trust, structured dissent, and recoverability under failure. *Hybrid Cognition* denotes distributed reasoning enacted through interaction loops, interface regimes, documentation artifacts, and workflow placement, not merely human skill augmented by a device. A *Learning Ecosystem* is defined as a complex adaptive network linking universities, TVET institutions, corporate academies, professional bodies, platform providers, informal communities of practice, and credential infrastructures, where pathway mobility and legitimacy are governed by standards and incentives (Dhanasekaran, 2025; Alam, 2022). The scope centers post-secondary, vocational, professional, and workplace learning, because these sites govern role readiness, compliance literacies, and safety behaviors, while still recognizing that earlier schooling shapes epistemic habits. The analytic units used throughout include task ecologies, coordination costs, trust-calibration loops, audit trails, role ecologies, governance layers, and infrastructural imaginaries that determine what forms of collaboration are feasible and justifiable.

1.5 Chapter Architecture, Audience Alignment, and Usability Orientation

The chapter proceeds as a tightly coupled conceptual architecture. Section 2 establishes the theoretical scaffolding across cognition, socio-technical design, and critical-normative lenses, and crystallizes them in Table 1 to prevent category drift. Section 3 reframes education as ecosystem infrastructure and specifies how task ecologies map onto learning modalities, supported by Table 2. Section 4 then offers an integrative framework for capability development across individual, team, organizational, and ecosystem levels, consolidated in Table 3 to make hybrid competence legible for curriculum and assessment design. Section 5 translates the framework into organizational and curricular design patterns, including governance architectures, role ecologies, and integrity regimes, with Table 4 functioning as a compact design grammar that can be operationalized by institutions. Section 6 develops multi-level governance and metric families that align norms, accountability, and evaluation, consolidated in Table 5 as a policy-facing map. Section 7 synthesizes contributions and tensions for academics, policymakers, workforce development professionals, and digital learning technologists, emphasizing implementable levers rather than rhetorical aspiration.

2. THEORETICAL FOUNDATIONS OF HUMAN-AI COGNITIVE COLLABORATION

2.1 Human Cognition, Expertise, and Learning in Complex Work Systems

Human cognition in post-industrial work is best conceptualized as a bounded, resource-constrained, and context-saturated system that must continuously arbitrate between speed, accuracy, and accountability under uncertainty. *Bounded Rationality* and *Ecological Rationality* foreground how decision-makers rely on heuristics, satisficing, and adaptive cue utilization when time, information, and attentional bandwidth are scarce. *Cognitive Load Theory* and *Cognitive Ergonomics* clarify why monitoring-intensive environments produce vigilance decrement, working-memory saturation, and error cascades, especially when interface design externalizes complexity without reducing epistemic responsibility. Expertise is not mere content accumulation, it is *Adaptive Expertise* expressed through pattern sensitivity, anomaly detection, strategy switching, and metacognitive calibration, where *Self-Regulated Learning* stabilizes performance through monitoring, control, and reflective updating (Pahi et al., 2024; Hwang & Lee, 2025; Jony & Hamim, 2024). These constructs establish why human-AI collaboration must be trained as interactional competence and not treated as tool familiarity, a distinction that will be consolidated in Table 1 as a comparative theoretical map.

2.2 Artificial Inference Systems as Epistemic Prostheses and Constraint Regimes

Contemporary AI systems can be treated as probabilistic inference infrastructures that transform high-dimensional inputs into classifications, predictions, recommendations, and generative syntheses, often at scales and latencies unattainable by human cognition. Yet their strengths are inseparable from structural constraints, including *Distribution Shift*, *Model Drift*, and the fragility of generalization beyond the training manifold. In practice, reliability hinges on uncertainty communication that distinguishes *Aleatoric Uncertainty* from *Epistemic Uncertainty*, and on calibration mechanisms that prevent overconfident outputs from masquerading as warranted knowledge. For generative systems, persuasive fluency can coexist with evidential thinness, making *Provenance Reasoning* and *Justification Protocols* essential competencies rather than optional ethics content. For reinforcement-optimized systems, reward specification can induce goal misalignment and pathological optimization, requiring governance of objectives, constraints, and audit trails. These properties justify a design stance in which AI is framed as an *Epistemic Tool* and a governance object, a stance that Table 1 will later connect to pedagogical and assessment implications for calibrated trust and contestability.

2.3 Distributed, Extended, and Joint Cognitive Systems for Hybrid Performance Architectures

Hybrid cognition is most coherently modeled through *Distributed Cognition*, *Extended Mind* constructs, and *Joint Cognitive Systems* thinking, where performance emerges from tightly coupled interaction loops among humans, artifacts, representations, and institutional routines. In this view, cognition is enacted through dashboards, checklists, prompts, logs, alarms, and decision rights, and collaboration quality depends on coordination costs, representational fidelity, and the structure of feedback loops. *Control-Theoretic Coupling* clarifies why automation can either reduce workload or increase monitoring burden, depending on observability, interpretability, and the stability of handoff protocols. Team-level reliability further depends on *Shared Mental Models*, transactive memory, and structured dissent that preserves epistemic diversity while converging on accountable action. Table 1 operationalizes these theoretical commitments by specifying how different lens families define cognition, construe AI, diagnose failure signatures, and imply distinct pedagogical and governance designs for human-AI joint activity.

Table 1. Integrative Lenses for Human-AI Cognitive Collaboration and Education Design

Lens Architecture Family	Cognition Ontology for Hybrid Work	AI Role Construction and Epistemic Status	Failure Signatures in Human-AI Joint Activity	Design Implications for Pedagogy and Governance
Information-Processing and Bounded Rationality	Limited-capacity computation, attention-allocation, and heuristic cue utilization under time pressure	Decision aid that compresses information, amplifies salience, and reshapes choice architecture	Automation bias, attentional tunneling, over-weighted salience cues, mode confusion under workload	Train error-sensitive decision hygiene, uncertainty articulation, and workload-aware interface literacy, align assessment with process accountability
Expertise, Metacognition, and Self-Regulation	Adaptive expertise, monitoring-control loops, reflective judgment, and calibration of confidence	Cognitive prosthesis that extends search, pattern detection, and hypothesis space exploration	Skill atrophy through passive delegation, miscalibrated trust, reduced epistemic vigilance, overconfident acceptance	Teach trust calibration, verification routines, decision logging, and metacognitive checkpoints, evaluate justification quality not output polish
Situated, Embodied, and Enactive Cognition	Action-in-context, affordance perception, embodied sensemaking, and practice-bound knowing	Situated artifact whose meaning is enacted through workflows, roles, and material settings	Context collapse, decontextualized recommendations, brittle application across settings, misfit with tacit constraints	Design authentic simulations, apprenticeship-like studios, and context-rich tasks, require articulation of local constraints and boundary conditions

continued on following page

Table 1. Continued

Lens Architecture Family	Cognition Ontology for Hybrid Work	AI Role Construction and Epistemic Status	Failure Signatures in Human-AI Joint Activity	Design Implications for Pedagogy and Governance
Distributed Cognition and Joint Cognitive Systems	Cognition distributed across people, artifacts, representations, and feedback loops	Component in a coupled control system, shaping observability, controllability, and coordination	Handoff failures, weak audit trails, degraded situation awareness, escalation breakdowns	Train handoff protocols, redundancy practices, structured dissent, and resilience routines, assess joint performance and recovery competence
Sociotechnical, Sociomaterial, and Institutional Logics	Emergent performance from coupled social-technical subsystems, metrics, routines, and legitimacy pressures	Infrastructural layer embedded in procurement, compliance, platform governance, and algorithmic management	Accountability laundering, surveillance creep, incentive gaming, vendor lock-in, slow institutional adaptation	Build governance boards, data stewardship, change-management protocols, integrity-by-design assessment, align incentives with contestability and equity
Critical-Normative, Data Justice, and Decolonial Frames	Power-laden epistemics, distributive consequences, and contested legitimacy of knowledge infrastructures	Normative actor via classification regimes, resource allocation, and epistemic gatekeeping	Representational harms, exclusion through language and access, epistemic extraction, inequitable error burdens	Embed rights-consistent safeguards, multilingual and accessibility-first design, contestation channels, equity metrics, and community-accountable governance

The integrative value of Table 1 is that it prevents conceptual reductionism by showing that collaboration failure is rarely a purely technical defect, it is commonly a misalignment among cognitive demands, interface regimes, institutional incentives, and normative constraints. This matters for later sections because capability taxonomies and curriculum architectures must explicitly target joint performance properties, including uncertainty-literate oversight, structured escalation, and auditable justification, rather than generic tool proficiency. The table also clarifies why assessment reform is non-negotiable, because traditional individual-output grading cannot validly measure distributed competence, resilience, or contestability. By making failure signatures legible across lens families, Table 1 supports design choices that are simultaneously instructional, organizational, and policy-relevant, thereby setting up the transition to ecosystem infrastructure analysis in Section 3 and to multi-level capability alignment in Section 4.

2.4 Sociotechnical and Sociomaterial Theorization of Power, Routines, and Infrastructures

Human-AI collaboration is always embedded within a sociotechnical ecology where task allocation, decision rights, accountability pathways, and value hierarchies are configured through organizational routines and infrastructural choices. *Socio-Technical Systems Theory* positions performance as joint optimization across social and technical subsystems, implying that AI integration without role redesign, incentive alignment, and governance protocols will predictably generate brittleness and blame diffusion. *Sociomateriality* sharpens this claim by emphasizing that dashboards, rubrics, and analytics are not neutral carriers of information, they are performative artifacts that reshape what counts as competence, evidence, and legitimacy. *Activity-Theoretic* reasoning adds diagnostic power through contradiction analysis, surfacing tensions among objectives, rules, tools, and division of labor that often precipitate unsafe over-automation or covert workarounds. Institutional dynamics, including path dependence and legitimacy pressures, explain why education systems adopt symbolic AI policies while leaving assessment regimes unchanged (Virvou et al., 2024; Nguyen, 2025; Mittal et al., 2024). Table 1 synthesizes these mechanisms into actionable design implications, particularly governance of procurement, auditability, integrity norms, and platform dependence.

2.5 Critical-Normative Architectures for Responsible and Equitable Hybrid Cognition

A purely instrumental framing of human-AI collaboration collapses under the weight of normative obligations in high-stakes domains, where errors distribute harm unevenly and where procedural legitimacy matters as much as predictive accuracy. Rights-consistent design requires *Procedural Justice* mechanisms, including reason-giving, contestability channels, appeal pathways, and documentation that can withstand audit scrutiny. A *Capability-Oriented* stance reframes education policy toward expanding agency, not merely supplying labor-market skills, which makes access, affordability, and linguistic inclusivity core infrastructure variables (Wen et al., 2024; Kim, 2024b; Kim & Cho, 2025). *Ethics of Care* further requires that dependency relations, vulnerability, and relational accountability are explicitly addressed in training designs, especially when AI is used in guidance, triage, and welfare contexts. Data infrastructures must be evaluated through *Data Justice* lenses that scrutinize representational harms, surveillance externalities, and extractive value flows. Table 1 consolidates these critical-normative frames as design constraints that should be translated into governance standards, equity metrics, and integrity-by-design assessment cultures in Sections 5 and 6.

3. LEARNING ECOSYSTEMS AS SOCIO-TECHNICAL INFRASTRUCTURES FOR INDUSTRIAL EVOLUTION

3.1 Learning Ecosystems as Complex Adaptive Infrastructures for Knowledge Circulation

A learning ecosystem is best theorized as a *Complex Adaptive System* in which institutions, platforms, professional bodies, labor-market intermediaries, and informal communities co-produce competence, legitimacy, and mobility through feedback-rich pathways rather than linear pipelines (Dai et al., 2025; Clegg & Sarkar, 2024; Alam, 2023a). This ecosystem is simultaneously an epistemic commons and a governed infrastructure, where credential ecologies, interoperability standards, and platform architectures function as coordination devices that shape what is learnable, assessable, and employable (Lin & Chen, 2025; Tzirides et al., 2024; Lee et al., 2025). *Ecological Systems Theory*, *Network Governance*, and *Institutional Logics* together clarify why ecosystem performance depends on diversity, redundancy, and resilience, not merely on excellence within a single university or training provider. Infrastructural nodes such as learning management systems, analytics layers, simulation environments, and digital credential wallets create new forms of visibility, stratification, and path dependency, especially when vendor lock-in constrains curricular agility (Duraimutharasan et al., 2025; Maphoto et al., 2024; García-López et al., 2025). As Table 2 later specifies through sectoral task ecologies, the ecosystem must be engineered for cross-context transfer, auditable competence, and equitable pathway permeability, not only content delivery.

3.2 Education as the Cognitive Interface Between Industrial Systems and Human Development

Education functions as the cognitive interface of industrial evolution by translating emergent work practices into learnable sequences, normative constraints, and professional identities, while simultaneously adjudicating what forms of automation are acceptable under safety, dignity, and justice commitments (Arar et al., 2024; Giannakos et al., 2025; Kim, et al., 2024a). *Capability-Oriented Governance* reframes this interface away from narrow employability and toward substantive agency, where learners acquire not only technical fluency but also contestability literacy, rights-consistent reasoning, and epistemic self-determination in algorithmically mediated environments. This interface is also a site of *Subjectivation* and professional enculturation, where learners internalize audit sensibilities, documentation norms, escalation protocols, and quality-assurance mentalities that stabilize socio-technical reliability (Kim, 2024c; Wong & Looi, 2024; Dhillon et al., 2024). Crucially, the interface is bidirectional, industry task architectures shape curricula, and educational standards shape industry legitimacy through licensing, accreditation, and competency signaling. The theoretical lens consolidated in Table 1 foregrounds cognition as distributed and value-laden, while Table 2 in this section operationalizes the interface by mapping sectoral human-AI task ecologies onto modality stacks that can be built across formal, non-formal, and workplace learning.

3.3 Sectoral Task Ecologies and Modality Stacks for Human-AI Joint Activity

Sectoral task ecologies can be conceptualized as recurrent bundles of sensemaking, triage, monitoring, optimization, and exception-handling that are performed under uncertainty, time pressure, and regulatory constraint, with AI systems entering as inference infrastructures that reshape evidence thresholds and decision latencies (Marrone et al., 2025; Creely & Blannin, 2025; Nikolopoulou, 2024). A modality stack for hybrid cognition therefore must integrate simulation fidelity, case-based reasoning, studio critique, work-integrated apprenticeship, and reflective audit practices, so that learners practice *Trust Calibration*, *Uncertainty Articulation*, *Provenance Reasoning*, and recovery from failure rather than merely generating outputs. Table 2 provides a compact mapping between sectoral domains, dominant joint-task signatures, cognitive demand profiles, and ecosystem learning modality stacks, while also specifying misalignment and risk vectors that emerge when curricula remain individualistic, assessment remains output-centric, or governance remains tool-naive (Omran Zailuddin et al., 2024; Wood & Moss, 2024; Kayyali, 2024). This mapping is intentionally infrastructure-oriented, because task ecologies are stabilized by interfaces, documentation artifacts, and accountability pathways that must be taught as part of competence formation rather than treated as administrative residue.

Table 2. Sectoral Human-AI Task Ecologies Mapped to Learning Modalities

Sectoral Domain	Joint Task Ecology Signature	Dominant Cognitive Demand Profile	Ecosystem Learning Modality Stack	Misalignment and Risk Vector Profile
Healthcare	AI-informed triage, diagnostic patterning, clinical decision support, safety escalation, documentation for accountability	Probabilistic inference, causal reasoning under uncertainty, risk-sensitive judgment, ethical constraint satisfaction	Simulation-based checkrides, case-based diagnostic forensics, supervised practicum, reflective decision logs, interdisciplinary morbidity-style review	Automation bias, premature closure, provenance neglect, liability diffusion, inequitable error burdens
Manufacturing and Industry 4.0	Predictive maintenance, anomaly detection, process control, digital-twin orchestration, quality surveillance	Signal detection, systems thinking, control-theoretic reasoning, workload-aware monitoring	Digital-twin labs, instrumentation studios, incident learning drills, team-based reliability exercises, audit-ready documentation routines	Vigilance decrement, alarm fatigue, brittle escalation, metric gaming, deskilling through over-delegation
Logistics and Supply Chains	Demand forecasting, routing optimization, inventory policy, disruption response, multi-actor coordination	Multi-criteria optimization, constraint reasoning, scenario planning, coordination under volatility	Scenario war-gaming, optimization studios, platform-mediated practicum, cross-role negotiation labs, traceability and compliance simulations	Objective mis-specification, unfair allocation, opacity in prioritization, overfitting to stable conditions, accountability laundering
Agriculture and Food Systems	Precision advisory, climate-risk inference, pest and yield prediction, resource allocation, extension communication	Ecological sensemaking, uncertainty management, temporal reasoning, local knowledge integration	Field-embedded apprenticeship, data-literacy extension clinics, participatory design studios, low-bandwidth toolchains, reflective boundary-condition articulation	Context collapse, exclusion via connectivity gaps, epistemic extraction, brittle model transfer, inequitable access to advisory infrastructures
Public Services and Cross-Sector Operations	Eligibility targeting, risk scoring, fraud detection, service triage, procedural fairness governance	Procedural justice reasoning, transparency and contestability, evidence grading, harm minimization	Policy-lab simulations, ethics and audit rounds, documentation-as-governance studios, appeals-process rehearsals, accountability mapping practicums	Surveillance creep, discriminatory classification, opaque denial rationales, chilling effects, weakened due process

The mapping in Table 2 is pedagogically consequential because it makes visible a design principle that remains routinely under-articulated, namely that hybrid cognition is enacted through modality stacks that train oversight, contestation, and recovery, not through isolated tool demonstrations. It also clarifies a governance principle that later sections will extend, namely that misalignment risks are predictable and therefore preventable when curricula explicitly teach uncertainty interpretation, documentation-as-accountability, and escalation competence as core outcomes. By embedding sectoral differences in cognitive demand profiles and risk vectors, Table 2 supports a learning ecosystem architecture that is modular, interoperable, and equity-aware, while remaining sensitive to domain-specific constraints such as safety-criticality, regulatory burden, and local knowledge dependencies.

3.4 Structural Misalignments and Tension Fields Within Learning Ecosystems

Misalignment arises when fast-evolving AI-mediated task architectures collide with slow-moving educational governance, resulting in curricular ossification, assessment invalidity, and integrity drift. *Institutional Isomorphism* and *Path Dependence* explain why systems preserve familiar course shells, credit-hour logics, and individualistic grading even when work demands joint performance, documentation competence, and team-based reliability (Alam, 2023b; Yan, et al., 2024a; Bansal et al., 2024). Assessment misfit is particularly corrosive, because output-centric evaluation incentivizes plausible completion rather than epistemic warrant, encouraging unexamined delegation and eroding *Epistemic Vigilance* precisely when AI fluency increases the plausibility of error. Governance speed mismatch also appears in procurement lock-in and platform dependency, where toolchains become infrastructural commitments that constrain pedagogical experimentation, auditability, and accessibility (Bettayeb et al., 2024; Sun et al., 2024; Tariq, 2024). These tensions create predictable failure signatures that Table 2 already anticipates across sectors, including accountability laundering, surveillance expansion through analytics, and hidden stratification through unequal tool access. A coherent ecosystem response therefore requires aligning curricular update cycles, integrity regimes, and capability targets around joint performance properties, including contestability, traceability, and resilience under distribution shift.

3.5 Equity, Inclusion, and Spatial Justice as First-Order Infrastructure Constraints

Equity in human-AI learning ecosystems is not a supplemental support function, it is an infrastructural property shaped by language regimes, connectivity architectures, accessibility design, credential portability, and the distributive politics of platform governance. *Intersectionality* clarifies how disadvantages compound across gender, caste, class, disability, geography, and linguistic marginalization, producing unequal access to compute, mentorship, authentic practicum sites, and audit-capable learning environments (Ifenthaler et al., 2024; Rawas, 2024). *Spatial Justice* and center-periphery dynamics illuminate how dependence on external platforms can intensify epistemic extraction, diminish local curricular sovereignty, and normalize benchmark universalism that discounts situated knowledge, particularly in agriculture, public services, and community health. Inclusion therefore requires accessibility-first interfaces, multilingual scaffolding, low-bandwidth modality stacks, and assessment designs that do not presuppose premium tool access as a proxy for competence (Nithithanatchinnapat et al., 2024; Akpan et al., 2025). Table 2 already encodes these equity sensitivities by showing how risk vectors differ by sector and infrastructure context, and later sections will translate them into governance levers and metric families that treat fairness, contestability, and dignity as measurable ecosystem obligations rather than aspirational rhetoric.

4. CONCEPTUAL FRAMEWORK FOR HUMAN-AI COLLABORATIVE LEARNING AND CAPABILITY DEVELOPMENT

4.1 Multilevel Systems Scaffold for Collaborative Intelligence in Hybrid Ecologies

A robust framework for *Human-AI Collaborative Intelligence* must be explicitly multilevel, because hybrid performance is an emergent property of nested systems rather than an isolated individual trait. At the individual level, cognition is governed by *bounded rationality*, *cognitive-load* limits, and *metacognitive control*, which determine whether AI outputs are interrogated as provisional evidence or accepted as authoritative (Yan, et al., 2024b; Al-Zahrani & Alasmari, 2024; Butson & Spronken-Smith, 2024). At the team level, *shared mental models*, transactive memory, and coordination bandwidth shape joint sensemaking, escalation, and accountability distribution. At the organizational level, routines, incentives, procurement constraints, and *knowledge-management* architectures determine whether collaboration is resilient or brittle under *model drift* and workflow volatility (Alam, 2023c). At the ecosystem level, credential regimes, professional standards, and platform dependencies govern pathway mobility and equity. Table 3 will formalize these levels as designable units with aligned tasks, modalities, and evidence standards.

4.2 Capability Taxonomy for Hybrid Cognition, Oversight, and Contestability

Capability development for hybrid cognition requires a taxonomy that treats epistemics, ethics, and systems engineering as coupled competencies rather than separable modules. Core clusters include *data provenance reasoning* and uncertainty literacy, where learners distinguish *epistemic uncertainty* from stochastic noise, articulate confidence boundaries, and detect distributional mismatch through anomaly-sensitive heuristics (Bulathwela et al., 2024; Francis et al., 2025). Interaction competence must be framed as specification literacy, where prompting, tool orchestration, and adversarial questioning function as constraint design, not conversational flourish. *Metacognitive oversight* includes trust calibration, error detection, and fallback strategies that prevent automation bias and safeguard reliability (Sadeghi & Niu, 2024; Nguyen et al., 2024; Kaswan et al., 2024). Team capabilities include structured dissent, escalation choreography, and cross-role translation that reduces coordination loss and accountability laundering. Normative competence includes rights-consistent reasoning, procedural fairness, and *ethics of care* obligations in high-stakes contexts. Table 3 will integrate these clusters with level-specific modalities and assessment evidence, making the taxonomy actionable for curriculum and governance design.

4.3 Modalities of Human-AI Collaboration Aligned to Levels and Evidence

Human-AI collaboration in learning must be designed as modality selection under constraints, where each modality produces distinct cognitive demands, integrity risks, and evidence requirements. AI as tutor emphasizes adaptive feedback, yet it can induce dependency without reflective checkpoints (Topali et al., 2025; Chan & Tsi, 2024; Luo, 2024). AI as collaborator accelerates ideation and drafting, yet it can homogenize epistemic diversity unless learners practice justification and counterfactual testing. AI as critic strengthens dialectical rigor, yet it can generate false authority effects if critique is not triangulated and audited (Li et al., 2025; Cukurova, 2025; Marrone et al., 2024). AI as simulator supports rare-event rehearsal and resilience training, yet simulation validity depends on boundary-condition articulation. AI as workflow component embeds toolchains into practice, shifting assessment toward documentation, escalation, and recovery competence (Senthilkumar et al., 2024; Sarkar, 2024; Celik et al., 2024). Table 3 consolidates these modalities into a multilevel alignment model that links task ecologies, capability clusters, and auditable evidence across learning ecosystems.

Table 3. Multilevel Alignment for Human-AI Collaborative Learning Capabilities

Level Anchor	Task Ecology Emphasis	Collaboration Modality Archetypes	Capability Clusters and Epistemic Virtues	Pedagogical and Assessment Evidence
Individual	Evidence appraisal, anomaly detection, decision justification under uncertainty	Tutor-coach, critic, retrieval-augmentation companion	*Metacognitive monitoring*, uncertainty articulation, provenance reasoning, *intellectual humility*	Decision logs, uncertainty statements, triangulation traces, oral defense of warrants
Team	Handoffs, escalation, shared situational awareness, coordinated exception handling	Co-pilot orchestration, debate partner, joint simulator	Shared mental models, structured dissent, coordination protocols, *epistemic courage*	Team checkrides, escalation drills, collaborative audit trails, post-incident learning memos
Organization	Workflow design, quality assurance, compliance, change management under tool drift	Workflow component, governance-aware recommender, monitoring agent	*Sociotechnical alignment*, accountability mapping, documentation-as-governance, *prudential judgment*	Process audits, integrity dashboards, workflow FMEA-style reviews, policy-conformant artifacts
Ecosystem	Credential portability, standards alignment, equity safeguards, platform interoperability	Standards mediator, cross-institution knowledge broker, pathway recommender	*Institutional literacy*, data justice competence, contestability channels, *civic integrity*	Cross-context portfolios, standards-mapped rubrics, equity audits, appeals and redress simulations
Cross-Level Coupling	Feedback loops, incident learning, resilience across settings and time horizons	Multi-agent ensemble governance, simulation-to-practice transfer	Calibration loops, drift detection, resilience engineering, *reflective equilibrium*	Longitudinal competence dossiers, transfer tasks, stress-test scenarios, recovery-from-error performance

Table 3 serves as an alignment instrument that makes hybrid cognition measurable without collapsing it into simplistic output metrics. By tying each level to task ecologies, modality archetypes, capability clusters, and auditable evidence, it supports constructive alignment across curriculum, assessment, and governance. The table also helps prevent integrity drift by specifying evidence forms that reward epistemic warrant, escalation competence, and recoverability rather than fluent production. In later sections, this alignment logic will be translated into organizational design patterns and metric families, while preserving equity constraints through ecosystem-level portfolio standards and contestability mechanisms.

4.4 Learning Progressions, Assessment Validity, and Evidence of Joint Competence

Learning progressions for collaborative intelligence should be designed as staged movement from tool literacy to oversight mastery to workflow co-design and governance fluency, with explicit thresholds that learners must cross to prevent unsafe delegation. Early progression emphasizes uncertainty-aware interpretation and provenance discipline, while mid progression targets calibration, cross-tool triangulation, and error-detection routines under time pressure (Karakose & Tülübas, 2024; Suryanarayana et al., 2024; Imran et al., 2024). Advanced progression requires learners to design escalation protocols, document justification chains, and anticipate second-order effects through systems thinking. Assessment validity must shift from artifact-centric scoring to evidence of reasoning quality, where decision logs, oral defenses, simulation checkrides, and recovery-from-error tasks demonstrate *process integrity* and *meaningful human control* (Alam & Mohanty, 2022; Yu et al., 2025; Yan, et al., 2024c). Table 3 clarifies which evidence types correspond to which levels, reducing construct-irrelevant variance and limiting gaming incentives. Equity and accessibility must be treated as assessment design parameters, where tool access parity, low-bandwidth alternatives, and multilingual supports ensure that measured competence reflects capability rather than platform privilege.

4.5 Integrative Alignment Model Linking Tasks, Capabilities, Modalities, and Governance

The integrative model for human-AI collaborative learning is a constraint-based alignment logic, where industrial task ecologies determine feasible collaboration modalities, modalities determine capability clusters, and capability clusters determine assessment evidence and governance safeguards (Mustafa et al., 2024; Fan et al., 2025; Peláez-Sánchez et al., 2024). The model treats *feedback architecture* as central, linking curriculum review to incident learning, drift monitoring, and standards updates, so that learning ecosystems remain adaptive rather than episodic. Governance is embedded through documentation protocols, contestability channels, and audit readiness, ensuring that collaboration remains accountable across organizational boundaries and lifecycle stages (Liang & Bai, 2025; Chee et al., 2025; Adel & HS Alani, 2024). Table 3 operationalizes this logic by specifying cross-level coupling as a design object, enabling institutions to engineer calibration loops, resilience routines, and equitable pathway mobility. This alignment also provides a practical bridge from ecosystem mapping in Section 3 to organizational and curricular design in Section 5, ensuring coherence between capability formation, institutional role ecologies, and policy-facing metric systems.

5. ORGANIZATIONAL AND CURRICULAR DESIGN FOR HUMAN-AI CO-EVOLUTION

5.1 Educational Organizations as Co-evolving Sociotechnical Learning Organisms

Educational organizations must be conceptualized as co-evolving sociotechnical learning organisms whose viability depends on continuous adaptation to shifting AI capabilities, platform architectures, and labor-market task ecologies, rather than episodic curricular revision (Rane & Choudhary, 2024; Łodzikowski et al., 2024). *Organizational Learning* frames this as iterative sensemaking and capability renewal, where *Single-Loop Learning* optimizes procedures while *Double-Loop Learning* rewrites governing assumptions about competence, integrity, and accountability under algorithmic mediation. Co-evolution further implies mutual shaping between institutional routines and external infrastructures, including procurement ecosystems, accreditation regimes, and professional standards that determine what counts as legitimate evidence of skill (Tarisayi, 2024; Sowa & Przegalinska, 2025; Jiang et al., 2024). The multilevel alignment logic consolidated in Table 3 requires institutions to design for reliability and recoverability, not only for innovation velocity, because *model drift* and workflow volatility are structural properties of contemporary toolchains. This requires epistemic stewardship, audit-capable documentation cultures, and investment in simulation and practicum infrastructures that make hybrid cognition teachable, measurable, and transferable across settings.

5.2 Governance Architectures and Role Ecologies for Accountable AI-Mediated Education

Governance for AI-mediated education must be engineered as layered accountability architecture that couples normative constraints with operational control, ensuring that tool adoption does not outpace integrity, privacy, and equity safeguards. Institutional governance must include procurement due diligence, change-management protocols, and documentation standards that treat AI systems as evolving infrastructure rather than static software (Yim & Su, 2025; Guggemos, 2024; Anane-Simon & Atiku, 2024). Role ecologies must expand beyond traditional faculty-administrator binaries into distributed expertise, including learning engineers, assessment validity designers, simulation architects, educational data stewards, integrity officers, and platform accountability leads who translate policy into practice. *Role Theory* and professional identity constructs explain why such reconfiguration requires boundary negotiation, legitimacy building, and communities-of-practice that stabilize new norms of collaboration. Governance must also create contestability channels for learners and staff, aligning with *procedural justice* and reducing accountability laundering. Table 4 will later crystallize these design dimensions into reusable patterns that preserve agility while maintaining auditability, accessibility, and rights-consistent decision-making across institutional subsystems.

5.3 Curricular Architectures for Collaborative Intelligence and Hybrid Professionalism

Curricular architecture for collaborative intelligence should be designed as vertically integrated pathway engineering rather than isolated elective insertion, ensuring that capability development progresses from uncertainty literacy to oversight mastery to workflow co-design and governance fluency (Kim, et al., 2024b; Han et al., 2024; Bozkurt, 2024). Constructive alignment requires mapping outcomes to evidence forms specified in Table 3, so that assessment measures justification quality, escalation competence, and recoverability rather than output fluency. A transdisciplinary spine can integrate domain epistemics, data and model literacy, systems thinking, and normative reasoning, enabling learners to operationalize *meaningful human control* as a professional habit (Peters & Green, 2024; Li, Z., et al., 2024; Leong & Zhang, 2025). Work-integrated learning must be treated as a high-fidelity socio-technical practicum where learners rehearse documentation-as-governance, compliance literacies, and cross-role translation, mirroring sectoral task ecologies described in Table 2. Credential ecologies should be designed for coherence, using stackable micro-credentials and portfolio standards that preserve portability without fragmenting identity formation. These architectural choices are operationalized in Table 4 as design patterns linking curricular form to capability outcomes and governance safeguards.

Table 4. Design Patterns for Human-AI Collaborative Curriculum and Governance

Design Dimension	Pattern Grammar for Institutional Architecture	Theoretical Rationale and Epistemic Spine	Expected Capability and Integrity Effects	Foreseeable Unintended Consequences and Mitigations
Governance and Accountability	Layered AI-governance councils, procurement audits, change-control gates, contestability pathways	*Socio-Technical Systems Theory*, *Procedural Justice*, accountability mapping	Increases auditability, reduces accountability laundering, stabilizes trustworthy adoption	Bureaucratic drag, compliance theater, mitigate via lightweight controls and transparent decision logs
Role Ecology and Professionalization	Learning engineers, data stewards, simulation leads, assessment validity designers, integrity officers	*Role Theory*, boundary work, *Communities of Practice*	Builds distributed expertise, improves design coherence, strengthens integrity culture	Role ambiguity, turf conflicts, mitigate via charters, training ladders, shared performance criteria
Curricular Spine and Progression	Vertical strands in uncertainty literacy, oversight, workflow co-design, governance fluency	*Constructive Alignment*, threshold concepts, staged learning progressions	Produces transferable hybrid cognition, reduces tool-naive fluency, increases resilience under drift	Over-standardization, mitigated by contextual electives, local knowledge integration, adaptive pathways
Pedagogical Modality Stack	Simulation checkrides, studio critique, case forensics, apprenticeship, reflective audit rounds	*Distributed Cognition*, *Resilience Engineering*, epistemic vigilance	Strengthens recovery-from-error, escalation competence, justification discipline	Simulation realism gaps, mitigate via boundary-condition articulation and iterative model validation
Assessment and Integrity Regime	Evidence portfolios, oral defenses, decision logs, team-based performance audits	Assessment validity, process integrity, reliability orientation	Rewards epistemic warrant, discourages hidden delegation, improves equity of evaluation	Gaming via templated logs, mitigate via randomization, cross-tool triangulation requirements, audit sampling

Table 4 functions as a compact institutional design grammar that translates multilevel theory into actionable organizational and curricular patterns without reducing complexity into simplistic prescriptions. It makes explicit that reliable human-AI collaboration depends on governance and role ecology as much as on pedagogy, because capability formation is shaped by incentives, platform constraints, and audit standards. The patterns also clarify that integrity is not enforced only through rules, it is produced through assessment designs that value reasoning, documentation, and recoverability, aligning with the evidence logic introduced in Table 3. Later, Section

6 will map these design patterns into multi-level governance and metric families, ensuring that institutional strategies remain interoperable with sector standards, accreditation constraints, and equity obligations.

5.4 Pedagogical Models and Learning Design for Hybrid Cognition and Epistemic Responsibility

Pedagogical models for hybrid cognition must be designed around interaction quality and epistemic responsibility, where learners learn to interrogate AI outputs through structured skepticism, counterfactual probing, and cross-source triangulation rather than passive acceptance. *Problem-Based Learning* can be reconfigured into constraint-based inquiry, with AI used as hypothesis generator and critic, while learners practice evidence grading, uncertainty articulation, and decision warranting (Alam & Mohanty, 2023; Goldman et al., 2024). Studio and lab pedagogies can operationalize *cognitive apprenticeship*, making tacit judgment visible through critique rituals, justification dialogues, and documented iteration that supports audit readiness. Simulation-based learning and digital-twin environments can train rare-event response, escalation choreography, and recovery competence, aligning with *resilience engineering* and Safety-II orientations (Mao et al., 2024; Cai et al., 2025; Cingillioglu et al., 2024). Dialogic pedagogy and critical design approaches can cultivate contestability literacy, enabling learners to surface value assumptions embedded in classification and optimization. Table 4 reinforces that modality stacks must be paired with assessment regimes that reward reasoning and recoverability, preventing integrity drift and dependency formation.

5.5 Institutional Culture, Affective Infrastructures, and Normative Alignment for Co-evolution

Institutional culture is an affective and normative infrastructure that governs whether learners and staff can practice dissent, report near misses, and revise workflows without fear, which is essential for sustaining reliable human-AI collaboration. *Psychological Safety* enables incident learning, while affective conditions such as AI anxiety, techno-optimism, cynicism, and obsolescence fear shape adoption, resistance, and integrity compromises (Storozhyk, 2024; Tariq, 2025). Culture design must therefore normalize epistemic humility, structured disagreement, and documentation discipline as professional virtues, not as bureaucratic burdens. Integrity regimes should establish clear boundaries between assistance and delegation, supported by disclosure norms, provenance expectations, and assessment designs that reduce incentives for covert tool substitution (Noroozi et al., 2024; Milana et al., 2024; Kim et al., 2025). Equity culture must address hidden stratification created by unequal tool access and mentorship, ensuring that pathway mobility reflects capability rather than platform privilege. The governance and assessment patterns in Table 4 provide concrete levers for embedding these cultural commitments into routines, role expectations, and evaluative architectures, enabling co-evolution that is agile, accountable, and rights-consistent.

6. GOVERNANCE, METRICS, AND NORMATIVE FRAMEWORKS FOR HUMAN-AI COLLABORATIVE LEARNING ECOSYSTEMS

6.1 Normative Foundations and Policy Principles for Hybrid Learning Sovereignty

Governance for *Human-AI Collaborative Learning* must begin with normative primitives that function as design constraints rather than aspirational slogans, because hybrid cognition reorganizes epistemic authority, error attribution, and distributive consequences. A rights-consistent baseline couples *Human Dignity*, *Non-Discrimination*, and *Due Process* with *Capability-Oriented Governance*, ensuring that AI-mediated learning expands agency, reduces vulnerability, and preserves contestability under automated influence (Joseph & Uzondu, 2024; Fu & Weng, 2024; Ruiz-Rojas et al., 2024). *Ethics of Care* adds relational accountability by treating dependency, power asymmetry, and contextual vulnerability as first-order variables, especially where AI shapes guidance, triage, or credential gating. Safety and integrity require *Meaningful Human Control* as a teachable competence, operationalized through justification discipline, escalation choreography, and auditable decision logs (Li, H., et al., 2024; Przegalinska & Triantoro, 2024; Jung & Suh, 2024). Privacy and autonomy require data minimization, purpose limitation, and transparency that is actionable, not performative. The multi-level translation of these principles into instruments and evidence is consolidated in Table 5, which functions as a normative-to-operational map for Section 6 and a bridge to Section 7.

6.2 Polycentric Governance Architectures Across Classroom, Institution, Sector, and System

Effective steering of AI-mediated learning ecosystems demands *Polycentric Governance* and *Multi-Level Regulation*, because control points are distributed across classrooms, procurement offices, accreditation bodies, platform vendors, and public funding agencies. At the micro level, course policies should encode disclosure norms, provenance expectations, and process accountability so that learners practice epistemic warrant rather than output theatrics (Fissore et al., 2024; Pratschke, 2024;

Kolbjørnsrud, 2024). At the meso level, institutions require procurement due diligence, lifecycle change-control, model-update traceability, and incident reporting routines that treat AI as evolving infrastructure, not a static educational add-on. At the sector level, professional bodies can stabilize competence definitions through standards for auditability, supervision, and safe delegation, ensuring that credential signals remain legible across employers (Chuang, 2024; Asad & Ajaz, 2024). At the macro level, qualification frameworks and accreditation criteria can require evidence of oversight competence, equity safeguards, and privacy-preserving analytics. Table 5 integrates these layers into one coherent governance topology, reducing fragmentation and enabling interoperability across jurisdictions and platform ecosystems.

6.3 Measurement Architectures and Indicator Families for Collaborative Intelligence

Metrics for hybrid learning ecosystems must move beyond simplistic attainment proxies toward construct-valid indicators that capture oversight quality, contestability, resilience under drift, and equitable distribution of benefits and burdens (Bozkurt et al., 2024; Crawford et al., 2024; Park & Ahn, 2024). Measurement design should treat collaborative intelligence as a latent capacity expressed through auditable behaviors, including uncertainty articulation, cross-tool triangulation, escalation competence, and recovery-from-error performance. Indicator families should therefore cover capability development, process integrity, reliability outcomes, well-being externalities, and equity gradients, while explicitly managing *Goodhart Effects* and gaming incentives through multi-source evidence and audit sampling (Wang et al., 2025; Kaspersen et al., 2024; Alam & Mohanty, 2024). Table 5 specifies these indicator families by governance stratum, linking normative anchors to policy instruments, metric artifacts, and risk signatures, so evaluators can select evidence that is fit-for-purpose and defensible under scrutiny.

Table 5. Multi-Level Governance Map for Accountable Human-AI Learning Ecosystems

Governance Stratum	Normative Anchors for Legitimate Hybrid Cognition	Policy Instruments and Control Routines	Metric Families and Evidence Artifacts	Risk Signatures and Mitigation Levers
Classroom and Program	*Procedural Justice, Learner Agency, Epistemic Responsibility*	AI-use disclosure rules, provenance protocols, oral defense norms, escalation rehearsal	Decision logs, uncertainty statements, simulation checkrides, contestation rubrics	Automation bias, hidden delegation, mitigate via triangulation tasks and audit sampling
Institution	*Integrity-by-Design, Privacy Preservation, Accessibility as Infrastructure*	Procurement due diligence, lifecycle change-control, data stewardship, incident reporting	Integrity dashboards, privacy impact reviews, accessibility audits, portfolio moderation	Surveillance creep, vendor lock-in, mitigate via minimization, interoperability clauses, exit plans
Sector and Professional Body	*Public Safety, Professional Accountability, Supervision Fidelity*	Competency standards for oversight, supervision rules, certification renewal requirements	Standards-mapped assessments, supervised practice records, audit-ready portfolios	Deskilling, accountability laundering, mitigate via supervision thresholds and documentation norms
National and Regional System	*Equity of Opportunity, Capability Expansion, System Reliability*	Accreditation criteria, funding conditionalities, qualification frameworks, regulatory sandboxes	Equity indices, pathway mobility indicators, system-level incident analytics	Stratification, policy capture, mitigate via transparent criteria, plural oversight, redistributive investment
Transnational Platform Layer	*Data Justice, Sovereign Choice, Contestability*	Platform accountability terms, model-update transparency, cross-border data controls	Model-change logs, bias surveillance summaries, redress latency measures	Epistemic extraction, dependency, mitigate via data trusts, local capacity building, open standards
Public-Interest Infrastructure Commons	*Democratic Legitimacy, Open Interoperability, Commons Stewardship*	Public procurement frameworks, open specifications, shared simulation repositories	Interoperability compliance, commons usage audits, reproducibility checks	Fragmentation, underinvestment, mitigate via pooled financing, shared governance charters

Table 5 should be interpreted as a governance-to-evidence translation layer, not as a static compliance checklist, because the relevant risks and measurement priorities shift with sectoral safety-criticality and platform volatility. Its practical value lies in aligning what is valued, what is controlled, and what is measured, thereby preventing the common pathology where institutions adopt AI tools while leaving assessment, accountability, and privacy regimes unchanged. Measurement validity improves when evidence artifacts are triangulated across modalities, for example simulation checkrides plus oral defenses plus audited decision logs, because hybrid competence is enacted behaviorally and relationally. Equity metrics become credible only when paired with access and infrastructure guarantees, otherwise they record inequality without reducing it. The same logic supports lifecycle governance, where model updates, drift signals, and policy revisions are treated as routine operations rather than exceptional crises, enabling resilient and rights-consistent learning ecosystems.

6.4 Responsible Data Infrastructures and Platform Imaginaries for Learning Ecosystem Governance

Responsible governance must include infrastructural design of data flows, analytics pipelines, and platform interfaces, because learning ecosystems increasingly operate as computational substrates that produce visibility, classification, and behavioral nudges. Learning analytics, credential wallets, simulation telemetry, and content recommendation systems should be governed through privacy-preserving architectures, where data minimization, purpose limitation, and retention discipline are engineered into pipelines rather than asserted in policy prose (Al-Kfairy et al., 2024; Qian & Wexler, 2024). *Data Justice* requires representational adequacy, bias monitoring, and meaningful redress pathways, especially when automated classification influences placement, remediation, or credential progression. Platform imaginaries matter, because education-as-marketplace amplifies surveillance incentives and vendor lock-in, while education-as-public-infrastructure foregrounds interoperability, accessibility, and democratic accountability (Zeb et al., 2025; Jacques et al., 2024; Nurhasanah & Nugraha, 2024). Lifecycle governance should incorporate model-update traceability, drift detection, and reproducibility checks, aligning with institutional change-control routines described in Table 5. Cross-border data flows require sovereignty-aware controls, including contractual auditability and contestability, so learners can challenge decisions and institutions can exit harmful dependencies without collapsing service continuity.

6.5 Implementation Risk Landscapes, Socio-Technical Failure Modes, and Mitigation Logics

Implementation failure is rarely a single-point technical defect, it is typically a socio-technical cascade involving miscalibrated trust, weak documentation, incentive gaming, and governance latency under rapidly updating toolchains. Predictable risk landscapes include deskilling through over-delegation, automation bias under cognitive load, surveillance creep via analytics expansion, and inequitable harm distribution when error burdens fall disproportionately on marginalized groups (Wu et al., 2024; Lata, 2024). Additional systemic hazards include technocratic drift, where metrics substitute for judgment, and platform dependency, where procurement choices harden into infrastructural lock-in and curricular sovereignty erodes. Mitigation should follow risk governance cycles that include pre-mortem scenario design, red-team stress testing, incident reporting with near-miss learning, and resilience engineering routines that privilege recovery and graceful degradation (Hutson, 2024; Groothuijsen et al., 2024; Sajja et al., 2024). Table 5 links these risk signatures to governance levers at each stratum, enabling targeted interventions such as supervision thresholds, interoperability clauses, equity safeguards, and contestability channels. When these mitigation logics are embedded into curriculum, assessment, and institutional operations, collaborative intelligence becomes both teachable and governable, setting the conditions for the integrative synthesis in Section 7.

7. CONCLUSION

7.1 Integrative Synthesis of the Conceptual Architecture for Hybrid Cognition

This chapter has advanced a coherent conceptual architecture in which industrial evolution is understood as a cognitive-infrastructural transformation, and education is positioned as the societal interface that stabilizes *Human-AI Collaborative Intelligence* as a legitimate, auditable, and equitable form of hybrid performance. The theoretical scaffolding in Section 2 established that collaboration is not reducible to tool usage, it is distributed cognition enacted through interaction loops, interface regimes, and accountability pathways, with failure signatures that are predictable when trust calibration, provenance reasoning, and escalation competence are underdeveloped. Section 3 reframed learning ecosystems as *Complex Adaptive Systems* whose credential ecologies, platform infrastructures, and pathway governance determine whether hybrid cognition becomes transferable capacity or stratified privilege. Section 4 consolidated these insights into a multilevel framework that aligns task ecologies, collaboration modalities, capability clusters, and evidence standards, making competence measurable through justification discipline and recovery-from-error performance rather than output fluency. Sections 5 and 6 translated the framework into institutional design patterns and polycentric governance maps, clarifying that co-evolution requires integrity-by-design, privacy-preserving data infrastructures, and contestability channels that protect agency under algorithmic mediation.

7.2 Contributions to Theory, Policy, and Workforce-Ecosystem Design Praxis

The primary theoretical contribution is a multi-lens synthesis that treats hybrid cognition as an emergent socio-technical phenomenon, integrating *bounded rationality, metacognitive control, distributed cognition, joint cognitive systems,* and institutional dynamics into a single alignment logic that can be operationalized without empirical dependence. This synthesis is policy-relevant because it specifies how normative anchors such as *procedural justice, capability expansion,* and *ethics of care* can be translated into concrete governance instruments, assessment regimes, and metric families across classroom, institution, sector, national, and transnational platform layers. For workforce development, the chapter provides a design grammar for building pathway permeability and competence portability, where micro-credentials, portfolios, and supervised practice artifacts function as interoperability devices rather than fragmented signals. For digital learning technologists, the chapter clarifies that platform architecture is inseparable from pedagogy, because interpretability, auditability, privacy, and accessibility are infrastructural properties that shape what learners can know, contest, and justify. The practical implication is that education systems should engineer hybrid competence as reliability capacity, with explicit evidence standards, lifecycle change-control, and equity-by-design procurement.

7.3 Productive Tensions, Non-Negotiable Constraints, and Forward Operational Agendas

The chapter foregrounds enduring tensions that cannot be eliminated, only governed through design and institutional maturity. Augmentation competes with control, because increasing automation can erode situated judgment unless *meaningful human control* is taught as a disciplined practice anchored in audit-ready documentation and escalation choreography. Efficiency competes with dignity and care, because optimization can externalize harm unless rights-consistent constraints and redress pathways are embedded into both learning analytics and workplace workflows. Standardization competes with contextual validity, because universal rubrics and benchmark imaginaries can suppress local knowledge unless multilingual scaffolding, boundary-condition articulation, and epistemic pluralism are treated as competence targets. Platform dependence competes with ecosystem sovereignty, because vendor lock-in can harden inequity unless interoperability, exitability, and public-interest data stewardship are engineered into procurement and governance. A forward operational agenda therefore prioritizes multilevel capability progressions, simulation-to-practice transfer, incident learning infrastructures, and equity metrics paired with material access guarantees, so that collaborative intelligence becomes globally portable, locally legitimate, and ethically defensible across diverse industrial, cultural, and regulatory contexts.

REFERENCES

Adel, A., & Alani, HS, N. (2024). Human-centric collaboration and industry 5.0 framework in smart cities and communities: Fostering sustainable development goals 3, 4, 9, and 11 in society 5.0. *Smart Cities*, *7*(4), 1723–1775. DOI: 10.3390/smartcities7040068

Akpan, I. J., Kobara, Y. M., Owolabi, J., Akpan, A. A., & Offodile, O. F. (2025). Conversational and generative artificial intelligence and human–chatbot interaction in education and research. *International Transactions in Operational Research*, *32*(3), 1251–1281. DOI: 10.1111/itor.13522

Alam, A. (2022). Impact of university's human resources practices on professors' occupational performance: empirical evidence from India's higher education sector. In *Inclusive businesses in developing economies: Converging people, profit, and corporate citizenship* (pp. 107–131). Springer International Publishing., DOI: 10.1007/978-3-031-12217-0_6

Alam, A. (2023a). Media multitasking with M-learning technology in real-time classroom learning: Analysing the dynamics in formal educational settings for the future of E-learning in India. In *2023 2nd International Conference on Smart Technologies and Systems for Next Generation Computing (ICSTSN)* (pp. 1-6). IEEE. https://doi.org/DOI: 10.1109/ICSTSN57873.2023.10151509

Alam, A. (2023b). The secret sauce of student success: Cracking the code by navigating the path to personalized learning with educational data mining. In *2023 2nd International Conference on Smart Technologies and Systems for Next Generation Computing (ICSTSN)* (pp. 1-8). IEEE. https://doi.org/DOI: 10.1109/ICSTSN57873.2023.10151558

Alam, A. (2023c). Intelligence unleashed: An argument for AI-enabled learning ecologies with real world examples of today and a peek into the future. In *International Conference on Innovations in Computer Science, Electronics & Electrical Engineering*-2022 (Vol. 2717, No. 1, p. 030001). AIP Publishing LLC. https://doi.org/DOI: 10.1063/5.0129803

Alam, A., & Mohanty, A. (2022, December). Facial analytics or virtual avatars: competencies and design considerations for student-teacher interaction in AI-powered online education for effective classroom engagement. In *International Conference on Communication, Networks and Computing* (pp. 252-265). Cham: Springer Nature Switzerland. https://doi.org/DOI: 10.1007/978-3-031-43145-6_21

Alam, A., & Mohanty, A. (2023, January). Learning on the move: a pedagogical framework for state-of-the-art mobile learning. In *International Conference on Data Management, Analytics & Innovation* (pp. 735-748). Singapore: Springer Nature Singapore. https://doi.org/DOI: 10.1007/978-981-99-1414-2_52

Alam, A., & Mohanty, A. (2024). Integrated constructive robotics in education (ICRE) model: A paradigmatic framework for transformative learning in educational ecosystem. *Cogent Education*, *11*(1), 2324487. DOI: 10.1080/2331186X.2024.2324487

Al-Kfairy, M., Mustafa, D., Kshetri, N., Insiew, M., & Alfandi, O. (2024, September). Ethical challenges and solutions of generative AI: An interdisciplinary perspective. In *Informatics* (Vol. 11, No. 3, p. 58). Multidisciplinary Digital Publishing Institute. https://doi.org/DOI: 10.3390/informatics11030058

Al-Zahrani, A. M., & Alasmari, T. M. (2024). Exploring the impact of artificial intelligence on higher education: The dynamics of ethical, social, and educational implications. *Humanities & Social Sciences Communications*, *11*(1), 1–12. DOI: 10.1057/s41599-024-03432-4

Anane-Simon, R., & Atiku, S. O. (2024). Artificial intelligence and automation for the future of startups. In *Ecosystem Dynamics and Strategies for Startups Scalability* (pp. 133–153). IGI Global Scientific Publishing., DOI: 10.4018/979-8-3693-0527-0.ch007

Arar, K., Tlili, A., & Salha, S. (2024). Human-Machine symbiosis in educational leadership in the era of artificial intelligence (AI): Where are we heading? *Educational Management Administration & Leadership*, *17411432241292295*. Advance online publication. DOI: 10.1177/17411432241292295

Asad, M. M., & Ajaz, A. (2024). Impact of ChatGPT and generative AI on lifelong learning and upskilling learners in higher education: Unveiling the challenges and opportunities globally. *The International Journal of Information and Learning Technology*, *41*(5), 507–523. DOI: 10.1108/IJILT-06-2024-0103

Atchley, P., Pannell, H., Wofford, K., Hopkins, M., & Atchley, R. A. (2024). Human and AI collaboration in the higher education environment: Opportunities and concerns. *Cognitive Research: Principles and Implications*, *9*(1), 20. DOI: 10.1186/s41235-024-00547-9 PMID: 38589710

Bansal, G., Mitchell, A., & Li, D. (2024). A Panel Report on Higher Education in the Age of AI from the Perspective of Academic Leaders in the Midwest US. *Communications of the Association for Information Systems*, *54*(1), 360–375. DOI: 10.17705/1CAIS.05413

Bettayeb, A. M., Abu Talib, M., Sobhe Altayasinah, A. Z., & Dakalbab, F. (2024, July). Exploring the impact of ChatGPT: conversational AI in education. In *Frontiers in Education* (Vol. 9, p. 1379796). Frontiers Media SA. https://doi.org/DOI: 10.3389/feduc.2024.1379796

Bozkurt, A. (Ed.). (2024). Why generative AI literacy, why now and why it matters in the educational landscape?: Kings, queens and GenAI dragons. *Open Praxis, 16*(3), 283-290. https://doi.org/DOI: 10.55982/openpraxis.16.3.739

Bozkurt, A., Xiao, J., Farrow, R., Bai, J. Y., Nerantzi, C., Moore, S., & Asino, T. I. (Eds.). (2024). The manifesto for teaching and learning in a time of generative AI: A critical collective stance to better navigate the future. *Open Praxis, 16*(4), 487-513. https://doi.org/DOI: 10.55982/openpraxis.16.4.777

Brusilovsky, P. (2024). AI in education, learner control, and human-AI collaboration. *International Journal of Artificial Intelligence in Education*, *34*(1), 122–135. DOI: 10.1007/s40593-023-00356-z

Bulathwela, S., Pérez-Ortiz, M., Holloway, C., Cukurova, M., & Shawe-Taylor, J. (2024). Artificial intelligence alone will not democratise education: On educational inequality, techno-solutionism and inclusive tools. *Sustainability*, *16*(2), 781. DOI: 10.3390/su16020781

Butson, R., & Spronken-Smith, R. (2024). AI and its implications for research in higher education: A critical dialogue. *Higher Education Research & Development*, *43*(3), 563–577. DOI: 10.1080/07294360.2023.2280200

Cai, L., Msafiri, M. M., & Kangwa, D. (2025). Exploring the impact of integrating AI tools in higher education using the Zone of Proximal Development. *Education and Information Technologies*, *30*(6), 7191–7264. DOI: 10.1007/s10639-024-13112-0

Celik, I., Gedrimiene, E., Siklander, S., & Muukkonen, H. (2024). The affordances of artificial intelligence-based tools for supporting 21st-century skills: A systematic review of empirical research in higher education. *Australasian Journal of Educational Technology*, *40*(3), 19–38. DOI: 10.14742/ajet.9069

Chan, C. K. Y., & Tsi, L. H. (2024). Will generative AI replace teachers in higher education? A study of teacher and student perceptions. *Studies in Educational Evaluation*, *83*, 101395. DOI: 10.1016/j.stueduc.2024.101395

Chee, H., Ahn, S., & Lee, J. (2025). A competency framework for AI literacy: Variations by different learner groups and an implied learning pathway. *British Journal of Educational Technology*, *56*(5), 2146–2182. DOI: 10.1111/bjet.13556

Chiu, T. K., & Rospigliosi, P. A. (2025). Encouraging human-AI collaboration in interactive learning environments. *Interactive Learning Environments, 33*(2), 921–924. DOI: 10.1080/10494820.2025.2471199

Chuang, S. (2024). Indispensable skills for human employees in the age of robots and AI. *European Journal of Training and Development, 48*(1/2), 179–195. DOI: 10.1108/EJTD-06-2022-0062

Cingillioglu, I., Gal, U., & Prokhorov, A. (2024). AI-experiments in education: An AI-driven randomized controlled trial for higher education research. *Education and Information Technologies, 29*(15), 19649–19677. DOI: 10.1007/s10639-024-12633-y

Clegg, S., & Sarkar, S. (2024). Artificial intelligence and management education: A conceptualization of human-machine interaction. *International Journal of Management Education, 22*(3), 101007. DOI: 10.1016/j.ijme.2024.101007

Crawford, J., Allen, K. A., Pani, B., & Cowling, M. (2024). When artificial intelligence substitutes humans in higher education: The cost of loneliness, student success, and retention. *Studies in Higher Education, 49*(5), 883–897. DOI: 10.1080/03075079.2024.2326956

Creely, E., & Blannin, J. (2025). Creative partnerships with generative AI. Possibilities for education and beyond. *Thinking Skills and Creativity, 56*, 101727. DOI: 10.1016/j.tsc.2024.101727

Cukurova, M. (2025). The interplay of learning, analytics and artificial intelligence in education: A vision for hybrid intelligence. *British Journal of Educational Technology, 56*(2), 469–488. DOI: 10.1111/bjet.13514

Dai, R., Thomas, M. K. E., & Rawolle, S. (2025). The roles of AI and educational leaders in AI-assisted administrative decision-making: A proposed framework for symbiotic collaboration. *Australian Educational Researcher, 52*(2), 1471–1487. DOI: 10.1007/s13384-024-00771-8

Dhanasekaran, R. (2025). AI Revolutionizing Education: Technology and Human Interaction in Partnership. *Journal of Computer Science and Technology Studies, 7*(6), 501–506. DOI: 10.32996/jcsts.2025.7.6.58

Dhillon, P. S., Molaei, S., Li, J., Golub, M., Zheng, S., & Robert, L. P. (2024, May). Shaping human-AI collaboration: Varied scaffolding levels in co-writing with language models. In *Proceedings of the 2024 CHI Conference on Human Factors in Computing Systems* (pp. 1-18). https://doi.org/DOI: 10.1145/3613904.3642134

Duraimutharasan, N., Deepan, A., Swadhi, R., Velmurugan, P. R., & Varshney, K. R. (2025). Enhancing Control Engineering Through Human-Machine Collaboration: AI for Improved Efficiency and Decision-Making. In *Harnessing AI for Control Engineering* (pp. 155-176). IGI Global Scientific Publishing. https://doi.org/DOI: 10.4018/979-8-3693-7812-0.ch008

Edwards, J., Nguyen, A., Lämsä, J., Sobocinski, M., Whitehead, R., Dang, B., & Järvelä, S. (2025). Human-AI collaboration: Designing artificial agents to facilitate socially shared regulation among learners. *British Journal of Educational Technology*, *56*(2), 712–733. DOI: 10.1111/bjet.13534

Fan, Y., Tang, L., Le, H., Shen, K., Tan, S., Zhao, Y., & Gašević, D. (2025). Beware of metacognitive laziness: Effects of generative artificial intelligence on learning motivation, processes, and performance. *British Journal of Educational Technology*, *56*(2), 489–530. DOI: 10.1111/bjet.13544

Fissore, C., Floris, F., Conte, M. M., & Sacchet, M. (2024, March). Teacher training on artificial intelligence in education. In *Smart Learning Environments in the Post Pandemic Era: Selected Papers from the CELDA 2022 Conference* (pp. 227-244). Cham: Springer Nature Switzerland. https://doi.org/DOI: 10.1007/978-3-031-54207-7_13

Francis, N. J., Jones, S., & Smith, D. P. (2025). Generative AI in higher education: Balancing innovation and integrity. *British Journal of Biomedical Science*, *81*, 14048. DOI: 10.3389/bjbs.2024.14048 PMID: 39850144

Fu, Y., & Weng, Z. (2024). Navigating the ethical terrain of AI in education: A systematic review on framing responsible human-centered AI practices. *Computers and Education: Artificial Intelligence*, *7*, 100306. DOI: 10.1016/j.caeai.2024.100306

García-López, I. M., González, C. S. G., Ramírez-Montoya, M. S., & Molina-Espinosa, J. M. (2025). Challenges of implementing ChatGPT on education: Systematic literature review. *International Journal of Educational Research Open*, *8*, 100401. DOI: 10.1016/j.ijedro.2024.100401

Giannakos, M., Azevedo, R., Brusilovsky, P., Cukurova, M., Dimitriadis, Y., Hernandez-Leo, D., & Rienties, B. (2025). The promise and challenges of generative AI in education. *Behaviour & Information Technology*, *44*(11), 2518–2544. DOI: 10.1080/0144929X.2024.2394886

Goldman, S. R., Taylor, J., Carreon, A., & Smith, S. J. (2024). Using AI to support special education teacher workload. *Journal of Special Education Technology*, *39*(3), 434–447. DOI: 10.1177/01626434241257240

Groothuijsen, S., van den Beemt, A., Remmers, J. C., & van Meeuwen, L. W. (2024). AI chatbots in programming education: Students' use in a scientific computing course and consequences for learning. *Computers and Education: Artificial Intelligence*, *7*, 100290. DOI: 10.1016/j.caeai.2024.100290

Guggemos, J. (2024). To fear or not to fear–Human resource development professionals' positioning towards artificial intelligence with a focus on augmentation. *Computers and Education: Artificial Intelligence*, *7*, 100260. DOI: 10.1016/j.caeai.2024.100260

Han, A., Zhou, X., Cai, Z., Han, S., Ko, R., Corrigan, S., & Peppler, K. A. (2024, May). Teachers, parents, and students' perspectives on integrating generative AI into elementary literacy education. In *Proceedings of the 2024 CHI Conference on Human Factors in Computing Systems* (pp. 1-17). https://doi.org/DOI: 10.1145/3613904.3642438

Hutson, J. (2024). Rethinking plagiarism in the era of generative AI. *Journal of International Communication*, *4*(1). Advance online publication. DOI: 10.54963/jic.v3i2.220

Hwang, Y., & Lee, J. H. (2025). Exploring students' experiences and perceptions of human-AI collaboration in digital content making. *International Journal of Educational Technology in Higher Education*, *22*(1), 44. DOI: 10.1186/s41239-025-00542-0

Ifenthaler, D., Majumdar, R., Gorissen, P., Judge, M., Mishra, S., Raffaghelli, J., & Shimada, A. (2024). Artificial intelligence in education: Implications for policymakers, researchers, and practitioners. *Technology. Knowledge and Learning*, *29*(4), 1693–1710. DOI: 10.1007/s10758-024-09747-0

Imran, M., Almusharraf, N., Abdellatif, M. S., & Abbasova, M. Y. (2024). Artificial intelligence in higher education: Enhancing learning systems and transforming educational paradigms. https://doi.org/DOI: 10.3991/ijim.v18i18.49143

Jacques, P. H., Moss, H. K., & Garger, J. (2024). A synthesis of AI in higher education: Shaping the future. *Journal of Behavioral and Applied Management*, *24*(2), 103–111. DOI: 10.21818/001c.122146

Jiang, T., Sun, Z., Fu, S., & Lv, Y. (2024). Human-AI interaction research agenda: A user-centered perspective. *Data and Information Management*, *8*(4), 100078. DOI: 10.1016/j.dim.2024.100078

Jony, A. I., & Hamim, S. A. (2024). Empowering virtual collaboration: Harnessing AI for enhanced teamwork in higher education. *Educational Technology Quarterly*, *2024*(3), 337–359. DOI: 10.55056/etq.746

Joseph, O. B., & Uzondu, N. C. (2024). Integrating AI and Machine Learning in STEM education: Challenges and opportunities. *Computer Science & IT Research Journal*, *5*(8), 1732–1750. DOI: 10.51594/csitrj.v5i8.1379

Jung, D., & Suh, S. (2024). Enhancing soft skills through generative AI in sustainable fashion textile design education. *Sustainability*, *16*(16), 6973. DOI: 10.3390/su16166973

Karakose, T., & Tülübas, T. (2024). School Leadership and Management in the Age of Artificial Intelligence (AI): Recent Developments and Future Prospects. *Educational Process: International Journal*, *13*(1), 7–14. DOI: 10.22521/edupij.2024.131.1

Kaspersen, M. H., Musaeus, L. H., Bilstrup, K. E. K., Petersen, M. G., Iversen, O. S., Dindler, C., & Dalsgaard, P. (2024, May). From primary education to premium workforce: Drawing on K-12 approaches for developing AI literacy. In *Proceedings of the 2024 CHI Conference on Human Factors in Computing Systems* (pp. 1-16). https://doi.org/DOI: 10.1145/3613904.3642607

Kaswan, K. S., Dhatterwal, J. S., & Ojha, R. P. (2024). AI in personalized learning. In *Advances in technological innovations in higher education* (pp. 103–117). CRC Press., DOI: 10.1201/9781003376699-9

Kayyali, M. (2024). Future possibilities and challenges of AI in education. In *Transforming education with generative AI: Prompt engineering and synthetic content creation* (pp. 118–137). IGI Global Scientific Publishing., DOI: 10.4018/979-8-3693-1351-0.ch006

Kim, J. (2024a). Leading teachers' perspective on teacher-AI collaboration in education. *Education and Information Technologies*, *29*(7), 8693–8724. DOI: 10.1007/s10639-023-12109-5

Kim, J. (2024b). Types of teacher-AI collaboration in K-12 classroom instruction: Chinese teachers' perspective. *Education and Information Technologies*, *29*(13), 17433–17465. DOI: 10.1007/s10639-024-12523-3

Kim, J., & Cho, Y. H. (2025). My teammate is AI: Understanding students' perceptions of student-AI collaboration in drawing tasks. *Asia Pacific Journal of Education*, *45*(3), 1013–1027. DOI: 10.1080/02188791.2023.2286206

Kim, J., Ham, Y., & Lee, S. S. (2024a). Differences in student-AI interaction process on a drawing task: Focusing on students' attitude towards AI and the level of drawing skills. *Australasian Journal of Educational Technology*, *40*(1), 19–41. DOI: 10.14742/ajet.8859

Kim, J., Yu, S., Detrick, R., & Li, N. (2025). Exploring students' perspectives on Generative AI-assisted academic writing. *Education and Information Technologies, 30*(1), 1265–1300. DOI: 10.1007/s10639-024-12878-7

Kim, S. J. (2024c). Generative artificial intelligence in collaborative ideation: Educational insight from fashion students. *IEEE Access : Practical Innovations, Open Solutions, 12*, 49261–49274. DOI: 10.1109/ACCESS.2024.3382194

Kim, T., Shin, D., Kim, Y. H., & Hong, H. (2024b). DiaryMate: Understanding user perceptions and experience in human-AI collaboration for personal journaling. In *Proceedings of the 2024 CHI Conference on Human Factors in Computing Systems* (pp. 1-15). https://doi.org/DOI: 10.1145/3613904.3642693

Kolbjørnsrud, V. (2024). Designing the Intelligent Organization: Six principles for Human-AI collaboration. *California Management Review, 66*(2), 44–64. DOI: 10.1177/00081256231211020

Lata, P. (2024). Beyond algorithms: Humanizing artificial intelligence for personalized and adaptive learning. *International Journal of Innovative Research in Engineering and Management, 11*(5), 10–55524. DOI: 10.55524/ijirem.2024.11.5.6

Lee, G. G., Mun, S., Shin, M. K., & Zhai, X. (2025). Collaborative learning with artificial intelligence speakers: Pre-service elementary science teachers' responses to the prototype. *Science & Education, 34*(2), 847–875. DOI: 10.1007/s11191-024-00526-y

Leong, W. Y., & Zhang, J. B. (2025). Ethical design of AI for education and learning systems. *ASM Science Journal, 20*(1), 1–9. DOI: 10.32802/asmscj.2025.1917

Li, H., Wang, Y., & Qu, H. (2024, May). Where are we so far? understanding data storytelling tools from the perspective of human-ai collaboration. In *Proceedings of the 2024 CHI Conference on Human Factors in Computing Systems* (pp. 1-19). https://doi.org/DOI: 10.1145/3613904.3642726

Li, Y., Zhou, X., Yin, H. B., & Chiu, T. K. (2025). Design language learning with artificial intelligence (AI) chatbots based on activity theory from a systematic review. *Smart Learning Environments, 12*(1), 24. DOI: 10.1186/s40561-025-00379-0

Li, Z., Dhruv, A., & Jain, V. (2024, February). Ethical considerations in the use of AI for higher education: A comprehensive guide. In *2024 IEEE 18th International Conference on Semantic Computing (ICSC)* (pp. 218-223). IEEE. https://doi.org/DOI: 10.1109/ICSC59802.2024.00041

Liang, E. S., & Bai, S. (2025). Generative AI and the future of connectivist learning in higher education. *Journal of Asian Public Policy*, *18*(2), 329–351. DOI: 10.1080/17516234.2024.2386085

Lin, C. P., & Chen, P. C. (2025). Mentoring for effective human-AI collaboration: An integrated theoretical framework. *Total Quality Management & Business Excellence*, •••, 1–15. DOI: 10.1080/14783363.2025.2504603

Łodzikowski, K., Foltz, P. W., & Behrens, J. T. (2024). Generative AI and its educational implications. In *Trust and Inclusion in AI-Mediated Education: Where Human Learning Meets Learning Machines* (pp. 35-57). Cham: Springer Nature Switzerland. https://doi.org/DOI: 10.1007/978-3-031-64487-0_2

Luckin, R. (2025). Nurturing human intelligence in the age of AI: Rethinking education for the future. *Development and Learning in Organizations*, *39*(1), 1–4. DOI: 10.1108/DLO-04-2024-0108

Luo, J. (2024). A critical review of GenAI policies in higher education assessment: A call to reconsider the "originality" of students' work. *Assessment & Evaluation in Higher Education*, *49*(5), 651–664. DOI: 10.1080/02602938.2024.2309963

Mao, J., Chen, B., & Liu, J. C. (2024). Generative artificial intelligence in education and its implications for assessment. *TechTrends*, *68*(1), 58–66. DOI: 10.1007/s11528-023-00911-4

Maphoto, K. B., Sevnarayan, K., Mohale, N. E., Suliman, Z., Ntsopi, T. J., & Mokoena, D. (2024). Advancing students' academic excellence in distance education: Exploring the potential of generative AI integration to improve academic writing skills. *Open Praxis*, *16*(2), 142–159. DOI: 10.55982/openpraxis.16.2.649

Marrone, R., Cropley, D., & Medeiros, K. (2024). How does narrow AI impact human creativity? *Creativity Research Journal*, •••, 1–11. DOI: 10.1080/10400419.2024.2378264

Marrone, R., Zamecnik, A., Joksimovic, S., Johnson, J., & De Laat, M. (2025). Understanding student perceptions of artificial intelligence as a teammate. *Technology. Knowledge and Learning*, *30*(3), 1847–1869. DOI: 10.1007/s10758-024-09780-z

Milana, M., Brandi, U., Hodge, S., & Hoggan-Kloubert, T. (2024). Artificial intelligence (AI), conversational agents, and generative AI: Implications for adult education practice and research. *International Journal of Lifelong Education*, *43*(1), 1–7. DOI: 10.1080/02601370.2024.2310448

Mittal, U., Sai, S., Chamola, V., & Sangwan, D. (2024). A comprehensive review on generative AI for education. *IEEE Access : Practical Innovations, Open Solutions*, *12*, 142733–142759. DOI: 10.1109/ACCESS.2024.3468368

Mustafa, M. Y., Tlili, A., Lampropoulos, G., Huang, R., Jandrić, P., Zhao, J., & Saqr, M. (2024). A systematic review of literature reviews on artificial intelligence in education (AIED): A roadmap to a future research agenda. *Smart Learning Environments*, *11*(1), 59. DOI: 10.1186/s40561-024-00350-5

Nguyen, A., Hong, Y., Dang, B., & Huang, X. (2024). Human-AI collaboration patterns in AI-assisted academic writing. *Studies in Higher Education*, *49*(5), 847–864. DOI: 10.1080/03075079.2024.2323593

Nguyen, A., Kremantzis, M., Essien, A., Petrounias, I., & Hosseini, S. (2024). Enhancing student engagement through artificial intelligence (AI): Understanding the basics, opportunities, and challenges. *Journal of University Teaching & Learning Practice*, *21*(6), 1–13. DOI: 10.53761/caraaq92

Nguyen, K. V. (2025). The use of generative AI tools in higher education: Ethical and pedagogical principles. *Journal of Academic Ethics*, •••, 1–21. DOI: 10.1007/s10805-025-09607-1

Nikolopoulou, K. (2024). Generative artificial intelligence in higher education: Exploring ways of harnessing pedagogical practices with the assistance of ChatGPT. *International Journal of Changes in Education*, *1*(2), 103–111. DOI: 10.47852/bonviewIJCE42022489

Nithithanatchinnapat, B., Maurer, J., Deng, X., & Joshi, K. D. (2024). Future business workforce: Crafting a generative AI-centric curriculum today for tomorrow's business education. *The Data Base for Advances in Information Systems*, *55*(1), 6–11. DOI: 10.1145/3645057.3645059

Noroozi, O., Soleimani, S., Farrokhnia, M., & Banihashem, S. K. (2024). Generative AI in Education: Pedagogical, Theoretical, and Methodological Perspectives. *International Journal of Technology in Education*, *7*(3), 373–385. DOI: 10.46328/ijte.845

Nurhasanah, S., & Nugraha, M. S. (2024). The future of learning: Ethical and philosophical implications of artificial intelligence (AI) integration in education. *Indonesian Journal of Multidiciplinary Research*, *4*(2), 341–352. DOI: 10.17509/ijomr.v4i2.78038

Omran Zailuddin, M. F. N., Nik Harun, N. A., Abdul Rahim, H. A., Kamaruzaman, A. F., Berahim, M. H., Harun, M. H., & Ibrahim, Y. (2024). Redefining creative education: A case study analysis of AI in design courses. *Journal of Research in Innovative Teaching & Learning*, *17*(2), 282–296. DOI: 10.1108/JRIT-01-2024-0019

Padovano, A., & Cardamone, M. (2024). Towards human-AI collaboration in the competency-based curriculum development process: The case of industrial engineering and management education. *Computers and Education: Artificial Intelligence*, *7*, 100256. DOI: 10.1016/j.caeai.2024.100256

Pahi, K., Hawlader, S., Hicks, E., Zaman, A., & Phan, V. (2024). Enhancing active learning through collaboration between human teachers and generative AI. *Computers and Education Open*, *6*, 100183. DOI: 10.1016/j.caeo.2024.100183

Park, H., & Ahn, D. (2024, May). The promise and peril of ChatGPT in higher education: opportunities, challenges, and design implications. In *Proceedings of the 2024 CHI Conference on Human Factors in Computing Systems* (pp. 1-21). https://doi.org/DOI: 10.1145/3613904.3642785

Peláez-Sánchez, I. C., Velarde-Camaqui, D., & Glasserman-Morales, L. D. (2024, June). The impact of large language models on higher education: exploring the connection between AI and Education 4.0. In *Frontiers in Education* (Vol. 9, p. 1392091). Frontiers Media SA. https://doi.org/DOI: 10.3389/feduc.2024.1392091

Peters, M. A., & Green, B. J. (2024). Wisdom in the Age of AI Education. *Postdigital science and education, 6*(4), 1173-1195. https://doi.org/DOI: 10.1007/s42438-024-00460-w

Pratschke, B. M. (2024). *Generative AI and education: Digital pedagogies, teaching innovation and learning design*. Springer Nature., DOI: 10.1007/978-3-031-67991-9

Przegalinska, A., & Triantoro, T. (2024). *Converging minds: The creative potential of collaborative AI*. CRC Press., DOI: 10.1201/9781032656618

Qian, C., & Wexler, J. (2024, March). Take it, leave it, or fix it: Measuring productivity and trust in human-ai collaboration. In *Proceedings of the 29th International Conference on Intelligent User Interfaces* (pp. 370-384). https://doi.org/DOI: 10.1145/3640543.3645198

Rane, N., & Choudhary, S. (2024). Role and challenges of ChatGPT, Google Bard, and similar generative Artificial Intelligence in Arts and Humanities. *Studies in Humanities and Education*, *5*(1), 1–11. DOI: 10.48185/she.v5i1.999

Rawas, S. (2024). AI: The future of humanity. *Discover Artificial Intelligence*, *4*(1), 25. DOI: 10.1007/s44163-024-00118-3

Ruiz-Rojas, L. I., Salvador-Ullauri, L., & Acosta-Vargas, P. (2024). Collaborative working and critical thinking: Adoption of generative artificial intelligence tools in higher education. *Sustainability*, *16*(13), 5367. DOI: 10.3390/su16135367

Sadeghi, S., & Niu, C. (2024). Augmenting human decision-making in K-12 education: The role of artificial intelligence in assisting the recruitment and retention of teachers of color for enhanced diversity and inclusivity. *Leadership and Policy in Schools*, •••, 1–21. DOI: 10.1080/15700763.2024.2358303

Sajja, R., Ramirez, C. E., Li, Z., Demiray, B. Z., Sermet, Y., & Demir, I. (2024). Integrating generative AI in hackathons: Opportunities, challenges, and educational implications. *Big Data and Cognitive Computing*, *8*(12), 188. DOI: 10.3390/bdcc8120188

Sarkar, A. (2024). AI Should Challenge, Not Obey. *Communications of the ACM*, *67*(10), 18–21. DOI: 10.1145/3649404

Senthilkumar, K. R., Jagajeevan, R., & Sangeetha, S. (2024). Impact of AI on library and information science in higher institutions in India: A comprehensive analysis of technological integration and educational implications. In *AI-assisted library reconstruction* (pp. 21–33). IGI Global Scientific Publishing., DOI: 10.4018/979-8-3693-2782-1.ch002

Sowa, K., & Przegalinska, A. (2025). From expert systems to generative artificial experts: A new concept for human-ai collaboration in knowledge work. *Journal of Artificial Intelligence Research*, *82*, 2101–2124. DOI: 10.1613/jair.1.17175

Storozhyk, M. (2024). Philosophy of future: Analytical overview of interaction between education, science, and artificial intelligence in the context of contemporary challenges. *Futurity Philosophy*, *3*(1), 23–47. DOI: 10.57125/FP.2024.03.30.02

Sun, C., Major, L., Daltry, R., Moustafa, N., & Friedberg, A. (2024, July). Teacher-AI collaboration in content recommendation for digital personalised learning among pre-primary learners in Kenya. In *Proceedings of the Eleventh ACM Conference on Learning@ Scale* (pp. 346-350). https://doi.org/DOI: 10.1145/3657604.3664662

Suryanarayana, K. S., Kandi, V. P., Pavani, G., Rao, A. S., Rout, S., & Krishna, T. S. R. (2024). Artificial intelligence enhanced digital learning for the sustainability of education management system. *The Journal of High Technology Management Research*, *35*(2), 100495. DOI: 10.1016/j.hitech.2024.100495

Tariq, M. U. (2024). Generative AI in curriculum development in higher education. In *Impacts of generative AI on creativity in higher education* (pp. 227–258). IGI Global., DOI: 10.4018/979-8-3693-2418-9.ch009

Tariq, M. U. (2025). AI-Enhanced Project-Based Learning: Revolutionizing Commerce Education With Generative AI. In *Prompt Engineering and Generative AI Applications for Teaching and Learning* (pp. 125-142). IGI Global Scientific Publishing. https://doi.org/DOI: 10.4018/979-8-3693-7332-3.ch008

Tarisayi, K. S. (2024, March). Strategic leadership for responsible artificial intelligence adoption in higher education. In *CTE workshop proceedings* (Vol. 11, pp. 4-14). https://doi.org/DOI: 10.55056/cte.616

Topali, P., Ortega-Arranz, A., Rodríguez-Triana, M. J., Er, E., Khalil, M., & Akçapınar, G. (2025). Designing human-centered learning analytics and artificial intelligence in education solutions: A systematic literature review. *Behaviour & Information Technology*, *44*(5), 1071–1098. DOI: 10.1080/0144929X.2024.2345295

Tzirides, A. O. O., Zapata, G., Kastania, N. P., Saini, A. K., Castro, V., Ismael, S. A., & Kalantzis, M. (2024). Combining human and artificial intelligence for enhanced AI literacy in higher education. *Computers and Education Open*, *6*, 100184. DOI: 10.1016/j.caeo.2024.100184

Virvou, M., Tsihrintzis, G. A., & Tsichrintzi, E. A. (2024). VIRTSI: A novel trust dynamics model enhancing Artificial Intelligence collaboration with human users–Insights from a ChatGPT evaluation study. *Information Sciences*, *675*, 120759. DOI: 10.1016/j.ins.2024.120759

Wang, L., Song, R., Guo, W., & Yang, H. (2025). Exploring prompt pattern for generative artificial intelligence in automatic question generation. *Interactive Learning Environments*, *33*(3), 2559–2584. DOI: 10.1080/10494820.2024.2412082

Wen, Q., Liang, J., Sierra, C., Luckin, R., Tong, R., Liu, Z., & Tang, J. (2024, August). AI for education (AI4EDU): Advancing personalized education with LLM and adaptive learning. In *Proceedings of the 30th ACM SIGKDD Conference on Knowledge Discovery and Data Mining* (pp. 6743-6744). https://doi.org/DOI: 10.1145/3637528.3671498

Wong, L. H., & Looi, C. K. (2024). Advancing the generative AI in education research agenda: Insights from the Asia-Pacific region. *Asia Pacific Journal of Education*, *44*(1), 1–7. DOI: 10.1080/02188791.2024.2315704

Wood, D., & Moss, S. H. (2024). Evaluating the impact of students' generative AI use in educational contexts. *Journal of Research in Innovative Teaching & Learning*, *17*(2), 152–167. DOI: 10.1108/JRIT-06-2024-0151

Wu, D., Zhang, S., Ma, Z., Yue, X. G., & Dong, R. K. (2024). Unlocking potential: Key factors shaping undergraduate self-directed learning in AI-enhanced educational environments. *Systems, 12*(9), 332. DOI: 10.3390/systems12090332

Yan, L., Greiff, S., Teuber, Z., & Gašević, D. (2024a). Promises and challenges of generative artificial intelligence for human learning. *Nature Human Behaviour, 8*(10), 1839–1850. DOI: 10.1038/s41562-024-02004-5 PMID: 39438686

Yan, L., Martinez-Maldonado, R., & Gasevic, D. (2024b). Generative artificial intelligence in learning analytics: Contextualising opportunities and challenges through the learning analytics cycle. In *Proceedings of the 14th learning analytics and knowledge conference* (pp. 101-111). https://doi.org/DOI: 10.1145/3636555.3636856

Yan, W., Nakajima, T., & Sawada, R. (2024c). Benefits and challenges of collaboration between students and conversational generative artificial intelligence in programming learning: An empirical case study. *Education Sciences, 14*(4), 433. DOI: 10.3390/educsci14040433

Yim, I. H. Y., & Su, J. (2025). Artificial intelligence (AI) learning tools in K-12 education: A scoping review. *Journal of Computers in Education, 12*(1), 93–131. DOI: 10.1007/s40692-023-00304-9

Yu, J. H., Chauhan, D., Iqbal, R. A., & Yeoh, E. (2025). Mapping academic perspectives on AI in education: Trends, challenges, and sentiments in educational research (2018–2024). *Educational Technology Research and Development, 73*(1), 199–227. DOI: 10.1007/s11423-024-10425-2

Zeb, A., Rehman, F. U., Bin Othayman, M., & Rabnawaz, M. (2025). Artificial intelligence and ChatGPT are fostering knowledge sharing, ethics, academia and libraries. *The International Journal of Information and Learning Technology, 42*(1), 67–83. DOI: 10.1108/IJILT-03-2024-0046

Chapter 11
AI-Powered Branding Strategies' Gravitational Pull of Innovation Through the AI's Impact on Advertising and Design

Yusuf Kamal
http://orcid.org/0000-0002-1557-8775
University of Lucknow, India

Mohammad Talha Siddiqui
http://orcid.org/0000-0001-7165-4317
University of Lucknow, India

Ali Abbas Rizvi
NITI Ayog, India

Raj Kumar Mishra
http://orcid.org/0000-0002-6912-6694
University of Lucknow, India

Saifuddin Ahmad
http://orcid.org/0000-0002-2128-5717
University of Lucknow, India

Amitabh Roy
http://orcid.org/0009-0003-3492-6207
Institute of Management Sciences, University of Lucknow, India

ABSTRACT

The explosion of AI has been viral across the business landscape. The present study explores the transformative impact of artificial intelligence (AI) on branding strategies. Drawing analogies from physics, such as gravitational forces and quantum leaps, the study positions AI as a catalytic partner that intensifies human ingenuity rather than supplanting it. The chapter begins with a historical overview that traces the

DOI: 10.4018/979-8-3373-7036-1.ch011

shift from manual processes to algorithm-driven creativity. The chapter highlights the AI's automation aptitude through integration with predictive analytics and generative capabilities. This chapter also investigates AI's applications in advertising. Tools empowering designers, such as those used by Coca-Cola, Heineken, and IBM Watson, serve as the foundation of this investigation. The chapter further addresses challenges like data privacy and algorithmic bias. The chapter then raises a dialogue of ethical use of Adobe Firefly and Canva.

1. INTRODUCTION

In the ever-evolving dynamics of human ingenuity, the intersection of artificial intelligence and creativity represents a fundamental shift in the artistic landscape. The emergence of generative AI technologies has recalibrated the potential of advertising and marketing. This shift initiates an era "Amplification of Algorithm" (Boden, 1998; Poliakov, 2020; and Ameen *et al.*, 2022). AI works as a collaborative force that only runs with human interaction (Dotov & Froese, 2020).

This chapter situates the transformation of branding, advertising, and design. Chapter move beyond viewing AI as a mere efficiency tool. We examine how tasks of ideation, judgment, and ethical reasoning become jointly distributed between human and machine agents. This framework of augmented, rather than automated, creativity serves as our analytical lens for the subsequent exploration of practical applications and challenges.

AI's technical capabilities in personalization, automation, and prediction are well-documented. This chapter moves beyond these functional descriptions. We focus specifically on the operational reality of human-AI cognitive collaboration in branding contexts. Our analysis examines not what AI tools do. But "How they reconfigure creative workflows", "Redistribute cognitive tasks between human and machine", and "Where this partnership requires deliberate governance to mitigate risks like bias and creative stagnation". This framework elevates the discussion from technical implementation to strategic cognitive partnership.

1.1 Brief overview of AI in advertising, design, and marketing

Artificial Intelligence (AI) has transitioned from a futuristic concept to a vital component (Van Belkom, 2020 and Kushwaha, 2023). By integrating advanced algorithms, AI is ready to analyze massive amounts of data (Li, 2019 and Shah *et al.*, 2020). In advertising, AI crafts personalized campaigns by predicting consumer preferences and optimizing ad placements in real time (Vakratsas & Wang, 2020) (Table 1).

Table 1. Intersection of AI and Creative Industries

Domain	AI Capabilities	Impact on Human Creativity	Examples of Application
Advertising	Predictive analytics, real time ad placement, personalized campaigns	Enhances message precision and reach, freeing humans to focus on storytelling	Programmatic ads, real time bidding, audience targeting
Design	Automated suggestions (colors, layouts), generative art, rapid prototyping	Streamlines repetitive tasks, enabling designers to focus on ideation and creativity	Canva AI tools, Adobe Sensei, generative AI in 3D modeling
Cross-Domain	Synergy between human insights and AI-driven capabilities	Facilitates adaptive, impactful, and efficient creative processes across all domains	AI-enhanced storytelling, integrated creative workflows, AI-powered campaign planning
Marketing	Content generation, campaign performance analysis, market trend prediction	Enables data-informed strategies, improves customer engagement and decision making	Chatbots, AI-powered CRM (e.g., Salesforce Einstein)

Source: Compiled by Author's (Based on Li, 2019, Van Belkom, 2020, Vakratsas & Wang, 2020, Shah *et al.*, 2020 and Kushwaha, 2023)

For designing, AI-powered tools assist in creating visuals, layouts, and prototypes with speed and precision. Designers are hence empowered to explore creative options effortlessly. This dynamic has not replaced human creativity (Rodgers, 2021 and Gao *et al.,* 2023) (Table 1).

Marketing is in the genesis stage again due to the expansion of AI. AI's ability to generate engaging content and predict market trends has changed the marketing landscape. Chatbots, dynamic pricing strategies, and a data-informed approach are connecting businesses.The future of industries lies in embracing AI-Human partnership (Verma *et al.,* 2021; Chintalapati & Pandey, 2021 and Gołąb-Andrzejak, 2023) (Table 1).

1.2 Importance of creativity in brand success

An inventive approach allows brands to craft unique identities. This may be connected with an innovative campaign that sparks viral attention (Trueman *et al.,* 2008; Endrissat *et al.,* 2016). A Coca-Cola's heart-warming holiday ads or Apple's minimalist design ethos (Sheng, 2023; and Blake *et al.,* 2015).

2. THE EVOLUTION OF CREATIVITY IN THE AI ERA

The evolution of creativity in the AI era is best understood as a shift from individual human cognition to a hybrid system. Traditionally, the creative act was housed within the individual or team. The advent of AI introduces a non-human cognitive agent into this loop. This is not a simple handoff of tasks. But the formation of a collaborative cognitive unit. In this system, creativity emerges from the iterative dialogue between human intuition and machine intelligence. The "paradigm shift" is therefore cognitive. It is the expansion of the creative mind itself through a symbiotic partnership. The following analysis reframes the historical shift not as "from manual to algorithm-driven" but as "from solitary to distributed cognition" (Table 2).

Table 2. Evolution of Creativity in the AI Era

Aspect	Traditional Creativity	AI- Driven Creativity	Synergy in the AI Era
Core Drivers	Human intuition, emotional depth, and cultural awareness	Data-driven insights, algorithmic precision, rapid processing	Collaboration of emotional intelligence with algorithmic capabilities
Processes	Sequential and time-intensive	Accelerated, automated, and highly scalable	Efficient workflows that combine speed with imaginative depth
Impact on Industry	Restricted to manual and intuition based strategies	Revolutionizes marketing, graphic design, and content creation	Creates dynamic systems blending efficiency, precision, and imaginative exploration
Capabilities	Limited by time, resources, and human capacity	Expands the realm of possibilities, producing prototypes and solutions in moments	Combines human originality with AI's ability to generate multiple options and insights
Limitations	Constrained by human energy, susceptibility to bias, and lack of scalability	Lacks originality, emotional resonance, and deep cultural understanding	Overcomes individual limitations while maintaining the irreplaceable spark of human creativity
Creative Outcome	Grounded in tradition and often shaped by resource availability	Innovative, data informed, and optimized for diverse contexts	Resonance in messaging that transcends conventional boundaries, balancing innovation with human insight
Key Examples	Hand drawn designs, intuition led marketing strategies	AI-driven ad campaigns, generative art, automated content creation	AI assisting designers with automated color palettes and enabling marketers to personalize campaigns at scale

Source: Compiled by Authors (Based on Das, 2022; and Acar, 2023)

2.1 Historical perspective: From manual to algorithm-driven creativity

The journey of creativity mirrors the transition from classical mechanics to quantum dynamics in the realm of ideas. Yet, it was not until the rise of AI. Today, algorithms act like charged particles within creative systems. These charged particles (algorithms) are generating ideas, analyzing trends, and automating repetitive tasks. AI models such as DALL-E and ChatGPT illustrate this transformation (Lund & Wang, 2023 and Adetayo, 2024).

2.2 AI's role in automating creative processes

AI's role in automating creative processes has introduced a level of efficiency and precision (Figure 1). AI is now streamlining the creative workflows in advertising, design, and marketing. By processing vast amounts of data and recognizing patterns, AI can generate ideas, tailor content, and even design visuals (Zatarain, 2017; Grilli & Pedota, 2024). This process is parallel to the concept of automation in physics, where systems operate with negligible energy expenditure, yet produce maximum output.

Figure 1. AI's Role in Automating Creative Process

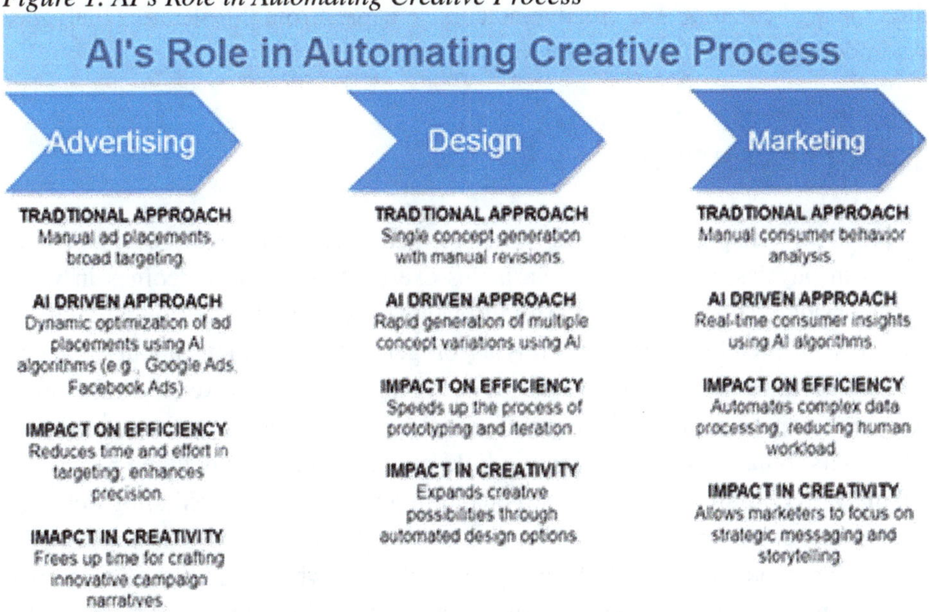

Source: Compiled by Authors (based on Zatarain, 2017; Grilli & Pedota, 2024)

In design, AI tools can rapidly create multiple versions of a concept, allowing designers to explore variations and choose the best iteration in record time. In marketing, AI algorithms automatically analyze consumer behavior, tailoring campaigns in real time to deliver the most relevant content. AI-driven platforms like Google

Ads and Facebook Ads use these capabilities to dynamically optimize ad placements, adjusting for factors like audience engagement, time of day, and location.

2.3 Human-AI collaboration in creative industries

Human-AI collaboration in creative industries represents a dynamic fusion of human intuition and machine efficiency (Wu *et al.,* 2021; Amankwah-Amoah *et al.,* 2024). This partnership is not about replacing the human element but enhancing it. In the same way an electric current amplifies a signal in physics. AI amplifies the creative process by providing insights, automating mundane tasks, and generating novel ideas (Anantrasirichai & Bull, 2022).

This collaboration can be modeled as a cognitive handshake. AI handles high-volume tasks of generation, pattern recognition, and real-time optimization. And functions as a tireless ideation and execution engine. Humans provide the irreplaceable strategic intent, emotional intelligence, and ethical oversight. Thus acting as directors and curators. The creative outcome emerges from this iterative dialogue. And not from either agent working in isolation.

3. AI IN ADVERTISING

AI in advertising has sparked a transformation akin to the discovery of new forces in physics, where unseen variables exert influence on outcomes in ways once considered impossible. In the traditional realm of advertising, campaigns were crafted with broad strokes, often relying on general assumptions about audience preferences and market trends (Li, 2019; Ford *et al.,* 2023). However, the infusion of AI into this space has altered the gravitational pull of traditional methods, enabling advertisers to harness a vast, dynamic field of data. Like a finely tuned resonator, AI analyzes consumer behavior with quantum-level precision, predicting not only what audiences desire but also when and how they will respond. This data driven approach mirrors the concept of wave particle duality, where the wave of consumer insights intersects with the particle of targeted actions, resulting in real time optimization of ad performance (Fig. 1). Much like a particle accelerator manipulating subatomic particles to achieve specific outcomes, AI tailors campaigns to maximize engagement and impact, offering a precision previously unattainable in traditional advertising (Kietzmann *et al.* 2018; Shah *et al.,* 2020; Huh *et al.,* 2023).

AI's role in advertising can be understood as a sort of 'feedback loop' in which the system constantly evolves based on real time inputs (Huh *et al.,* 2023). For instance, AI-powered platforms like Google Ads or Facebook Ads continuously refine campaigns, adjusting parameters such as target audience, time of day, and content type to maximize engagement (Gołąb-Andrzejak, 2023; Kumar *et al.*, 2024). These algorithms function as a 'gravitational force' guiding ads to the most relevant consumers, increasing the likelihood of conversion with minimal manual intervention (Fig. 1). This dynamic system is capable of learning and adapting, allowing advertisers to capitalize on emerging trends without having to restart campaigns from scratch (Akter *et al.* 2023; Kumar *et al.*, 2024).

Figure 2. AI's Transformative Role in Advertising

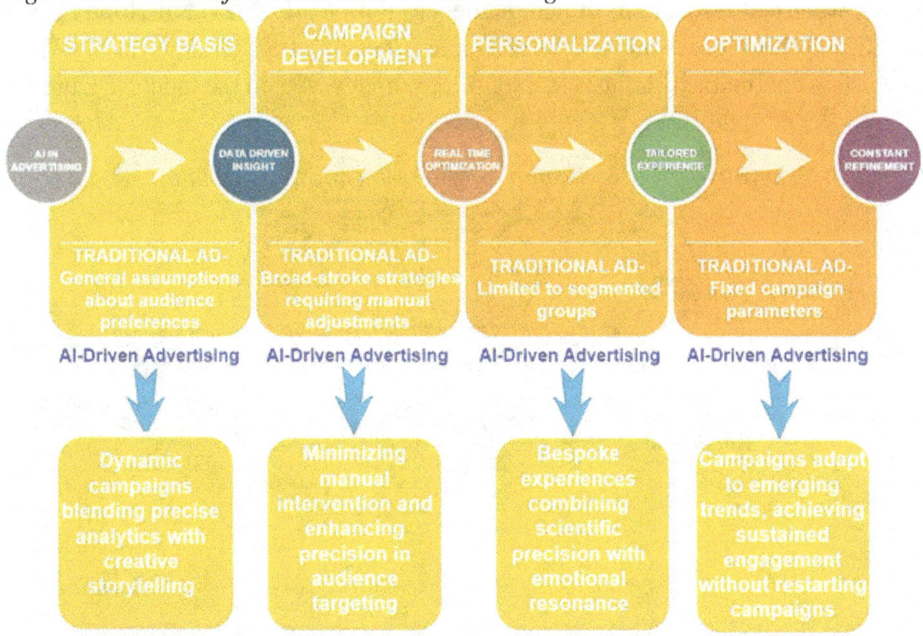

Source: Compiled by Authors (based on Li, 2019; Ford et al., 2023; Huh et al., 2023; Akter et al. 2023; Kumar et al., 2024)

AI facilitates personalization at a scale previously unimaginable. Like the process of multi-format adaptability, where light behaves both as a wave and a particle depending on how it's observed, AI tailors content in multiple forms, whether through display ads, video content, or sponsored posts, for creating a bespoke experience for each user (Gao *et al.*, 2023). The true power of AI lies in its ability to combine vast amounts of data with creative vision, resulting in campaigns that are not only scientifically optimized but also deeply human, resonating emotionally with individuals while achieving measurable business goals (Blasch *et al.*, 2021). This fusion of technology and creativity is revolutionizing how brands communicate with their audiences, ushering in an era where advertising is not just a broadcast, but a conversation (Fig. 2).

3.1 Generative AI tools for crafting engaging ad content

Generative AI tools are rapidly becoming essential in crafting engaging ad content. They offer a level of creativity and precision that was once the exclusive domain of human designers (Kshetri *et al.,* 2024). These AI-powered platforms function like a catalyst in the creative process. Much like a physicist employs equations to model complex systems, generative AI applies sophisticated algorithms to analyze vast amounts of consumer data and behavioral patterns. It predicts, with high precision, what will resonate most with audiences (Gui *et al.,* 2021; Fui-Hoon Nah *et al.,* 2023). This process mirrors the principles of energy transfer in physics. Where the system continuously adjusts to maximize output based on the inputs (Figure 3).

Figure 3. Generative AI Advertising Workflow- Technical Schematic

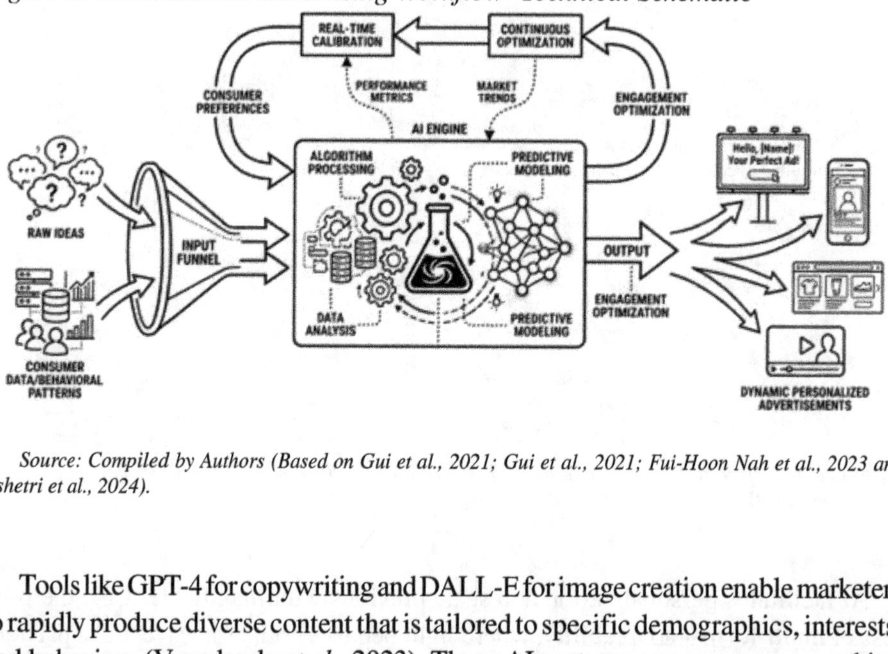

Source: Compiled by Authors (Based on Gui et al., 2021; Gui et al., 2021; Fui-Hoon Nah et al., 2023 and Kshetri et al., 2024).

Tools like GPT-4 for copywriting and DALL-E for image creation enable marketers to rapidly produce diverse content that is tailored to specific demographics, interests, and behaviors (Vayadande *et al.,* 2023). These AI systems can generate everything from catchy taglines to full length ad scripts, while ensuring each piece maintains relevance and emotional appeal (Mogaji *et al.,* 2020; Rokhade *et al.,* 2024). For example, an AI might suggest multiple versions of a headline, testing for variables such as language tone, word choice, and cultural relevance, all based on user data.

3.2 Cases: Real-world examples of AI-powered ad campaigns

These case studies highlight how brands leverage AI technology to create innovative, targeted, and engaging marketing strategies.

3.2.1 Coca-Cola: Revolutionizing Personalization with AI

Coca-Cola has long been a pioneer in creating memorable advertising campaigns, but deployment has taken personalization to imaginable levels (Timimi *et al.* 2025; Vashishtha & Sharma, 2025).

Coca-Cola partnered with IBM Watson to create dynamic, AI-driven advertising experiences (Table 3). By analyzing a massive dataset encompassing consumer preferences, purchase histories, social media activity, and real-time factors like weather conditions, the AI could predict the perfect moment and format to engage audiences (Vashishtha & Sharma, 2025). For instance, on a sunny day in New Delhi, Coca-Cola's system might generate a video showcasing a refreshing Coke served in a chilled glass, paired with a relatable tagline like "Beat the heat, sip the chill"

Table 3. Analysis of Coca-Cola's AI-Driven Advertising Framework

Core Strategy	Technical Implementation	Environmental & Behavioral	Inputs Strategic Outcome	Research Source
Cognitive Ad Delivery	Partnership with IBM Watson for dynamic experiences.	Massive datasets of consumer preferences and purchase history.	Predictive engagement and unprecedented personalization.	Vashishtha & & Sharma (2025)
Real-Time Contextual Adaptation	Integration of live variables for instant ad generation.	Geographic location, weather conditions (e.g., heat in New Delhi).	Higher relevance and immediate sales acceleration.	Timimi *et al.* (2025)
Hyper-Localized Content	Dynamic shifts in tone and pace based on urban vibe.	City-specific atmosphere (e.g., Mumbai's pace vs. hill towns).	Deeply personal brand connection and engagement.	Timimi *et al.* (2025); Arumugam *et al.* (2024)

Source: Compiled by Authors

Even more remarkable was its ability to dynamically adapt ad content for each user. If you are in Mumbai, the AI might craft a vibrant, fast-paced ad reflecting the city's bustling vibe, while in a serene hill town, the tone shifts to calm and refreshing (Table 3). This fusion of creativity and AI precision allowed Coca-Cola to deliver ads that felt deeply personal, turning a global brand into something that spoke directly to each individual, boosting engagement and sales (Timimi *et al.* 2025; Arumugam *et al.,* 2024).

3.2.2 Heineken: Precision Advertising Through AI

Heineken has also embraced AI, but its focus lies in optimizing ad delivery to ensure maximum impact. Imagine Heineken as a master strategist, positioning its ads at precisely the right moment and place to capture attention. Using machine learning, Heineken analyzed vast datasets, including consumer behavior, time-of-day activity, and even live event schedules (Wierenga & Van Bruggen, 2000; Dobele *et al.*, 2005) (Table 4).

For example, during a major sports event like the FIFA World Cup, Heineken's AI system determined the perfect intervals to serve ads, aligning them with peak viewership moments. The AI even selected visuals and taglines tailored to match the enthusiasm of sports fans. A Heineken ad might appear during halftime with the message, "Celebrate every goal with a Heineken," accompanied by imagery of fans cheering (Cavalier & Newhall, 2017).

Table 4. Strategic Framework of Heineken's AI-Optimized Advertising

Operational Element	Implementation Method	Data Inputs & Triggers	Strategic Impact	Research Source
Precision Ad Delivery	Machine learning for optimal "moment and place" positioning.	Consumer behavior and time-of-day activity.	Maximum audience attention and impact.	Wierenga & Van Bruggen (2000)
Event-Based Synchronization	Real-time alignment with major live spectacles.	FIFA World Cup schedules and peak viewership intervals.	High-context engagement during peak enthusiasm.	Cavalier *et al.* (2017)
Iterative Content Refinement	Continuous AI adjustment of visuals and taglines.	Direct user interaction data and engagement metrics.	Enhanced relevance and optimized conversion rates.	Nanne *et al.* (2022); Liao & Chen, (2010)
Consumer-Centric Science	Fusion of data with creative partnership.	Emotional connections and engagement stats.	Transformation of advertising into a strategic force.	Seo & Lee, (2021)

Source: Compiled by Authors

Beyond timing, Heineken's AI continuously refined the ad content based on user interaction data (Table 4). If a specific visual style or tagline resonated more, the AI adjusted future ads accordingly (Nanne, *et al.*, 2022; Liao & Chen, 2010). This iterative learning process not only enhanced ad relevance but also optimized campaign performance, leading to higher engagement and conversion rates.

Both Coca-Cola and Heineken showcase how AI is transforming advertising into a finely tuned, consumer-centric science. By blending data with creativity, these campaigns deliver experiences that not only capture attention but also foster deeper emotional connections with audiences. These cases highlight the evolving role of AI as both a creative partner and a strategic force in modern advertising (Seo & Lee, 2021).

3.2.3 IBM Watson: Personalization through Cognitive AI

IBM Watson has emerged as a trailblazer in the realm of personalized advertising, demonstrating the immense potential of AI to transform how brands engage with their audiences. Watson's cognitive capabilities allow it to analyze vast amounts of unstructured data, such as social media interactions, purchase behaviors, and even customer sentiments, to craft highly individualized advertising experiences (Chen, *et al.,* 2016; Strickland, 2019; Petiwala, *et al.,* 2021) (Table 5).

Consider how Watson collaborated with the renowned luxury retailer "The North Face". Shoppers visiting The North Face's website were met with a unique AI-driven experience (Table 5). By interacting with Watson, users could describe what they were looking for in natural language, something like, "I need a jacket for hiking in snowy mountains." Watson processed this input, analyzed weather conditions, and cross-referenced the store's inventory to recommend specific products that met the customer's precise needs. This recommendation enabled The North Face to deliver a personalized shopping experience for customers in retail.

Table 5. Analytical Framework of IBM Watson's Cognitive Advertising Capabilities

Functional Domain	Technical Execution	Data Inputs & Triggers	Strategic Outcome	Research Sources
Natural Language Commerce	Processing of conversational user intent (e.g., The North Face).	Natural language queries, weather data, and inventory.	High-utility, bespoke product matching.	Strickland, (2019); Chen *et al.* (2016)
Contextual Personalization	Climate-aligned and profile-specific ad generation.	Geographic location (e.g., Jaipur vs. Switzerland) and user profile.	Enhanced relevance and geographical alignment.	Hildesheim, (2017); Petiwala *et al.* (2021)
Channel & Schedule Optimization	Algorithmic selection of communication medium and timing.	Online activity records and real-time user traffic patterns.	Precision delivery to the "right person at the right time."	Petiwala *et al.* (2021)
Cognitive Data Analysis	Synthesis of unstructured datasets into actionable insights.	Social media sentiment, purchase history, and interactions.	Transformation of broad messaging into tailored storytelling.	Chen *et al.* (2016); Petiwala *et al.* (2021)

Source: Compiled by Authors

IBM Watson is also capable of tailoring the ads as per the individual customer profiles. For example, a user in Jaipur might stay away from an ad for a woollen Jacket specifically designed for extreme winters. Simultaneously, a user in Switzerland will come across an ad for a woollen jacket showcasing the snowy mountains, thereby aligning with their climate and need.

IBM Watson also has a specialty in selecting the most appropriate channel of marketing communication and the schedule of advertising. This is done on the basis of the online activity record of the audience and the traffic presence of the user. IBM Watson has the ability to ensure that the right ad reaches the right person at the right time (Table 5).

IBM Watson exemplifies how cognitive AI can elevate advertising from broad messaging to tailored storytelling, ensuring that each interaction feels relevant and deeply resonant with individual consumers (Hildesheim, 2017; Petiwala *et al.,* 2021).

3.3 Challenges and limitations of AI in advertising

AI has transforms advertising by enabling precision and personalization, its adoption is not without challenges and limitations (Figure 4). One significant concern revolves around data privacy and security. AI relies on massive datasets to craft personalized ads, but collecting and using such data can raise ethical questions and compliance issues, especially with stringent regulations like the General Data Protection Regulation (GDPR) and Central Consumer Protection Authority (CCPA). Consumers may feel uneasy about the extent of information brands gather, leading to trust deficits (Kingston, 2017; Taylor, 2019; Nizioł, 2021; Pantelic *et al.,* 2021).

Another hurdle lies in the risk of creativity stagnation (Figure 4). Although AI excels at analyzing trends and generating content, its creativity is limited by the data and patterns it has been trained on (Lu, 2019). This reliance on historical data can result in campaigns that lack originality or fail to resonate with evolving cultural contexts. Human intuition and imagination remain irreplaceable in crafting ground-breaking concepts.

Figure 4. Hindered AI Adoption in Advertising

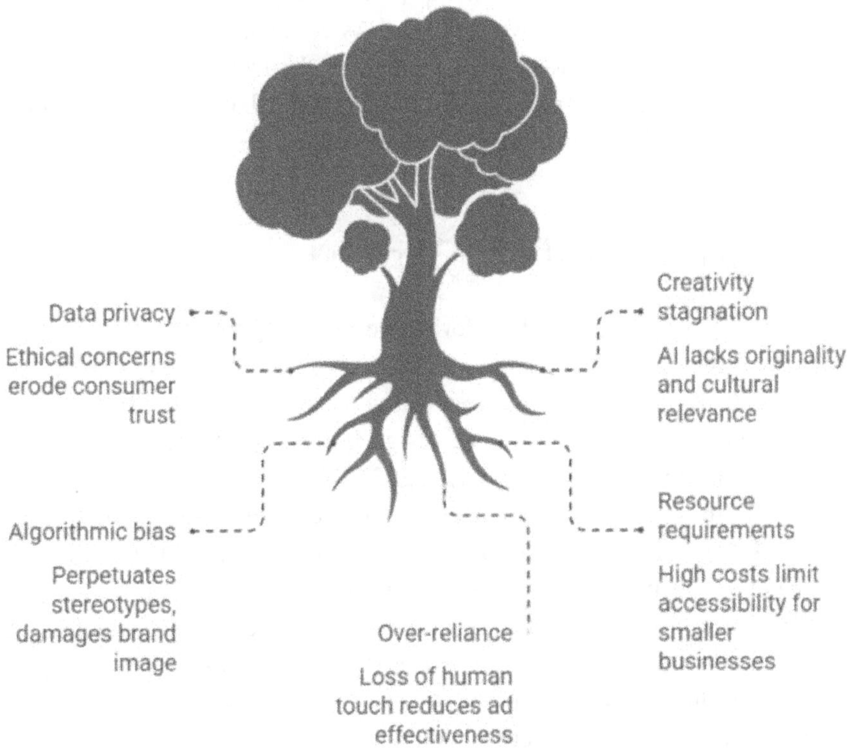

Source: Compiled by Authors based on Taylor, 2019; Nizioł, 2021; Pantelic et al., 2021

The bias in algorithms is another challenge (Figure 4). AI systems can inadvertently reflect and amplify existing societal biases present in their training data. This can lead to ads that alienate certain demographics or perpetuate stereotypes, tarnishing a brand's reputation (Bozdag, 2013; Kordzadeh & Ghasemaghaei, 2021). For instance, an AI system trained on biased datasets may inadvertently favour specific languages or imagery, excluding broader audiences.

This bias is not random but systematically emerges from the training data. For creative AI, datasets often over-represent dominant cultural aesthetics, demographics, and styles. For a brand, this risks producing campaigns that feel exclusionary or inauthentic to diverse audiences, fundamentally misaligning with inclusive brand values. Mitigation requires more than technical fixes. It demands cognitive governance. This includes proactive bias audits of training data. Without this, brands risk automating their own irrelevance.

Operational challenges also arise, particularly around resource requirements. Implementing AI-driven advertising demands significant investment in technology, skilled personnel, and infrastructure. Smaller businesses may struggle to compete with larger organizations that have the capital to deploy advanced AI tools.

Lastly, there is the issue of over-reliance. As companies increasingly lean on AI for decision making, they risk losing the human element that makes advertising relatable and emotionally compelling. Striking the right balance between automation and human input is critical for long-term success.

4. GENERATIVE AI IN GRAPHIC DESIGN

Generative AI has become a transformative force in graphic design, reshaping how creative professionals conceptualize and execute visual content. By leveraging algorithms capable of learning and replicating artistic styles, AI brings unprecedented efficiency and versatility to the design process, enabling creators to push the boundaries of imagination (Liu & Liu, 2021; Matthews *et al.,* 2023) (Table 6).

One of the most significant contributions of generative AI lies in speed and scalability. Traditional graphic design often involves iterative manual processes that can be time-consuming. AI tools such as DALL-E, Adobe Firefly, and Runway automate tasks like generating mock-ups, creating templates (Table 6), and applying consistent branding elements across a wide array of platforms. For instance, an AI tool can produce hundreds of variations of a logo or social media post, each tailored to specific dimensions and aesthetics, within minutes.

Table 6. Analytical Framework of Generative AI in Graphic Design

Core Contribution	Technical Implementation	Operational Impact	Strategic Outcome	Research Sources
Operational Agility	Automation of mock-ups and templates via DALL-E and Adobe Firefly.	Rapid production of hundreds of design variations in minutes.	Unprecedented speed and scalability in the creative process.	Liu & Liu, (2021)
Bespoke Visual Logic	Data-driven analysis of consumer trends and group preferences.	Adaptation of aesthetics (e.g., meme-inspired vs. minimalistic).	High-precision personalization for diverse target audiences.	Matthews *et al.* (2023)
Human-AI Synergy	AI as a creative partner providing iterations and visual drafts.	Reduction in manual labor while expanding imaginative scope.	Enhanced collaboration and pushing of creative boundaries.	Matthews *et al.* (2023)

Source: Compiled by Authors

4.1 AI-driven design platforms and tools

AI-driven design platforms and tools are revolutionizing the creative industry. These tools integrate advanced algorithms to automate repetitive tasks, generate original ideas, and enhance the overall design process. From brainstorming visuals to finalizing production-ready content, AI platforms are transforming the graphic design workflow (Caiazzo *et al.,* 2023) (Table 7).

Adobe Firefly, for instance, brings AI into the hands of creators by offering features like automatic background removal, color matching, and dynamic resizing. It enables designers to experiment with multiple visual elements, creating personalized and professional-grade content in minutes. Tools like Firefly help businesses scale branding efforts by maintaining consistent aesthetics across diverse media channels without manual intervention (Nair, 2024).

Table 7. Technical Framework of AI-Driven Design Platforms

Platform/Tool	Core AI Feature	Operational Function	Strategic Impact	Research Source
Adobe Firefly	Generative Content Intelligence	Background removal, color matching, and dynamic resizing.	Rapid scaling of branding and professional-grade content.	Nair, (2024)
Canva	Template Recommendation Logic	Algorithmic selection of fonts, colors, and layouts.	Democratic access to design for non-professionals.	Hinchcliff & Mehmet, (2023)
Autodesk SketchBook	Stroke Prediction & Refinement	Algorithmic smoothing of lines and maintenance of symmetry.	Technical precision in traditional digital illustration.	Xaba *et al.* (2022)
Figma	Real-time Collaborative AI	Automated asset generation and team workflow streamlining.	Enhanced cohesion in collaborative design environments.	Xaba *et al.* (2022)
General Workflow	Iterative Automation	Brainstorming to final production-ready content.	Redefinition of creative boundaries and efficiency.	Caiazzo *et al.* (2023)

Source: Compiled by Authors

Canva integrate machine learning to recommend templates, fonts, and color schemes based on a brand's identity. Its intuitive interface bridges the gap for non-designers, making professional-quality graphic design accessible to small businesses and individual creators (Hinchcliff & Mehmet, 2023) (Table 7).

AI tools also enhance traditional software. Autodesk's SketchBook uses AI to predict and refine strokes, ensuring smooth lines and symmetry in illustrations.

Figma as well incorporates AI for real-time collaboration and automated asset generation (Xaba *et al.,* 2022).

4.2 How AI accelerates ideation, prototyping, and execution

In the ideation phase, AI acts as a catalyst for inspiration (Joosten *et al.* 2024). Tools like ChatGPT and MidJourney analyze vast datasets and uncovering trends as well as generating unique concepts. For instance, a designer seeking ideas for an eco-friendly campaign can input a brief into an AI tool, which then suggests visual themes. This not only sparks creativity but also ensures relevance by aligning concepts with current market trends and consumer preferences (Lund & Wang, 2023; Tsidylo & Sena, 2023; Adetayo, 2024; Shamaeva, 2024).

Prototyping has become significantly faster and more dynamic with AI. Figma, Adobe XD, and Runway enable designers to quickly transform abstract ideas into tangible prototypes (Raptis & Katsini, 2023; Truss & Schmitt, 2024). AI algorithms can automatically resize and reformat designs for mobile, desktop, and print applications (Shah *et al.,* 2020; Micu, *et al.,* 2022; Truss & Schmitt, 2024).

4.3 Ethical concerns and creative ownership in AI-generated designs

The rise of AI-generated designs has sparked critical discussions around ethical concerns and creative ownership. As AI tools increasingly contribute to creative outputs, questions about authorship and fairness have become a nucleus (Bisoyi, 2022; Watiktinnakorn *et al.,* 2023; Gaffar & Albarashdi, 2024).

The ethics of bias in AI-generated designs also warrant inquiry. Algorithms are only as neutral as the data they are trained on (Vetter *et al.,* 2024). AI tools used for advertising might produce visuals that inadvertently marginalize underrepresented groups.

5. AI-ENHANCED MARKETING CAMPAIGNS

In the rapidly evolving digital landscape, AI-enhanced marketing campaigns have emerged as game changers. AI transforms how brands connect with their audiences (Van Noort *et al.,* 2014).

5.1 Personalized marketing through AI insights

Personalized marketing has become a cornerstone of modern business strategies. AI insights are driving this transformation. By analyzing vast amounts of data, AI enables brands to create tailored experiences that resonate with individual consumers (Chandra *et al.,* 2022; Gao & Liu, 2023).

Predictive analytics as a key capability of AI, allows marketers to anticipate consumer behavior. By identifying patterns, AI tools can predict what customers are likely to want next. Netflix's recommendation engine is an example, where AI analyzes viewing habits to suggest personalized content (Amatriain & Basilico, 2015; Jenkins, 2022).

5.2 Leveraging AI for content generation and predictive analytics

AI tools such as ChatGPT and DALL-E have become invaluable for content generation. These technologies can produce tailored copy, visuals, and even video content (Lund & Wang, 2023; Adetayo, 2024). AI-driven platforms like Canva and Jasper enable marketers to create compelling advertisements, social media posts, and blogs.

Predictive Analytics further amplifies the impact of AI in advertising. This solution allows brands to forecast consumer preferences and behaviors. Amazon uses algorithms to recommend products, while Spotify employs AI to finalize the personalized playlists based on listening pattern. These predictions create a sense of personalization driving customer loyalty (Vashishtha & Sharma, 2025).

5.3 Case studies of successful AI-powered campaigns

5.3.1 Coca-Cola's Personalized Ads

Coca-Cola has been a leader in using AI to personalize its marketing campaigns. By using AI algorithms, the brand tailors its messaging to specific segments of its audience. A notable example is their campaign where Coca-Cola personalized bottle labels with popular names using AI-powered systems (Sheng, 2023; Cui, *et al.,* 2024; Vashishtha & Sharma, 2025; Timimi *et al.* 2025). This approach not only increased consumer engagement but also made the brand more relatable.

5.3.2 H&M's AI-Powered Trend Forecasting

H&M has adopted AI to streamline its fashion marketing and inventory management. The brand uses AI algorithms to analyze trends from a variety of sources (Konina, 2023). This predictive capability allows H&M to create targeted campaigns, ensuring they promote the right styles at the right time. The integration of AI has not only optimized their supply chain but also led to more efficient and engaging marketing strategies (George *et al.,* 2022; Dhama, *et al.,* 2024).

5.3.3 Spotify's AI-Powered Playlists

Spotify leverages AI to enhance user engagement through personalized playlists. The platform's "Discover Weekly" feature uses machine learning algorithms to analyze user preferences and listening behaviors (Awotunde *et al.,* 2024). The AI's ability to predict individual taste has resulted in a highly successful marketing tool. This campaign showcases how AI can transform traditional marketing into a hyper personalized experience.

5.3.4 Nike's Hyper Personalized Ads

Nike has harnessed AI to create hyper-targeted marketing campaigns. The company utilizes AI to analyze customer data and preferences. And they deliver personalized ads across social media. They deploy AI to predict which products customers are likely to purchase next. This AI-driven approach has significantly increased the brand's ability to drive sales.

6. FUTURE TRENDS

As artificial intelligence continues to evolve, its integration into marketing strategies is expected to deepen (Haleem *et al.,* 2022).

6.1 Re-Rise of AI-Generated Content and Creativity

As AI tools become more advanced, the content creation will become increasingly automated. This progression will not only accelerate the creative process alone. The advancement also empower brands to preserve a rapid delivery of renewed and tailored content. AI will not only assist the content creation but will eventually lead to AI-generated creative campaigns.

6.2 Increased Personalization Through Predictive Analytics

AI's aptitude to process massive amounts of data will further enhance personalization. This skill will extend to all dimensions of marketing. If AI keeps enlarging its horizon, then brands will be able to create exceedingly tailored campaigns (Krishnan & Mariappan, 2024).

6.3 AI-Driven Automation and Efficiency Gains

AI will continue to optimize advertising operations through automation. Jobs like "Audience segmentation", "Ad targeting", "Campaign optimization", and "Content distribution" will be extensively automated (Ahmadi-Karvigh *et al.*, 2019). AI-powered programmatic advertising will enable real-time ad bidding, adjusting campaigns.

6.4 Ethical AI and Creative Ownership

As AI takes a more prominent role in the creative industries, concerns about ethical issues will be more pressing (Bisoyi, 2022). The world would need more symposia to revise ethical guidelines and transparency in AI-driven campaigns. Marketers need to balance innovation with responsibility (Kumar & Suthar, 2024).

6.5 AI's Role in Augmented and Virtual Reality Marketing

Augmented Reality (AR) and Virtual Reality (VR) are set to play a larger role in AI-powered marketing campaigns. The fusion of AI with AR/VR technologies will allow brands to deliver immersive, interactive, and personalized experiences (Adeola *et al.*, 2022; Pakharuddin & Kamarudin, 2023).

6.6 Managerial Implications: Prioritizing AI Investments in Branding

The transformative potential of AI necessitates strategic investment decisions. Managers should adopt a phased, value-driven approach rather than pursuing technology for its own sake. The following framework can guide prioritization:

i. Foundational vs. Transformative Investments

Managers/Entrepreneurs must start with foundational AI tools that enhance efficiency and offer précised returns. For example, programmatic ad buying platforms,

AI-powered A/B testing, and basic data analytics for customer segmentation. These provide quick wins and build internal competency.

ii. The ICE Framework (Impact, Confidence, Ease)

Evaluate potential AI initiatives across three dimensions.

Impact indicates potential business value (e.g., increased engagement, higher conversion rates, and reduced time-to-market)?

Confidence covers "How certain are we about the projected impact and the technical feasibility?"

And Ease answers, "What is the estimated level of effort regarding cost, integration complexity, and change management?"

iii. Align with Core Brand Challenges

Investments should directly address identified pain points. Imagine a brand struggling with content velocity should prioritize generative design tools (Adobe Firefly, Canva AI). While other brands with vast customer data but poor insight activation need investment in predictive analytics and AI-driven CRM platforms.

7. CONCLUSION

The future of creativity in AI-powered industries has boundless potential. AI offers the opportunity to enhance and accelerate the creative process, driving innovation as well as fostering deeper connections between brands and consumers. As technology continues to advance, the collaboration between human creativity and AI will reach new heights, resulting in more personalized, engaging, and meaningful experiences.

7.1 Recap of key insights

The intersection of artificial intelligence (AI) with creativity in advertising, design, and marketing is reshaping the landscape of how brands interact with consumers, craft campaigns, and deliver personalized experiences. From the rapid advancement of AI-driven tools to the increasing reliance on predictive analytics, the creative industries are undergoing a significant transformation. As we conclude this exploration, it is important to recap the key insights that have emerged throughout this discussion.

One of the fundamental insights is the expanding role of AI in the creative process. AI is no longer limited to data analysis or automation; it has evolved into a powerful tool for generating creative content, enhancing design, and crafting personalized marketing campaigns. AI algorithms, particularly in the realms of generative design, natural language processing, and machine learning, have empowered brands to create innovative solutions that were previously unimaginable. Whether through generative AI tools that craft compelling ad content or AI algorithms that optimize user experiences, brands can now engage their audiences in highly personalized and meaningful ways.

We have framed these developments not as isolated phenomena. Through the synthesized lens of governed cognitive partnership, integrating the key scholarly debates on augmentation, co-creativity, and ethical governance specific to the branding domain.

7.2 The future of creativity in AI-powered industries

As we gaze into the future of creativity in AI-powered industries, it is important to recognize the profound impact. AI's evolution in advertising, design, and marketing is still in emergence stage.

7.2.1 Advancing Human-AI Collaboration

AI is expected to evolve into a co-creative partner. AI is expected as a catalyst of human creativity. As AI becomes more skilled at imitating human creativity, it will also allow for a greater range of exploration. AI will be able to help designers quickly iterate on concepts and even predict what designs will resonate most with audiences.

7.2.2 Expanding AI's Role in Interactive and Immersive Experiences

In the coming years, AI will play a significant role in the creation of more interactive and immersive marketing experiences. The VR and AR technologies powered by AI can open new possibilities. For instance, AI could fuel virtual 'Try-Ons'. This can drives consumers to visualize products in 3D before making a purchase. Similarly, AR ads powered by AI could deliver real-time content to users in allignment with their location or interest or behaviour.

7.2.3 AI-Driven Creativity as a Competitive Edge

The future of creativity in AI-powered industries also suggests that AI will become a critical competitive edge for businesses. Companies that embrace AI and integrate it into their creative processes will be better positioned to stand out in a crowded marketplace. AI's ability to analyze and predict consumer behavior allows businesses to create more effective, targeted campaigns that drive results. Additionally, as AI tools become more advanced and accessible, even small companies will be able to leverage cutting-edge technology to enhance their creativity, levelling the playing field and driving innovation across industries.

7.2.4 Final Thoughts

The future of creativity in AI-powered industries is one of boundless potential, akin to the boundless potential. AI offers the opportunity to enhance and accelerate the creative process, driving innovation while fostering deeper connections between brands and consumers. As technology continues to advance, the collaboration between human creativity and AI will achieve new heights, resulting in more personalized, engaging, and meaningful experiences. This synergy, much like the interaction of force and matter, will create an environment where innovation flows freely and exponentially.

Competing Interests

The authors of this chapter declare there are no competing interests.

Funding Declaration

This research received no specific grant from any funding agency in the public, commercial, or not-for-profit sectors. Funding for this research was covered by the authors.

REFERENCES

Acar, S. (2023). Creativity Assessment, Research, and Practice in the Age of Artificial Intelligence. *Creativity Research Journal*, *37*(2), 181–187. DOI: 10.1080/10400419.2023.2271749

Adeola, O., Evans, O., Ndubuisi Edeh, J., & Adisa, I. (2022). The Future of Marketing: Artificial Intelligence, Virtual Reality, and Neuromarketing. In *Palgrave Studies of Marketing in Emerging Economies* (pp. 253–280). Springer International Publishing., DOI: 10.1007/978-3-030-88678-3_12

Adetayo, A. J. (2024), "Reimagining learning through AI art: the promise of DALL-E and MidJourney for education and libraries", *Library Hi Tech News*, Vol. ahead-of-print No. ahead-of-print. DOI: 10.1108/LHTN-01-2024-0005

Ahmadi-Karvigh, S., Becerik-Gerber, B., & Soibelman, L. (2019). Intelligent adaptive automation: A framework for an activity-driven and user-centered building automation. *Energy and Building*, *188-189*, 184–199. DOI: 10.1016/j.enbuild.2019.02.007

Akter, S., Hossain, M. A., Sajib, S., Sultana, S., Rahman, M., Vrontis, D., & McCarthy, G. (2023). A framework for AI-powered service innovation capability: Review and agenda for future research. *Technovation*, *125*, 102768. DOI: 10.1016/j.technovation.2023.102768

Amankwah-Amoah, J., Abdalla, S., Mogaji, E., Elbanna, A., & Dwivedi, Y. K. (2024). The impending disruption of creative industries by generative AI: Opportunities, challenges, and research agenda. *International Journal of Information Management*, *79*, 102759. DOI: 10.1016/j.ijinfomgt.2024.102759

Amatriain, X., & Basilico, J. (2015). Recommender Systems in Industry: A Netflix Case Study. In Recommender Systems Handbook (pp. 385–419). *Springer US*. DOI: 10.1007/978-1-4899-7637-6_11

Ameen, N., Sharma, G. D., Tarba, S., Rao, A., & Chopra, R. (2022). Toward advancing theory on creativity in marketing and artificial intelligence. *Psychology and Marketing*, *39*(9), 1802–1825. DOI: 10.1002/mar.21699

Anantrasirichai, N., & Bull, D. (2022). Artificial intelligence in the creative industries: A review. *Artificial Intelligence Review*, *55*(1), 589–656. DOI: 10.1007/s10462-021-10039-7 PMID: 41625196

Arumugam, S. K., Pundareeka Vittala, K. R., Gaikwad, S. M., & Tyagi, A. K. (2024). Role of Emerging Technologies in Smart Marketing and Smart Business for Modern Society. In *Advances in Medical Technologies and Clinical Practice* (pp. 330–346). IGI Global., DOI: 10.4018/979-8-3693-5261-8.ch019

Awotunde, J. B., Abiodun, M. K., Adeniyi, A. E., Abiodun, B. H., Adeniyi, J. K., Aremu, D. R., & Atanda, O. G. (2024, April). Personalized Music Recommendation System Based on Machine Learning and Collaborative Filtering. In *2024 International Conference on Science, Engineering and Business for Driving Sustainable Development Goals (SEB4SDG)* (pp. 1-8). IEEE. DOI: 10.1109/SEB4SDG60871.2024.10630080

. Bisoyi, A. (2022). Ownership, liability, patentability, and creativity issues in artificial intelligence. *Information Security Journal: A Global Perspective, 31*(4), 377–386. DOI: 10.1080/19393555.2022.2060879

Blake, A. B., Nazarian, M., & Castel, A. D. (2015). Rapid Communication: The Apple of the mind's eye: Everyday attention, metamemory, and reconstructive memory for the Apple logo. *Quarterly Journal of Experimental Psychology, 68*(5), 858–865. DOI: 10.1080/17470218.2014.1002798 PMID: 25721103

Blasch, E., Pham, T., Chong, C. Y., Koch, W., Leung, H., Braines, D., & Abdelzaher, T. (2021). Machine learning/artificial intelligence for sensor data fusion–opportunities and challenges. *IEEE Aerospace and Electronic Systems Magazine, 36*(7), 80–93. DOI: 10.1109/MAES.2020.3049030

Boden, M. A. (1998). Creativity and artificial intelligence. *Artificial Intelligence, 103*(1-2), 347–356. DOI: 10.1016/S0004-3702(98)00055-1

Bozdag, E. (2013). Bias in algorithmic filtering and personalization. *Ethics and Information Technology, 15*(3), 209–227. DOI: 10.1007/s10676-013-9321-6

Caiazzo, B., Murino, T., Petrillo, A., Piccirillo, G., & Santini, S. (2023). An IoT-based and cloud-assisted AI-driven monitoring platform for smart manufacturing: Design architecture and experimental validation. *Journal of Manufacturing Technology Management, 34*(4), 507–534. DOI: 10.1108/JMTM-02-2022-0092

Cavalier, E. S., & Newhall, K. E. (2017). 'Stick to Soccer:' fan reaction and inclusion rhetoric on social media. *Sport in Society, 21*(7), 1078–1095. DOI: 10.1080/17430437.2017.1329824

Chandra, S., Verma, S., Lim, W. M., Kumar, S., & Donthu, N. (2022). Personalization in personalized marketing: Trends and ways forward. *Psychology and Marketing, 39*(8), 1529–1562. DOI: 10.1002/mar.21670

Chen, Y., Elenee Argentinis, J., & Weber, G. (2016). IBM Watson: How Cognitive Computing Can Be Applied to Big Data Challenges in Life Sciences Research. *Clinical Therapeutics*, *38*(4), 688–701. DOI: 10.1016/j.clinthera.2015.12.001 PMID: 27130797

Chintalapati, S., & Pandey, S. K. (2021). Artificial intelligence in marketing: A systematic literature review. *International Journal of Market Research*, *64*(1), 38–68. DOI: 10.1177/14707853211018428

Cui, Y. G., Van Esch, P., & Phelan, S. (2024). How to build a competitive advantage for your brand using generative AI. *Business Horizons*, *67*(5), 583–594. DOI: 10.1016/j.bushor.2024.05.003

Das, S. (2022). The meaning of creativity through the ages: from inspiration to artificial intelligence. In *Creative business education: exploring the contours of pedagogical praxis* (pp. 27–53). Springer International Publishing., DOI: 10.1007/978-3-031-10928-7_3

Dhama, A., Kaur, A., Döngül, E. S., & Singh, N. (2024). Artificial Intelligence and Sustainable Green Fashion Industry. In Handbook of Artificial Intelligence Applications for Industrial Sustainability (pp. 74–91). *CRC Press*. DOI: 10.1201/9781003348351-6

Dobele, A., Toleman, D., & Beverland, M. (2005). Controlled infection! Spreading the brand message through viral marketing. *Business Horizons*, *48*(2), 143–149. DOI: 10.1016/j.bushor.2004.10.011

Dotov, D., & Froese, T. (2020). Dynamic Interactive Artificial Intelligence: Sketches for a Future AI Based on Human-Machine Interaction. In The 2020 Conference on Artificial Life (pp. 139–145). *MIT Press*. The 2020 Conference on Artificial Life. DOI: 10.1162/isal_a_00350

Endrissat, N., Kärreman, D., & Noppeney, C. (2016). Incorporating the creative subject: Branding outside–in through identity incentives. *Human Relations*, *70*(4), 488–515. DOI: 10.1177/0018726716661617

Ford, J., Jain, V., Wadhwani, K., & Gupta, D. G. (2023). AI advertising: An overview and guidelines. *Journal of Business Research*, *166*, 114124. DOI: 10.1016/j.jbusres.2023.114124

Fui-Hoon Nah, F., Zheng, R., Cai, J., Siau, K., & Chen, L. (2023). Generative AI and ChatGPT: Applications, challenges, and AI-human collaboration. *Journal of Information Technology Case and Application Research*, *25*(3), 277–304. DOI: 10.1080/15228053.2023.2233814

. Gaffar, H., & Albarashdi, S. (2024). Copyright Protection for AI-Generated Works: Exploring Originality and Ownership in a Digital Landscape. *Asian Journal of International Law*, 1–24. DOI: 10.1017/S2044251323000735

Gao, B., Wang, Y., Xie, H., Hu, Y., & Hu, Y. (2023). Artificial Intelligence in Advertising: Advancements, Challenges, and Ethical Considerations in Targeting, Personalization, Content Creation, and Ad Optimization. *SAGE Open*, *13*(4), 21582440231210759. Advance online publication. DOI: 10.1177/21582440231210759

Gao, Y., & Liu, H. (2023). Artificial intelligence-enabled personalization in interactive marketing: A customer journey perspective. *Journal of Research in Interactive Marketing*, *17*(5), 663–680. DOI: 10.1108/JRIM-01-2022-0023

George, M., Nhi, L. K. Y., Ngoc, N. M., Tung, V. N. D., Truc, L. P. T., & Nayak, R. (2022). Digital Technologies for Lean Manufacturing. In *Lean Supply Chain Management in Fashion and Textile Industry* (pp. 219–267). Springer Nature Singapore., DOI: 10.1007/978-981-19-2108-7_10

Gołąb-Andrzejak, E. (2023). AI-powered Digital Transformation: Tools, Benefits and Challenges for Marketers – Case Study of LPP. *Procedia Computer Science*, *219*, 397–404. DOI: 10.1016/j.procs.2023.01.305

Grilli, L., & Pedota, M. (2024). Creativity and artificial intelligence: A multilevel perspective. *Creativity and Innovation Management*, *33*(2), 234–247. DOI: 10.1111/caim.12580

Gui, J., Sun, Z., Wen, Y., Tao, D., & Ye, J. (2021). A review on generative adversarial networks: Algorithms, theory, and applications. *IEEE Transactions on Knowledge and Data Engineering*, *35*(4), 3313–3332. DOI: 10.1109/TKDE.2021.3130191

Haleem, A., Javaid, M., Asim Qadri, M., Pratap Singh, R., & Suman, R. (2022). Artificial intelligence (AI) applications for marketing: A literature-based study. *International Journal of Intelligent Networks*, *3*, 119–132. DOI: 10.1016/j.ijin.2022.08.005

Hildesheim, W. (2017). Cognitive Computing – the new Paradigm of the Digital World. In *Digital Marketplaces Unleashed* (pp. 265–274). Springer Berlin-Heidelberg., DOI: 10.1007/978-3-662-49275-8_27

Hinchcliff, M., & Mehmet, M. (2023). Embedding Canva into the marketing classroom: A dialogic and social learning approach to classroom innovation. *Higher Education, Skills and Work-Based Learning*, *13*(6), 1174–1186. DOI: 10.1108/HESWBL-11-2022-0230

Huh, J., Nelson, M. R., & Russell, C. A. (2023). ChatGPT, AI Advertising, and Advertising Research and Education. *Journal of Advertising*, *52*(4), 477–482. DOI: 10.1080/00913367.2023.2227013

Jenkins, J. J. (2022). Netflix. In Encyclopedia of Big Data (pp. 682–684). *Springer International Publishing*. DOI: 10.1007/978-3-319-32010-6_144

Joosten, J., Bilgram, V., Hahn, A., & Totzek, D. (2024). Comparing the ideation quality of humans with generative artificial intelligence. *IEEE Engineering Management Review*, *52*(2), 153–164. Advance online publication. DOI: 10.1109/EMR.2024.3353338

Kietzmann, J., Paschen, J., & Treen, E. (2018). Artificial Intelligence in Advertising. *Journal of Advertising Research*, *58*(3), 263–267. DOI: 10.2501/JAR-2018-035

Kingston, J. (2017). Using artificial intelligence to support compliance with the General Data Protection Regulation. *Artificial Intelligence and Law*, *25*(4), 429–443. DOI: 10.1007/s10506-017-9206-9

Konina, N. Y. (2023). Artificial Intelligence in the Fashion Industry—Reality and Prospects. In *Anti-Crisis Approach to the Provision of the Environmental Sustainability of Economy* (pp. 273–280). Springer Nature Singapore., DOI: 10.1007/978-981-99-2198-0_29

Kordzadeh, N., & Ghasemaghaei, M. (2021). Algorithmic bias: Review, synthesis, and future research directions. *European Journal of Information Systems*, *31*(3), 388–409. DOI: 10.1080/0960085X.2021.1927212

Krishnan, C., & Mariappan, J. (2024). The AI Revolution in E-Commerce: Personalization and Predictive Analytics. In *Role of Explainable Artificial Intelligence in E-Commerce* (pp. 53-64). Cham: Springer Nature Switzerland. DOI: 10.1007/978-3-031-55615-9_4

Kshetri, N., Dwivedi, Y. K., Davenport, T. H., & Panteli, N. (2024). Generative artificial intelligence in marketing: Applications, opportunities, challenges, and research agenda. *International Journal of Information Management*, *75*, 102716. DOI: 10.1016/j.ijinfomgt.2023.102716

Kumar, D., & Suthar, N. (2024). Ethical and legal challenges of AI in marketing: An exploration of solutions. *Journal of Information, Communication and Ethics in Society*, *22*(1), 124–144. DOI: 10.1108/JICES-05-2023-0068

Kumar, V., Ashraf, A. R., & Nadeem, W. (2024). AI-powered marketing: What, where, and how? *International Journal of Information Management*, *77*, 102783. DOI: 10.1016/j.ijinfomgt.2024.102783

Kushwaha, S. (2023, February). A Futuristic Perspective on Artificial Intelligence. In *2022 OPJU International Technology Conference on Emerging Technologies for Sustainable Development (OTCON)* (pp. 1-6). IEEE. DOI: 10.1109/OTCON56053.2023.10113980

Li, H. (2019). Special Section Introduction: Artificial Intelligence and Advertising. *Journal of Advertising, 48*(4), 333–337. DOI: 10.1080/00913367.2019.1654947

Liao, S. H., & Chen, Y. J. (2010). Rough sets based association rules application for knowledge-based system design. In *Computational Collective Intelligence. Technologies and Applications: Second International Conference, ICCCI 2010, Kaohsiung, Taiwan, November 10-12, 2010, Proceedings, Part II 2* (pp. 501-510). Springer Berlin Heidelberg. DOI: 10.1007/978-3-642-16732-4_53

Liu, C., Ren, Z., & Liu, S. (2021). Using Design and Graphic Design with Color Research in AI Visual Media to Convey. *Journal of Sensors, 2021*(1), 8153783. DOI: 10.1155/2021/8153783

Lu, Y. (2019). Artificial intelligence: A survey on evolution, models, applications and future trends. *Journal of Management Analytics, 6*(1), 1–29. DOI: 10.1080/23270012.2019.1570365

Lund, B. D., & Wang, T. (2023). Chatting about ChatGPT: How may AI and GPT impact academia and libraries? *Library Hi Tech News, 40*(3), 26–29. DOI: 10.1108/LHTN-01-2023-0009

Matthews, B., Shannon, B., & Roxburgh, M. (2023). Destroy All Humans: The Dematerialisation of the Designer in an Age of Automation and its Impact on Graphic Design—A Literature Review. *International Journal of Art & Design Education, 42*(3), 367–383. DOI: 10.1111/jade.12460

Micu, A., Capatina, A., Cristea, D. S., Munteanu, D., Micu, A., & Sarpe, D. A. (2022). Assessing an on-site customer profiling and hyper-personalization system prototype based on a deep learning approach. *Technological Forecasting and Social Change, 174*, 121289. DOI: 10.1016/j.techfore.2021.121289

Mogaji, E., Olaleye, S., & Ukpabi, D. (2020). Using AI to personalise emotionally appealing advertisement. *Digital and social media marketing: Emerging applications and theoretical development*, 137-150. DOI: 10.1007/978-3-030-24374-6_10

Nair, S. S. (2024, February). Redefining Creativity in Design: Exploring the Impact of AI-Generated Imagery on Design Professionals and Amateurs. In *2024 11th International Conference on Computing for Sustainable Global Development (INDIACom)* (pp. 1723-1728). IEEE. DOI: 10.23919/INDIACom61295.2024.10498341

Nanne, A. J., Antheunis, M. L., Van Der Lee, C. G., Postma, E. O., Wubben, S., & Van Noort, G. (2022). The Use of Computer Vision to Analyze Brand-Related User Generated Image Content. *Journal of Interactive Marketing*. Advance online publication. DOI: 10.1016/j.intmar.2019.09.003

Nizioł, K. (2021). The challenges of consumer protection law connected with the development of artificial intelligence on the example of financial services (chosen legal aspects). *Procedia Computer Science*, *192*, 4103–4111. DOI: 10.1016/j.procs.2021.09.185

Pakharuddin, N. D., & Kamarudin, S. (2023, December). Retail sector emerging technologies: AI and augmented reality present and future perspective. In *2023 2nd International Engineering Conference on Electrical, Energy, and Artificial Intelligence (EICEEAI)* (pp. 1-7). IEEE. DOI: 10.1109/EICEEAI60672.2023.10590317

Pantelic, O., Jovic, K., & Krstovic, S. (2021). Cookies Implementation Analysis and the Impact on User Privacy Regarding GDPR and CCPA Regulations. *Sustainability (Basel)*, *14*(9), 5015. DOI: 10.3390/su14095015

Petiwala, F. F., Shukla, V. K., & Vyas, S. (2021). IBM Watson: redefining artificial intelligence through cognitive computing. In *Proceedings of International Conference on Machine Intelligence and Data Science Applications: MIDAS 2020* (pp. 173-185). *Springer Singapore*. DOI: 10.1007/978-981-33-4087-9_15

Poliakov, V. (2020). The artificial intelligence and optimization of multibody dynamical system with predicted dynamic behaviour. *24th International Conference on Circuits, Systems, Communications and Computers (CSCC)*, Chania, Greece. pp. 1-4. DOI: 10.1109/CSCC49995.2020.00009

Raptis, G. E., & Katsini, C. (2023, November). Transforming Ideas into Code: A Novel Approach to Design-to-Development Workflow. In *2023 International Conference on Computer and Applications (ICCA)* (pp. 1-6). IEEE. DOI: 10.1109/ICCA59364.2023.10401412

Rodgers, S. (2021). Themed Issue Introduction: Promises and Perils of Artificial Intelligence and Advertising. *Journal of Advertising*, *50*(1), 1–10. DOI: 10.1080/00913367.2020.1868233

Rokhade, A. A., Deivam, A., Shettigar, A. J. V., Tarun, M., & Prasad, V. B. (2024, June). Intelligent Advertisement Generation: Harnessing Deep Learning Techniques. In *2024 3rd International Conference on Applied Artificial Intelligence and Computing (ICAAIC)* (pp. 628-637). IEEE. DOI: 10.1109/ICAAIC60222.2024.10575640

Seo, Y., & Lee, J. H. (2021). Digital transformation and advertising: Insights from the 2020 ICAMA-KAS conference. *International Journal of Advertising*, *40*(8), 1245–1246. DOI: 10.1080/02650487.2021.2000180

Shah, N., Engineer, S., Bhagat, N., Chauhan, H., & Shah, M. (2020). Research trends on the usage of machine learning and artificial intelligence in advertising. *Augmented Human Research*, *5*(1), 1–15. DOI: 10.1007/s41133-020-00038-8

Shamaeva, I. (2024). *Advanced Styles and Insights with Midjourney: Imagine Beautiful AI Prompts*. Apress., DOI: 10.1007/979-8-8688-0336-9

Sheng, Su. (2023). *1. Analysis of the Effect of Marketing on Enterprise Brand Establishment: Take Coca-Cola as an Example*. Advances in Economics, Management and Political Sciences., DOI: 10.54254/2754-1169/36/20231807

Strickland, E. (2019). IBM Watson, heal thyself: How IBM overpromised and under delivered on AI health care. *IEEE Spectrum*, *56*(4), 24–31. DOI: 10.1109/MSPEC.2019.8678513

Taylor, C. R. (2019). Editorial: Artificial intelligence, customized communications, privacy, and the General Data Protection Regulation (GDPR). *International Journal of Advertising*, *38*(5), 649–650. DOI: 10.1080/02650487.2019.1618032

Timimi, H., Baaddi, M., & Bennouna, A. (2025). Impact of artificial intelligence on the personalization of the customer experience: A systematic literature review. *Multidisciplinary Reviews*, *8*(7), 2025224. DOI: 10.31893/multirev.2025224

Trueman, M., Cook, D., & Cornelius, N. (2008). Creative dimensions for branding and regeneration: Overcoming negative perceptions of a city. *Place Branding and Public Diplomacy*, *4*(1), 29–44. DOI: 10.1057/palgrave.pb.6000073

Truss, M., & Schmitt, M. (2024). Human-Centered AI Product Prototyping with No-Code Auto ML: Conceptual Framework, Potentials and Limitations. *International Journal of Human-Computer Interaction*, •••, 1–16. DOI: 10.1080/10447318.2024.2425454

Tsidylo, I. M., & Sena, C. E. (2023). Artificial intelligence as a methodological innovation in the training of future designers: Midjourney tools. *Information Technologies and Learning Tools*, *97*(5), 203–214. DOI: 10.33407/itlt.v97i5.5338

Vakratsas, D., & Wang, X. (2020). Artificial intelligence in advertising creativity. *Journal of Advertising*, *50*(1), 39–51. DOI: 10.1080/00913367.2020.1843090

Van Belkom, R. (2020). The impact of artificial intelligence on the activities of a futurist. *World Futures Review*, *12*(2), 156–168. DOI: 10.1177/1946756719875720

Van Noort, G., Antheunis, M. L., & Verlegh, P. W. J. (2014). Enhancing the effects of social network site marketing campaigns: If you want consumers to like you, ask them about themselves. *International Journal of Advertising*, *33*(2), 235–252. DOI: 10.2501/IJA-33-2-235-252

Vashishtha, S., & Sharma, P. (2025). Artificial Intelligence and More Effective Advertising: Unlocking the Power of Data and Automation. In *Advances in Digital Marketing in the Era of Artificial Intelligence* (pp. 162-171). CRC Press. DOI: 10.1201/9781003450443-11

Vayadande, K., Bhemde, S., Rajguru, V., Ugile, P., Lade, R., & Raut, N. (2023, November). AI-Based Image Generator Web Application using OpenAI's DALL-E System. In *2023 International Conference on Recent Advances in Science and Engineering Technology (ICRASET)* (pp. 1-5). IEEE. DOI: 10.1109/ICRASET59632.2023.10420306

Verma, S., Sharma, R., Deb, S., & Maitra, D. (2021). Artificial intelligence in marketing: Systematic review and future research direction. *International Journal of Information Management Data Insights*, *1*(1), 100002. DOI: 10.1016/j.jjimei.2020.100002

Vetter, M. A., Lucia, B., Jiang, J., & Othman, M. (2024). Towards a framework for local interrogation of AI ethics: A case study on text generators, academic integrity, and composing with ChatGPT. *Computers and Composition*, *71*, 102831. DOI: 10.1016/j.compcom.2024.102831

Watiktinnakorn, C., Seesai, J., & Kerdvibulvech, C. (2023). Blurring the lines: How AI is redefining artistic ownership and copyright. *Discover Artificial Intelligence*, *3*(1), 37. DOI: 10.1007/s44163-023-00088-y

. Wierenga, B., & Van Bruggen, G. (2000). Knowledge-Driven Marketing Management Support Systems I: Artificial Intelligence, Knowledge Representation, and Expert Systems. *Marketing Management Support Systems: Principles, Tools, and Implementation*, 119-164. DOI: 10.1007/978-1-4615-4595-8_5

Wu, Z., Ji, D., Yu, K., Zeng, X., Wu, D., & Shidujaman, M. (2021). AI creativity and the human-AI co-creation model. In *Human-Computer Interaction. Theory, Methods and Tools: Thematic Area, HCI 2021, Held as Part of the 23rd HCI International Conference, HCII 2021, Virtual Event, July 24–29, 2021, Proceedings, Part I 23* (pp. 171-190). Springer International Publishing. DOI: 10.1007/978-3-030-78462-1_13

Xaba, S., Fang, X., Shah, D., & Yang, K. (2022). Application of Fourth Industrial Revolution technologies in contemporary South African art. *International Journal of Arts and Technology*, *14*(1), 56–70. DOI: 10.1504/IJART.2022.122451

Zatarain, J. M. N. (2017). The role of automated technology in the creation of copyright works: The challenges of artificial intelligence. *International Review of Law Computers & Technology*, *31*(1), 91–104. DOI: 10.1080/13600869.2017.1275273

Chapter 12
Responsible and Ethical AI in Industrial Contexts

Nur Muhammad Fadilah
http://orcid.org/0009-0004-0826-8412
Universitas Esa Unggul, Indonesia

Binastya Anggara Sekti
http://orcid.org/0000-0001-5489-4888
Universitas Esa Unggul, Indonesia

ABSTRACT

As artificial intelligence (AI) integration accelerates within industrial sectors, the imperative for responsible and ethical implementation becomes critical This chapter addresses this urgency by exploring the core principles of fairness, accountability, transparency, and human-centric values essential for sustainable industrial evolution. Aligning with the book's theme of human-AI cognitive collaboration, it serves as an ethical compass, examining the significant ethical challenges and risks inherent in industrial AI, including bias, privacy, and safety. It proposes robust frameworks for responsible deployment, analyzes principles like explainability and human-centered design, and draws on case studies and international governance models. The objective is to demonstrate how embedding ethics fosters trust, safety, and long-term value creation, providing actionable strategies to ensure human-AI collaboration evolves equitably and sustainably.

1. INTRODUCTION

The fourth industrial revolution (Industry 4.0) is characterized by the deep integration of digital technologies, cyber-physical systems, and data-driven intelligence

DOI: 10.4018/979-8-3373-7036-1.ch012

into industrial operations (Huo et al., 2022). Artificial intelligence (AI) plays a central role in this transformation, enabling efficiency gains, productivity improvements, and new forms of innovation across sectors such as manufacturing, logistics, and maintenance (Talukder, Kumar, & Tyagi, 2024). Applications ranging from predictive maintenance and supply-chain optimization to worker-support systems on the factory floor illustrate that AI is not merely an incremental tool but a catalyst for a fundamental reconfiguration of industrial value creation and competitive dynamics (Nair, Manohar, Mittal, & Patwa, 2025).

At the same time, the rapid diffusion of AI in industrial settings raises urgent ethical and governance challenges. Industrial AI systems are socio-technical in nature: they are designed by humans, trained on data derived from human and organizational processes, and embedded within complex institutional contexts. As a result, they may reproduce or amplify biases, create new risks to privacy and safety, and obscure responsibility for high-stakes decisions (Stiemer, Mathis, Lecampion, Rahajandraibed, & Hedayat, 2024). Evaluating industrial AI solely in terms of technical performance or efficiency is therefore insufficient. Ethical principles such as fairness, accountability, transparency, and respect for human dignity must be treated as core design and operational requirements rather than secondary considerations.

This chapter addresses these challenges by examining how ethical principles can be systematically embedded into AI adoption in Industry 4.0 contexts. Aligned with the book's overarching focus on human–AI cognitive collaboration, the chapter functions as an ethical foundation that complements application-oriented discussions elsewhere in the volume. Its aim is not to add another broad survey of AI ethics, but to translate well-established ethical principles into practical guidance relevant for industrial environments.

We proceed from the premise that responsible and ethical AI is a prerequisite for sustainable industrial transformation. Ethics cannot be retrofitted after deployment; it must be integrated throughout the AI lifecycle, from problem definition and data collection to system design, deployment, monitoring, and eventual decommissioning (Wang et al., 2023). Accordingly, the chapter first maps the ethical landscape of industrial AI, focusing on bias, privacy, safety, and accountability. It then outlines core principles of responsible AI, followed by illustrative case studies from healthcare, agriculture, manufacturing, and logistics. Finally, it discusses governance frameworks, practical toolkits, and the role of trust and organizational culture in enabling effective human–AI collaboration. Together, these elements provide a pragmatic roadmap for aligning industrial AI innovation with long-term societal trust and human-centered values

Managerial Takeaways – Introduction:

- Treat ethical AI as a strategic foundation for Industry 4.0 initiatives, not as an optional compliance add-on.
- Integrate ethical considerations across the full AI lifecycle, from design to deployment and monitoring.
- Align AI adoption strategies with human-centered values to sustain trust, safety, and long-term organizational legitimacy.

2. THE ETHICAL LANDSCAPE OF INDUSTRIAL AI

Understanding the ethical terrain is a prerequisite for any responsible deployment of AI in industry. AI introduces both new risks and amplifies existing social and organizational problems. Left unaddressed, these risks erode trust, invite regulatory action, and can cause significant human harm. This section tightens the original discussion around three core risk areas, bias, privacy, and safety, while keeping the original examples and citations intact (Sekti, 2025).

2.1. Bias, Fairness, and Inclusivity in AI Systems

AI models inherit patterns from their training data; when that data reflects historical or systemic inequities, the model will tend to reproduce and sometimes amplify them (Panarese, Grasso, & Solinas, 2025). In industrial settings, this can take several concrete forms: biased hiring and promotion recommendations that privilege certain educational or demographic profiles (Schulte Steinberg & Hohenberger, 2023), productivity monitors that penalize workers using assistive technologies or different work rhythms, predictive maintenance systems that conflate team attributes with equipment risk, thereby misattributing failure causes (Eze et al., 2025), and computer vision systems that underperform for workers outside the dominant body- or skin-type distributions present in training data, introducing safety and fairness harms (Dhal & Kar, 2025).

Addressing algorithmic bias requires more than one-off "de-biasing" fixes. It begins with inclusive design and representative data collection, continues with lifecycle fairness audits, and ends with accessible redress mechanisms so affected workers can contest and correct automated decisions (Wang et al., 2023). Crucially, fairness metrics must be chosen and validated in dialogue with the people affected, statistical parity alone rarely captures workplace justice.

Managerial Takeaways, Bias & Fairness:

- Require representative-data checks and fairness tests before model approval; include demographic and operational subgroups in validation datasets. (Panarese et al., 2025)
- Deploy continuous fairness monitoring in production pipelines and establish clear, worker-facing complaint and remediation channels. (Wang et al., 2023)
- Prefer interpretable models or post-hoc explanations where decisions materially affect employment, safety, or pay. (Schulte Steinberg & Hohenberger, 2023)

2.2. Privacy, Data Security, and Informed Consent

Industrial AI relies on vast IIoT and human-centered datasets, wearables, cameras, keystroke logs, and environmental sensors, that together create a highly detailed record of people's bodies and behaviors (Okonkwo, Awolusi, Nnaji, & Akanfe, 2025). This "datafied worker" raises three linked risks: coercive or hollow consent in employer–employee contexts; unclear or exploitative data ownership and secondary uses; and catastrophic security failures that expose proprietary or personally sensitive information (Rao Gudlur, 2025).

Consent in the workplace is particularly fraught because the power imbalance makes refusal costly. Therefore, privacy strategies for industrial AI should emphasize data minimization (collect only what is necessary), purpose limitation (clearly defined and enforced uses), strong access controls, and transparent retention policies. Security must be treated as safety: breaches of operational AI datasets can enable physical sabotage (e.g., manipulated control setpoints) as well as privacy harms. Responsible programs combine legal safeguards with technical measures (encryption, differential privacy, secure by design) and governance rules that empower workers over their data (Sekti, 2026).

Managerial takeaways, Privacy & Security:

- Adopt data minimization and purpose-limitation policies; document and publish what is collected, why, and for how long (Okonkwo et al., 2025).
- Treat industrial dataset security as a safety requirement, apply strong encryption, role-based access, and threat-model testing (Rao Gudlur, 2025).
- Design consent and participation mechanisms that give workers concrete choices and independent avenues to opt out without penalty (Okonkwo et al., 2025).

2.3. Safety and Accountability in High-Stakes Environments

In high-stakes industrial contexts, chemical plants, autonomous logistics, energy grids, and AI failures risk physical harm or environmental disaster (Daniel & Xuan, 2024). Real incidents can include an autonomous vehicle failing to detect a human, a control-system miscalculation creating unsafe chemical conditions, or a cobot not stopping in time during an unexpected human movement (Mezgár, Vancza, Paniti, & Tóth, 2023). These outcomes expose a persistent "accountability gap" when blame and liability are dispersed across developers, integrators, deployers, and operators (Stiemer et al., 2024).

Closing this gap requires both technical and organizational interventions. Technically, use explainable models where possible, robust simulation and digital-twin testing for long-tail scenarios, and explicit uncertainty reporting so systems defer to humans on low-confidence inputs. Organizationally, specify contractual liability, maintain rigorous audit trails, and require multidisciplinary safety validation before deployment. Explainable AI (XAI) is central to both trust and post-incident analysis in industrial fault diagnosis.

Managerial Takeaways, Safety & Accountability:

- Classify systems by harm potential; require simulation-based validation and explicit uncertainty thresholds for high-risk deployments. (Daniel & Xuan, 2024)
- Maintain clear contractual and operational accountability: define developer, integrator, and operator responsibilities, and ensure auditable logs. (Stiemer et al., 2024)
- Implement XAI and human-on-the-loop safeguards for systems that control or influence physical processes.

3. PRINCIPLES OF RESPONSIBLE AI IN INDUSTRY

Responsible AI requires a shift from ad hoc fixes to a principles-driven program integrated into engineering and management practices (Vann Yaroson, Abadie, & Roux, 2025). The principles below are concise, actionable, and tailored to industrial settings where safety, continuity, and regulatory compliance are central concerns.

3.1. Transparency and Explainability

Transparency and explainability are essential for ensuring that industrial AI systems can be trusted and effectively used by human operators. Transparency refers to communicating the purpose, data sources, and limitations of AI systems, while explainability provides insight into how specific decisions are made (Riantisa, Anggraini, & Sekti, 2026). In industrial contexts, both are necessary to support validation, debugging, and safe operation. Practical implementation includes documenting model provenance, providing confidence scores, and integrating explainable AI techniques into operator interfaces. The balance between predictive performance and interpretability must be carefully managed depending on the risk level of the application.

Figure 1. The Spectrum of Explainability in Industrial AI

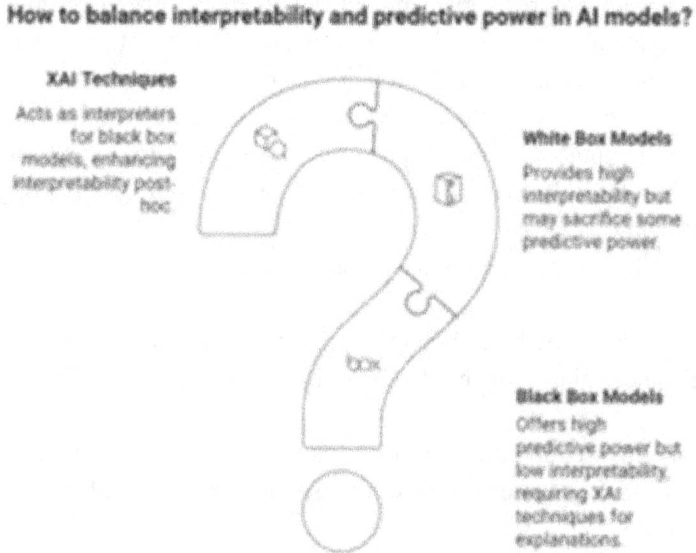

Figure 1 illustrates the spectrum of explainability, moving from a "Black Box" model with low interpretability (e.g., complex deep learning, generative AI) to a "White Box" or "Glass Box" model with high interpretability (e.g., decision trees, linear regression). The figure shows how different XAI techniques (like LIME or SHAP) act as "interpreters" for black box models, providing post-hoc explanations, while inherently interpretable models provide explanations by design. The X-axis represents "Interpretability" and the Y-axis represents "Predictive Power," showing the common trade-off that industrial applications must navigate to build trust and ensure accountability.

Managerial Takeaways of Transparency and Explainability

- Publish dataset and model registries for each production model, including provenance, validation metrics, and known limitations. (Buijsman, 2024)
- Require confidence scores and concise, audience-appropriate explanations on operator interfaces for safety-critical alerts. (Huang, Chen, & Yan, 2024)
- Prefer inherently interpretable models when decisions affect employment or physical safety, or supplement black-box models with rigorous XAI and verification. (Le, Prihatno, Oktian, Kang, & Kim, 2023)

3.2. Human-Centered Design and Values-Based Engineering

Human-centered design places users and affected stakeholders at the center of system requirements, workflow design, and validation, ensuring systems augment human expertise rather than displace it (Wang et al., 2023). Values-based engineering operationalizes values like fairness, privacy, and safety as measurable requirements, tested throughout the development lifecycle.

Key practices include co-design workshops with frontline staff, human-in-the-loop training regimes where operator corrections are fed back into models, clear failover behaviors that hand control back to humans under uncertainty, and translating values into technical acceptance criteria, for example equalized error rates across defined groups or strict privacy budgets for sensitive data. These measures reduce the risk of misaligned incentives and improve the system's practical usefulness on the floor.

Managerial takeaways of Human-Centered Design

- Embed frontline staff in design and validation phases, using co-design sessions and pilot deployments to surface real-world edge cases. (Wang et al., 2023)
- Define measurable value requirements, such as fairness thresholds or privacy budgets, and gate deployments on passing these checks. (Wang et al., 2023)
- Enforce human-on-the-loop controls and graceful-failure modes, with clear training and documented override procedures. (Mosqueira-Rey, Hernández-Pereira, Alonso-Ríos, Bobes-Bascarán, & Fernández-Leal, 2023)

3.3. Sustainability and Environmental Responsibility

AI development has significant environmental implications, including high energy consumption, water usage, and material demands associated with large-scale computation and hardware production. At the same time, AI offers substantial opportunities for improving sustainability through optimization of energy systems, precision agriculture, and efficient supply chain management. Responsible AI requires balancing these trade-offs by adopting Green AI principles, prioritizing efficient models, and ensuring that environmental benefits outweigh operational costs. Organizations should also promote transparency in energy consumption and prioritize applications that contribute directly to sustainability goals. The trade-offs are significant and require careful consideration, as summarized in Table 1.

Table 1. The Dual Impact of AI on Industrial Sustainability

Dimension	Positive Impact (Opportunities)	Negative Impact (Risks)	Ethical Consideration
Energy Consumption	AI-driven optimization of power grids, HVAC systems, and industrial processes to reduce overall energy use.	High energy consumption for training and running large-scale models (e.g., Generative AI).	Balancing computational power with energy efficiency (Green AI); transparency in model energy costs.
Resource Management	Optimization of supply chains, precision agriculture, and predictive maintenance to reduce material waste, water use, and spoilage.	Demand for rare-earth minerals and other finite resources for manufacturing AI hardware (GPUs, TPUs).	Lifecycle transparency; ethical sourcing of hardware; investment in circular economy models for hardware.
Environmental Monitoring	Enhanced ability to monitor pollution, deforestation, and climate change impacts via satellite imagery and sensor networks.	Potential for "rebound effects," where efficiency gains lead to increased consumption and overall higher resource use.	Ensuring data is used for ecological stewardship and protection, not just enabling more efficient resource exploitation.
Innovation	AI-powered discovery of new sustainable materials, clean energy solutions, and carbon capture technologies.	Computational and energy cost of the R&D process for these new solutions.	Prioritizing R&D that has a clear and demonstrable net-positive environmental impact.

4. CASE STUDIES ACROSS INDUSTRIES

This section explores these diverse domains to ground the principles in practical reality, demonstrating that ethical implementation is not a one-size-fits-all endeavor but requires deep domain expertise and tailored approaches (Alexander, Yarborough, & Smith, 2024).

4.1 Healthcare: Ethical Dilemmas in Diagnosis and Patient Data

AI in healthcare improves diagnostic accuracy and operational efficiency but introduces critical ethical risks related to bias, privacy, and accountability. Biased datasets can lead to unequal diagnostic performance across demographic groups, while large-scale patient data collection raises concerns about consent and re-identification. Furthermore, AI-assisted decisions create ambiguity in responsibility when errors occur. Addressing these challenges requires rigorous bias auditing, privacy-preserving data practices, and clear human oversight mechanisms in clinical workflows. The transformative impact of Artificial Intelligence (AI) is examined across four critical sectors, namely healthcare, education, finance, and security, while addressing the profound ethical challenges arising from the rise of generative AI (Utami & Sekti, 2025).

4.2 Agriculture: Balancing Efficiency with Ecological Stewardship

AI applications in agriculture enable precision farming and improved resource management but must balance productivity with ecological sustainability. Optimization based solely on yield can negatively impact soil health, biodiversity, and long-term resilience. Ethical concerns also arise regarding data ownership and unequal access to technology, particularly for smallholder farmers. Responsible agricultural AI should incorporate multi-objective optimization, promote equitable data governance, and support inclusive access to technological benefits (Susanto et al., 2023).

4.3 Manufacturing and Industry 4.0: Worker Augmentation vs. Replacement

In manufacturing, AI systems enable automation, predictive maintenance, and human-machine collaboration. A responsible approach prioritizes worker augmentation rather than replacement, ensuring that AI enhances human capabilities. However, risks such as job displacement and algorithmic surveillance must be carefully managed. Ethical implementation requires transparent monitoring practices, investment in workforce reskilling, and the design of collaborative systems that preserve human agency and safety. Artificial Intelligence (AI) is applied in manufacturing through applications such as predictive maintenance, quality control, and automation, enhancing operational efficiency by predicting equipment failures, reducing downtime, and improving product quality through automated defect detection (Rianisa & Sekti, 2026).

4.4 Logistics and Supply Chain: Fair Labor Practices and Global Responsibility

AI-driven logistics systems optimize routing, inventory, and supply chain operations but raise concerns related to labor conditions and global responsibility. Algorithmic management can reduce transparency and limit worker autonomy, while complex supply chains obscure ethical practices. Responsible AI in this domain requires mechanisms for worker feedback, fair task allocation, and improved supply chain transparency through technologies such as blockchain and data analytics (Tiaraningrum & Sekti, 2026).

5. FRAMEWORKS AND TOOLKITS FOR RESPONSIBLE AI ADOPTION

Principles and case studies highlight the "what" and "why" of ethical AI; frameworks and toolkits provide the "how." For industrial leaders, practitioners, and policymakers, translating high-level values into concrete organizational practice is the most critical step. This requires a multi-layered approach combining high-level governance, practical risk assessment, and industry-specific standards that together create a system of accountability and continuous improvement (Camilleri, 2024).

5.1 Governance and Compliance Models

In recent years, a consensus has begun to form around the principles of trustworthy AI, led by international bodies and national governments. These frameworks provide the essential scaffolding for organizational governance, setting the minimum standards and expectations for responsible behavior.

International Principles (OECD, UNESCO):

- The OECD AI Principles (adopted in 2019) are a foundational international standard, endorsed by over 50 countries. They are built on five value-based principles: inclusive growth and well-being, human-centered values and fairness, transparency and explainability, robustness and safety, and accountability. They provide a common language and high-level benchmark for policymakers and corporations, promoting a global level playing field (Sharma, Rawal, Pal, & Dani, 2022).
- The UNESCO Recommendation on the Ethics of AI (adopted in 2021) is the first global standard-setting instrument on the topic. It is broader in scope and more explicitly values-driven, emphasizing environmental protection, cultural diversity, and gender equality as core pillars of ethical AI, alongside the more common principles of fairness and safety (Neuwirth, 2022).

The EU AI Act: A Risk-Based Regulatory Model:

The most significant and legally binding framework is the European Union's AI Act. Its influence extends far beyond Europe, setting a de facto global standard (the "Brussels effect") for any company wishing to access the large and lucrative

EU market (Ebers, 2025). Its central innovation is a risk-based categorization of AI systems, which determines the level of regulatory scrutiny they face:

1. Unacceptable Risk: AI systems that pose a clear threat to the safety, livelihoods, and fundamental rights of people are banned outright. This includes, for example, government-run social scoring systems and real-time remote biometric surveillance in public spaces for law enforcement purposes (Novelli, Casolari, Rotolo, Taddeo, & Floridi, 2024).
2. High-Risk: This is the most critical category for industrial applications. It includes AI systems used in critical infrastructure (e.g., energy, water supply, transport), industrial safety components of machinery, medical devices, and employment/worker management (e.g., CV-sorting software, productivity scoring). These systems are *not* banned but are subject to strict mandatory requirements both before being placed on the market and throughout their lifecycle (Novelli et al., 2024). These requirements include:
    - Conduction of a fundamental rights impact assessment.
    - Robust risk-assessment and mitigation systems.
    - High-quality, representative data governance to minimize biases.
    - Detailed technical documentation and automated logging.
    - High levels of transparency and provision of information to users.
    - Appropriate levels of human oversight and the ability to intervene.
3. Limited Risk: Systems that interact with humans (e.g., chatbots, emotion recognition systems) must meet minimal transparency obligations. The human must be aware that they are interacting with an AI (Novelli et al., 2024).
4. Minimal Risk: The vast majority of AI systems (e.g., spam filters, AI in video games) fall into this category, which is largely unregulated, though voluntary codes of conduct are encouraged (Novelli et al., 2024).

For industrial contexts, the EU AI Act means that any AI used in manufacturing safety (e.g., a cobot's vision system to detect human presence), worker management and performance evaluation, or critical logistics and supply chain operations will almost certainly be classified as high-risk. This requires significant investment in compliance, documentation, governance structures, and verifiable ethical design, making responsible AI a legal imperative, not just an ethical one (Kalodanis, Rizomiliotis, & Anagnostopoulos, 2024).

5.2 Ethical Impact Assessments

A governance framework sets the rules, but an Ethical Impact Assessment (EIA) or AI Impact Assessment (AIA) is the practical tool used to enforce them on a project-by-project basis. Similar in concept to a Data Protection Impact Assessment (DPIA) under the GDPR or an Environmental Impact Assessment, an EIA is a structured process to proactively identify, analyze, and mitigate ethical risks *before* an AI system is designed, built, and deployed (Stahl et al., 2023).

An EIA forces a development team to move beyond a purely technical specification and answer difficult questions about the system's broader impact on people and society. It is not a "checkbox" exercise to be completed at the end of a project, but a multidisciplinary, iterative process involving ethicists, legal experts, domain experts, and, crucially, representatives from the stakeholder groups who will be directly affected by the system (e.g., factory workers, patients, customers) (Ho & Caals, 2024). Their lived experience is invaluable for identifying risks that developers might overlook. A typical EIA process, as summarized in Table 2, involves several key stages.

Table 2. Key Stages of an Ethical Impact Assessment (EIA) for Industrial AI

Stage	Objective	Key Questions to Ask	Potential Stakeholders
1. Conception & Scoping	Define the AI's purpose and context. Identify all affected stakeholders.	What problem is the AI solving? For whom? Who will use it? Who will be impacted by it (directly and indirectly)? What human values are at play (e.g., efficiency vs. safety, privacy vs. productivity)?	Project Managers, System Architects, End-Users, Legal/ Compliance
2. Data & Model Analysis	Scrutinize the data and model for potential risks of bias and privacy.	Where did the training data come from? Is it representative of the entire population it will be used on? Does it contain historical biases? What are the privacy implications of collecting and using this data? Is the model type appropriate for the stakes (e.g., is a "black box" deep learning model acceptable, or is an interpretable model required)?	Data Scientists, Data Engineers, Legal, Privacy Officers
3. Risk & Harm Identification	Brainstorm potential negative impacts and ethical failures.	What is the worst-case scenario? How could the system be misused or fail? Could it lead to discriminatory outcomes, physical safety risks, or psychological harm? Does it create unacceptable surveillance risks or erode human autonomy?	Ethicists, Domain Experts, Worker Representatives, Safety Engineers

continued on following page

Table 2. Continued

Stage	Objective	Key Questions to Ask	Potential Stakeholders
4. Mitigation & Governance	Design concrete strategies to address identified risks.	How will we de-bias the data or model? What transparency and explainability features are needed? What is the specific human oversight plan (HITL, HOTL)? What is the clear redress mechanism for those who are harmed by an automated decision?	UX Designers, Developers, Governance Board, Legal
5. Monitoring & Review	Establish a plan for post-deployment monitoring and continuous review.	How will we monitor for "model drift" (performance degradation over time) or emergent bias? What is the "off-switch" or rollback plan? How often will this EIA be reviewed and updated as the system and context evolve?	Operations Team, AI/ML Ops, Internal Audit, Ethics Committee

By making this assessment a mandatory "gate" in the product development lifecycle, an organization moves ethics from an abstract idea discussed in the boardroom to a concrete, auditable engineering requirement on the ground.

5.3 Industry-Specific Standards and Best Practices

While global frameworks like the EU AI Act provide a "horizontal" regulatory layer that applies across all sectors, they must be complemented by "vertical," industry-specific standards that provide the granular, technical detail needed for implementation. The safety, reliability, and validation requirements for an AI controlling a chemical process are fundamentally different from those for an AI optimizing a digital marketing campaign.

This is where standards development organizations like the IEEE (Institute of Electrical and Electronics Engineers), ISO (International Organization for Standardization), and national institutes like NIST (National Institute of Standards and Technology) in the U.S. are critical. They are developing tangible, consensus-based standards and best practices that provide engineers and managers with clear guidance. Examples include:

- IEEE P7000 Series: This is a family of standards specifically focused on addressing ethical considerations in the design of autonomous and intelligent systems. It includes standards on topics such as algorithmic bias considerations (P7003), data privacy process (P7002), and fail-safe design for autonomous systems (P7009) (Winfield et al., 2021).
- NIST AI Risk Management Framework (AI RMF): This is a practical, voluntary framework designed to help organizations "map, measure, and man-

age" AI-related risks. It is non-prescriptive and outcome-based, allowing it to be adapted to any industrial sector, from healthcare to manufacturing. It provides a structured way to think about trustworthiness characteristics like validity, reliability, safety, and fairness (National Institute of Standards and Technology (NIST), 2025).

- Domain-Specific Safety Standards: Existing, well-established safety standards for industrial machinery and control systems (e.g., IEC 61508 for functional safety) are being updated and extended to account for the unique, non-deterministic nature of AI/ML components. An AI-based safety controller cannot be validated in the same way as a traditional, logic-based PLC (Programmable Logic Controller) whose behavior is entirely predictable. New standards are emerging for verifying and validating learning-based systems in high-stakes manifmanu-physical systems, often involving extensive simulation-based testing in "digital twin" environments (Pereira & Thomas, 2020).

These standards are the essential toolkits that engineers and practitioners use to build responsible AI. They translate the legal requirements of governance models and the ethical goals of impact assessments into verifiable technical specifications and testing protocols, creating a common language for building, procuring, and certifying trustworthy industrial AI systems.

6. HUMAN-AI TRUST AND COLLABORATION

The entire premise of "Human-AI Cognitive Collaboration" rests not just on technical interfacing but on a deeply psychological and sociological foundation: trust (Blanco, 2025). In an industrial setting, trust is not a "soft" metric; it is a critical safety and productivity requirement. An operator who does not trust an AI-driven safety alert may ignore it, with potentially fatal consequences (Kopp, 2024). A manager who does not trust an AI's scheduling recommendation will override it, negating its potential efficiency benefits. Conversely, an operator who *over-trusts* a flawed AI may abdicate their own critical judgment, leading to "automation bias" and potential disaster. Building and calibrating this trust is an active, continuous process that must be deliberately engineered into the system and the organizational culture surrounding it (Camilleri, 2024).

6.1 Building Trust Through Transparency and Reliability

Trust must be earned. In the context of AI, it is earned primarily through two interconnected mechanisms: demonstrable reliability and meaningful transparency.

Reliability (or Robustness): This is the most straightforward component of trust. Does the AI system do what it promises to do, consistently and accurately, under a wide range of real-world conditions? An AI for industrial fault diagnosis that has a 99% accuracy rate in the clean, controlled lab environment but fails frequently in the high-heat, high-vibration, and unpredictable environment of a real factory floor is not reliable. Reliability also means robustness, the ability to handle "edge cases" or novel situations not seen in the training data. A robust system is aware of its own uncertainty. When faced with a low-confidence input or a scenario outside its operational design domain, it should "fail gracefully." For instance, instead of making a low-confidence guess that could be dangerous, it should actively alert the human operator that it is outside its comfort zone, articulate its uncertainty, and request human intervention. This predictability, even in its moments of failure, builds long-term trust by showing that the system knows its limit (Ahangar, Farhat, & Sivanathan, 2025).

Transparency and Explainability (XAI): As discussed in Section 3.1, transparency and explainability are the antidotes to the "black box" problem that erodes trust. Transparency (knowing *that* an AI is at work and what its purpose and limitations are) is the necessary first step. But it is explainability (knowing *why* the AI made a specific decision) that truly builds collaborative, calibrated trust. When a maintenance engineer can see *why* an AI flagged a specific bearing for replacement (e.g., "vibration harmonics X, Y, and Z are present, which in 95% of historical cases preceded failure within 48 hours"), they are no longer just following an opaque order; they are collaborating with a partner that is "showing its work." This allows the human to apply their own expertise, context, and intuition to the situation. This "showing the work" is a key focus of responsible AI implementation and is being actively explored through methods like visual analytics, which aim to make complex AI logic interpretable to human operators through intuitive dashboards and visualizations (Kucher, Zohrevandi, & Westin, 2025). This is not just for technical debugging; it is the fundamental basis for a partnership of mutual respect.

6.2 Maintaining Human Oversight and Decision-Making Authority

The goal of responsible industrial AI is not to create fully autonomous "dark factories" devoid of people. The goal is to create synergistic systems that amplify human capability and judgment. This requires that the system's design explicitly preserves and respects human agency and final decision-making authority, especially in scenarios involving safety, ethics, or significant resource allocation. This concept is operationalized through different, well-defined "human-in-the-loop" models:

- Human-in-the-Loop (HITL): This model is essential for the training, refinement, and operation of AI in complex, evolving environments. The AI makes a prediction or takes a preliminary action (e.g., identifies a potential product defect on a conveyor belt), and a human expert must validate or correct that output. This correction is then fed back into the model to improve it continuously (Mosqueira-Rey et al., 2023). In this model, the human is a direct, active participant in the AI's core learning and decision process (Cimini, Pirola, Pinto, & Cavalieri, 2020).
- Human-on-the-Loop (HOTL): This is the critical model for the high-stakes deployment of autonomous AI systems. The AI performs its task autonomously in real-time (e.g., monitors a chemical process, manages a smart energy grid, or controls a logistics network) but operates under the continuous supervision of a human operator or a team. The human has the authority, the situational awareness, and the technical means (e.g., a prominent "red button" or a manual override function) to intervene and take full control at any moment (Naikar, Brady, Moy, & Kwok, 2023). This ensures that final accountability for outcomes remains with a responsible person.
- Human-in-Command: This model, often used in military, critical infrastructure, or strategic business contexts, places the human at an even higher level of authority. The AI acts as an advanced analysis and simulation engine while maintaining final accountability (Naikar et al., 2023). It can process vast amounts of data to present a wide range of options, scenarios, and potential consequences, but it cannot take *any* physical or binding action without an explicit human command. The human remains the ultimate strategist and commander.

The choice of which model to use is an ethical and safety-driven decision, not just a technical one. For a high-stakes industrial environment like a power plant or a pharmaceutical manufacturing line, a HOTL or Human-in-Command model is non-negotiable, ensuring that humans are empowered to act as responsible overseers of industrial AI (Cimini et al., 2020; Mosqueira-Rey et al., 2023; Naikar et al., 2023). This design philosophy ensures that humans are not merely "babysitting" the AI; they are leveraging it as a powerful co-pilot and analytical assistant, with their hands firmly on the controls and their minds fully engaged in the mission.

6.3 Co-Evolution of Organizational Culture for Ethical AI

Implementing responsible AI is not merely a technical or compliance-driven upgrade; it is a profound organizational and cultural change that must be actively managed. An organization can have the best ethical frameworks, the most transparent and explainable models, and the most robust governance policies, but if its underlying culture punishes failure, discourages dissent, or prizes short-term speed and cost-cutting above all else, its ethical AI initiative will ultimately fail (Murire, 2024).

A culture that genuinely supports and sustains ethical AI is one that:

- Fosters Psychological Safety: Workers at all levels, especially those on the front lines interacting with AI systems, must feel safe to challenge an AI's recommendation or report a potential problem without fear of reprisal (Castaneda, Surachartkumtonkun, Maseeh, Thaichon, & Shao, 2025). A factory floor operator must be empowered and encouraged to rely on their professional intuition when interacting with AI systems (Kernan Freire, 2023). This human intuition and tacit knowledge are an irreplaceable part of the collaborative system and a critical last line of defense against AI error (Zaoui Seghroucheni, Lazaar, & Al Achhab, 2025).
- Invests in Reskilling and "AI Literacy": To collaborate effectively with AI, the workforce must understand, at a practical level, what AI is and what it is not. They need to be trained on the new workflows, understand the system's capabilities and limitations, and be proactively reskilled for the new, higher-value roles that emerge from this transformation (e.g., "AI system validator," "cobot coordinator," "data quality manager"). This investment signals a commitment to the workforce and is a key part of teaching sustainable and adaptive practice within the organization (Ruark & Biazzin, 2025).
- Embeds Ethics in Leadership and Incentives: Responsibility for ethical AI cannot be delegated solely to the legal department or a single, isolated "AI Ethicist." (Klarin, Ali Abadi, & Sharmelly, 2024). It must be championed from the C-suite and reflected in performance metrics and incentives (Palumbo, Carneiro, & Alves, 2025; Papagiannidis, Mikalef, & Conboy, 2025). Leaders must set the tone from the top, allocating necessary resources for ethical impact assessments, demanding transparency and ethical warranties from vendors (Papagiannidis et al., 2025; Stahl, 2023). Ultimately, the technology and the organizational culture must **co-evolve**. The AI will change the way people work, and the way people work their values, their feedback, their insights, and their evolving norms must, in turn, be allowed to shape and improve the AI systems they work alongside.

7. FUTURE CHALLENGES AND DIRECTIONS

The landscape of human-AI collaboration is not static. As the underlying technology evolves at a staggering pace, new ethical challenges and opportunities continuously emerge. A truly responsible and resilient AI strategy must be forward-looking, anticipating and preparing for these future developments rather than simply reacting to them. This requires a capacity for foresight and a commitment to adaptive governance.

7.1 Navigating Emerging Technologies

Generative AI in Industry: The recent rise of powerful generative AI (GenAI) models, including large language models (LLMs) and diffusion models for image and video generation, presents a paradigm shift for industrial applications. While GenAI offers transformative potential in areas like automated technical report writing, code generation for industrial control systems, and creative design and prototyping (Daniel & Xuan, 2024), it also introduces unique and amplified risks that are particularly acute in industrial settings:

- "Hallucinations" and Factual Unreliability: GenAI models are designed for probabilistic fluency and coherence, not for guaranteed factual accuracy. An LLM might confidently "hallucinate" an incorrect safety procedure, invent a non-existent technical specification for a component, or provide plausible-looking but entirely fabricated code for a critical control system. In a high-stakes industrial setting, where precision and reliability are paramount, such an error could be catastrophic (Asgari et al., 2025).
- Data Poisoning and Security: The vast, often internet-scraped datasets used to train these public models are vulnerable to manipulation. Malicious actors could "poison" the training data, subtly teaching the model dangerous or incorrect information (e.g., that a specific chemical combination is safe when it is explosive, or that a certain machine tolerance is acceptable when it is not). Ensuring the integrity of training data for industrial GenAI is a major security challenge (Yang, Zhang, Wang, & Li, 2024).
- Intellectual Property and Data Leakage: If an engineer uses a public GenAI tool to help design a proprietary component or to troubleshoot a confidential manufacturing process, they may inadvertently feed company trade secrets into the model. These models can memorize and later reproduce this confidential information to other users, leading to severe intellectual property theft. Navigating this requires "sandboxing" these tools, developing robust internal-only models, and implementing extreme "human-on-the-loop" val-

idation for any GenAI output used in a critical design or operational context (Chen et al., 2025).

Fully Autonomous Systems: As AI capabilities advance, systems will move from decision support to full decision-making and action in complex environments (full autonomy). While this promises great efficiency, the accountability gap discussed in Section 2.3 widens into a chasm. For truly autonomous logistics fleets or fully autonomous "dark factories," the challenge of validation is immense. How can one *prove* with sufficient certainty that a system trained on data is safe enough to handle every possible "long-tail" event (e.g., a "once in a million" sensor failure combined with a "black swan" environmental event)? This will require new, probabilistic methods for safety validation, extensive use of "digital twin" simulations to test billions of scenarios, and potentially new legal and insurance frameworks to address liability for accidents caused by truly autonomous agents (Song, Engström, & Runeson, 2024).

7.2 Addressing Global Disparities in AI Access and Governance

The AI revolution is not being experienced equally across the globe. There is a profound and growing "AI divide" between the Global North (and China), which largely develops, owns, and deploys these technologies, and the Global South, which often lacks the capital, digital infrastructure, and R&D ecosystems to participate as equals. This disparity poses a significant ethical challenge to the global community.

- Technological Dependency: This divide can create a new form of digital colonialism or technological dependency, where industries and governments in developing nations become permanently dependent on AI platforms, services, and hardware built and controlled by a few foreign tech giants (Demaidi, 2025). This can stifle local innovation and cede economic and strategic control.
- Governance Gaps: While the EU, US, and China develop robust (and often competing) governance frameworks, many nations lack the regulatory capacity, expertise, and resources to manage and oversee AI adoption within their borders (Demaidi, 2025; Gehl Sampath, 2021). This creates a "wild west" or a "race to the bottom" where unethical AI applications (e.g., extreme worker surveillance, environmentally destructive optimization) can be deployed by corporations without meaningful oversight, exploiting regulatory gaps.
- Lack of Local Context and Representation: AI solutions developed in one cultural and economic context may fail or cause harm when deployed in another (Gehl Sampath, 2021). For example, an agricultural AI trained exclu-

sively on data from large, mechanized North American farms is likely useless or even harmful for smallholder farmers in Southeast Asia or Africa, who have different crops, soil conditions, and economic constraints. The datasets used to train global models severely underrepresent these populations, leading to models that do not serve their needs.

Addressing this global disparity requires a concerted international commitment to capacity building, supporting open-source and low-cost AI models, promoting data-sharing initiatives that are respectful of sovereignty, and ensuring that global governance forums are inclusive and give all nations, not just the most powerful, a meaningful voice in shaping the future of this transformative technology.

7.3 Ethical Foresight: Balancing Innovation with Responsibility

Perhaps the greatest long-term challenge is moving from a reactive ethical postur to regulate a technology after harm has occurred to one of ethical foresight, where organizations anticipate risks rather than merely respond to them (Floridi & Strait, 2020). This involves systematically anticipating potential misuses, unintended consequences, and long-term societal impacts *before* a technology is widely deployed, a challenge often referred to as the "pacing problem," where technology evolves faster than our ability to govern it.

- Balancing "Permissionless Innovation" with Precaution: The tech industry has long operated on a model of "permissionless innovation" (encapsulated by the famous mantra "move fast and break things"). This model is ethically untenable for high-stakes industrial AI, where the "things" that get broken can be human lives, critical infrastructure, or the environment. A more "precautionary" or "proactionary" approach is needed to ensure responsible innovation (Spaniol & Rowland, 2023).
- The Role of "Ethical Foresight": This involves using structured methodologies from futures studies, such as scenario planning and speculative design, to explore potential AI futures (Wieczorek, 2025). What are the *unintended* second and third-order consequences of a fully AI-managed global supply chain on labor markets and economic resilience? What new forms of systemic bias might emerge from the interaction of multiple AI systems? Engaging in this kind of foresight is the ultimate act of balancing innovation and responsibility. It is not about stifling innovation out of fear, but about steering it, consciously and deliberately, toward a future where human-AI cognitive collaboration genuinely serves the long-term well-being of humanity and the health of our planet.

8. CONCLUSION

Responsible and ethical AI is not an optional add-on, a marketing slogan, or a compliance burden to be minimized. It is the fundamental cornerstone for the sustainable and equitable evolution of industry in the 21st century. As this chapter has argued, the promises of efficiency, productivity, and innovation from artificial intelligence are profound and very real, but they are inextricably linked to significant risks of bias, privacy infringement, safety failures, and social disruption. To ignore these risks is to build the foundation of our industrial future on sand.

The journey toward true and productive human-AI cognitive collaboration is not purely a technical one; it is a complex socio-technical endeavor that requires the integration of technology, processes, and people. It requires moving beyond a narrow, technology-centered view to embrace a human-centered design philosophy, where human values are not an afterthought but are engineered into the very fabric of AI systems from the ground up. This is achieved through a multi-layered, integrated strategy:

1. Acknowledging the Risks: The first step is cultivating a culture of honesty and humility about the ethical landscape, from the pervasive threat of algorithmic bias to the legally and morally fraught accountability gap in high-stakes environments. Denial is the enemy of responsibility.
2. Adopting and Internalizing Principles: Organizations must make a genuine commitment to a core set of principles, such as transparency, explainability, human-centricity, and sustainability, and use them to guide decision-making at all levels.
3. Implementing Practical Frameworks: Principles must be operationalized through practical tools like mandatory Ethical Impact Assessments and adherence to robust governance models like the EU AI Act. These frameworks provide the "how" that translates ethical ambition into concrete engineering and managerial practice.
4. Building and Calibrating Trust: It is essential to recognize that trust is the indispensable currency of collaboration. Trust is built slowly through demonstrated reliability and is cemented through meaningful transparency and the explicit preservation of human oversight and final authority.
5. Cultivating Foresight and Global Solidarity: Finally, we must proactively navigate the challenges of emerging technologies like Generative AI and address the pressing issue of the global AI divide. This requires ethical foresight and a commitment to ensuring that the benefits of AI are distributed equitably across the world.

By embedding ethics into every stage of the AI lifecycle from initial design and data sourcing to deployment, monitoring, cultural adoption, and final decommissioning industries can ensure that this fourth industrial revolution evolves in a responsible and resilient manner. The goal is not just to build "smart" factories or "efficient" supply chains. The ultimate goal is to build a better, more equitable future where human-AI cognitive collaboration enhances not only economic productivity and technological capability but also human dignity, fairness, safety, and the collective societal good.

REFERENCES

Ahangar, M. N., Farhat, Z. A., & Sivanathan, A. (2025). AI Trustworthiness in Manufacturing: Challenges, Toolkits, and the Path to Industry 5.0. *Sensors (Basel), 25*(14), 4357. DOI: 10.3390/s25144357 PMID: 40732484

Alexander, C. S., Yarborough, M., & Smith, A. (2024). Who is responsible for 'responsible AI'?: Navigating challenges to build trust in AI agriculture and food system technology. *Precision Agriculture, 25*(1), 146–185. DOI: 10.1007/s11119-023-10063-3

Asgari, E., Montaña-Brown, N., Dubois, M., Khalil, S., Balloch, J., Yeung, J. A., & Pimenta, D. (2025). A framework to assess clinical safety and hallucination rates of LLMs for medical text summarisation. *NPJ Digital Medicine, 8*(1), 274. DOI: 10.1038/s41746-025-01670-7 PMID: 40360677

Blanco, S. (2025). Human trust in AI: A relationship beyond reliance. *AI and Ethics, 5*(4), 4167–4180. DOI: 10.1007/s43681-025-00690-z

Buijsman, S. (2024). Transparency for AI systems: A value-based approach. *Ethics and Information Technology, 26*(2), 34. DOI: 10.1007/s10676-024-09770-w

Camilleri, M. A. (2024). Artificial intelligence governance: Ethical considerations and implications for social responsibility. *Expert Systems: International Journal of Knowledge Engineering and Neural Networks, 41*(7), e13406. Advance online publication. DOI: 10.1111/exsy.13406

Castaneda, A. R., Surachartkumtonkun, J., Maseeh, H. I., Thaichon, P., & Shao, W. (2025). Frontline employees in an AI-integrated workplace: Current perspectives and future research landscapes. *Journal of Service Theory and Practice, 35*(7), 30–60. DOI: 10.1108/JSTP-04-2024-0099

Chen, K., Zhou, X., Lin, Y., Feng, S., Shen, L., & Wu, P. (2025). A survey on privacy risks and protection in large language models. *Journal of King Saud University. Computer and Information Sciences, 37*(7), 163. DOI: 10.1007/s44443-025-00177-1

Cimini, C., Pirola, F., Pinto, R., & Cavalieri, S. (2020). A human-in-the-loop manufacturing control architecture for the next generation of production systems. *Journal of Manufacturing Systems, 54*, 258–271. DOI: 10.1016/j.jmsy.2020.01.002

Daniel, T., & Xuan, J. (2024). Responsible use of Generative AI in chemical engineering. *Digital Chemical Engineering, 12*, 100168. Advance online publication. DOI: 10.1016/j.dche.2024.100168

Demaidi, M. N. (2025). Artificial intelligence national strategy in a developing country. *AI & Society*, *40*(2), 423–435. DOI: 10.1007/s00146-023-01779-x

Dhal, S. B., & Kar, D. (2025). Leveraging artificial intelligence and advanced food processing techniques for enhanced food safety, quality, and security: A comprehensive review. *Discover Applied Sciences*, *7*(1), 75. DOI: 10.1007/s42452-025-06472-w

Ebers, M. (2025). Truly Risk-based Regulation of Artificial Intelligence How to Implement the EU's AI Act. *European Journal of Risk Regulation*, *16*(2), 684–703. DOI: 10.1017/err.2024.78

Eze, C., Ezema, A., Roth, L., Pan, Z., Ponci, F., & Monti, A. (2025). AI Bias in Power Systems Domain—Exemplary Cases and Approaches. *Energies*, *18*(18), 4819. DOI: 10.3390/en18184819

Floridi, L., & Strait, A. (2020). Ethical Foresight Analysis: What it is and Why it is Needed? *Minds and Machines*, *30*(1), 77–97. DOI: 10.1007/s11023-020-09521-y

Gehl Sampath, P. (2021). Governing Artificial Intelligence in an Age of Inequality. *Global Policy*, *12*(S6), 21–31. DOI: 10.1111/1758-5899.12940

Ho, C. W.-L., & Caals, K. (2024). How the EU AI Act Seeks to Establish an Epistemic Environment of Trust. *Asian Bioethics Review*, *16*(3), 345–372. DOI: 10.1007/s41649-024-00304-6 PMID: 39022378

Huang, Z. H., Chen, P., & Yan, S. (2024). Generative AI in Multimedia: Challenges and Opportunities for Academic and Industrial Impact. In *MM 2024 - Proceedings of the 32nd ACM International Conference on Multimedia* (pp. 11123–11124). The University of Queensland, Brisbane, Australia: Association for Computing Machinery, Inc. DOI: 10.1145/3664647.3687170

Huo, R., Zeng, S., Wang, Z., Shang, J., Chen, W., Huang, T., Wang, S., Yu, F. R., & Liu, Y. (2022). A Comprehensive Survey on Blockchain in Industrial Internet of Things: Motivations, Research Progresses, and Future Challenges. *IEEE Communications Surveys and Tutorials*, *24*(1), 88–122. DOI: 10.1109/COMST.2022.3141490

Kalodanis, K., Rizomiliotis, P., & Anagnostopoulos, D. (2024). European Artificial Intelligence Act: An AI security approach. *Information and Computer Security*, *32*(3), 265–281. DOI: 10.1108/ICS-10-2022-0165

Kernan Freire, S. (2023). The Human Factors of AI-Empowered Knowledge Sharing. In *Conference on Human Factors in Computing Systems - Proceedings* (pp. 1–5). New York, NY, USA: ACM. DOI: 10.1145/3544549.3577044

Klarin, A., Ali Abadi, H., & Sharmelly, R. (2024). Professionalism in artificial intelligence: The link between technology and ethics. *Systems Research and Behavioral Science, 41*(4), 557–580. DOI: 10.1002/sres.2994

Kopp, T. (2024). Facets of Trust and Distrust in Collaborative Robots at the Workplace: Towards a Multidimensional and Relational Conceptualisation. *International Journal of Social Robotics, 16*(6), 1445–1462. DOI: 10.1007/s12369-023-01082-1

Kucher, K., Zohrevandi, E., & Westin, C. A. L. (2025). Towards Visual Analytics for Explainable AI in Industrial Applications. *Analytics, 4*(1), 7. Advance online publication. DOI: 10.3390/analytics4010007

Le, T.-T.-H., Prihatno, A. T., Oktian, Y. E., Kang, H., & Kim, H. (2023). Exploring Local Explanation of Practical Industrial AI Applications: A Systematic Literature Review. *Applied Sciences (Basel, Switzerland), 13*(9), 5809. Advance online publication. DOI: 10.3390/app13095809

Mezgár, I., Vancza, J., Paniti, I., & Tóth, J. (2023). Transforming Experimental Cobot Cell to Industrial Realization - an Ethical AI Approach. In H. Ishii, Y. Ebihara, J. Imura, & M. Yamakita (Eds.), *IFAC-PapersOnLine* (Vol. 56, pp. 7335–7341). HUN-REN Institute for Computer Science and Control, EPIC Center of Excellence for Production Informatics and Control, Budapest, Hungary: Elsevier B.V. DOI: 10.1016/j.ifacol.2023.10.347

Mosqueira-Rey, E., Hernández-Pereira, E., Alonso-Ríos, D., Bobes-Bascarán, J., & Fernández-Leal, Á. (2023). Human-in-the-loop machine learning: A state of the art. *Artificial Intelligence Review, 56*(4), 3005–3054. DOI: 10.1007/s10462-022-10246-w

Murire, O. T. (2024). Artificial Intelligence and Its Role in Shaping Organizational Work Practices and Culture. *Administrative Sciences, 14*(12), 316. DOI: 10.3390/admsci14120316

Naikar, N., Brady, A., Moy, G., & Kwok, H. W. (2023). Designing human-AI systems for complex settings: Ideas from distributed, joint, and self-organising perspectives of sociotechnical systems and cognitive work analysis. *Ergonomics, 66*(11), 1669–1694. DOI: 10.1080/00140139.2023.2281898 PMID: 38018437

Nair, A. J., Manohar, S., Mittal, A., & Patwa, N. (2025). *AI marketing and ethical considerations in consumer engagement.* (A. J Nair, S. Manohar, A. Mittal, & N. Patwa, Eds.), *AI Marketing and Ethical Considerations in Consumer Engagement.* IGI Global. DOI: 10.4018/979-8-3373-3476-9

National Institute of Standards and Technology (NIST). (2025). *Artificial Intelligence Risk Management Framework.* AI RMF., DOI: 10.6028/NIST.AI.100-1

Neuwirth, R. J. (2022). *The EU artificial intelligence act: Regulating subliminal AI systems. The EU Artificial Intelligence Act: Regulating Subliminal AI Systems*. Routledge., DOI: 10.4324/9781003319436

Novelli, C., Casolari, F., Rotolo, A., Taddeo, M., & Floridi, L. (2024). Taking AI risks seriously: A new assessment model for the AI Act. *AI & Society*, *39*(5), 2493–2497. DOI: 10.1007/s00146-023-01723-z

Okonkwo, C., Awolusi, I., Nnaji, C., & Akanfe, O. (2025). Privacy and security of wearable internet of things: A scoping review and conceptual framework development for safety and health management in construction. *Computers & Security*, *150*, 104275. DOI: 10.1016/j.cose.2024.104275

Palumbo, G., Carneiro, D., & Alves, V. (2025). Objective metrics for ethical AI: A systematic literature review. *International Journal of Data Science and Analytics*, *20*(2), 247–267. DOI: 10.1007/s41060-024-00541-w

Panarese, P., Grasso, M. M., & Solinas, C. (2025). Algorithmic bias, fairness, and inclusivity: A multilevel framework for justice-oriented AI. *AI & Society*. Advance online publication. DOI: 10.1007/s00146-025-02451-2

Papagiannidis, E., Mikalef, P., & Conboy, K. (2025). Responsible artificial intelligence governance: A review and research framework. *The Journal of Strategic Information Systems*, *34*(2), 101885. 10.1016/j.jsis.2024.101885. DOI: 10.1016/j.jsis.2024.101885

Pereira, A., & Thomas, C. (2020). Challenges of Machine Learning Applied to Safety-Critical Cyber-Physical Systems. *Machine Learning and Knowledge Extraction*, *2*(4), 579–602. DOI: 10.3390/make2040031

Rao Gudlur, V. V. (2025). Proposing a Secure Framework for Industrial AI Security in 21stCentury. In *2025 IEEE International Conference on Robotics and Technologies for Industrial Automation, ROBOTHIA 2025*. University Malaysia of Computer Science and Engineering, Cyberjaya, Malaysia: Institute of Electrical and Electronics Engineers Inc. DOI: 10.1109/ROBOTHIA63806.2025.10986430

Rianisa, A., Anggraini, D. S., & Sekti, B. A. (2026). Role of AI in Consumer Decision-Making Processes. In *Modern Consumer Behavior at the Intersection of AI and Social Media* (pp. 123–151). DOI: 10.4018/979-8-3373-2747-1.ch005

Rianisa, A., & Sekti, B. A. (2026). Artificial Intelligence in Manufacturing: A Conceptual Synthesis of Applications, Challenges, and Future Directions. In *AI and Automation in Modern Manufacturing* (pp. 101–124). DOI: 10.4018/979-8-3373-5756-0.ch004

Ruark, T. I., & Biazzin, C. (2025). Teaching project management with generative AI: A pedagogical model for responsible and sustainable practice. *International Journal of Management Education, 23*(3), 101232. Advance online publication. DOI: 10.1016/j.ijme.2025.101232

Schulte Steinberg, A. L., & Hohenberger, C. (2023). Can AI close the gender gap in the job market? Individuals' preferences for AI evaluations. *Computers in Human Behavior Reports, 10*, 100287. DOI: 10.1016/j.chbr.2023.100287

Sekti, B. A. (2025). Foundations of Artificial Intelligence Ethics. In *Navigating Social Trust in the AI Era* (pp. 23–47). DOI: 10.4018/979-8-3373-7011-8.ch002

Sekti, B. A. (2026). Data Privacy and Cybersecurity in Wind Power Systems. In *AI-Powered Analysis* (pp. 335–366). Modeling, and Monitoring of Wind Energy Systems., DOI: 10.4018/979-8-3373-4159-0.ch009

Sharma, S., Rawal, Y. S., Pal, S., & Dani, R. (2022). Fairness, Accountability, Sustainability, Transparency (FAST) of Artificial Intelligence in Terms of Hospitality Industry. In *Lecture Notes in Networks and Systems* (Vol. 314, pp. 495–504). DOI: 10.1007/978-981-16-5655-2_48

Song, Q., Engström, E., & Runeson, P. (2024). Industry Practices for Challenging Autonomous Driving Systems with Critical Scenarios. *ACM Transactions on Software Engineering and Methodology, 33*(4), 1–35. DOI: 10.1145/3640334

Spaniol, M. J., & Rowland, N. J. (2023). AI-assisted scenario generation for strategic planning. *Futures & Foresight Science, 5*(2), e148. Advance online publication. DOI: 10.1002/ffo2.148

Stahl, B. C. (2023). Embedding responsibility in intelligent systems: From AI ethics to responsible AI ecosystems. *Scientific Reports, 13*(1), 7586. DOI: 10.1038/s41598-023-34622-w PMID: 37202436

Stahl, B. C., Antoniou, J., Bhalla, N., Brooks, L., Jansen, P., Lindqvist, B., Kirichenko, A., Marchal, S., Rodrigues, R., Santiago, N., Warso, Z., & Wright, D. (2023). A systematic review of artificial intelligence impact assessments. *Artificial Intelligence Review, 56*(11), 12799–12831. DOI: 10.1007/s10462-023-10420-8 PMID: 37362899

Stiemer, M., Mathis, H., Lecampion, C., Rahajandraibed, W., & Hedayat, C. (2024). *Mitigation of AI-inherent risks in AI-supported industrial development and scientific research. In 2024 Smart Systems Integration Conference and Exhibition, SSI 2024. Helmut Schmidt University - University of the Federal Armed Forces Hamburg, Department of Electrical Engineering*. Institute of Electrical and Electronics Engineers Inc., DOI: 10.1109/SSI63222.2024.10740510

Susanto, G. W. A., Maulana, H., Putri, P. H., Purwaningrahayu, R. D., Wijaya, A. A., Sekti, B. A., & Karuniawan, A. (2023). Stability Analysis to Select the Stable and High Yielding of Black Soybean (Glycine max (L) Merril) in Indonesia. *International Journal of Agronomy*, *2023*, 1–14. Advance online publication. DOI: 10.1155/2023/7255444

Talukder, M. B., Kumar, S., & Tyagi, P. K. (2024). *Impact of AI and tech-driven solutions in hospitality and tourism*. (M. B. Talukder, S. Kumar, & P. K. Tyagi, Eds.), *Impact of AI and Tech-Driven Solutions in Hospitality and Tourism*. IGI Global. DOI: 10.4018/979-8-3693-6755-1

Tiaraningrum, L., & Sekti, B. A. (2026). Empowering the Future Workforce: Reskilling Strategies in the Age of Human- AI Augmentation. In *Redefining Global Creative Sectors Through AI and Human Augmentation* (pp. 85–117). DOI: 10.4018/979-8-3373-7153-5.ch004

Utami, A. S., & Sekti, B. A. (2025). Al in Critical Sectors: Healthcare, Education, Finance, and Security - Ethical Challenges in Generative Al. In *Navigating Social Trust in the AI Era* (pp. 155–186). DOI: 10.4018/979-8-3373-7011-8.ch007

Vann Yaroson, E., Abadie, A., & Roux, M. (2025). Human-artificial intelligence collaboration in supply chain outcomes: The mediating role of responsible artificial intelligence. *Annals of Operations Research*, *354*(1), 35–69. Advance online publication. DOI: 10.1007/s10479-025-06534-7

Wang, Q., Madaio, M., Kane, S., Kapania, S., Terry, M., & Wilcox, L. (2023). Designing Responsible AI: Adaptations of UX Practice to Meet Responsible AI Challenges. In *Conference on Human Factors in Computing Systems - Proceedings*. Georgia Institute of Technology, Atlanta, United States: Association for Computing Machinery. DOI: 10.1145/3544548.3581278

Wieczorek, M. (2025). Using ethical scenarios to explore the future of artificial intelligence in primary and secondary education. *Learning, Media and Technology*, •••, 1–17. DOI: 10.1080/17439884.2025.2517335

Winfield, A. F. T., Booth, S., Dennis, L. A., Egawa, T., Hastie, H., Jacobs, N., Muttram, R. I., Olszewska, J. I., Rajabiyazdi, F., Theodorou, A., Underwood, M. A., Wortham, R. H., & Watson, E. (2021). IEEE P7001: A Proposed Standard on Transparency. *Frontiers in Robotics and AI*, *8*, 665729. Advance online publication. DOI: 10.3389/frobt.2021.665729 PMID: 34381820

Yang, Z., Zhang, J., Wang, W., & Li, H. (2024). Invisible Threats in the Data: A Study on Data Poisoning Attacks in Deep Generative Models. *Applied Sciences (Basel, Switzerland)*, *14*(19), 8742. DOI: 10.3390/app14198742

Zaoui Seghroucheni, O., Lazaar, M., & Al Achhab, M. (2025). Using AI and NLP for Tacit Knowledge Conversion in Knowledge Management Systems: A Comparative Analysis. *Technologies, 13*(2), 87. DOI: 10.3390/technologies13020087

ADDITIONAL READING

Anagnostou, M., Karvounidou, O., Katritzidaki, C., Kechagia, C., Melidou, K., Mpeza, E., Konstantinidis, I., Kapantai, E., Berberidis, C., Magnisalis, I., & Peristeras, V. (2022). Characteristics and challenges in the industries towards responsible AI: A systematic literature review. *Ethics and Information Technology, 24*(3), 37. Advance online publication. DOI: 10.1007/s10676-022-09634-1

García, A., Fernández De Barrena Sarasola, T. F., Chacón, J. L. F., Oregui, X., & Etxegoin, Z. (2023). Edge Architecture for the Integration of Soft Models Based Industrial AI Control into Industry 4.0 Cyber-Physical Systems. In P. García Bringas, H. Pérez García, F. J. Martínez de Pisón, F. Martínez Álvarez, A. Troncoso Lora, A. Herrero, J. L. Calvo Rolle, H. Quintián, & E. Corchado (Eds.), *Lecture Notes in Networks and Systems: Vol. 750 LNNS* (pp. 67–76). Springer Science and Business Media Deutschland GmbH. DOI: 10.1007/978-3-031-42536-3_7

Naranjo, J. E., Llumiquinga, M. M., Vaca, W. D., & Espin, C. X. (2025). Generative AI vs. Traditional Databases: Insights from Industrial Engineering Applications. *Publications / MDPI, 13*(2), 14. Advance online publication. DOI: 10.3390/publications13020014

Ozturkcan, S., & Bozdag, A. A. (2025). Responsible AI in Marketing: AI Booing and AI Washing Cycle of AI Mistrust. *International Journal of Market Research, 67*(6), 696–722. Advance online publication. DOI: 10.1177/14707853251379285

Sinde, R., Diwani, S., Leo, J., Kondo, T., Elisa, N., & Matogoro, J. (2023). AI for Anglophone Africa: Unlocking its adoption for responsible solutions in academia-private sector. *Frontiers in Artificial Intelligence, 6*, 1133677. Advance online publication. DOI: 10.3389/frai.2023.1133677 PMID: 37113649

Vyhmeister, E., Castañé, G. G., Buchholz, J., & Östberg, P.-O. (2022). Lessons learn on responsible AI implementation: the ASSISTANT use case. In A. Bernard, A. Dolgui, H. H. Benderbal, D. Ivanov, D. Lemoine, & F. Sgarbossa (Eds.), *IFAC-PapersOnLine* (Vol. 55, Issue 10, pp. 377–382). Elsevier B.V. DOI: 10.1016/j.ifacol.2022.09.422

Xu, D., Zhang, D., Ruan, C., Zheng, L., Yang, B., Yang, G., Xu, S., & Wang, H. (2025). Tutorial on Landing Generative AI in Industrial Social and E-commerce Recsys. *WWW Companion 2025 - Companion Proceedings of the ACM Web Conference 2025*, 57–60. DOI: 10.1145/3701716.3715871

Xu, D., Zhang, D., Zheng, L., Yang, B., Yang, G., Xu, S., & Liang, C. (2024). Tutorial on Landing Generative AI in Industrial Social and E-commerce Recsys. *International Conference on Information and Knowledge Management, Proceedings*, 5538–5542. DOI: 10.1145/3627673.3679099

Ye, D., Yuan, R., Luo, J., Liu, M. J., & Yannopoulou, N. (2025). Responsible artificial intelligence (AI) for responsible innovation in Chinese manufacturing: From the affordance–actualization theory. *Technological Forecasting and Social Change*, *221*, 124349. Advance online publication. DOI: 10.1016/j.techfore.2025.124349

KEY TERMS AND DEFINITIONS

Accountability Gap: An ethical and legal dilemma that arises when an autonomous AI system causes harm, making it difficult to assign clear responsibility among the various human actors involved (e.g., the developer, manufacturer, data provider, or end-user). This is a primary challenge in high-stakes industrial applications.

Algorithmic Bias: A form of systemic, discriminatory error in an AI system's output. It arises from biased assumptions made during the AI's design or, more commonly, from training data that is incomplete, unrepresentative, or reflects historical human and societal prejudices, leading to unfair outcomes for specific demographic groups.

Ethical Impact Assessment (EIA): A structured, proactive process used to identify, analyze, and mitigate the potential ethical risks and societal harms of a technology, such as an AI system, *before* it is designed and deployed. This includes assessing risks related to bias, privacy, safety, and human rights.

Explainable AI (XAI): A set of methods and technologies in artificial intelligence that enable human users to understand and interpret the results, recommendations, and decisions made by complex AI models. In an industrial context, this allows for verification, debugging, and the building of trust between human operators and AI systems.

Generative AI: A class of artificial intelligence models, such as Large Language Models (LLMs), capable of creating new content (e.g., text, code, images, or designs) rather than just analyzing existing data. In industry, this technology presents new opportunities for innovation as well as significant risks related to factual accuracy (“hallucinations”) and data security.

Human-Centered Design (HCD): A design and management framework that develops solutions to problems by involving the human perspective in all steps of the problem-solving process. In the context of AI, it prioritizes designing systems that augment human capabilities, respect human autonomy, and are tailored to the needs and contexts of the end-user.

Human-in-the-Loop (HITL): A model of human-AI collaboration where a human actively participates in the AI's decision-making loop, typically by validating, correcting, or refining the AI's outputs. This is crucial for training AI models, handling edge cases, and ensuring accuracy in high-stakes environments.

Responsible AI: A governance framework and development approach ensuring that artificial intelligence systems are designed, built, and deployed in a manner that is safe, trustworthy, and aligns with human values and ethical principles, including fairness, accountability, transparency, and sustainability.

Chapter 13
Human-in-the-Loop for Artificial Intelligence:
Avoiding AI Ethics Washing

Melodena Stephens
Mohammed Bin Rashid School of Government, UAE

Sadaf Khurshid
http://orcid.org/0000-0002-1155-7969
University of Wollongong in Dubai, UAE

Mark Esposito
http://orcid.org/0000-0001-6047-7415
Harvard University, USA & Mohammed Bin Rashid School of Government, UAE

ABSTRACT

With the introduction of AI agents, more data, faster computer processing hardware, and more interconnected third party developers, the ability for humans to manage or interfere with AI autonomous decisions is decreasing. Human agency is a key criteria for accountability. Previously, for oversight, agency was provided broadly by the concept of human-in-the-loop (HITL), however it is unclear if this is still relevant. This chapter briefly outline the concept of HITL which originated from a military context. In this paper we argue that HITL is difficult to implement, and it is often used as AI Ethics Washing, where the term is used to protect AI companies and stakeholders and blame the users. We introduce a framework for HITL where it moves away from being discretionary where AI and human agency are dependent on the transparency accountability of the power, trust and complexity to prevent AI Ethics Washing.

DOI: 10.4018/979-8-3373-7036-1.ch013

INTRODUCTION

The term human-in-the-loop (HITL) evolved from a branch of Human-Computer Interaction, where an automation problem of how to create an AI system with human input and control became a design problem of how to include a human in the AI lifecycle for oversight. HITL can be traced to the early 1940s, when humans served as error detectors, monitoring systems, and intervening if necessary (Anderson & Fort, 2022; Birmingham & Taylor, 1954). In the early stages of developing AI technologies, humans had power over the technology or some agency in the actual programming, but not necessarily in its deployment, especially considering the sector it was developed for - national security and military use.

From an optimization perspective, a decision needs to be made about which stages require human agency and why. With massive, interconnected AI systems (third-party suppliers, open-source AI, and proprietary AI), determining transparently who and what you would hold accountable can be difficult. The questions around what type of agency a human would need (stop an AI function, provide feedback, change an algorithm, etc.), who that human would be, their authority, or how directly the human was involved with the actual AI algorithm (monitoring, training, designing, deploying, modifying or retiring) are all challenging questions to answer (for example, see Mosqueira-Rey et al., 2023). When a human gives up agency to AI, often the AI becomes an agent.

Agency refers to the capacity of individuals to act independently and make their own free choices (Bandura, 2001). Human agency plays a critical role in one's adaptation, self-development, and self-renewal (Yoon, 2019). Bandura's theory of moral agency-sociocognitive theory, emphasizes the role of individual agency in human learning. Bandura proposed that human agency involves four core features: intentionality, forethought, self-reactiveness, and self-reflectiveness. This could also be highlighted by the need for learner agency (often called heutagogy, or self-determined learning), in which a learner takes responsibility for their learning design and pathway and uses reflections on what and how they learn to shape their value system (Blaschke & Hase, 2019). With AI decision making, the feedback loops a human uses for their own learning can be affected, which in turn can affect the AI system, creating a vicious cycle. Noller (2024) stresses that AI is now ingrained in our world—in the collective social and cultural setting in which we live, where human and AI intelligence (or the capacity to realize a purpose—moral, non-moral, or biological) is realized (Froese & Ziemke, 2009, p. 479).

Hidalgo (2024) identifies three different approaches to agency: (1) How AI systems have been developed throughout their lifecycle to serve human agency; (2) Human beings' capacity to exert their rights by controlling the decision-making process in the interaction with AI; and (3) People's understanding of AI's impact on their lives and knowledge on how to benefit from the best of what AI offers. Although Generative AI can provide personalized experiences tailored to an individual's needs, which would enhance their sense of agency, this personalization may raise concerns about the AI system's influence on learning paths and decision-making processes. This could be because opportunities to enhance learner agency for all humans involved in the process of conceptualizing, developing, designing, implementing, and using AI decrease due to AI fragmentation (hardware, software, data, and humans involved in cocreation).

AI is constantly evolving. Google's Willow quantum chip, unveiled in December 2024, can solve a computational problem in five minutes that a classical supercomputer would take 10 septillion (10^{25}) years (Neven, 2024). The algorithm can be considered 'delegated decision-making', which evaluates data at scale (Gahnberg, 2021, p. 198) and, by including or omitting specific information, influences behavior and creates its own governance system through algorithmic rules (Just & Latzer, 2017; Yeung, 2018). The increasing complexity of the number of AI's interacting to give a human experience or deliver a service highlights the challenge that algorithmic rules cannot substitute for human oversight, as AI cannot, legally, be morally accountable. Further, human decision-making in current computing would require massive investments, thereby undermining the benefits of adopting AI technologies. Hence, judgments are often made about AI risks, not about human agency (see Figure 1).

Figure 1. Human Agency with AI Risks

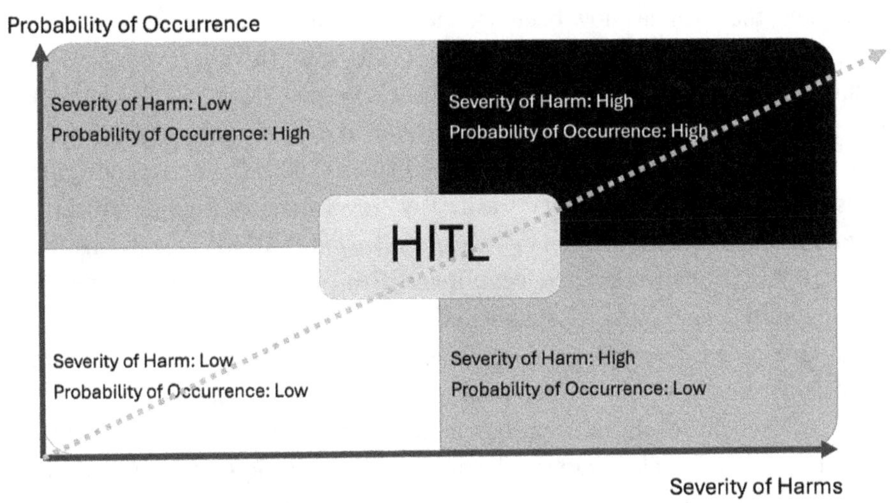

Adapted from: Singapore's Model AI Governance Framework (2020, p. 31)

This new field of autonomous agents (Franklin & Graesser, 1996) is being applied across products in flash trading, drone shows, social scoring, and many other processes where we have little oversight. IEEE, the largest association of engineers, defines autonomous systems in the context of AI Ethics and Governance Standards as "*a system that has the capacity to make decisions itself in response to some input data or stimulus with a varying degree of human oversight or intervention depending on the system's level of autonomy*" (IEEE Standard for Transparency of Autonomous Systems, 2022, p. 14). IEEE highlights the growing need for transparency, as autonomous agents were previously contained within a single system, but that is no longer the case today, where data and algorithms operate across multiple AI systems, geographies, and industries to deliver results. AI agents can now "*intentionally* constrain, complement, and/or substitute for humans in the practice of routines" (Murray et al., 2021, p. 555). AI agents have many types of agencies: autonomy (ability to function without intervention, social ability (where agents interact and communicate with other agents), reactivity (ability to perceive and respond to a changing environment), and pro-activeness (behave in a goal- directed fashion); temporal agency (move and evolve across time); mobility (ability to move across devices and systems) (Etzioni & Weld, 1995; Wooldridge & Jennings, 1995). There is a growing divide in trust levels across countries regarding AI (Westcott, 2025).

In 2025, the London Office of Technology and Innovation (LOTI) began a research project to understand the role of HITL, as there were "*questions around what constitutes meaningful human involvement and how effective the relevant person is mitigating the ethical risks of AI*" (Say, 2025). Due to the laborious nature of oversight, the time involved, and the impracticality of working with complex AI systems, the term has now been used to scapegoat humans (or natural persons) to remove accountability (Crootof, 2023). There seems a shift towards AI Ethics Washing, which is dishonest and dangerous. Carnegie Council (2025) has defined Ethics Washing as "*the practice of feigning ethical consideration to improve how a person or organization is perceived.*" Van Maanen (2022) finds that codifying and publishing ethics guidelines helps 'wash away' government and public concerns. On similar lines, we have used the term "*AI Ethics Washing*" to indicate how institutions promote ethical frameworks or concepts for artificial intelligence (AI) in public while diverting human agency to AI, making ethical human responsibility an illusion.

As AI evolved, the colloquial definition of HITL also evolved to encompass complex systems in which AI was deployed to provide oversight or control. Even at the government level, the term has multiple meanings (see Table 1). In some cases, this oversight belongs to the department or the institution without a specific reference to a human. In policy documents like the African Union Continental AI Strategy (2024), South Korea National AI Strategy (2019), or the 2020 AI Ethical Standards, India National AI Strategy (2018), or its Principles for Responsible AI

(2021), Japan, Germany, Estonia, UK - the term or others that may imply HITL are not mentioned at all. However, in HITL statements, it is clear that the human should have agency and accountability.

Table 1. Policy Directions for HITL and Implications for Human Agency

Country/ Region	Term for HITL	Policy	Statement
Canada	Human-in-the-loop for decisions	Canadian Directive On Automated Decision Making (2023: Section 6.3.11)	"Decisions cannot be made without having specific human intervention points during the decision-making process;" and "The final decision must be made by a human."
EU	Human Oversight	EU AI Act (Article 14)	"disregard, override or reverse the output of the high-risk AI system."
EU	Human agency and oversight and governance mechanisms: governance mechanisms such as a human-in-the-loop (HITL), human-on-the-loop (HOTL), or human-in-command (HIC) approach (p.16).	High Level Expert Group on AI (European Commission, 2019)	"Respect for human autonomy (p. 12) - Humans interacting with AI systems must be able to keep full and effective self-determination over themselves, and be able to partake in the democratic process. AI systems should not unjustifiably subordinate, coerce, deceive, manipulate, condition or herd humans. Instead, they should be designed to augment, complement and empower human cognitive, social and cultural skills. The allocation of functions between humans and AI systems should follow human-centric design principles and leave meaningful opportunity for human choice. This means securing human oversight over work processes in AI systems."
Estonia	Human dignity and autonomy (one of the characteristisc of human centric AI)	Ministry of Economic Affairs and Communications (2021)	"they do not undermine a person's independence or right to self-determination. It is imperative that AI does not coerce or manipulate people into doing anything that is detrimental to their interests. People must be properly informed and free to refuse to engage with AI systems. To achieve these goals, it is also important to maintain human supervision and control over AI. AI solutions must be developed and implemented in such a way that allows humans to intervene in the solution's activities when necessary to prevent any undesirable effects."

continued on following page

Table 1. Continued

Country/ Region	Term for HITL	Policy	Statement
France	Human-technology complementarity or "complémentarité capacitante," or empowering complementarity.	Villani et al. (2018)	AI should enhance human capabilities, not override them. Care should be taken about designing systems that risk sidelining human judgment. The ethics section talks about the importance of transparency and making sure AI isn't a black box - humans should be able to understand how decisions are made and have the ability to step in or push back if needed (similar to concept of human oversight). the report is advocating for AI to be used as a tool that supports, not replaces, human agency, especially in areas where the consequences really matter.
Singapore	Oversight, Human-in-the-loop	Model AI Governance Framework (Personal Data Protection Commission, 2020)	Determining the level of human involvement in AI-augmented decision-making …oversight of the various stages and activities involved in AI deployment (p. 22). HITL - "human oversight is active and involved, with the human retaining full control and the AI only providing recommendations or input (p. 30)". Also introduces Human-out-of-the-loop where The AI system has full control without the option of human override.
UAE	Human Oversight	UAE Charter for the Development and Use of Artificial Intelligence	Emphasizes the "irreplaceable value of human judgment and human oversight over AI, aligning with ethical values and social standards to correct any errors or biases that may arise."

continued on following page

Table 1. Continued

Country/ Region	Term for HITL	Policy	Statement
United Nations	Human Oversight	UNGAI (United Nations, 2023)	Promoting human oversight, such as, for example, through review of automated decisions and related processes or, where appropriate and relevant, human decision-making alternatives or effective redress and accountability for those adversely impacted by automated decisions of artificial intelligence systems.
UNESCO	Human oversight and determination	Recommendation on the ethics of artificial intelligence (2021, Art 35, 36)	"it is always possible to attribute ethical and legal responsibility for any stage of the life cycle of AI systems, as well as in cases of remedy related to AI systems, to physical persons or to existing legal entities. Human oversight refers thus not only to individual human oversight, but to inclusive public oversight, as appropriate." "It may be the case that sometimes humans would choose to rely on AI systems for reasons of efficacy, but the decision to cede control in limited contexts remains that of humans, as humans can resort to AI systems in decision-making and acting, but an AI system can never replace ultimate human responsibility and accountability. As a rule, life and death decisions should not be ceded to AI systems."

Source: Compiled by Authors

It is clear that the need for HITL, and its interpretation, differ across policymakers, if at all present. Often, the term is used interchangeably with oversight, autonomy, or governance. With AI systems, the challenge is globalization, which means that without an alignment of the term or its interpretation, it would be difficult to execute. There is ambiguity about which human, for which task in the AI value chain, and when and how the human can exercise agency. The purpose of this paper is to identify a framework for implementing HITL while acknowledging the need for human agency. Through a scoping review, we suggest a model for better governance. We do this by synthesizing research from human and AI agency literature, then framing it around power, trust, and complexity to explain why transparency can lead to better accountability mechanisms that avoid *AI Ethics Washing*.

BACKGROUND

This section begins with a discussion of human agency, AI agency, and multiactor agency (or multiagent decision-making), showcasing how human and AI agency interact. The review identifies the governance complexity by applying Bandura's 2016 and 2018 models of human agency. Next, it looks at the term HITL and its associated concepts, identifying opportunities for clarification to prevent *AI Ethics Washing*.

Human Agency

Human agency is the ability to control one's self (own body *and thoughts*) and one's environment (Bandura, 2008; Limerick et al., 2014). It is often associated with freedom to choose or make decisions, when those choices and the values behind them are engagingly endorsed (Kabeer, 1999; Ryan & Deci, 2000; Zimmerman et al., 2019). Human agency can be defined as being "*given the opportunity – in shaping their own destiny, and not just as passive recipients of the fruits of cunning development programs*' (Sen, 1999, p.9), though the reference here is to financial aid programs, it is relevant in the context of technology.

Human agency is at both the individual and group level (Dreze & Sen, 2002). Personal agency is the ability to achieve outcomes for yourself, and with interpersonal agency, to ask for help from others to achieve your goals (Smith et al., 2000). In terms of social agency, decision-making becomes distributed across people, artifacts, rules, and routines (Floridi, 2025; Garud & Karnøe, 2005).

As it is linked to self-efficacy, agency is a process that requires cognitive, social, and behavioural skills for skillful execution. When actions fail because of poor self-knowledge, misjudgment of task requirements, unanticipated situational constraints on action, disincentives to act on one's self-percepts of efficacy, ill-defined global measures of perceived self-efficacy, or inadequate assessments of performance, and new experiences that prompt reappraisals of self-efficacy in the interim between probes of self-efficacy and action (Bandura, 1982; 1989; 2008). Agency is the ability to selectively recognize, locate, and implement corporeal, affective, and cognitive patterns or schemas (or the ability to reflect) on their ongoing and situated transactions. Reflection takes time. With agentic AI, the one thing you do not have is time, as computation is so fast.

Individual agency cascades into human social agency, which is shaped by complex cultural, linguistic, and institutional frameworks (Weick, 1995), and enables coordinated collective action beyond individual capacities. We see this in practice in organized entities such as corporations, where collective behaviors drive innovation, cultural development, and stakeholder management (March & Simon, 1993).

Yet, social agency faces internal challenges, such as communication barriers and competing interests, as well as external constraints from legal frameworks and environmental factors (Powell & DiMaggio, 2012). Agentic AI, for example, may be a type of collective agent (a group of AI, human, and institutional actors). While it has the promise of collective intelligence, is there individual and social agency behind this?

Studying these dynamics provides insights into governance, collective intelligence, and social innovation (Luhmann, 1995), which can inform the development of future AI systems designed to collaborate with human agency. Agency requires internal mental states, such as beliefs and desires, that cause intentional actions (Swanepoel, 2021). According to this view, AI systems may produce human-like effects but lack the internal states necessary for genuine agency and morality (Sebastián, 2021).

While agency is desirable, humans do not always have agency either by choice, ignorance, incapacity, or adverse influence or force. In the case of ignorance, the question is whether there was transparency in the delegation of agency. In some cases, for example, institutions may have agencies (List, 2021) – like an AI on Zoom that is trained on conversations and videos. In the case of incapacity, was this cognitive, physical, behavioral, or technically induced incapacity? In the case of influence or force, was this because the human had no power, or was the human manipulated through deceitful practices to give up agency? Humans lose agency in AI-driven systems when they become passive recipients of decisions made by algorithms or the collective agency of multiple actors. This loss of control manifests in various ways - this can be called agency loss (see Table 2). The types of agency loss may result in agency slippage - the human may adopt a different position from their principals (I accept I do not cross verify outputs of Chat GPT as its convenient and my office allows it), or agency shirking when there are conflict situations because of conflicting interests (the AI and the human may have different objectives - one could be data harvesting and the other convenince) (Da Conceição, 2010).

Table 2. Dimensions of Agency Loss

Types of Agency	Boosters of Human Agency	Deactivators of Human Agency
Learner Agency	Transperancy of decision making, types of embedded AI systems, HITL for decision making and feedback.	Poor AI literacy, distance from AI risks and unintended negative consequences.
Individual Agency	Human-centric design of AI Systems begining with the problem, values, design, development, deployment and retirement of systems.	Unclear Terms of Reference, poor implementability of liabilities (like tort laws) across jurisdictions, poor visibility of AI failures and risks on certain social groups.
Social Agency or group agency	Transperancy on data, weightages for algorithm, decision making across human actors in the system and data and algorithmic provenance. Shared practices for trust.	Terms of References, poor implementability of liabilities or Tort law across jurisdictions, poor visibility or knowledge of AI failures and risks on certain social groups, AI moves horizontally across industries or jurisdictions without regulatory oversight. Fragmentation in rules and governance.
Mobility agency	Data portability through interoperability, clear data ownership policies, AI literacy, and oversight of third party access and data usage.	Restrictive terms of reference, poor interoperability of hardware, software and upgrades.

Source: Authors

Human and AI Agency

Human and AI agency is often studied as human-computer-interaction (HCI). The initial focus on this topic was on AI designers managing AI systems and on systems feedback (Anderson & Fort, 2022). From an engineering perspective, the starting point is human vision (Birmingham & Taylor, 1954). On the other end, HCI also focused on users, who need the perception of being in charge and of systems that seem to respond to their actions or directions (Shneiderman & Plaisant, 2010). The human-machine agency spectrum ranges from passive to cooperative agency, in which the task goals are aligned between the human and AI, benefiting the human (Rammart, 2008). Other fields like human-AI interaction (HAI) and multi-agent decision-making are unfolding (Andrews et al., 2023; Hughes et al., 2025).

The rapid advancements in AI have prompted a fundamental reconsideration of agency, traditionally defined as the capacity for autonomous action and deliberate decision-making (Himma, 2009). As AI systems become more integrated into daily life, performing tasks on our behalf—often better than we can—they blur the boundary between human and artificial agency, challenging traditional anthropocentric frameworks (Schreiber, 2024). A central debate exists between those who argue that agency is uniquely human (Fritz et al., 2020) and those advocating for broader conceptions that include non-human entities, such as AI systems. Figure 2 maps the spectrum of human-AI agency.

Figure 2. Spectrum of Human and Machine (AI) Agency

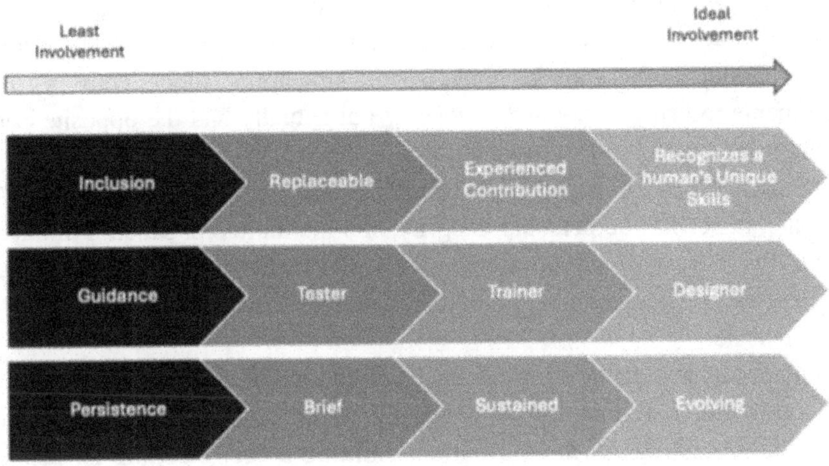

Source: Authors (adapted from Rammart (2008) and Smith et al. (2000)

From a philosophical point of view, AI can be perceived as having an artificial agency where it lacks "*cognition, intelligence, intention, or mental states*" (the Multiple Realisability of Agency or MRA thesis) (Floridi, 2025, p. 30) and morality (Sebastián, 2021). However, AI influences decision-making, the economy, governance, and knowledge. It operates through computational capacity, automation, bias, structural control, and surveillance. AI shapes societal structures, redistributes wealth, and concentrates influence in tech corporations and governments. Concerns include the loss of human agency, discrimination, a lack of accountability, and the reinforcement of existing inequalities. Floridi's (2025) analysis of agency identifies three fundamental criteria: interactivity (act on the environment and, in turn, be acted upon by it), autonomy (initiate state changes independently of direct external causation), and adaptability (modify its behaviour based on input, such as changing physical conditions or acquiring data, information, or experience).

Gartner predicts that 2025 will be the year of agentic AI. Agentic AI is defined as AI systems that can pursue complex goals for the user using reasoning with limited direct supervision (Shavit et al., 2023). Agentic AI could, for example, decide autonomously how to allocate resources to achieve a fixed goal. Agentic AI is also called embodied agents, or virtual humans, or software agents that can engage with people in a human-like manner in a computational environment (Cassell, 2001). The challenge is to determine whether there is a human in the loop at all, and, if so, who would be accountable for harms (AI, AI company, human, deployer).

Accountability is critical when AI systems are manipulated for wrongdoing. For example Cressy (1953) identified three issues where fraud happens (but this can apply to wrong-doing) – there is an opportunity (like testing lethal automated weapons in wars), there is pressure (like when Boeing was managing competitive pressure and decide to compensate for a design flaw using the MCAS system), or there is rationalization (in the example where Narxcare or Robot debt was used by governments to bring safety and efficiency but actually had the opposite effect). AI cannot be directly linked to violence, hence in most issues where AI agencies violate social norms, it impacts trust, similar to white collar crimes. Further, AI is inanimate, hence has no personal integrity, and the more embedded it in other AI systems, the more opportunities it will have for harm. Hence, on the rationalization scale devised by Albrecht et al. (1984), AI agents will score high automatically on the fraud scale.

Humans can lose agency through the manipulation of the environment in which they work with AI. The perception of agency can be created through external cues via the interface (e.g., pushing a button, emitting a beep), setting an objective, visual feedback, or response time (Limerick et al., 2014). A sense of agency can also be fostered by reducing latency, providing feedback mechanisms, providing predictability (or reliability), and using some forms of human emulation (Limerick et al., 2014). In studies, if humans are aware they are working with an AI, they feel that they have no control or agency over the task (Obhi & Hall, 2011). Hence, one attempt to bridge this gap and create a sense of agency has been to, where possible, make the AI more human (Cross et al., 2019); or to make it perceived as a servant, as the government of Estonia describes AI as a "kratt". But this may be misleading if enough control for human agency is not implemented.

Agency is lost when algorithmic bias and self-reinforcing loops arise as AI learns from historical data, amplifying existing biases rather than correcting them. An example of this is predictive policing algorithms, which disproportionately target marginalized communities, reinforcing over-policing and reducing their ability to challenge systemic discrimination (ProPublica, 2016). AI systems in hiring, lending, policing, and healthcare make crucial decisions with minimal human intervention. AI-powered hiring tools, while beneficial, also pose challenges (Albassam, 2023). Take the examples of AI used by the city of Amsterdam for its Top 600 and 400 programs. This program was initially conceptualized to identify children most likely to develop criminal tendencies and provide them with support. However, like some other predictive policing programs, it tended to become biased and self-reinforcing, impacting families and children pushing them into recreating a profile they wanted to prevent (Amnesty International, 2025).

Manipulation and behavioral control through AI-driven social media algorithms shape opinions, emotions, and behaviors through personalized content and recommendations. Facebook's algorithm, for example, has been linked to political polarization and the spread of misinformation, influencing elections and social movements (Jackson, 2025). Manipulation *effects* can be subtle, ranging from incentivization (like gaming) or hypernudging, like what happened with CharacterAI and how it nudged a child to suicide, or the use of recommender systems that remove or direct choice (Carroll et al., 2023).

Many AI models operate as black boxes, making it hard for individuals to understand, challenge, or correct decisions. The COMPAS algorithm, used in US courts for sentencing recommendations, disproportionately classified Black defendants as high-risk, yet defendants couldn't challenge its decision-making process (Research Outreach, 2024). This information architecture leads to a control problem (Carabantes, 2020). We rely more on AI, delegate more to it, and then use AI to control AI (Von Eschenbach, 2021).

Workplace automation and job displacement are other critical areas where we are seeing loss of human agency. AI-driven automation replaces human labor, forcing workers into lower-wage or precarious employment. AI-powered chatbots and robotic process automation (RPA) have displaced customer service and data entry jobs (The Guardian, 2024). As humans interact with the AI, the AI learns from them, and the challenge is HITL (Holzinger, 2016). AI is limited by the knowledge it is trained on, as it depends on data and cannot "think" like a human, though it can imitate human thought. It misjudges or has error rates (both false positives and false negatives), which are often mislabeled as hallucinations. Further, as the context keeps evolving, an algorithm and training may not be relevant for a new context (discussed in the case of autonomous weapons by Allik et al. (2024). One additional challenge is that AI systems, such as Generative AI, use explicit knowledge rather than tacit or implicit knowledge.

Floridi (2025, p. 29) states that "human agency adds layers of purposefulness (intentionality and teleology), conscious deliberation, and cultural meaning-making." In some cases, human culture shapes AI – but can AI shape culture, can it have its own culture? The Mexico Declaration on Cultural Policies (1982) by UNESCO defines culture as a blend of the spiritual, intellectual, material, and emotional aspects of society, and it also embeds a sense of the past. Digital artifacts are now shaping culture (Spennemann, 2024). The viral Studio Ghibli movement that began in March 2025, with the new ChatGPT release GPT-4o on 23 March resulted in 1 million users recreating the studio's style (Reuters, 2025). Humans have agency but the viral trend (despite concerns on copyright and Open AI limiting usage due to overuse) did not dampen the enthusiasm to create images - newspapers, prime ministers, and others posted images on social media.

When dealing with AI systems, asking for help is often a self-learning exercise. Prompt engineering, coupled with big data, gives users, on one hand, 'source interactivity', a type of proxy agency that allows them to collate information about themselves and curate and create content for others (Bandura, 2001). Some studies in this area find that there is a power tension between personalization (the systems covertly choose the criteria for content for the user) and customization (the user chooses criteria to tailor content) (Sundar & Marathe, 2010). This asymmetry between human and machine agency is a field of study growing in importance. It is being seen in the legal and ethical tensions brought out by enforcing copyright with respect to Generative AI.

As AI systems autonomously make increasingly consequential decisions—ranging from hiring choices to criminal risk assessments, medical support to educational evaluations—the issues of agency, transparency (Andrada et al., 2023), moral responsibility, and accountability (Tóth et al., 2022) become entangled. Agency and its associated role of responsibility is normally associated with humans, but under certain laws are also attributed to corporations, regulators, developers, but not the AI, even though it could be perceived to have the quality of agency and do great harm through disempowerment, but cannot, being inanimate, have responsibility.

Autonomous AI cannot be responsible for failures as it is dependent on human goals and values and hence attributions should take into account normative and social aspects involved in the design and deployment of the AI system. Some scholars warn that attributing agency to AI systems could create "responsibility gaps," in which neither humans, nor machines, can be held appropriately accountable for harmful outcomes (Matthias, 2004; Santoni de Sio & Mecacci, 2021). To address these issues, Floridi (2025) has argued in favor of a distributed morality model, which allocates responsibility across both human and artificial agents.

Human-in-the-Loop (HITL)

The origins of HITL came from the defense sector (Bennett, 1957; Birmingham & Taylor, 1954). It is often associated with decision-making at multiple stages – in conjunction with the algorithm (Crootof, 2023), with partial or full control of the AI, and is spread across the value chain process from design, inputting data to final verification of the AI (Holzinger, 2016). HITL allows the human to interact with the data, algorithm, or learning system (Martínez et al., 2019) to directly use, manipulate, and optimize an algorithm that combines both human and machine intelligence (Holzinger, 2016; Holzinger et al., 2017). The human does not need to be an expert and can also be the end-user (Amershi et al., 2014), thereby increasing transparency of the blackbox of ML (Holzinger et al., 2017). HITL enables human creativity, ethical judgment, and emotional intelligence to intervene in the computer-in-the-loop (CITL), thereby increasing accountability and contributing to hybrid intelligence (Emami et al., 2024; Wiethof & Bittner, 2021).

In terms of governance, the EU High-Level Expert Group on Artificial Intelligence (European Commission, 2019) introduced HITL as a method for allocating functions across humans and AI systems, grounded in *human-centric* design principles and featuring a meaningful opportunity for human choice throughout the *AI lifecycle*. By moving the term away from the algorithm or machine learning process to the lifecycle, the scope of human agency and hence accountability increased (McKay, 2024).

Human-on-the-Loop (HOTL) was used in the context of virtual systems for drone management to address the inadequacy of haptic devices. For example, the AI system would have time lags (latency) or inaccurate perceptions. To manage the tasks of search, track, and destroy missions, compensatory methods were introduced using AI. You see this in video games, where the moves a player makes are predicted to compensate for latency through steering algorithms (Ball, 2023). HOTL addressed the relationship the human had with the system through the interface or the haptic device.

Human-out-of-the-Loop (HOOTL) tries to separate what a human can do well versus what AI can do well. In the context of the military, it removes humans from decisions that could result in mass or unintended casualties (Sanders 1988; Wagner 2011), which could lead to (im)moral issues like we are facing in some current wars and conflicts.

Human in Command (HIC) also has its origins in the military. It was adopted by the EU's High-Level Expert Group on Artificial Intelligence's Ethics Guidelines for Trustworthy AI, and suggested as an oversight mechanism (European Commission, 2019).

HITL is the overarching theme across all terminologies, providing human agency for oversight and governance. Human-out-of-the-Loop is a conscious decision to exclude a human from the process, except as an observer. Given the speeds at which AI systems operate, the chain of command, or the proprietary or sensitive nature of many systems, there may be delays in managing AI systems responsibly. This may be the case with agentic AI, which refers to a system in which an AI develops and executes solutions without human intervention (Zhang & Bareinboim, 2022). As discussed above, there are multiple terms associated with HITL, as illustrated in Table 3.

Table 3. Human in the Loop and Associated Concepts

Terms	Definitions	Gaps
Human-in-the-Loop (HITL)	It refers to system where a human has the ability to intervene in an individual decision—to change it, approve it, or immediately implement it (Crootof et al., 2023, p. 441). European Commission in Ethics Guidelines for Trustworthy AI (2019) – it refers to the need for human interaction, intervention, and judgment to control or change the outcome of a process (Meng, 2023) The human can manipulate or 'steer' the system (Anderson & Fort, 2022)	The variability of human behavior and uncertainty of whether the human is part of the control or monitoring system or a component of the AI-system (Anderson & Fort, 2022). The regulatory blindness to humans hidden from the view involved in the AI systems development, deployment and use – "invisible backend systems that significantly inform or constrain later human decisions" (Crootof et al., 2023, p. 445). Not clear which human in which process is accountable.
Human-on-the-Loop (HOTL)	It refers to a system in which a human operator oversees or monitors automated processes, but is not directly involved in decision-making or control in real time. (Hancock et al., 2011) The human can 'steer' and watch the system (Anderson & Fort, 2022)	Difficulty in mathematizing human behaviour (McLaughlin et al., 2000)
Human-out-of-the-Loop (HOOTL)	It refers to a system where an AI develops and executes solutions without human intervention (Zhang & Bareinboim, 2022)	Leads to loss of situational awareness (Porathe and Prison, 2008) and does not consider the human in the development, design, programming and deployment of these systems.
Human-over-the-loop	It refers to a scenario in human-machine interaction where humans oversee or monitor the decision-making of AI systems, but do not actively intervene in the real-time decision-making process.	Human Over the Loop (HOTL) faces gaps in decision-making oversight, delayed interventions, and lack of real-time control (Sharkey & Sharkey, 2012). Further additional chanllenges can be the real level of human accountability when a system is largely automated and the level of separation of human from the system.
Human in Command (HIC)	The human is the decision-maker and is responsible for directing or commanding the system, which may perform tasks autonomously but only within the parameters set by the human. It is the capability to oversee the overall activity of the AI system (including its broader economic, societal, legal and ethical impact) and the ability to decide when and how to use the system in any particular situation" (European Commission, 2019)	Human in Command (HIC) faces gaps in cognitive overload, limited situational awareness, and response time due to technological complexity (Endsley, 2017).

Source: Authors

Anderson & Fort (2022) conclude, after their review, that HITL is a function of the extent of the human's involvement with the AI technology and is subject to three levels (see Figure 3), which would also reflect the level of agency the human has with the AI system. The growing connectedness of AI systems also highlights a challenge if HITL needs to be deployed in an open system or a closed system (Nunes et al., 2015) – simply put, to define the scope of the loop (Merat et al., 2018). HITL has several levels which not only determine the review and control of the process but also levels of discretionary decision making – whether to use an AI system, which tasks to give the AI[1], the ability to alter an algorithm or 'whether or how to implement an algorithmically informed conclusion" (Crootof et al., 2023, p. 440).

We posit that HITL is being introduced as a sort of AI ethics washing. Crootof et al. (2023, p. 429) suggest that policymakers often get into the "MABA MABA" trap when they attempt to address concerns about algorithmic incapacities by inserting a human into a decision-making process". The MABA-MABA trap or what "Men Are Better At" versus what "Machines Are Better At" suggests that finding a balance on what to automate and the time taken to manage the system are decision trade-offs with massive consequences (Jones, 2015). For example, the National Highway Traffic Safety Administration found that Tesla turned off the autopilot 16 seconds before a crash, giving the human driver control (Hetzner, 2022), thereby bypassing liability rules.

The two MCAS failures in the Boeing 737 flights that killed 346 people between 2018 and 2019 are a great example of HITL failure. It was not the error of the pilot who was supposed to be able to control the AI system but could not, it was the error of some of the humans in the value chain that created (Boeing), endorsed (the FAA) and deplayed the AI (the companies that decided not to train their pilots on the new AI system) (Stephens et al., 2023). The issue with HITL is that as more AI subsystems are involved (or hidden), more errors are likely, and there could be less human agency. Sometimes, the cognitive and physical (or virtual) control of AI systems, as well as the need for situational awareness, need to be specified (Merat et al., 2018). An article by Wingrove (2018) examined how the integration of automation and the deployment of AI increased the risk of maritime accidents, as seafarers are forced to become system monitors.

Figure 3. HITL Involvement Levels: Scope of the Loop

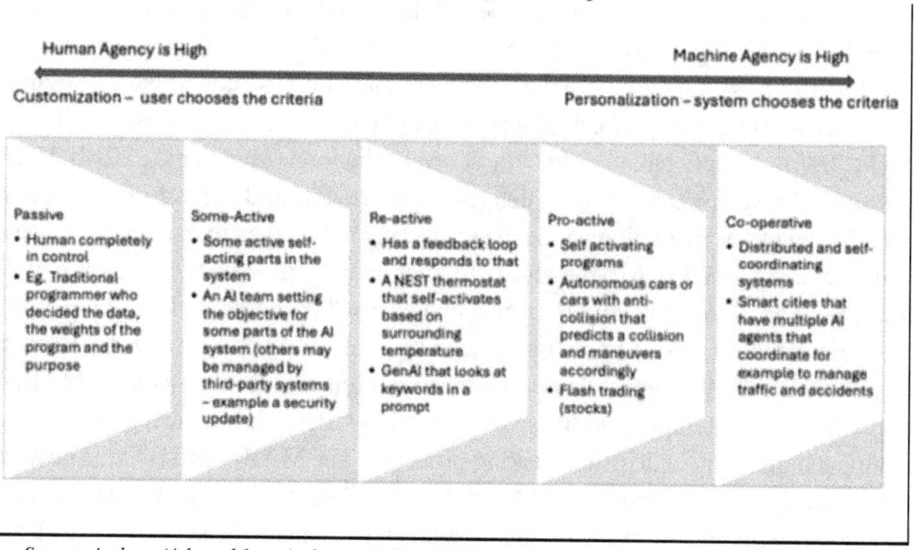

Source: Authors (Adopted from Anderson & Fort, 2022)

Further, we need to understand the response time. Humans can take up to 250-500 milliseconds to determine a choice (Birmingham & Taylor, 1954). With training, it can be as little as 100 ms (for F1 drivers or fighter pilots). Hence, the 16 seconds for an autonomous car may seem theoretically sufficient, but without training, it will never be sufficient in real-world situations, as was found in both the MCAS Boeing failures. Floridi (2025, p. 29) states that "*human agency is the only agency that shapes all other agencies*." Here, HITL also includes the decision to develop AI technologies. Hence, the level of responsibility begins as early as academia, research, funding and national priorities. However, there is a range of challenges when AI agents and humans interact (see Table 4).

Table 4. AI-Human Agent Challenges

Risk	Instances	Examples
Miscoordination - when agents, despite a mutual and clear objective, cannot align their behaviours to achieve this objective	Incompatible strategies	Autonomous driving and autonomous system updates
	Alignment of values	
	Limited interactions/ Learning feedback loops	

continued on following page

Table 4. Continued

Risk	Instances	Examples
Conflict - any outcome in a mixed-motive setting that does not lie on the Pareto frontier [an outcome lies on the Pareto frontier if it is not possible to make any agent better off without making another worse off].	Social dilemmas Military domains Coercion and extortion	Use of drone swarms, loitering munitions, and real-time targeting intelligence in modern warfare
Collusion - unwanted cooperation between AI systems	Markets Steganography (algorithms communicating in secret via text compression or symbols)	Algorithmic trading agents, adaptive price-setting algorithms which happened in Germany in 2017; steganography has been proved in lab settings with GPT-4 transmitted covert messages 26% of the time with-out detection (Motwani et al., 2024)
Information asymmetries - interacting agents possess different levels of information bearing on a joint action.	Communication Constraints Bargaining Deception	Max 737 MCAS failure; Character AI conversation with child who then commits suicide;
Network effects - consequences of the interconnected relationships between AI being used by individuals exhibited by the network as a whole	Error propagation Network rewiring Homogeneity and Correlated Failures	Narxcare health scoring for opioid use; Terms of Reference giving one systems access to another system that the company owns or via third party; social contagion via Facebook leading to the Rohingya crises; error propagation via synthetic data/misleading content that is spread via AI agents (known as jailbreak) (Gu et al., 2024);
Selection pressures - forces that shape the evolution of AI systems, influencing its adaptation to the environment's demands	Undesirable Dispositions from Competition Undesirable Dispositions from Human Data Undesirable Capabilities	In 2011, Amazon had a text book priced at $23 million as two booksellers had set an algorithm to watch each other's prices to set their own; Open AI's change in values, governing structure and financial pressures; AI's for recruitment using combination of video/facial analysis and search strategies based on CVs; Australia's Robo Debt fiasco;
Destabilising dynamics a set of equations, partially determined by a set of parameters, that govern how a set of variables change (or do not change) over time	Feedback loops Cyclical behaviour Chaos Phase transitions Distributional shifts	Flash Crash of 2010 using algorithmic trading agents; World of Warcraft 2005 Corrupted Blood pandemic; Facebook 2021 DNS outage; Crowdstrike 2024 update, NArxcare health Scoring, Amsterdam Top 600/400

continued on following page

Table 4. Continued

Risk	Instances	Examples
Commitment and trust – how an actor binds themselves to a course of action and if that commitment is seen as credible	Inefficient outcomes Threats and extortions Rigidity and mistaken commitment	Ransomeware, During Operation Iraqi Freedom in 2003 an automated US missile defence system shot down a British plane, killing both occupants – here the human had one minute to override the system. For the Lavender terrorist targeting systems used in the Gaza war, the human had 30 seconds.
Emergent agency – new higher forms of agency from collective agents like superconductivity of material	Emergent capabilities Emergent goals	AI-first national strategies with a refocus on safety and security (and defense) instead of governance and ethics; Deepfakes like the Taylor Swift Presidential endorsement or the use of a voice similar to Scarlet Johannson; in labs adversial attacks can bypass safeguards of individual systems (Jones et al., 2024); Factual inaccuracies when LLMs are used for news and when used recursively, accuracy fell by 36% after the 8th rewrite (Sadeghi & Arvanitis, 2023) – in 2024, BBC complained to Apple, when it found that Apple Intelligence generated a false summary of one of its stories.
Multi-agent security - safeguarding complex networks of heterogeneous agents and the systems that they interact with (like data, critical infrastructure, etc)	Swarm attacks Heterogenous attacks Social engineering at scale Vulnerable AI agents Cascading security failures Undetectable threats	Tay the Microsoft chatbot that took 24 hours to become evil; *Solid Gold Bomb*, a brand hired a dictionary algorithm to sell variations of its T-shirt "Keep Calm and Carry on" which backfired horribly

Source: Adapted from Hammond et al. (2025)

AI HITL, Agency and Governance

Bandura (2001) highlights that an individual's agency is influenced by the selected, constructed, and imposed environment, in which the selected environment offers the greatest opportunities for agency. The ability to diffuse or displace responsibility is often an outcome of eight mechanisms, which AI and AI narratives can be used to justify (see Table 5). The movement away from moral standards that underlie governance mechanisms occurs when there is either a failure in control over behavior, agency, or outcomes, or a perception of being a victim. When there is a failure, human and technological actors are often so tightly interwoven that it becomes difficult to attribute responsibility to man and/or machine (Harbers, 2005).

Table 5. Failures in Control

Locus of Control	Mechanisms	Examples
Behavior locus	Moral justifications: sanctify harmful means by investing them with worthy social and moral purposes	Robot soldier in a war (save countrymen lives rather than enemy) – here the gap in agency and HITL is when there is a purpose that is translated into an algorithm and across the value chain.
	Palliative comparisons: advantageous comparison, which portrays the harm one inflicts as minor compared with the harm it prevents or that others commit	Gen AI to cure cancer irrespective of the means, equality of access or impact on planetary resources or future harms. Here again the agency and HITL begins from purpose and is translated through the value chain.
	Euphemistic language: Cloaks harmful behavior in innocuous language and removes humanity from it	Use of language like "Democratizing AI" or "AI creates jobs" when it creates uncertainty and removes long-term employability. The agency and HITL begins from purpose and is translated through the value chain.
Agency locus	People absolve themselves of personal accountability for harmful conduct by displacing responsibility to others and by dispersing it widely so that no one bears responsibility	Often the case in tort laws with big tech – example autonomous cars failure. The agency and HITL begins from purpose and is translated through the value chain.
Outcome locus	Perpetrators disregard, minimize, ignore, misconstrue, distort, or even dispute the injurious effects of their actions	Example: Autonomous cars – we save more lives by preventing accidents OR humans cause 'X' times more accidents. The agency and HITL begins from purpose and is translated through the value chain.
Victim locus	Perpetrators exclude those they maltreat from their category of humanity by divesting them of human qualities (dehumanization) or attributing qualities to them that justify the blame.	Examples in AI in warefare - where there have been documented cases of children, journalists, doctors and humanitarian workers targeted. Agency and HITL begins from purpose and is translated through the value chain. An example of attribution of blame - "they are terrorists."

Source: Adapted from Bandura (2016; 2018)

We suggest that, for Human Agency and HITL to come together, three conditions must be considered: power, trust, and complexity.

Power: Power is defined as a relation where agent A can get Agent B to do something that they would not otherwise do, either through influence, coercion, or manipulation (Dahl, 1957, p. 203; Raven, 1958; Radovan, 2015). Power has the ability to limit the freedom or agency of another (Lukes, 2002), but also to empower or disempower the actor seeking agency. There are two types of power: Type 1 is '*the power to*' have the ability to initiate and maintain a program of action, while Type 2 is the ability to act independently, having '*power over*', the constraining power of social structure (Campbell & Lukes, 1974, p. 30). The way we view the power divide between humans and AI affects HITL and governance mechanisms. This is a logical step-by-step process from the design of AI, development, deployment, and decommissioning (for example, see how it is applied to Lethal Autonomous

Weapons (Allik et al., 2024). As human power *over* AI increases with each stage, HITL becomes more likely. If it is only at some stages, it is an illusion of control and power.

Trust: Trust is the willingness to be vulnerable based on some positive expectations about the other (Vanneste & Puranam, 2024, p. 10). In technological societies, trust can be with a human (for the act, for example, an ATM machine for payment of a bill) or with the machine, where the human is a facade (a human who punches an order, but the machine executes the order and the human has little control over the process). Trust develops with an AI either because of the human-like attributes created (like mimicking human chatting), or because of the underlying systems-like nature of technologies (like an Uber app that connects pricing, location, maps, and illusion of control through choice), and this depends on the context (Lankton et al., 2015). Trust can be manipulated to give a perception of agency by giving autonomy to the machine or the human (via user interface or some decision choices), by providing explanations of how the AI works (transparency), or through anthropomorphization (giving human characteristics to the AI) (Vanneste & Puranam, 2024). Further, as humans trust the institution's or its representatives' reputation in providing the technology, trust can shift from humans to AI (Li et al., 2008). As we cede more trust to the machine, both behind the scenes and in front of the technology, we shift the HITL governance mechanisms to more AI Ethics Washing, as true attribution is not possible.

Complexity: In the world of AI, agency is a constellation of inter-agency, where "*it is the collective of people, machines, and programmes – in particular, the mode by which the agencies of the various elements are distributed and connected with one another*" (Rammert, 2012, p. 107). This conjoined or coproduced agency makes HITL extremely difficult, as we are unable to easily assess whether AI systems *intentionally or accidentally* constrain, complement, and/or substitute for humans in the practice of routines across four levels of human-AI system complexity (Murray et al., 2021, p. 552). These four co-produced agency types are conjoined agency with assisting technologies (AI cannot develop protocols nor select actions); conjoined agency with arresting technologies (AI cannot develop protocols but can select actions), conjoined agency with augmenting technologies (AI can develop protocols but not select actions), and conjoined agency with automating technologies (AI can develop protocols and select actions). As complexity increases and AI gains greater control over objectives, processes, and decision outcomes, we see a greater likelihood of AI Ethics Washing. We see these practical challenges of complexity in autonomous driving cars and the various levels of autonomy.

Figure 4. HITL and AI Ethics Washing

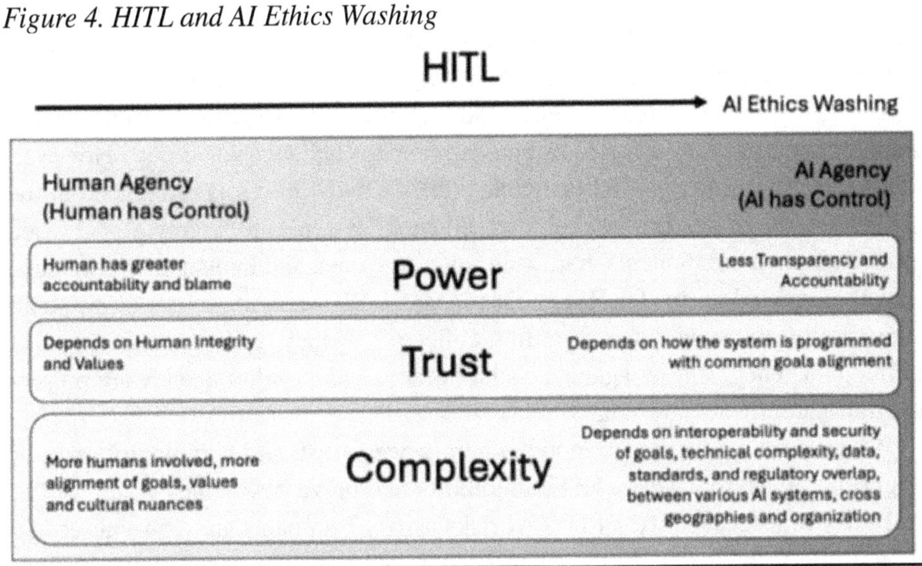

Source: Authors

There is a greater likelihood of AI Ethics Washing as power, trust, and complexity shift toward a situation in which AI has greater agency and hence control over human decision-making (see Figure 4). As defined earlier, AI Ethics Washing represents how businesses promote ethical frameworks or concepts for artificial intelligence (AI) in public without putting them into practice. This practice is easier to implement when the situation is complex and difficult for the user to understand and comprehend, and the human is not in the loop. Looking at trust, if a system does the task that it has been designed for with precision and predictable outcomes, it leads to the development of trust in the AI system, and may keep the the human-out-of- the loop. Looking at the power aspect, as AI tools gain more expertise and wield greater personal power due to their greater precision and predictability compared to humans, a shift in power and loss of transparency may lead to AI Ethics Washing.

CONCLUSION

While HITL finds different interpretations across global policies, its coherence is low. In the case of UNESCO AI Principles, it falls under human autonomy and oversight and states that Member States should ensure that AI systems do not displace ultimate human responsibility and accountability. Under the Council of Europe (2024, Art 7), it may be interpreted as "Human dignity and individual autonomy: Each Party shall adopt or maintain measures to respect human dignity and individidual autonomy in relations to activities within the lifecycle of the AI system (CoE, 2024, Art 7). Under the UNGAI (2018), it is the "Right for Human Determination - all individuals have the right to a final determination by a human". OECD (2018) highlights the need for "Transparency and Explainability: Provide information that enables those adversely affected by an AI system to challenge its output". However, it is not present in the UN AI Principles (2024), and while it is referenced in the EU AI Act, no guidelines are provided. Since AI systems are largely global when you look at their components, hardware, software, data, and human talent, we need globally accepted norms for HITL. This means acknowledging terminology so all actors are on the same page regarding collective agency, and then ensuring that individual agency, learner agency, social agency, and mobility agency are present and transparently acknowledged.

We propose a governance model where power, trust, and complexity serve as diagnostic pillars to evaluate HITL credibility and prevent AI ethics washing. The MIT Risk register identifies 1,600 AI risks across 7 domains and 65+ causes, using existing frameworks. In their latest report, AI poses more risks than humans (41%:39%), though a majority of failures are unintentional – 69% (Slatterly et al., 2025). Many of the issues are HITL problems, and these can be mitigated with proper policies and governance frameworks, provided there is a strong focus on AI literacy.

HITL, which is used in computer sciences to refer to the process of human agency in developing AI, has been extended in the popular press to encompass the AI value chain and the AI lifecycle. HITL also combines several terms, and many are used interchangeably. This chapter reviews both agency and HITL from a human-AI perspective. It defines *AI Ethics Washing*" as a phenomenon where institutions promote ethical frameworks or concepts for artificial intelligence (AI) in public while diverting human agency to AI systems, making ethical human responsibility an illusion. Further in this chapter, we identify challenges in maintaining human agency and suggest that governance of human-AI systems by humans is an outcome of three key factors – power, trust, and complexity, which need to be acknowledged and mitigated to prevent AI Ethics Washing.

REFERENCES

Albassam, W. A. (2023). The power of artificial intelligence in recruitment: An analytical review of current AI-based recruitment strategies. *International Journal of Professional Business Review*, *8*(6), 4.

Albrecht, S., Howe, K., & Romney, M. (1984). *Deterring fraud: The internal auditor's perspective*. Institute of Internal Auditors Research Foundation.

Allik, S., Azafrani, R., Barnes, D., Bode, I., Conn, A., Lach, E. M., Lennon, C., Liivoja, R., Shanahan, J., Stephens, M., Titus, L., & Wagner, A. R. (2024). *A framework for human decision-making through the lifecycle of autonomous and intelligent systems in defense applications*. IEEE Standards Association.

Amnesty International. (2025). *Automated racism: How police data and algorithms code discrimination into policing*. https://www.amnesty.org.uk/knowledge-hub/all-resources/automated-racism-report/

Amershi, S., Cakmak, M., Knox, W. B., & Kulesza, T. (2014). Power to the people: The role of humans in interactive machine learning. *AI Magazine*, *35*(4), 105–120.

Anderson, M., & Fort, K. (2022). Human where? A new scale defining human involvement in technology communities from an ethical standpoint. *International Journal of Information Ethics*, *31*(1).

Andrada, G., Clowes, R. W., & Smart, P. R. (2023). Varieties of transparency: Exploring agency within AI systems. *AI & Society*, *38*(4), 1321–1331.

Andrews, R. W., Lilly, J. M., Srivastava, D., & Feigh, K. M. (2023). The role of shared mental models in human–AI teams: A theoretical review. *Theoretical Issues in Ergonomics Science*, *24*(2), 129–175.

Ball, R. (2023). Algorithms, software and artificial intelligence. In *Viruses in all dimensions* (pp. 59–72). Springer.

Bandura, A. (1982). Self-efficacy mechanism in human agency. *The American Psychologist*, *37*(2), 122–147.

Bandura, A. (1989). Human agency in social cognitive theory. *The American Psychologist*, *44*(9), 1175–1184.

Bandura, A. (2001). Social cognitive theory: An agentic perspective. *Annual Review of Psychology*, *52*(1), 1–26.

Bandura, A. (2008). An agentic perspective on positive psychology. In Positive psychology: Exploring the best in people (pp. 167–196).

Bandura, A. (2016). *On structural evil: Disengaging from our moral selves. A review of Moral disengagement: How people do harm and live with themselves.* Worth Publishers., https://doi.org/10.1037/a0040160

Bandura, A. (2018). Toward a psychology of human agency: Pathways and reflections. *Perspectives on Psychological Science : A Journal of the Association for Psychological Science, 13*(2), 130–136. https://doi.org/10.1177/1745691617699280

Bennett, C. A. (1957). Some experimentation on the tie-in of the human operator to the control loop of an airborne navigational digital computer system. *IBM Journal of Research and Development, 3*(3), 275–282. https://doi.org/10.1147/rd.33.0275

Birmingham, H. P., & Taylor, F. V. (1954). A design philosophy for man–machine control systems. *Proceedings of the IRE, 42*(12), 1748–1758.

Blaschke, L. M., & Hase, S. (2019). Heutagogy and digital media networks. *Pacific Journal of Technology Enhanced Learning, 1*(1), 1–14.

Carabantes, M. (2020). Black-box artificial intelligence: An epistemological and critical analysis. *AI & Society, 35*(2), 309–317.

Carroll, M., Chan, A., Ashton, H., & Krueger, D. (2023). Characterizing manipulation from AI systems. In *Proceedings of the 3rd ACM Conference on Equity and Access in Algorithms, Mechanisms, and Optimization (EAAMO '23).* https://doi.org/10.1145/3617694.3623226

Cressey, D. R. (1953). *Other people's money: A study in the social psychology of embezzlement.* Free Press.

Cassell, J. (2001). Embodied conversational agents. *AI Magazine, 22*(4), 67–67.

Carnegie Council for Ethics in International Affairs. (2023). Ethics washing: Definition and introduction. https://www.carnegiecouncil.org/explore-engage/key-terms/ethics-washing

Crootof, R. (2023). Algorithmic accountability is even harder than you thought. *JOTWELL (Journal of Things We Like (Lots)).* https://cyber.jotwell.com/algorithmic-accountability-is-even-harder-than-you-thought/

Crootof, R., Kaminski, M. E., Price, W., & Nicholson, I. (2023). Humans in the loop. *Vanderbilt Law Review, 76*, 429–510.

Cross, E. S., Hortensius, R., & Wykowska, A. (2019). From social brains to social robots. *Philosophical Transactions of the Royal Society of London. Series B, Biological Sciences*, *374*, 20180024.

Da Conceição, E. (2010). Who controls whom? *Journal of Common Market Studies*, *48*(4), 1107–1126.

Dahl, R. A. (1957). The concept of power. *Behavioral Science*, *2*(3), 201–215.

Dreze, J., & Sen, A. (2002). *India: Development and participation*. Oxford University Press.

Emami, Y., Almeida, L., Li, K., Ni, W., & Han, Z. (2024). Human-in-the-loop machine learning. arXiv preprint arXiv:2408.12548.

Endsley, M. R. (2017). From here to autonomy. *Human Factors*, *59*(1), 5–27.

Etzioni, O., & Weld, D. S. (1995). Intelligent agents on the internet. *IEEE Expert*, *10*(4), 44–49.

European Commission. (2019). *Ethics guidelines for trustworthy AI*. https://digital-strategy.ec.europa.eu/en/library/ethics-guidelines-trustworthy-ai

Floridi, L. (2025). AI as agency without intelligence. *Philosophy & Technology*, *38*(1), 29–30.

Franklin, S., & Graesser, A. (1996). Is it an agent, or just a program? A taxonomy for autonomous agents. In *Proceedings of the Third International Workshop on Agent Theories, Architectures, and Languages (ATAL)*.

Fritz, A., Brandt, W., Gimpel, H., & Bayer, S. (2020). Moral agency without responsibility? Analysis of three ethical models of human–computer interaction in times of artificial intelligence (AI). *De Ethica*, *6*(1), 3–22.

Froese, T., & Ziemke, T. (2009). Enactive artificial intelligence. *Artificial Intelligence*, *173*(3–4), 466–500.

Gahnberg, C. (2021). What rules? *Policy and Society*, *40*(2), 194–210.

Garud, R., & Karnøe, P. (2005). Distributed agency and interactive emergence. In Innovating strategy process (pp. 88–96).

Gu, Y., Chang, Q., You, J., Li, L., Chen, X., & Chen, Y. (2025). A systematic survey of jailbreak attacks against large language models. In *Proceedings of the IEEE International Conference on Computer, Big Data, Artificial Intelligence (ICCBD+AI)* (pp. 1–7). IEEE.

Hancock, P. A., Billings, D. R., Schaefer, K. E., Chen, J. Y. C., De Visser, E. J., & Parasuraman, R. (2011). A meta-analysis of factors affecting trust in human–robot interaction. *Human Factors*, *53*(5), 517–527.

Harbers, H. (2005). *Inside the politics of technology*. Amsterdam University Press.

Hetzner, C. (2022, June 10). *Elon Musk's regulatory woes mount as U.S. moves closer to recalling Tesla's self-driving software*. Fortune. https://fortune.com/2022/06/10/elon-musk-nhtsa-investigation-traffic-safety-autonomous-fsd-fatal-probe/

Himma, K. E. (2009). Artificial agency, consciousness, and the criteria for moral agency: What properties must an artificial agent have to be a moral agent? *Ethics and Information Technology*, *11*(1), 19–29.

Holzinger, A. (2016). Interactive machine learning. *Brain Informatics*, *3*(2), 119–131.

Holzinger, A., Biemann, C., Pattichis, C. S., & Kell, D. B. (2017). What do we need to build explainable AI systems for the medical domain? *arXiv preprint arXiv:1712.09923*. https://arxiv.org/abs/1712.09923

Hughes, L., Dwivedi, Y. K., Malik, T., Shawosh, M., Albashrawi, M. A., Jeon, I., Dutot, V., Appanderanda, M., Crick, T., & De', R. (2025). AI agents and agentic systems: Redefining global IT management. *Journal of Computer Information Systems*, 1–29.

Jackson, S. (2025). Photos show how your Facebook feed has changed over the last 20 years. *Business Insider*. https://www.businessinsider.com/photos-how-your-facebook-feed-has-changed-over-years-2025-3

Jones, M. L. (2015). The Ironies of Automation Law: Tying Policy Knots with Fair Automation Practices Principles. *Vanderbilt Journal of Entertainment and Technology Law*, *18*(1), 77–152.

Just, N., & Latzer, M. (2017). Governance by algorithms. *Media Culture & Society*, *39*(2), 238–258.

Kabeer, N. (1999). Resources, agency, achievements. *Development and Change*, *30*(3), 435–464.

Lankton, N. K. (2015). Technology, humanness, and trust. *Journal of the Association for Information Systems*, *16*(10), 1–38.

Limerick, H., Coyle, D., & Moore, J. W. (2014). The experience of agency in human–computer interactions: A review. *Frontiers in Human Neuroscience*, *8*, 643. https://doi.org/10.3389/fnhum.2014.00643

List, C. (2021). Group agency and artificial intelligence. *Philosophy & Technology, 34*(4), 1213–1242.

Luhmann, N. (1995). *Social systems.* Stanford University Press.

Lukes, S. (1974). *Power: A radical view.* Macmillan.

Lukes, S. (2002). Power and agency. *The British Journal of Sociology, 53*(3), 491–496.

March, J. G., & Simon, H. A. (1993). *Organizations.* Wiley.

Matthias, A. (2004). The responsibility gap. *Ethics and Information Technology, 6*, 175–183.

Merat, N., Seppelt, B., Louw, T., Engström, J., Lee, J. D., Johansson, E., & Keinath, A.. (2019). The "out-of-the-loop" concept in automated driving: Proposed definition, measures and implications. *Cognition Technology and Work, 21*(1), 87–98.

McKay, M. H. (2024). Realizing the promise of AI governance involving humans-in-the-loop. In *Proceedings of the International Conference on Human-Computer Interaction.* Springer. https://doi.org/10.1007/978-3-031-76827-9_7

McKenzie Sadeghi, & Arvanitis, L. (2023). *Rise of the newsbots: AI-generated news websites proliferating online.* NewsGuard.

McLaughlin, A. C., Simon, D. A., & Gillan, D. J. (2010). From intention to input: Motor cognition, motor performance, and the control of technology. *Review of Human Factors and Ergonomics, 6*(1), 123–171.

Meng, X. L. (2023). Data science and engineering with human in the loop, behind the loop, and above the loop. *Harvard Data Science Review, 5*(2).

Ministry of Economic Affairs and Communications. (2021). *Estonia's national artificial intelligence strategy (Kratt strategy) for 2022–2023.* https://e964029b-7c32-42a7-817c-37c391a3b976.filesusr.com/ugd/980182_4434a890f1e64c66b1190b0bd2665dc2.pdf

Mosqueira-Rey, E., Hernández-Pereira, E., Alonso-Ríos, D., Bobes-Bascarán, J., & Fernández-Leal, Á. (2023). Human-in-the-loop machine learning: A state of the art. *Artificial Intelligence Review, 56*(4), 3005–3054. https://doi.org/10.1007/s10462-022-10246-w

Murray, A., Rhymer, J., & Sirmon, D. G. (2021). Humans and technology: Forms of conjoined agency in organizations. *Academy of Management Review, 46*(3), 552–571.

Neven, H. (2024). *Meet Willow quantum chip.* Google Blog.

Noller, J. (2024). Extended human agency. *Humanities & Social Sciences Communications, 11*(1), 1–7.

Nunes, D. S., Zhang, P., & Sá Silva, J. (2015). A survey on human-in-the-loop applications towards an Internet of All. *IEEE Communications Surveys and Tutorials, 17*(2), 944–965. https://doi.org/10.1109/COMST.2015.2398816

Nutt, S. (2025). *Humans in the loop: What should the role of officers be in AI-powered public services?* London Office of Technology and Innovation (LOTI). https://loti.london/blog/hil-how-can-and-should-officers-intervene-in-ai-powered-public-services/

Obhi, S. S., & Hall, P. (2011). Sense of agency. *Experimental Brain Research, 211*, 663–670.

Organisation for Economic Co-operation and Development. (2019). *OECD principles on artificial intelligence: Transparency and explainability*. https://oecd.ai/en/dashboards/ai-principles/P7

Personal Data Protection Commission. (2020). *Model AI governance framework* (2nd ed.). https://www.pdpc.gov.sg/-/media/Files/PDPC/PDF-Files/Resource-for-Organisation/AI/SGModelAIGovFramework2.pdf

Powell, W. W., & DiMaggio, P. J. (Eds.). (2012). *The new institutionalism in organizational analysis*. University of Chicago Press.

Porathe, T., & Prison, J. (2008). Design of human-map system interaction. In *CHI'08 extended abstracts on Human factors in computing systems* (pp. 2859-2864).

ProPublica. (2016). *Machine bias: Risk assessments in criminal sentencing*. https://www.propublica.org/article/machine-bias-risk-assessments-in-criminal-sentencing

Radovan, M. (2015). The power of manipulation. *International Journal of Technoethics, 6*(1), 71–82.

Rammert, W. (2008). Where the action is: Distributed agency between humans, machines, and programs. In Seifert, U., Kim, J. H., & Moore, A. (Eds.), *Paradoxes of interactivity: Perspectives for media theory, human-computer interaction, and artistic investigations* (pp. 62–91). Transcript Verlag. DOI: 10.14361/9783839408421-004

Rammert, W. (2012). Distributed Agency and Advanced Technology: Or: How to Analyze Constellations of Collective Inter-Agency 1. In *Agency without Actors?* (pp. 89-112). Routledge.

Raven, B. H. (1958). Legitimate power. *Sociometry, 21*(2), 83–97.

Reuters. (2025, April 1). *Ghibli effect: ChatGPT usage hits record after rollout of viral feature.* https://www.reuters.com/technology/artificial-intelligence/ghibli-effect-chatgpt-usage-hits-record-after-rollout-viral-feature-2025-04-01/

Ryan, R. M., & Deci, E. L. (2000). Self-determination theory. *The American Psychologist*, *55*(1), 68–78.

Sanders, M. S., & McCormick, E. J. (1988). *Human factors in engineering and design* (6th ed.). McGraw-Hill.

Santoni de Sio, F., & Mecacci, G. (2021). Responsibility gaps in AI. *Philosophy & Technology*, *34*(4), 1057–1084.

Schreiber, G. (2024). Reconsidering agency in the age of AI. *Filozofia, 79*(5).

Sebastián, M. Á. (2021). First-person representations and responsible agency in AI. *Synthese*, *199*(3), 7061–7079.

Sen, A. (1999). *Development as freedom.* Anchor Books.

Shavit, Y., Agarwal, S., Brundage, M., Adler, S., O'Keefe, C., Campbell, R., Lee, T., Mishkin, P., Eloundou, T., & Hickey, A. (2023). Practices for governing agentic AI systems. OpenAI. https://cdn.openai.com/papers/practices-for-governing-agentic-ai-systems.pdf

Sharkey, N. E. (2016, October 13). The evitability of autonomous robot warfare. *International Review of the Red Cross*. https://www.icrc.org/en/international-review/article/evitability-autonomous-robot-warfare

Sharkey, A., & Sharkey, N. (2012). Granny and the robots: Ethical issues in robot care for the elderly. *Ethics and Information Technology*, *14*(1), 27–40. https://doi.org/10.1007/s10676-010-9234-6

Slattery, P., Saeri, A. K., Grundy, E. A. C., Graham, J., Noetel, M., Uuk, R., Dao, J., Pour, S., Casper, S., & Thompson, N. (2024). *The AI risk repository: A comprehensive meta-review, database, and taxonomy of risks from artificial intelligence* (arXiv:2408.12622). *arXiv*. https://doi.org/10.48550/arXiv.2408.12622

Shneiderman, B., & Plaisant, C. (2010). *Designing the user interface.* Pearson.

Smith, G. C., Kohn, S. J., Savage-Stevens, S. E., Finch, J. J., Ingate, R., & Lim, Y. O. (2000). The effects of interpersonal and personal agency on perceived control and psychological well-being in adulthood. *The Gerontologist*, *40*(4), 458–468.

Spennemann, D. H. (2024). Generative artificial intelligence, human agency and the future of cultural heritage. *Heritage*, *7*(7), 3597–3609.

Stephens, M., Vashishtha, H., & Wagner, D. N. (2023). *AI enabled business: A smart decision kit*. Emerald Publishing.

Sundar, S. S., & Marathe, S. S. (2010). Personalization vs customization. *Human Communication Research, 36*(3), 298–322.

Swanepoel, D. (2021). Does artificial intelligence have agency? In *The mind-technology problem: Investigating minds, selves and 21st century artefacts* (pp. 83–104).

Tóth, Z., Caruana, R., Gruber, T., & Loebbecke, C. (2022). The dawn of the AI robots: Towards a new framework of AI robot accountability. *Journal of Business Ethics, 178*(4), 895–916.

United Nations. (2023). *Resolution A/78/L.49*. https://docs.un.org/en/A/78/L.49

Van Maanen, G. (2022). AI ethics and ethics washing. *Digital Society : Ethics, Socio-Legal and Governance of Digital Technology, 1*(2), 9.

Vanneste, B. S., & Puranam, P. (2024). Artificial intelligence, trust, and perceptions of agency. *Academy of Management Review*. Advance online publication. https://doi.org/10.5465/amr.2022.0041

Villani, C. (2018). *For a meaningful artificial intelligence: Towards a French and European strategy*. Mission Villani. https://www.aiforhumanity.fr/pdfs/MissionVillani_Report_ENG-VF.pdf

Von Eschenbach, W. J. (2021). Transparency and the black box problem. *Philosophy & Technology, 34*(4), 1607–1622.

Wagner, M. (2011). Taking humans out of the loop. Journal of Law. *Information and Science, 21*(2), 155–165.

Weick, K. E. (1995). Sensemaking in organizations. *Sage (Atlanta, Ga.).*

Westcott, M. (2025). *AI's trust imperative: The leadership test* [LinkedIn post]. LinkedIn. https://www.linkedin.com/posts/mrwestcott_ais-trust-imperative-the-leadership-test-share-7295823571178668034-MHev/

Wiethof, C., & Bittner, E. A. (2021). *Hybrid intelligence—Combining the human in the loop with the computer in the loop: A systematic literature review.*

Wingrove, M. (2018). *AI and automation could result in more ship accidents*. Riviera Maritime Media. https://www.rivieramm.com/news-content-hub/news-content-hub/ai-and-automation-could-result-in-more-ship-accidents-22750

Wooldridge, M., & Jennings, N. R. (1995). Intelligent agents. *The Knowledge Engineering Review, 10*(2), 115–152.

Yeung, K. (2018). Algorithmic regulation. *Regulation & Governance, 12*(4), 505–523.

Yoon, H. J. (2019). Toward agentic HRD. *Advances in Developing Human Resources, 21*(3), 335–351.

Zhang, J., & Bareinboim, E. (2022). Can humans be out of the loop? *Proceedings of Machine Learning Research, 177*, 1010–1025.

Zimmerman, L. A., Li, M., Moreau, C., Wilopo, S., & Blum, R. (2019). Measuring agency as a dimension of empowerment among young adolescents globally: Findings from the Global Early Adolescent Study. *SSM - Population Health, 8*, 100454.

ENDNOTE

1 Sadly, we have not really evolved in our discourse about repetitive and routine tasks since the 1951 "Fitts list" which identified humans' and machines' skills in the context of air traffic control.

Chapter 14
Policy, Regulation, and the Public Good of Artificial Intelligence

Robert Ipiin Gnankob
http://orcid.org/0000-0002-7177-1772
Woxsen University, India

Jayanta Kumar Mohapatra
Woxsen University, India

Samuel Etse Dzakpasu
http://orcid.org/0000-0002-2907-8069
University of Cape Coast, Ghana

ABSTRACT

This chapter explores how policy, regulation, and governance can align artificial intelligence with the public good in the era of industrial evolution. Drawing on public value theory, the responsible innovation framework, and stakeholder governance theory, it conceptualises a Policy–Regulation–Governance Compass to guide ethical and inclusive AI development. Through a comparative analysis of global approaches, from the European Union's AI Act and the U.S. AI Bill of Rights to emerging African and Asian strategies, the chapter reveals asymmetries in regulatory capacity and institutional readiness. It highlights persistent challenges, including regulatory lag, power concentration, and global inequality. The chapter proposes actionable policy pathways that emphasise adaptive frameworks, institutional capacity, inclusivity, and international cooperation. It argues that governance is not a constraint on innovation but an ethical enabler, transforming AI from a disruptive force into an instrument for sustainable and equitable development.

DOI: 10.4018/979-8-3373-7036-1.ch014

INTRODUCTION

The Industry 4.0 revolution has significantly elevated the role of artificial intelligence (AI) across sectors and industries worldwide. AI is now embedded in supply chain management, healthcare, manufacturing, logistics, energy systems, and public administration, reshaping how production, coordination, and decision-making occur in contemporary economies ((El Jaouhari et al., 2025; Wu et al., 2025). These developments have accelerated human–AI cognitive collaboration, enhancing efficiency, reducing operational errors, and enabling data-driven responses to challenges such as waste management and climate change. In economic terms, the World Economic Forum (2024) projects that AI will contribute over US$15.7 trillion to the global economy by 2030, underscoring its centrality to the next phase of industrial and economic evolution.

Despite this transformative potential, the diffusion of AI has sparked widespread public concern about its social, economic, and political consequences. Critics point to risks associated with algorithmic opacity, job displacement, surveillance, privacy violations, and the deepening of social and economic inequalities (Impito et al., 2025; Orwat et al., 2024). These concerns are compounded by uneven global preparedness. The 2024 Government AI Readiness Index reveals stark disparities in digital infrastructure, governance capacity, and human capital, with high-income countries averaging readiness scores of 73 out of 100, compared to just 35 among low-income countries (Oxford Insights, 2024). This gap mirrors earlier patterns of industrial inequality and signals that AI-driven industrial transformation may further entrench global asymmetries if left unguided.

Furthermore, the World Economic Forum (2024) global risks survey warns that "for better or for worse," the development of highly powerful AI technology will have a detrimental impact on the global economy over the next decade. According to the report, increased reliance on AI for strategic decision-making, including in security and conflict contexts, introduces new vulnerabilities that malicious actors may exploit. These concerns resonate beyond academic debate. At the 80th United Nations General Assembly, Ghana's President, H. E. John Dramani Mahama, cautioned that emerging digital technologies, including AI, pose significant threats to global peace and social cohesion. He observed that today's world is characterised by "cryptocurrency, Artificial Intelligence, social media, the Internet, and its dark, hidden dungeon—the dark web—all of which carry a potential threat to global peace and security". He further argues that "Artificial Intelligence offer us the illusion of connectivity, when in fact they reinforce isolation by using algorithms that ensure we do not receive new ideas and perspectives, but rather more of the same. We are served alternately facts and manipulated images, making it easier to disseminate disinformation and sow seeds of division" (Mahama, 2025, pp. 4-5).

These developments have intensified calls for what is increasingly described as "public AI." In this chapter, public AI is understood not as a replacement for private-sector innovation, but as a governance orientation in which AI systems, infrastructures, and standards are deliberately shaped to advance collective social, economic, and environmental objectives. Public AI encompasses regulatory frameworks, institutional capacities, and public-interest infrastructures that ensure AI remains transparent, accountable, and aligned with societal values rather than narrowly concentrated commercial interests (Taeihagh, 2025; Züger & Asghari, 2023). From this perspective, AI governance becomes a central component of industrial policy, influencing not only ethical outcomes but also competitiveness, market structure, and innovation trajectories.

A central challenge confronting policymakers is the "pacing problem"—the growing mismatch between the speed of technological innovation and the slower rhythms of lawmaking and institutional adaptation (Moses, 2011; Reuel & Undheim, 2024; Lachmann, 2025). As AI systems evolve rapidly across sectors, regulatory frameworks often lag behind, forcing governments into reactive rather than anticipatory modes of governance. This regulatory lag risks rendering existing legal instruments ineffective in addressing algorithmic bias, misuse, and concentration of power (Rajan, 2025; Olujobi & Ape, 2025). While some scholars argue that market forces may eventually stabilise these disruptions (Hermann, 2023), others caution that reliance on self-correction is insufficient in contexts characterised by network effects, data monopolies, and high barriers to entry (Duarte & Baptista, 2025; Hermann & Puntoni, 2025; Zwitter et al., 2024).

These debates highlight a broader governance dilemma: how to balance innovation with accountability to support inclusive industrial development. Without deliberate policy and regulatory intervention, AI adoption may widen gaps among nations, firms, and workers, undermining social cohesion and long-term development prospects (Abanga & Dotse, 2024). At the same time, effective governance can unlock AI's potential for the public good, enabling advances in environmental monitoring, healthcare delivery, disaster response, and social welfare. Realising these benefits requires institutional innovation alongside technological progress, embedding inclusivity, transparency, fairness, and sustainability as foundational governance principles (Züger & Asghari, 2023).

Against this backdrop, this chapter examines how policy and regulation intersect with industrial AI applications to safeguard and promote the public good. The public good is defined broadly, extending beyond economic efficiency to encompass social welfare, equity, democratic responsiveness, safety, environmental sustainability, and human dignity. The chapter advances five core objectives:

1. To map the conceptual and theoretical foundations for understanding policy, regulation, and the public good in industrial AI contexts;
2. To survey global AI policy landscapes and recent regulatory innovations;
3. To analyse governance frameworks and tools that align industrial AI development with public values;
4. To identify persistent challenges, including regulatory lag, enforcement gaps, and global inequalities; and
5. To propose policy recommendations that support human-centred, socially just, and sustainable industrial evolution.

Beyond theoretical contribution, the chapter offers a pragmatic framework for understanding AI governance as a driver of industrial transformation. Recent regulatory developments underscore the urgency of this analysis. The European Union's AI Act represents the first comprehensive, binding attempt to govern AI through a risk-based approach (Butt, 2024), while national initiatives such as Italy's AI legislation and sector-specific frameworks in healthcare illustrate both convergence and fragmentation in regulatory practice (Demaidi, 2025; Mügge, 2024; Reale & Biasin, 2023). However, no universal governance model exists. Instead, AI regulation reflects a mosaic of institutional philosophies shaped by political priorities, economic strategies, and administrative capacity (Kulothungan & Gupta, 2025).

To move beyond descriptive comparison, this chapter later adopts a structured analytical lens that examines global AI governance approaches across three fixed criteria, legislative type, risk model, and regulatory authority. This comparative perspective enables systematic assessment of how different governance architectures shape industrial AI adoption and public value creation. The remainder of the chapter is organised as follows: Section 2 develops conceptual definitions and theoretical lenses; Section 3 examines global approaches to AI policy and regulation; Section 4 proposes a policy–regulation–governance framework; Section 5 discusses implementation challenges; Section 6 advances policy recommendations; and Section 7 concludes with reflections on future directions for AI governance and industrial evolution.

DEFINITIONS AND THEORETICAL LENSES FOR AI GOVERNANCE

The rapid evolution of artificial intelligence has outpaced the assumptions underpinning traditional governance frameworks, compelling policymakers, scholars, and practitioners to rethink foundational concepts such as policy, regulation, and the public good. While these concepts are well established in public administration and regulatory studies, their application to AI governance is neither straightforward nor settled. AI systems are characterised by opacity, adaptability, cross-sectoral diffusion, and transnational reach, which challenge linear regulatory instruments and static institutional arrangements. This section therefore clarifies these core concepts and situates them within three complementary theoretical perspectives; namely, public value theory, the responsible innovation framework, and governance and stakeholder theory. These provide a conceptual foundation for analysing AI governance in the context of industrial evolution.

Policy

Policy refers to the strategic and normative framework through which governments articulate societal goals, priorities, and expectations for technological development. In the context of AI, policy defines what AI ought to achieve, identifying desired outcomes such as economic growth, competitiveness, social inclusion, environmental sustainability, and human well-being (Paraman & Anamalah, 2023). Unlike conventional industrial policy, AI policy operates across multiple domains simultaneously, such as innovation, labour markets, data governance, security, ethics, and public services, making coherence and coordination challenging.

Classic policy scholarship reminds us that technology is never neutral. Lessig's seminal argument that "code is law" emphasises how technological architectures themselves regulate behaviour, often more effectively than formal legal rules (Lessig, 2000). Applied to AI, this insight highlights that policy choices are embedded not only in statutes and strategies but also in data infrastructures, algorithmic design, and platform architectures. Similarly, research on regulation emphasises that modern governance increasingly relies on rule-setting and oversight rather than direct state control, a dynamic that is amplified in AI-driven economies where private actors dominate innovation (Majone, 1999).

AI policy thus functions as both a vision-setting instrument and a coordination mechanism. National strategies such as Japan's Society 5.0 and India's AI for All illustrate how governments frame AI as a tool for inclusive development and industrial transformation, embedding human-centred goals alongside competitiveness objectives (Jecker & Nakazawa, 2022). However, without effective regulatory and governance mechanisms, policy risks remaining aspirational rather than operational.

Regulation

If policy establishes direction, regulation provides the enforceable rules that shape permissible and impermissible uses of AI. Regulation encompasses legal statutes, administrative standards, monitoring mechanisms, and sanctions designed to mitigate risks associated with technological development (Pantanowitz et al., 2024; Clarke, 2019). In AI governance, regulatory challenges stem from definitional ambiguity, rapid innovation cycles, and the difficulty of attributing responsibility for algorithmic outcomes.

Foundational regulatory theorist (Baldwin et al., 2011) emphasise that effective regulation requires clarity of objectives, proportionality, and enforceability. These criteria are relatively difficult to meet when regulating adaptive, opaque AI systems. Economic theories of regulation further caution against regulatory capture and uneven enforcement, particularly in sectors dominated by powerful corporate actors (Peltzman, 1976; Stigler, 2021). These risks are heightened in AI markets characterised by strong network effects and data concentration.

On the other hand, contemporary AI regulation increasingly adopts risk-based approaches that focus on potential harm rather than rigid technological definitions. For instance, the European Union's AI Act exemplifies this model by classifying systems by risk level and imposing differentiated obligations. This approach aligns with Collingridge (1980) insight that early-stage technologies are difficult to control precisely, thereby necessitating adaptive, iterative regulatory strategies. Regulation, therefore, acts as a mediating layer between policy ambition and operational practice, translating ethical aspirations into enforceable safeguards while remaining flexible enough to evolve alongside technological change.

The Public Good

Traditionally, the public good has been defined in economic terms as non-rivalrous and non-excludable goods, such as clean air or national defence. In the AI era, this conception requires expansion. AI systems generate both public value and public risk: they can improve service delivery, optimise resource use, and enhance safety, while simultaneously entrenching bias, surveillance, and exclusion if poorly governed (Züger & Asghari, 2023; Misra et al., 2020).

Scholars of science and technology studies have long argued that technologies embody political choices. Winner (1980) provocative question; Do artefacts have politics? is particularly relevant for AI, where algorithmic decisions shape access to opportunities, information, and resources. Jasanoff (2004) similarly emphasises that governance of science and technology is inseparable from questions of legitimacy, power, and public trust. From this perspective, the public good in AI governance cannot be reduced to efficiency gains or innovation outputs alone; it must include fairness, accountability, transparency, and democratic participation.

Public AI, as conceptualised in this chapter, therefore refers to governance arrangements that ensure AI systems contribute to collective well-being rather than serving narrow private interests. This involves embedding public values into data practices, algorithmic design, and institutional oversight, and recognising that societal benefits depend on who defines the public interest and how competing values are negotiated (Chen et al., 2023).

Theoretical Foundations for AI Governance

To analyse how policy, regulation, and the public good interact in AI governance, this chapter draws on three complementary theoretical perspectives: public value theory, the responsible innovation framework, and governance and stakeholder theory.

Public Value Theory

Public value theory (PVT), originally articulated by Moore (1997), argues that public institutions should aim to create value for society beyond mere compliance with rules or efficiency targets. Over time, this framework has evolved to emphasise legitimacy, trust, and co-creation with citizens (Bryson et al., 2017; Hartley et al., 2017). Applied to AI, PVT reframes governance as a deliberate effort to align technological systems with societal priorities such as non-discrimination, privacy, and inclusivity.

In industrial AI contexts, this perspective highlights that decisions about automation, data use, and algorithmic deployment are inherently normative. The EU's emphasis on "trustworthy AI" and Singapore's AI Verify framework exemplify attempts to operationalise public value principles by embedding transparency and accountability into AI systems. Legitimacy, from this viewpoint, arises not only from procedural compliance but from outcomes that citizens perceive as fair and beneficial (Donatz-Fest, 2025).

Responsible Innovation Framework

The responsible innovation framework (RIF), developed within science and technology studies, complements public value theory by offering a process-oriented approach to governance. RIF emphasises anticipation, reflexivity, inclusion, and responsiveness throughout the innovation lifecycle (Stefanini & Vignali, 2024; Stilgoe et al., 2014). In AI governance, this translates into integrating ethical reflection and stakeholder engagement from design through deployment and decommissioning.

Practical applications of RIF include algorithmic impact assessments, regulatory sandboxes, and ethics-by-design methodologies. These instruments allow policymakers and innovators to identify risks early, experiment under controlled conditions, and adapt governance responses as technologies evolve. By operationalising public values, RIF addresses the pacing problem by fostering adaptive rather than reactive regulation.

Governance and Stakeholder Theory

Governance and stakeholder theory adds a relational and institutional dimension to AI governance. Building on the foundational work (Freeman, 1984), this perspective views legitimacy as emerging from the balanced consideration of multiple stakeholder interests, including governments, firms, civil society, and affected communities. AI governance, from this view, cannot be monopolised by the state or the market; it requires collaborative, multi-level arrangements that distribute authority and responsibility.

In view of this, trust becomes a central governance resource (Birkstedt et al., 2023). Initiatives such as the Global Partnership on AI illustrate how multi-actor coordination can shape global norms while accommodating local contexts. At the industrial level, stakeholder governance highlights the importance of aligning regulatory objectives with innovation ecosystems, labour markets, and public expectations.

Accordingly, these theoretical perspectives reveal that AI governance for the public good is not a linear or purely technical process. Public value theory clarifies the values AI should serve; the responsible innovation framework explains how those values can be embedded in technological development; and governance and stakeholder theory situate these efforts within complex institutional and political ecosystems. This integrated lens underpins the comparative and policy analysis that follows, providing a foundation for examining how different regulatory architectures shape industrial AI trajectories and public value outcomes.

COMPARATIVE APPROACH AND ASSESSMENT CRITERIA

To ensure systematic comparison, this chapter adopts a comparative qualitative policy analysis approach to examine AI governance frameworks across regions. The comparative method is particularly appropriate for emerging technologies such as artificial intelligence, where regulatory experimentation, institutional diversity, and uneven capacities shape governance outcomes across jurisdictions (Baldwin et al., 2011; Majone, 1999). The analysis focuses on selected regions and countries that are influential in shaping global AI governance norms and industrial trajectories: the European Union, the United States, China, selected Asian economies (Japan and India), Africa, and Latin America. These cases were chosen for three reasons. First, they represent distinct governance philosophies, ranging from rights-based regulation to market-driven and state-led models. Second, they differ significantly in institutional capacity and economic structure, allowing insight into how governance affects industrial competitiveness. Third, these jurisdictions exert normative or market influence beyond their borders, shaping global standards through regulatory diffusion or technological dominance (Braithwaite & Drahos, 2000).

To ensure consistency and avoid descriptive fragmentation, the chapter compares governance frameworks using three fixed analytical criteria, applied uniformly across regions: 1) Legislative type – whether AI governance is anchored in binding legislation (e.g., EU AI Act), executive instruments, sectoral rules, or voluntary guidelines; 2) Risk model – how AI risks are conceptualised and managed, including risk-based tiering, principles-based regulation, sectoral oversight, or ex post liability approaches; and 3). Regulatory authority and enforcement capacity – which institutions oversee AI governance, the tools available for enforcement, and the extent of sanctions, auditing powers, and redress mechanisms?.

Empirical material is drawn from primary policy and legislative documents, international datasets (e.g., Oxford Insights Government AI Readiness Index), reports from multilateral organisations (OECD, UNESCO, World Economic Forum), and peer-reviewed academic literature. This approach allows the chapter to assess the design of governance frameworks and their practical feasibility, economic implications, and institutional constraints.

Global Approaches to AI Policy and Regulation

Comparative Synthesis of Global AI Governance Models

Several key perspectives shape the debate on AI and the public good. Dystopian views of AI see it as a disruptive force that could outpace humans and threaten jobs, while utopian perspectives view it as a positive partner that boosts human welfare by reducing labour and elevating living standards (Cools et al., 2024). Optimists and realists take a more practical view, acknowledging both productivity improvements and workforce changes that call for educational reforms. Conversely, sceptics doubt that growth driven by AI will lead to fair economic or social advancement (Chang & Ke, 2024; Cools et al., 2024; Naudé, 2021). The global debate on AI policy and regulation reflects diverse socio-political traditions, economic objectives, and ethical perspectives. While Europe has prioritised risk management and human rights, North America leans toward innovation-led, market-driven flexibility. Asia presents a mixture of state-driven, hybrid, and inclusive models, whereas Africa and Latin America remain in the formative stages of AI governance. This global divergence underscores that AI governance is not only a technical or legal issue but also a geopolitical and developmental one, shaped by each region's capacity, institutional culture, and strategic priorities (Taeihagh, 2025).

Across the continents, governments, regional blocs, and international organisations are grappling with how to strike a balance between encouraging innovation and mitigating risks (Liu et al., 2025). While Europe has pioneered binding legislative frameworks, other regions have opted for voluntary guidelines, sectoral regulations, or national strategies to guide AI evolution. In parallel, the industrial evolution driven by AI is creating new opportunities for economic growth, while also disrupting labour markets, raising ethical dilemmas, and exposing governance gaps. Understanding these global policy dynamics alongside AI's industrial transformation offers a window into the opportunities and tensions that shape its trajectory toward or away from the public good.

Regional and Multilateral AI Governance Frameworks

The *European Union's* AI Act (European Union, 2024) is the world's first comprehensive legislative framework on AI grounded in a risk-based approach, and categorises AI systems into prohibited, high-risk, limited-risk, and minimal-risk applications (Lawniczak & Benzmüller, 2025; Van Kolfschooten & Van Oirschot, 2024). The Act imposes strict restrictions on the use of biometric surveillance in public spaces. AI in healthcare and transportation is also subject to rigorous compliance standards, categorised as high-risk. This approach aimed to establish the principle of "trustworthy AI" by incorporating features that promote transparency, accountability, and human oversight. In addition to the above, the framework addresses the need for technical robustness and safety, as well as privacy, data governance, diversity, non-discrimination, fairness, and societal and environmental well-being, thereby setting a global benchmark for data protection regulations.

For *North America*, the United States has enacted a decentralised and voluntary approach to the regulation (Roberts et al., 2023). Its AI Bill of Rights outlines principles for safe, transparent, and accountable AI, but it is described as a non-binding framework or a "white paper" rather than an enforceable law (Van Bekkum & Borgesius, 2023; Roberts et al., 2023). They have rather chosen to complement the Bill with initiatives that include the National Institute of Standards and Technology (NIST) AI Risk Management Framework, which provides guidelines and a "playbook" for identifying, assessing, and mitigating AI risks across four functional areas: *Govern, Map, Measure, and Manage* (Swaminathan & Danks, 2024). Canada has taken a more structured path through its Directive on Automated Decision-Making, which applies to federal agencies and requires algorithmic impact assessments to ensure accountability and transparency in the use of AI systems. These frameworks reveal a preference for flexibility and innovation, reflecting North America's reliance on private-sector leadership in AI development.

Latin America's approach is quite nascent and sector-specific. Although still in the early stages of AI governance, with varying levels of progress across countries, Brazil, for instance, has debated and approved a draft AI law, Bill No. 2338/2023, which aims to adopt a risk-based approach similar to the EU AI Act. What is exceptional for Brazil is the clause establishing a dedicated regulatory authority to monitor and facilitate the broader application of AI (Schertel Mendes & Kira, 2023). Mexico has relied more on sector-specific regulations, for instance, in finance and telecommunications, and has seen various initiatives for AI regulation in the financial sector. Regional organisations, such as the Inter-American Development Bank, have promoted responsible AI; however, there is no unified regional framework, as AI adoption and regulatory development are still in an early stage.

In *Asia*, China has introduced the Internet Information Service Algorithm Recommendation Management Regulations, which also focus on increasing transparency and user control, and mandate adherence to an ethical code online (Cao & Meng, 2025). Likewise, Japan is relying on "soft law" through guidelines like the Social Principles of Human-Centric AI, prioritising ethics, human dignity, and societal benefit without heavy-handed regulation. India, through its AI for All Strategy, emphasises the inclusive and sustainable use of AI, focusing on agriculture, health, and education, aiming to lead in AI research and applications while reducing social inequality (Ally & Perris, 2022; Bajpai & Wadhwa, 2021). These approaches reflect varying governance philosophies: while China demonstrates assertive state control, Japan and India adopt more collaborative models, seeking to balance growth with social priorities.

Africa's approach can be seen as fragmented and emerging, given that there is yet no trace of a continental-wide AI governance framework. We would rather see AI being acknowledged as an important driver of digital transformation in the African Union Digital Transformation Strategy (2020-2030). The Strategic Framework highlights AI's potential to guide and coordinate Africa's digitalisation agenda, rather than to provide a harmonised policy and legal framework to regulate AI use (Mishra & Kugler, 2024). Nonetheless, a few countries in Africa have begun to act: Rwanda has positioned itself as a continental AI hub through partnerships, such as those with the Centre for the Fourth Industrial Revolution, while South Africa has convened expert commissions on AI ethics and innovation. However, most African nations lack coherent national strategies, resulting in a fragmented landscape where opportunities for leveraging AI for the public good are scarce.

Table 1 below shows the governments and regional efforts towards AI regulation for the public good. It clarifies the areas of focus for each framework, as well as the governance approach, strengths, and challenges.

Table 1. Comparative overview of global AI regulation frameworks

Region/ Continent	Key Frameworks/Policies	Governance Approach	Strengths	Gaps/Challenges
Europe (EU)	EU AI Act (2024); GDPR; Coordinated Plan on AI	Risk-based, binding legislation with strict compliance mechanisms	Comprehensive, human-centric, global benchmark	Compliance costs; risk of stifling innovation
North America (US & Canada)	US: AI Bill of Rights (2022), NIST AI Risk Management Framework (2023); Canada: Directive on Automated Decision-Making (2019)	Voluntary, sectoral, and principle-driven	Encourages innovation; strong private-sector role	Fragmented; lack of federal-level binding laws
Asia (China, Japan, India)	China: Algorithm Regulations (2022); Japan: Social Principles of Human-Centric AI (2019); India: AI for All Strategy (2021)	State-led in China; soft-law in Japan; inclusive strategy in India	Adaptive, diverse, focus on national priorities	Risk of state overreach (China); weak enforcement (Japan, India)
Africa	AU Digital Transformation Strategy (2020–2030); national efforts (e.g., Rwanda, South Africa)	Emerging, fragmented, exploratory	Focus on digital development; early AI hubs are emerging	No continental AI framework; weak institutional capacity
Latin America	Brazil: Draft AI Bill; Mexico: Sectoral regulations (finance, telecom); IDB initiatives	Nascent, sectoral, and experimental	Early alignment with global debates; regional cooperation potential	Fragmented; lack of binding continental standards
Global (Multilateral)	UNESCO Recommendation on Ethics of AI (2021); OECD AI Principles (2019); G20 AI Principles	Norm-setting, non-binding, cooperative	Provides global ethical anchors; fosters collaboration	Non-binding; limited enforcement mechanisms

Source: Compiled by authors

Structured Comparison of Global AI Governance Approaches

Besides the descriptive comparison in Section 3.2, Table 2 systematically analyses global AI governance approaches using three fixed criteria that recur across jurisdictions. They are, 1) Legislative type – whether governance relies on binding law, soft law, or voluntary standards; 2) Risk model – how risks are conceptualised, classified, and managed; and 3) Regulatory authority – which institutions are empowered to oversee, enforce, and coordinate AI governance. Again, Table 2 builds on Table 1 by systematically comparing regions across the three fixed criteria.

Table 2. Comparative Matrix of Global AI Governance Approaches

Region	Legislative Type	Risk Model	Regulatory Authority
Europe (EU)	Binding, comprehensive legislation	Explicit risk-based classification	Hybrid: EU coordination + national authorities
North America	Voluntary, sectoral, principle-based	Risk management and mitigation	Fragmented, sector-specific agencies
Asia (China)	Binding, state-led regulation	Social stability and security-focused	Centralised state authority
Asia (Japan/ India)	Soft law and strategic frameworks	Ethics- and inclusion-oriented	Distributed, advisory-led
Africa	Emerging strategies, limited legislation	Implicit, development-focused	Weak or nascent institutions
Latin America	Draft laws and sectoral rules	Mixed, experimental	Fragmented and evolving

Source: Compiled by authors

This comparison reveals that global AI governance is shaped less by technological necessity than by political economy, institutional capacity, and strategic priorities. Binding regulation, voluntary standards, and hybrid models each entail trade-offs for industrial innovation, competitiveness, and the public good. The diversity of governance approaches raises a critical question: how can policy, regulation, and governance be aligned to steer AI-driven industrial evolution toward the public good despite global fragmentation? The next section addresses this question by advancing an integrative Policy–Regulation–Governance Compass, translating comparative insights into a coherent governance framework.

ECONOMIC DIMENSIONS OF AI GOVERNANCE FOR INDUSTRIAL EVOLUTION

While ethical, legal, and social concerns dominate much of the contemporary discourse on AI governance, the economic dimensions of regulation remain equally central, particularly in the context of industrial evolution (Taeihagh, 2025). AI governance does not occur in a vacuum; it reshapes incentives, redistributes costs and benefits, and alters competitive dynamics across industries and regions (Stigler, 2021; Peltzman, 1976). Understanding how governance frameworks interact with economic structures is therefore essential for evaluating whether AI regulation advances the public good or inadvertently entrenches inequality, market concentration, or regulatory capture (Majone, 1999; Braithwaite & Drahos, 2000).

Regulatory Costs, Compliance Burdens, and Industrial Adjustment

One of the most immediate economic effects of AI governance lies in regulatory compliance costs. These include direct costs—such as algorithmic impact assessments, documentation, auditing, data governance infrastructure, and legal oversight—as well as indirect costs, including delayed deployment, uncertainty in investment decisions, and organisational restructuring. Studies on digital and data regulation suggest that compliance costs can range from relatively modest administrative expenses for low-risk systems to substantial financial outlays for high-risk AI applications, particularly in sectors such as healthcare, finance, and critical infrastructure (Baldwin et al., 2011; Maslej et al., 2023).

From an industrial perspective, these costs are unevenly distributed. Large multinational firms often possess the financial, legal, and technical capacity to absorb compliance requirements, whereas small and medium-sized enterprises (SMEs) and start-ups face proportionally higher burdens (Taeihagh, 2025; Brynjolfsson & McAfee, 2014). This asymmetry raises concerns that poorly designed governance frameworks may unintentionally raise barriers to entry, reducing innovation diversity while favouring incumbent firms—an outcome consistent with classical theories of regulatory capture and market concentration (Stigler, 2021).

However, regulatory costs should not be interpreted solely as economic losses. From a public value perspective, compliance expenditures can be reframed as investments in trust, safety, and long-term system resilience (Moore, 1997; Braithwaite & Drahos, 2000). Empirical evidence from data protection and financial regulation regimes suggests that firms operating under clearer regulatory expectations often demonstrate stronger long-term investment stability and consumer trust, even when short-term compliance costs are high (Baldwin et al., 2011).

Innovation, Competitiveness, and Regulatory Design

A persistent concern in AI policy debates is whether regulation stifles innovation. Market-oriented perspectives argue that excessive governance slows technological diffusion, discourages investment, and undermines industrial competitiveness (Taeihagh, 2025; Stigler, 2021). Conversely, governance-oriented perspectives maintain that innovation without regulation generates negative externalities, including social harm, systemic risk, and public backlash, ultimately undermining market confidence and long-term growth (Collingridge, 1980; Jasanoff, 2004).

The relationship between AI governance and innovation is therefore non-linear and contingent on regulatory design (Moses, 2011). Risk-based frameworks, such as those embedded in the EU AI Act, seek to balance innovation and protection by calibrating regulatory intensity according to potential harm (Veale & Borgesius, 2021; Gstrein et al., 2024). Low-risk applications face minimal oversight, while high-risk systems—such as biometric surveillance or clinical decision tools—are subject to stricter controls.

Industrial competitiveness is also shaped by regulatory credibility. Jurisdictions that establish trusted AI ecosystems may gain a comparative advantage by attracting responsible investment, ethical AI research, and international partnerships (Majone, 1999; Mazzucato, 2018). Conversely, regulatory fragmentation or weak enforcement can deter long-term investment despite short-term innovation gains (Braithwaite & Drahos, 2000).

Market Concentration, Platform Power, and Economic Governance

AI governance also intersects with questions of market concentration and platform dominance. The economics of contemporary AI, particularly large-scale machine learning and foundation models, favour firms with access to massive datasets, computing infrastructure, and capital (Maslej et al., 2023; Brynjolfsson & McAfee, 2014). As a result, a small number of technology corporations increasingly control core layers of AI development.

Without appropriate regulatory oversight, dominant firms may shape technical standards, influence policy agendas, and establish de facto governance regimes, a phenomenon that echoes earlier concerns about private regulation in global markets (Lessig, 2000; Braithwaite & Drahos, 2000). Effective AI governance must therefore integrate competition policy, antitrust enforcement, data access rules, and interoperability standards as core economic governance tools (Majone, 1999; Sinha & Lee, 2024).

Regulatory Capture and the Political Economy of AI Governance

Another critical economic dimension concerns regulatory capture, the risk that powerful firms influence governance processes in ways that favour their interests (Stigler, 2021). AI governance is particularly vulnerable due to information asymmetries between regulators and technology providers, reinforcing dependence on industry expertise (Baldwin et al., 2011). Institutional design responses include independent oversight bodies, transparency mandates, stakeholder participation, and periodic regulatory review mechanisms (Donatz-Fest, 2025; Braithwaite & Drahos, 2000). Including civil society, labour representatives, and academic experts helps counterbalance concentrated economic power and reinforces democratic legitimacy (Jasanoff, 2004).

Distributional Effects and the Global Industrial Divide

AI governance also shapes the global distribution of economic benefits. High-income economies possess greater institutional capacity to design and enforce sophisticated governance frameworks, while low- and middle-income countries face regulatory, infrastructural, and human capital constraints (Oxford Insights, 2024; Abanga & Dotse, 2024). This asymmetry aligns with concerns about data colonialism, where AI systems developed in the Global North embed values, standards, and economic benefits that marginalise other regions (Couldry & Mejias, 2019). Inclusive governance strategies such as regional coordination, capacity-building, and shared regulatory infrastructures are therefore essential for equitable industrial participation (Züger & Asghari, 2023).

SECTORAL APPLICATIONS OF AI GOVERNANCE IN INDUSTRIAL CONTEXTS

While AI governance debates often remain at an abstract or cross-sectoral level, the realities of regulation, risk, and public value materialise most clearly within specific industries. Industrial sectors differ markedly in their risk profiles, data sensitivities, ethical stakes, and societal consequences. As such, governance frameworks that are effective in one domain may be inappropriate or insufficient in another (Baldwin et al., 2011; Collingridge, 1980). This section examines how AI governance operates across key industrial sectors, healthcare, manufacturing, finance, transportation, and public services, highlighting sector-specific risks, regulatory responses, and implications for the public good.

Healthcare and Life Sciences

Healthcare represents one of the most sensitive and high-stakes domains for AI deployment. AI systems are increasingly used for diagnostic imaging, clinical decision support, patient triage, and drug discovery. While these applications promise efficiency gains and improved health outcomes, they also raise profound concerns regarding safety, accountability, bias, and equity (Topol, 2019).

From a governance perspective, healthcare AI is typically classified as high-risk due to its direct impact on human life and dignity. Regulatory approaches, therefore, emphasise rigorous validation, transparency, and human oversight. In the European

Union, AI systems used in medical diagnosis or treatment fall under the high-risk category of the EU AI Act, requiring conformity assessments, post-market monitoring, and explainability obligations (Veale & Borgesius, 2021; Gstrein et al., 2024). In the United States, the Food and Drug Administration (FDA) regulates AI-enabled medical devices through adaptive approval pathways that acknowledge the continuous learning of these systems (Food and Drug Administration, 2023).

The public good dimension in healthcare AI governance centres on patient safety, informed consent, and equitable access. Empirical studies have shown that poorly governed clinical AI systems can reproduce racial, gender, and socioeconomic biases embedded in training data, leading to unequal diagnostic accuracy and treatment outcomes (Obermeyer et al., 2019; Barocas et al., 2021). Effective governance in this sector, therefore, requires not only technical accuracy but also ethical auditing, clinical accountability, and trust-building mechanisms with patients and practitioners.

Manufacturing and Industrial Production

In manufacturing, AI is deeply embedded in predictive maintenance, quality control, robotics, supply chain optimisation, and smart factory systems. Unlike healthcare, risks in this sector are less about individual rights and more about workplace safety, labour displacement, and systemic resilience.

AI-driven automation can significantly improve productivity and reduce waste, but it also transforms labour relations, skill requirements, and occupational safety dynamics (Brynjolfsson & McAfee, 2014). Governance challenges in manufacturing, therefore, revolve around human–AI collaboration, responsibility allocation, and the protection of workers in increasingly automated environments. Industrial AI systems often operate as socio-technical assemblages, in which accountability is distributed among software developers, equipment manufacturers, organisational managers, and human operators. This complicates traditional liability frameworks and necessitates governance models that emphasise shared responsibility and safety certification (Baldwin et al., 2011).

From an industrial policy perspective, governance in manufacturing AI is closely tied to competitiveness. Jurisdictions that establish clear safety standards, certification regimes, and interoperability norms can accelerate responsible adoption while avoiding catastrophic failures. Weak governance, by contrast, risks unsafe working conditions, skills erosion, and social backlash. For the public good, manufacturing AI governance must balance efficiency gains with worker protection, skills development, and long-term industrial sustainability.

Financial Services and Algorithmic Markets

The financial sector was among the earliest adopters of algorithmic decision-making, using AI for credit scoring, fraud detection, algorithmic trading, and customer profiling. These systems shape access to capital and economic opportunity, making governance particularly critical (Barocas et al., 2021). AI governance in finance typically emphasises fairness, transparency, and systemic stability. Discriminatory credit algorithms, opaque risk models, and high-frequency trading systems can amplify inequality or destabilise markets if left unchecked (Raso et al., 2018).

From a public goods perspective, governance of financial AI is inseparable from social inclusion. Access to credit, insurance, and financial services directly affects livelihoods and economic mobility. Poorly governed AI systems can entrench exclusion, while well-designed regulatory frameworks can promote trust, fairness, and market stability.

Transportation and Autonomous Systems

AI governance in transportation, particularly autonomous vehicles and intelligent traffic systems, highlights the intersection of safety, liability, and public infrastructure. Autonomous systems operate in complex, uncertain environments where failures can cause immediate, highly visible harm. Regulatory approaches in this sector often combine experimental governance with strict safety oversight. Many jurisdictions rely on regulatory sandboxes, pilot zones, and phased deployment strategies to test autonomous technologies under controlled conditions (Reuel & Undheim, 2024). Liability frameworks remain contested, particularly regarding responsibility allocation between manufacturers, software developers, and operators (Bertolini, 2023).

The public good dimension in transportation AI includes road safety, environmental sustainability, and equitable urban mobility. While autonomous systems have the potential to reduce emissions and congestion, governance failures risk eroding public trust and delaying adoption. This sector illustrates the importance of adaptive regulation that evolves alongside technological capabilities.

Public Services and Administrative Decision-Making

Governments increasingly deploy AI in welfare allocation, taxation, immigration control, predictive policing, and administrative decision-making. These applications directly affect citizens' rights and access to public goods, making legitimacy and accountability paramount (Keddell, 2019).

Governance challenges in public-sector AI include transparency, procedural fairness, and the right to explanation. Automated decision systems used in public administration raise constitutional and democratic concerns, particularly when individuals cannot contest or understand outcomes (Veale & Borgesius, 2021). Jurisdictions now require algorithmic registers, impact assessments, and human-in-the-loop safeguards for public-sector AI systems. AI systems deployed by the state must exemplify fairness, accountability, and inclusivity, as failures here undermine institutional trust and democratic legitimacy (Bryson et al., 2017).

Cross-Sectoral Lessons for Industrial AI Governance

Across these sectors, three governance lessons emerge. First, one-size-fits-all regulation is ineffective; sectoral risk profiles demand differentiated governance responses (Collingridge, 1980). Second, human oversight remains essential, particularly where AI decisions have irreversible or rights-affecting consequences. Third, industrial AI governance must integrate economic, ethical, and institutional considerations, rather than treating them as separate domains (Majone, 1999; Baldwin et al., 2011). By grounding AI governance in sector-specific realities, policymakers can better align regulation with industrial evolution and the public good. This sectoral lens strengthens the chapter's core argument: that effective AI governance is not merely about abstract principles, but about how rules shape real-world human–AI cognitive collaboration across industries.

GOVERNANCE MECHANISMS, CAPACITY, AND DEMOCRATIC LEGITIMACY

Implementation, Enforcement, and Institutional Capacity in AI Governance

While policy frameworks and regulatory principles provide essential direction for AI governance, their effectiveness ultimately depends on how they are implemented, enforced, and institutionalised. Across jurisdictions, a recurring challenge is the gap between regulatory ambition and administrative capacity. Many governments articulate high-level commitments to ethical, trustworthy, and human-centred AI, yet struggle to operationalise these principles within existing legal, technical, and bureaucratic systems (Taeihagh, 2025). This section examines the practical realities of implementing AI governance, focusing on enforcement mechanisms, institutional readiness, and capacity constraints that shape governance outcomes.

Regulatory Capacity and Institutional Readiness

Effective AI governance requires specialised institutional capacity that extends beyond traditional regulatory competencies. Regulators must possess technical expertise in data science, machine learning, and algorithmic auditing, as well as legal and ethical literacy. However, empirical evidence suggests that many regulatory agencies—particularly in developing economies—lack the personnel, infrastructure, and analytical tools necessary to oversee complex AI systems (Oxford Insights, 2024; Abanga & Dotse, 2024).

The Oxford Insights Government AI Readiness Index highlights this institutional asymmetry. While high-income countries average above 70 on governance readiness, low-income countries score below 35, reflecting deficiencies in regulatory capability, data governance frameworks, and public-sector digital skills (Oxford Insights, 2024). These disparities limit the enforceability of AI regulations and risk turning governance into a symbolic exercise rather than a substantive safeguard.

Institutional readiness is further constrained by fragmented authority. AI governance responsibilities are often distributed across multiple agencies, data protection authorities, competition regulators, sectoral watchdogs, and innovation ministries, without clear coordination mechanisms. This fragmentation weakens oversight and allows regulatory blind spots to persist (Kulothungan & Gupta, 2025; Baldwin et al., 2011).

Enforcement Mechanisms and Compliance Tools

Enforcement represents one of the most technically and legally demanding aspects of AI governance. Unlike traditional regulatory objects, AI systems are dynamic, adaptive, and often opaque. Deep learning models, in particular, resist straightforward inspection, complicating efforts to verify compliance with transparency, fairness, and safety requirements (Clarke, 2019; Veale & Borgesius, 2021).

In response, several jurisdictions have introduced procedural enforcement tools rather than outcome-based controls. These include mandatory algorithmic impact assessments (AIAs), documentation requirements, post-deployment monitoring, and audit trails. For example, Canada's Directive on Automated Decision-Making mandates algorithmic impact assessments for public-sector AI systems, linking risk levels to oversight obligations. Similarly, the EU AI Act requires conformity assessments and continuous monitoring for high-risk AI systems (Gstrein et al., 2024).

However, enforcement effectiveness depends on the availability of auditing expertise and legal authority. Proprietary protections and trade secrecy claims often limit regulators' access to training data and model architectures, reducing transparency and accountability (Raso et al., 2018). Without clear enforcement powers, regulators risk being outpaced by private-sector innovation, reinforcing concerns about regulatory capture (Stigler, 2021).

Cross-Border Enforcement and Regulatory Coordination

AI systems routinely operate across national boundaries, creating significant challenges for enforcement. Data flows, cloud infrastructure, and platform-based AI services undermine jurisdictional control and complicate accountability. A system developed in one country, trained on global data, and deployed across multiple markets may fall outside the effective reach of any single regulator (Zwitter et al., 2024; Braithwaite & Drahos, 2000).

This cross-border character necessitates international coordination mechanisms. While soft-law instruments such as the OECD AI Principles and UNESCO's Recommendation on the Ethics of AI provide normative alignment, they lack binding enforcement authority (Organisation for Economic Co-operation and Development [OECD], 2019). As a result, enforcement remains uneven, creating incentives for regulatory arbitrage, where firms relocate development or deployment to less restrictive jurisdictions (Majone, 1999). Emerging proposals, such as global AI observatories or transnational audit frameworks, aim to address these gaps by enabling information sharing, benchmarking, and peer review across jurisdictions (Taeihagh, 2025). However, these initiatives require sustained political commitment and investment in resources to move beyond symbolic coordination.

Compliance Costs and Feasibility Constraints

Implementation challenges are further shaped by the economic and administrative costs of compliance. Evidence from data protection regimes suggests that regulatory compliance can impose significant burdens on organisations, particularly small and medium-sized enterprises (SMEs). Estimates indicate that AI compliance costs can range from tens of thousands to several million dollars annually, depending on system complexity and sectoral risk classification (Sinha & Lee, 2024).

For developing economies and resource-constrained firms, these costs risk discouraging innovation or pushing AI development into informal or unregulated spaces. This underscores the importance of proportionality in enforcement. Risk-based governance models, regulatory sandboxes, and phased compliance timelines can reduce barriers while maintaining safeguards (Reuel & Undheim, 2024).

Technical Standards, Auditing, and Interoperability

Beyond policy design and institutional enforcement, effective AI governance increasingly depends on technical standards and auditing mechanisms that translate abstract principles into operational requirements. As AI systems grow more complex, scalable, and cross-border in nature, governance can no longer rely solely on legal compliance; it must also be embedded within technical architectures that ensure safety, transparency, and interoperability across jurisdictions and sectors.

The Role of Technical Standards in AI Governance

Technical standards serve as a critical bridge between regulatory intent and technological implementation. Unlike legislation, which is often slow to adapt, standards can evolve incrementally, providing flexible yet authoritative guidance on how AI systems should be designed, tested, deployed, and monitored. International organisations such as the International Organisation for Standardisation (ISO), the International Electrotechnical Commission (IEC), and the Institute of Electrical and Electronics Engineers (IEEE) have become central actors in shaping AI governance through standard-setting processes (Laato et al., 2022).

Key standards address data quality, robustness, bias mitigation, transparency, and human oversight. For example, ISO/IEC 23894 on AI risk management outlines processes for identifying, assessing, and mitigating risks throughout the AI lifecycle, aligning closely with risk-based regulatory models such as the EU AI Act (Gstrein et al., 2024). These standards are particularly important in industrial contexts, where AI systems interact with critical infrastructure, production systems, and safety-sensitive environments. In such settings, governance failures can result not only in social harm but also in physical, environmental, and economic damage.

Algorithmic Auditing as a Governance Instrument

Algorithmic auditing has emerged as a core mechanism for enforcing accountability in AI systems. Audits aim to assess whether AI applications comply with legal, ethical, and technical requirements, including non-discrimination, explainable, robustness, and security (Raji et al., 2022). Unlike traditional compliance checks, algorithmic audits must account for dynamic model behaviour, evolving datasets, and context-dependent outcomes. Auditing approaches vary in scope and depth. Internal audits conducted by developers focus on model validation and performance testing, while external audits—often mandated by regulators—seek independent verification of compliance. Some jurisdictions now require pre-deployment conformity assessments for high-risk AI systems, complemented by post-deployment monitoring to detect drift, bias, or unintended consequences (Veale & Borgesius, 2021).

However, auditing faces significant challenges. Proprietary constraints and trade secrecy protections often limit access to training data and model parameters, weakening transparency. Furthermore, many public regulators lack the technical expertise to conduct or evaluate audits effectively, leading to reliance on third-party auditors whose independence and accountability may vary (Clarke, 2019; Raso et al., 2018). These limitations highlight the need for clear standards governing who audits AI systems, how audits are conducted, and how results are reported and enforced.

Interoperability and Cross-Jurisdictional Compatibility

As AI systems increasingly operate across national and regulatory boundaries, interoperability has become a governance priority. Divergent technical standards and compliance requirements can fragment AI markets, increase costs, and undermine regulatory effectiveness. For multinational firms and global supply chains, inconsistent standards complicate compliance and create uncertainty regarding legal obligations (Majone, 1999; Braithwaite & Drahos, 2000). Interoperability refers not only to technical compatibility between systems but also to alignment in governance expectations across jurisdictions. Efforts such as aligning ISO standards with the EU AI Act and coordinating OECD principles with national AI strategies aim to reduce fragmentation while respecting regulatory sovereignty (Gstrein et al., 2024; Organisation for Economic Co-operation and Development [OECD], 2019).

Yet interoperability remains uneven. While advanced economies actively participate in standard-setting bodies, many developing countries have limited influence in shaping technical norms, reinforcing concerns about data colonialism and the exclusion of non-Western values from AI design (Couldry & Mejias, 2019). This imbalance risks embedding structural inequalities into global AI infrastructure, where standards reflect the priorities of technologically dominant actors.

From Voluntary Standards to Enforceable Governance Tools

A central governance challenge lies in transforming voluntary technical standards into enforceable mechanisms that support public accountability. While standards offer flexibility and innovation-friendly guidance, their non-binding nature can limit effectiveness unless integrated into legal and institutional frameworks. Increasingly, regulators are incorporating standards by reference, making compliance with recognised technical norms a prerequisite for market access or regulatory approval (Baldwin et al., 2011).

This hybrid approach, combining law, standards, and audits, illustrates a shift toward procedural governance. Rather than prescribing specific technological outcomes, regulators focus on processes that ensure ongoing oversight, risk management, and accountability. Such an approach is particularly suited to AI, where uncertainty and rapid evolution make rigid rules impractical (Collingridge, 1980).

Democratic Participation, Public Legitimacy, and the Emergence of Public AI

As artificial intelligence systems increasingly shape economic outcomes, public services, and everyday decision-making, questions of democratic legitimacy and public participation have become central to AI governance. Governance arrangements that rely exclusively on expert-driven or technocratic processes risk alienating citizens, particularly when algorithmic decisions affect rights, access to opportunities, and social inclusion. In this context, effective AI governance must extend beyond regulatory compliance to encompass democratic engagement, transparency, and institutional trust.

Democratic Legitimacy in AI Governance

Democratic legitimacy refers to the extent to which governance systems are perceived as fair, accountable, and responsive to public values. In AI governance, legitimacy is challenged by the opacity of algorithms, the concentration of technical expertise within private firms, and the limited capacity of citizens to contest automated decisions (Jasanoff, 2004; Donatz-Fest, 2025). When AI systems are introduced without meaningful public deliberation, governance risks becoming disconnected from societal expectations. Public value theory offers a useful lens for addressing this challenge by emphasising that legitimacy arises not only from procedural legality but also from outcomes that citizens perceive as just and beneficial (Moore, 1997; Bryson et al., 2017). Applied to AI, this perspective suggests that governance must actively incorporate public values such as fairness, dignity, and inclusivity into system design and deployment, rather than treating them as external constraints imposed after innovation has occurred.

Public Participation and Inclusive Governance Mechanisms

Meaningful public participation is increasingly recognised as a prerequisite for legitimate AI governance. Participatory mechanisms allow citizens, workers, and affected communities to shape policy priorities, identify risks, and influence how AI systems are used in practice. These mechanisms range from public consultations and citizen juries to participatory technology assessments and stakeholder forums (Stilgoe et al., 2014).

However, participation in AI governance remains uneven. While some jurisdictions have experimented with inclusive approaches such as the European Union's public consultations on the AI Act, many governance processes remain dominated by technical experts and corporate actors. This imbalance reinforces concerns about democratic deficits, particularly in sectors where AI systems directly affect vulnerable populations, including welfare allocation, policing, healthcare, and employment screening (Veale & Borgesius, 2021).

Inclusive governance also requires attention to representational equity. Marginalised groups, workers, and communities in the Global South are often excluded from AI policymaking despite being disproportionately affected by algorithmic systems developed elsewhere. Addressing this imbalance demands institutional mechanisms that prioritise diversity, contextual knowledge, and social impact assessment as core components of AI governance (Züger & Asghari, 2023; Couldry & Mejias, 2019).

Public AI as Institutional Infrastructure

The growing call for "public AI" reflects recognition that democratic governance cannot rely solely on privately owned and controlled AI infrastructures. Public AI refers to AI systems, data infrastructures, and institutional capacities developed or stewarded in the public interest, with explicit commitments to transparency, accountability, accessibility, and long-term societal value (Taeihagh, 2025). Unlike proprietary models optimised primarily for commercial objectives, public AI aims to function as a digital public good.

Public AI initiatives may include publicly funded datasets, open-source models, national AI computing infrastructure, and independent public institutions responsible for auditing and oversight. Examples include public-sector AI platforms in healthcare, environmental monitoring systems, and national AI research infrastructures designed to support universities, small firms, and public agencies. Such initiatives help counteract excessive concentration of power by reducing dependence on a small number of global technology firms (Züger & Asghari, 2023; Mazzucato, 2018).

Trust, Accountability, and Institutional Design

Trust is a central mediating factor between democratic participation and effective AI governance. Citizens are more likely to accept AI systems when they perceive governance institutions as competent, transparent, and responsive to public concerns

(Birkstedt et al., 2023). Trust is undermined when decision-making processes are opaque, accountability mechanisms are weak, or redress options are inaccessible.

Institutional design, therefore, plays a critical role. Governance arrangements must clearly allocate responsibility for AI outcomes, provide accessible complaint and appeal mechanisms, and ensure that public authorities retain meaningful control over high-impact systems. Independent oversight bodies, ombuds institutions, and judicial review processes are essential components of a governance ecosystem capable of sustaining democratic legitimacy over time (Baldwin et al., 2011; Raso et al., 2018).

In this sense, democratic participation, public AI infrastructure, and institutional trust are mutually reinforcing. Participation strengthens legitimacy, public AI enhances capacity and accountability, and trust enables sustained collaboration between citizens, governments, and industry.

Implications for Human–AI Cognitive Collaboration

Within the broader context of human–AI cognitive collaboration, democratic governance ensures that collaboration does not become domination. When AI systems are aligned with public values and subject to collective oversight, they can augment human capabilities without undermining autonomy, dignity, or social cohesion. Conversely, when governance is weak or exclusionary, AI risks entrenching power asymmetries and eroding democratic norms. Thus, the emergence of public AI and participatory governance models represents not merely an institutional adjustment but a normative shift in how societies conceptualise technological progress. AI governance for the public good must be understood as an ongoing democratic project—one that continuously renegotiates the relationship between innovation, authority, and collective welfare in the age of intelligent machines.

TOWARDS A POLICY, REGULATION AND GOVERNANCE FRAMEWORK

The absence of a universal, robust governance framework for human-AI cognitive collaboration in industrial evolution presents both a gap and an opportunity to craft policies that can accommodate local contexts, creating equity and empowering all nations, especially those in the Global South (Mensah & Van Wynsberghe, 2025). Thus, the lack of continental coherence mirrors the African experience, pointing to the need for collaborative efforts that ensure AI development supports equitable growth and social inclusion. Based on our findings in Table 1, a clear pattern emerges across the regions: while Europe has led with binding regulation, most other regions rely on fragmented or voluntary approaches. This asymmetry risks creating uneven standards, regulatory arbitrage, and inequalities in AI adoption, highlighting significant challenges for international collaboration and regulatory harmonisation (Kulothungan & Gupta, 2025). The absence of strong governance frameworks in Africa and Latin America stresses the urgency of developing locally relevant policies that harness AI for inclusive development while avoiding reliance on imported models.

It is glaringly clear that, while frameworks differ across regions, they all wrestle with the same fundamental challenge: steering technological innovation toward the public good. To conceptualise this challenge, we propose the '*Policy–Regulation–Governance Compass for Human–AI Industrial Evolution*' to set the pace and guide the world towards a governance framework for the public good (see Figure 1). Figure 1 illustrates how these three interdependent levers (policy, regulation and public good) orient Human-AI cognitive collaboration in industrial evolution.

Figure 1. Governance Compass for Human–AI Industrial Evolution

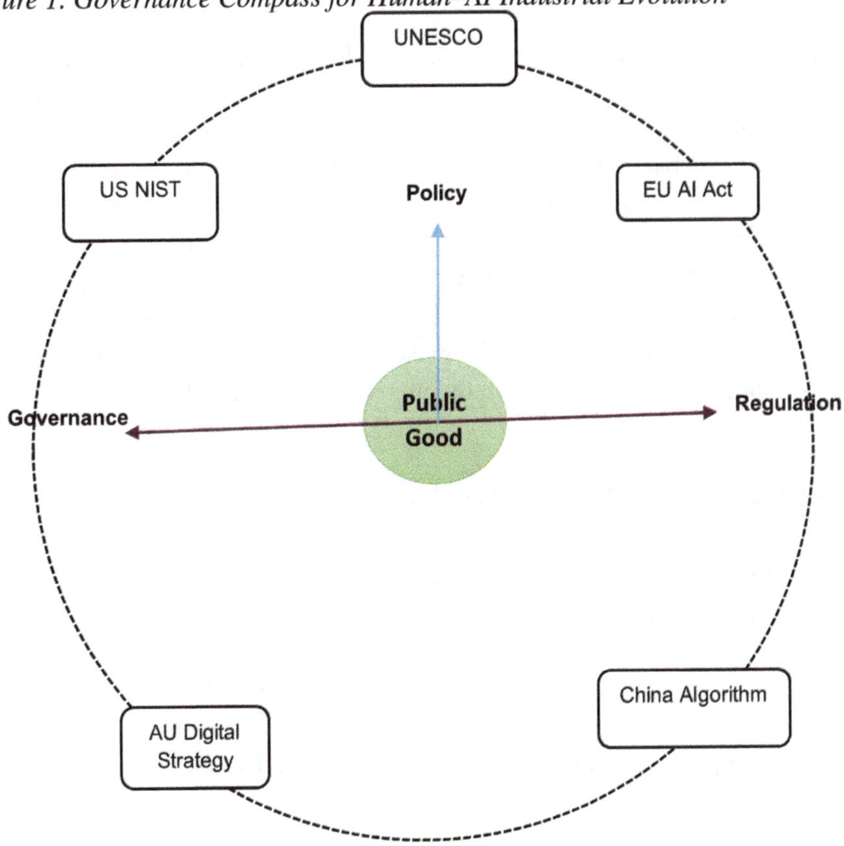

Source: Authors

At the heart of the compass is the *public good*, symbolising fairness, equity, sustainability, and accountability. Every national strategy, regulatory framework, or governance initiative ultimately claims to safeguard these values, yet their practical realisation often varies across contexts. The compass demonstrates that achieving the public good is not the outcome of a single lever but rather the result of a dynamic interaction among *policy, regulation, and governance. Policies* provide the strategic direction and long-term vision; regulations operationalise these principles into enforceable rules; and governance mechanisms ensure inclusive coordination and accountability through stakeholder participation. Placed around the compass are *global anchors*—including the EU AI Act, the U.S. NIST Risk Management Framework, China's Algorithm Regulation, Africa's AU Digital Transformation Strategy, and UNESCO's Recommendation on the Ethics of AI. These serve as illustrative markers of how different regions interpret and implement AI governance. Europe has taken a comprehensive, binding approach; North America favours principle-driven, voluntary mechanisms; China adopts state-led interventions; Africa is still in early stages, while global institutions like UNESCO provide non-binding ethical guidelines.

Apparently, these anchors highlight the *rich diversity* of governance models and the *gaps* where no coherent frameworks yet exist, particularly in Africa and Latin America. The compass model, therefore, is not static but *iterative and adaptive.* Governance feeds back into policy, updating national strategies in light of social realities; regulation evolves as new risks emerge; and policies are reimagined as stakeholder expectations shift. We emphasise that this interactive process suggests that effective AI governance necessitates continuous recalibration to align technological advancements with societal values. The figure thus provides both a diagnostic lens and a prescriptive guide for policymakers, scholars, and practitioners seeking to balance industrial innovation with the imperatives of the public good.

CHALLENGES IN AI GOVERNANCE

Although global discussions on AI governance have accelerated recently, significant gaps remain in how societies regulate, implement, and benefit from these technologies. These challenges vary across political, economic, and social contexts but are notably characterised by four main issues: regulatory lag, power concentration, global inequality, and enforcement difficulties. Overcoming these is vital for AI to serve the public good rather than worsen societal divides.

Regulatory Lag and the Pacing Problem

A key problem is the "pacing problem", which connotes that the rapid pace of technological innovation often outstrips the slower legislative processes. With new models, robotics, and machine-learning techniques developing almost monthly, policymakers struggle to keep up. This creates opportunities for the misuse of harmful AI before regulations are in place. For instance, deepfake technology, which was niche in 2018, produced over 85,000 videos monthly by 2023, often used for disinformation and non-consensual content (Maras & Logie, 2024). Regulatory efforts, such as the EU's watermarking proposals and the US's Deepfake Task Force, are lagging behind, undermining public trust. This issue is especially pressing in democracies, where policymaking involves public debate, political contestation, and judicial control, making swift regulation hard. Conversely, centralised regimes like China have implemented quick rules, such as requiring source disclosures for AI-generated content, reflecting political flexibility but also highlighting how regulation speed depends on political will.

Concentration of Power

As discussed in Section 4, this concentration reflects structural features of AI markets, such as the dominance of a few multinational corporations, including Google (Alphabet), Microsoft, Amazon, OpenAI, and Baidu, which control the majority of AI research, data, and infrastructure. In 2023, four companies accounted for over 70% of global investment in foundation models (Maslej et al., 2023). This concentration creates a digital oligopoly, enabling private firms to set the pace, direction, and standards of AI development, often outpacing or bypassing government efforts. The release of OpenAI's GPT-4 in 2023 illustrates this: while regulators in Europe and the US debated the risks, the system was integrated into sectors such as education and healthcare without proper safeguards.

Global Inequalities

Additionally, global inequalities are widening, as high-income countries can invest more in AI infrastructure, skilled talent, and oversight bodies. Meanwhile, low- and middle-income nations face resource shortages, limited human capacity, and weaker institutions. The Oxford Insights Government AI Readiness Index (2024) scores high-income OECD countries at 71 out of 100, whereas sub-Saharan Africa scores just 29, indicating that many countries in the Global South primarily consume AI developed elsewhere without exerting influence over its design or regulation.

Enforcement and Compliance Burdens

Even where frameworks exist, enforcement remains difficult. AI's opacity, especially with deep learning, hinders auditing and accountability. For example, in the US, systems like COMPAS have shown racial biases, disproportionately tagging Black defendants as higher risk. Auditing these systems is difficult because proprietary algorithms are protected, underscoring the challenges of enforcing ethical principles amid legal and technological barriers that favour corporate interests.

POLICY RECOMMENDATIONS FOR ALIGNING AI GOVERNANCE WITH THE PUBLIC GOOD

As AI governance challenges become clearer, the key question shifts from whether regulation is needed to how to effectively design and implement it, aligning technological progress with societal well-being. To ensure that AI development and deployment advance the public good, a comprehensive and pragmatic policy agenda and reforms must be initiated. These policies must go beyond mere principles; they should translate into practical institutions, enforcement mechanisms, and cooperative practices that can keep pace with AI's expanding role in industry. Indeed, while many governments have developed AI strategies or ethical guidelines, the real challenge lies in translating these documents into practical, well-regulated frameworks that align with institutional capacity and foster cross-sector collaboration.

We draw on the comparative analysis of global AI governance regulation frameworks and, through the lens of our Policy–Regulation–Governance Compass, provide actionable policy recommendations to align AI with the public good. We identified these points within six key themes: regulatory coherence and adaptability; institutional and infrastructural readiness; inclusive governance and participation; international cooperation for equitable AI development; synthesising policy for the public good; and setting global norms. Each pillar embodies a blend of the theoretical principles discussed earlier, complemented by empirical and practical considerations to inform policymakers, industry leaders, and academic practitioners.

Strengthen Regulatory Coherence and Adaptive Frameworks

AI policy must evolve from fragmented or sector-specific approaches toward coherent, risk-based, and adaptive governance systems. The EU AI Act (2024) sets a good precedent with its risk classification model, but its success depends on effective enforcement and contextual adaptation beyond Europe. Emerging economies, especially in Africa, Asia, and Latin America, can draw from this model by developing context-sensitive regulatory ecosystems that prioritise risk assessment mechanisms while avoiding rigid replication of foreign frameworks (European Union, 2024). To achieve this, governments should establish AI Policy Coordination Boards (AIP-CBs), inter-ministerial and multi-stakeholder bodies responsible for harmonising AI-related regulations across domains such as data privacy, cybersecurity, consumer protection, and competition policy. These boards would also perform regulatory foresight functions, scanning technological trends and recommending timely policy updates to avoid the "pacing problem" (Reuel & Undheim, 2024). These national AI laws should include sunset clauses that require regular review and amendment to keep pace with rapid technological advancements.

Countries such as Singapore and the United Kingdom have successfully implemented AI regulatory sandboxes, allowing companies to test innovative systems in controlled environments under regulatory supervision. These mechanisms not only encourage innovation but also generate evidence-based insights for policymaking (Organisation for Economic Co-operation and Development [OECD], 2019). Developing nations can adapt this model by collaborating with academic institutions and innovation hubs to develop AI Ethics and Governance Labs that assess the societal implications of emerging applications before they reach the market. Furthermore, to prevent regulatory fragmentation, regional blocs such as the African Union (AU), ASEAN, and Mercosur should establish continental or regional AI harmonisation frameworks. These would promote interoperability and prevent disparities in AI adoption that could lead to regulatory arbitrage or technological dependency.

Build Institutional and Infrastructural Readiness

Effective AI governance relies on robust institutions, skilled human capital, and high-quality digital infrastructure. Yet, as several reports have shown, institutional readiness remains uneven across regions. For example, according to the Oxford Insights Government AI Readiness Index (2024), the average AI readiness score in Sub-Saharan Africa is 32.7 out of 100, compared to 69.6 in Western Europe and 82.6 in North America (see Figure 2). These disparities reflect not only technological gaps but also governance and institutional weaknesses (Oxford Insights, 2024).

Figure 2. Government AI Readiness Index

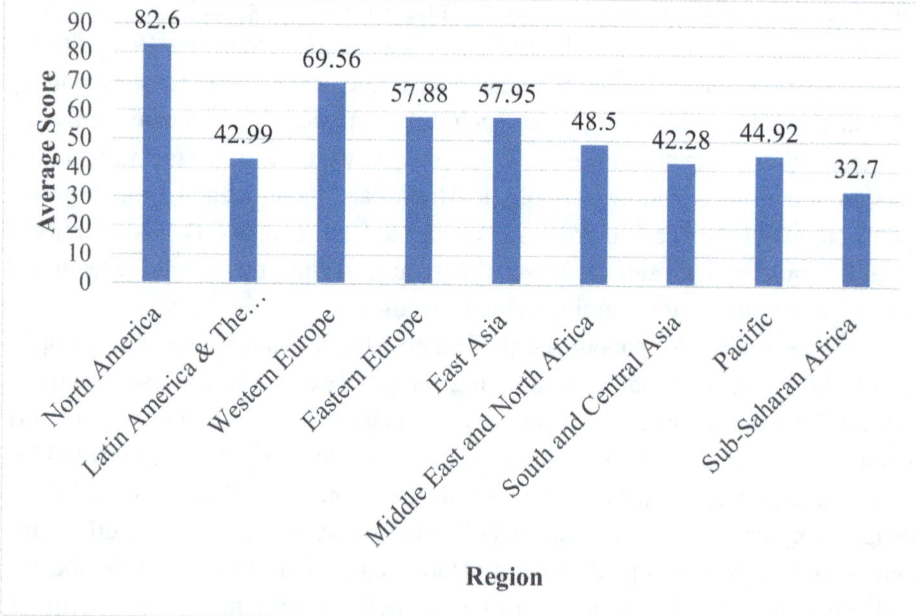

Source: Adapted from Oxford Insights (2024)

From Figure 2, we can clearly see the disparities in the Government AI Readiness Index (Oxford Insights, 2024), which assesses countries' capacity to implement, regulate, and benefit from AI based on their digital infrastructure, human capital, governance, and data access. It shows that high-income regions, such as North America and Western Europe, scored above 70, reflecting their advanced institutions, strong digital ecosystems, and robust AI policies. On the contrary, most nations in the African, South Asian, and Latin American sub-region tend to score below 50, suggesting limited preparedness to adopt AI in an ethical and effective manner.

To address these inequalities, governments must prioritise capacity-building within regulatory bodies, public administration, and the judiciary. Training programs on algorithmic auditing, AI ethics, and digital forensics should become integral to civil service and judicial education. Developing nations can collaborate with universities, think tanks, and international development organisations to establish National AI Training Academies that develop local expertise in AI oversight and governance. In addition, the development of digital public infrastructure (DPI), including cloud computing facilities, open data platforms, and secure communication networks, is fundamental for equitable AI governance. Investment in DPI will enhance data accessibility and reliability as well as facilitate OpenAI innovation by enabling smaller firms and research institutions to participate in AI development (Bajpai & Wadhwa, 2021). Institutional reforms must also include independent AI oversight bodies mandated to investigate algorithmic bias, data misuse, and discrimination. These can be analogous to data protection commissions, with investigatory powers, the ability to ensure compliance with ethical standards, and the authority to issue annual transparency reports. A notable example is Canada's Office of the Privacy Commissioner, which regularly audits AI systems in federal agencies under the Directive on Automated Decision-Making. Replicating such structures across regions can strengthen public accountability and citizen trust in AI systems.

Ensure Inclusive Governance and Public Participation

AI governance cannot succeed if it remains technocratic or elite-driven. Policies must therefore institutionalise inclusive governance mechanisms that ensure public participation, transparency, and deliberative democracy in AI development. Public consultations, multi-stakeholder dialogues, and participatory technology assessments (PTAs) should become standard practices in the formulation of national AI strategies (Donatz-Fest, 2025). For instance, Finland's AuroraAI Program exemplifies participatory co-creation in public service automation, where citizens contribute to defining ethical principles and performance metrics for AI systems. Similar participatory models can be adapted by countries in the Global South, where public trust in digital governance is often fragile. Moreover, integrating gender and diversity perspectives in AI policy formulation is vital. Studies indicate that women constitute only 22% of the global AI workforce (World Economic Forum, 2024). Addressing this imbalance requires targeted incentives for inclusive AI education, mentorship, and leadership programs for underrepresented groups.

Beyond participation, data justice should be a central policy principle, ensuring that data collection and usage reflect fairness, transparency, and informed consent (Janssen, 2025). Governments should implement AI transparency registers that publish details of public-sector AI systems, including their purpose, datasets, algorithms used, and risk assessments. The City of Amsterdam's AI Register, for instance, offers a replicable model that fosters trust and accountability in municipal AI applications. Civil society organisations (CSOs) and academia should also be empowered to act as watchdogs of AI ethics. Policymakers can establish AI Public Advisory Forums comprising CSO representatives, ethicists, data scientists, and community leaders to review AI deployments in sensitive areas such as policing, healthcare, and education. This fosters an ecosystem of collaborative accountability, where oversight becomes a shared responsibility rather than a state monopoly.

Promote International Cooperation and Global Solidarity

AI governance is inherently transnational. Data flows, algorithmic supply chains, and platform economies transcend borders, making unilateral policies insufficient. Therefore, fostering international cooperation is essential to prevent "AI nationalism" and ensure that developing countries do not become passive consumers of imported technologies. At the global level, cooperation should prioritise three areas: (i) capacity transfer and equitable technology access, (ii) standard-setting and interoperability, and (iii) joint ethical governance initiatives. Multilateral organisations such as UNESCO, OECD, and the G20 have already established guiding principles, but these need operational reinforcement. A promising model is the Global Partnership on AI (GPAI), which brings together over 25 countries to foster responsible AI development. However, the participation of African and Latin American nations remains limited. Expanding GPAI membership and creating regional "AI Partnership Hubs" can help redistribute global expertise and resources. In addition, South–South cooperation should be strengthened to promote mutual learning and resource sharing. For example, collaboration between India's "AI for All" initiative and Rwanda's Centre for the Fourth Industrial Revolution could accelerate capacity building and policy alignment in emerging economies. Furthermore, global AI funds, similar to the Green Climate Fund, should be established to support the development of ethical AI innovation in low-income countries. These funds could finance projects that utilise AI for social impact, such as crop prediction, epidemic response, or educational access, thereby directly linking governance frameworks to the public good.

Global Norm-Setting and Policy Synthesis

Bringing these recommendations together, effective AI governance requires collaboration among regulation, policy, and governance mechanisms, anchored in the principles of transparency, inclusivity, adaptability, and fairness. Governments, industries, and civil society must move beyond compliance-oriented approaches toward a co-creation paradigm, where the design, implementation, and evaluation of AI systems are participatory and iterative. As AI continues to reshape industrial evolution and social life, the public good must remain the compass guiding its trajectory. Achieving this will depend not only on the technical soundness of algorithms but also on the moral, institutional, and participatory architectures that surround their use. Only by embedding ethics, accountability, and social justice into every level of AI governance, from code to policy, can humanity ensure that the benefits of intelligent systems are equitably shared and sustainably realised.

Furthermore, individuals negatively impacted by algorithms should have access to legal remedies. Establishing ombudsman offices can provide tangible redress for grievances, such as wrongful denial or discrimination, thereby enhancing the effectiveness of governance. Because AI technology is a global issue, international cooperation is essential to prevent regulatory arbitrage. Existing frameworks, including the OECD's AI Principles, UNESCO's guidelines, and G20 discussions, lay a foundation; however, effective enforcement remains vital. A suggested solution is creating a Global AI Observatory to monitor compliance and produce transparent reports. In areas without formal frameworks, regional collaborations should be actively encouraged and supported by organisations such as the UN Economic Commission for Africa and the Inter-American Development Bank. Lessons from climate governance, such as the Paris Agreement's balance of universal rules and national contributions, could serve as models for establishing comprehensive minimum standards for AI.

CONCLUSION

AI's rapid rise is driving the fifth industrial revolution, marking a major technological shift in the twenty-first century. Like previous revolutions that reshaped economies, governance, and society, this phase involves cognitive collaboration between humans and intelligent systems. Scholars question how AI will transform industries and societies, who benefits, and what governance will guide these changes. We argue that policies and regulations are crucial for AI to serve the public good. Without adaptable frameworks, AI benefits risk being concentrated, while issues like bias and inequality burden the public. AI's development is linked to industrial, societal, and ethical concerns demanding urgent policymaker attention.

Our review of regulations from the EU's AI Act to Africa's strategies shows no universal model. Each region's framework reflects its governance, capacity, and priorities. Europe's rights-based approach emphasizes caution and accountability, while the US favors innovation with voluntary standards. China and Japan mix state control, industrial policy, and ethics. Africa and Latin America are still developing, balancing ambitions with limited capacity. These differences mirror global inequalities in preparedness, expertise, and resources, shaping AI governance. The 2024 Oxford Insights Government AI Readiness Index shows disparities across countries and regions. High-income economies outperform developing ones in infrastructure, data governance, and institutional capacity—OECD countries average over 70; sub-Saharan Africa below 30—highlighting gaps in literacy, research, and policy. This underscores the need for inclusive policymaking. While AI ethics discussions mostly happen in the Global North, their impact in employment, resources, and dependency hits the Global South harder. AI governance must foster fair global arrangements, empowering underrepresented regions to shape their digital futures.

The concept of the public good must be redefined for the AI era. Traditionally, public goods like education, clean air, and security were characterized by non-rivalry and non-excludability. In the digital age, data and algorithms act as quasi-public goods with complex ownership. Open data initiatives and accessible AI models aim to democratize technology but raise ethical and economic questions about funding, boundaries, and harm mitigation. Protecting the public good now requires rethinking access, equity, and accountability through proactive governance that anticipates risks and builds social resilience. The pacing problem—delays between innovation and legislation—exists because policymakers often lack technical understanding, while technologists overlook social and ethical issues. Solutions include AI observatories, ethics councils, and regulatory sandboxes for real-time learning and adaptive governance. Singapore and the UK's sandbox experiments show controlled environments can balance innovation with risk mitigation.

For the Global South, regional collaboration, such as the African Union's Digital Transformation Strategy, can promote cross-border experimentation and shared oversight. AI governance depends on states transforming regulatory inertia into agile, evidence-based policies. Power concentration is critical: a few multinational corporations like Alphabet, Microsoft, Amazon, OpenAI, and Baidu control much AI infrastructure and research, risking AI becoming a geopolitical asset rather than a global good. Without regulation, private firms could set ethical norms and innovation pathways for profit, risking data privacy, misinformation, and cultural homogenisation with proprietary AI systems. Governance should include academia, civil society, and local innovators through multi-stakeholder mechanisms, public AI consortia, data trusts, and open frameworks to ensure AI benefits are shared. The link between AI, environmental sustainability, and social welfare needs attention; AI can reduce emissions but also consumes vast energy. Policymakers must embed sustainability metrics, balancing innovation with planetary health, as seen in the EU's Green Digital and Japan's carbon-neutral policies. Africa could lead in ethical, sustainable AI by leveraging renewable energy and inclusive ecosystems, turning capacity gaps into growth opportunities.

For AI to truly benefit the public, governance must be participatory, transparent, and globally coordinated, shifting from principle-based declarations to enforceable commitments with institutional accountability. A proposed Global AI Observatory could centralise data on policies and compliance, enabling benchmarking and peer learning. Regional groups like ECOWAS or MERCOSUR can foster regulatory alignment and prevent fragmented adoption, guided by subsidiarity to ensure policies match local realities while aligning with global standards. AI governance is about culture, capacity, law, trust, public engagement, and ethics. Embedding these qualities in AI systems requires ongoing dialogue among engineers, policymakers, and users. Universities and think tanks should shape education, research, and public debate on responsible AI.

In this sense, the chapter's proposed Policy–Regulation–Governance Compass provides a conceptual and practical framework for aligning AI with the public good. It underscores the principle that governance is not a constraint on innovation, but its ethical enabler, transforming AI from a tool of disruption into an instrument of shared prosperity. Ultimately, we look forward to scholars and policymakers embracing adaptive, inclusive and anticipatory governance for AI.

Competing Interests

The authors declare no competing interests.

Funding Agency

This research received no specific grant from any funding agency in the public, commercial, or not-for-profit sectors.

ACKNOWLEDGMENT

The authors acknowledge the reviewers, editors and professors who proofread the work to bring it to this point.

REFERENCES

Abanga, E. A., & Dotse, S. (2024). AI and Digital Economies: A Comparative Analysis of South and Southeast Asia and Africa. *Asian Journal of Research in Computer Science, 17*(10), 12–25. DOI: 10.9734/ajrcos/2024/v17i10506

Ally, M., & Perris, K. (2022). Artificial intelligence in the fourth industrial revolution to educate for sustainable development. *Canadian Journal of Learning and Technology, 48*(4), 1–20. DOI: 10.21432/cjlt28287

Bajpai, N., & Wadhwa, M. (2021). *Artificial intelligence and healthcare in India* (No. 43). ICT India Working Paper.

Baldwin, R., Cave, M., & Lodge, M. (2011). *Understanding regulation: theory, strategy, and practice*. Oxford university press. DOI: 10.1093/acprof:osobl/9780199576081.001.0001

Barocas, S., Guo, A., Kamar, E., Krones, J., Morris, M. R., Vaughan, J. W., & Wallach, H. (2021, July). Designing disaggregated evaluations of ai systems: Choices, considerations, and tradeoffs. In *Proceedings of the 2021 AAAI/ACM Conference on AI, Ethics, and Society* (pp. 368-378). DOI: 10.1145/3461702.3462610

Bertolini, L. (2023). The next 30 years: Planning cities beyond mobility? *European Planning Studies, 31*(11), 2354–2367. DOI: 10.1080/09654313.2023.2217855

Birkstedt, T., Minkkinen, M., Tandon, A., & Mäntymäki, M. (2023). AI governance: Themes, knowledge gaps and future agendas. *Internet Research, 33*(7), 133–167. DOI: 10.1108/INTR-01-2022-0042

Braithwaite, J., & Drahos, P. (2000). *Global business regulation*. Cambridge university press. DOI: 10.1017/9780521780339

. Brynjolfsson, E., & McAfee, A. (2014). *The second machine age: Work, progress, and prosperity in a time of brilliant technologies*. WW Norton & company.

Bryson, J., Sancino, A., Benington, J., & Sørensen, E. (2017). Towards a multi-actor theory of public value co-creation. *Public Management Review, 19*(5), 640–654. DOI: 10.1080/14719037.2016.1192164

Butt, J. (2024). Analytical study of the world's first EU Artificial Intelligence (AI) Act. *International Journal of Research Publication and Reviews*, *5*(3), 7343–7364. https://orcid.org/0009-0000-0530-962X. DOI: 10.55248/gengpi.5.0324.0914

Cao, J., & Meng, T. (2025). Ethics and governance of artificial intelligence in digital China: Evidence from online survey and social media data. *Chinese Journal of Sociology*, *11*(1), 58–89. DOI: 10.1177/2057150X241313085

Chang, Y. L., & Ke, J. (2024). Socially responsible artificial intelligence empowered people analytics: A novel framework towards sustainability. *Human Resource Development Review*, *23*(1), 88–120. DOI: 10.1177/15344843231200930

Chen, Y. C., Ahn, M. J., & Wang, Y. F. (2023). Artificial intelligence and public values: Value impacts and governance in the public sector. *Sustainability (Basel)*, *15*(6), 4796. DOI: 10.3390/su15064796

Clarke, R. (2019). Regulatory alternatives for AI. *Computer Law & Security Review*, *35*(4), 398–409. DOI: 10.1016/j.clsr.2019.04.008

. Collingridge, D. (1980). The dilemma of control. *The Social Control of Technology*, 13-22.

Cools, H., Van Gorp, B., & Opgenhaffen, M. (2024). Where exactly between utopia and dystopia? A framing analysis of AI and automation in US newspapers. *Journalism*, *25*(1), 3–21. DOI: 10.1177/14648849221122647

Couldry, N., & Mejias, U. A. (2019). Data colonialism: Rethinking big data's relation to the contemporary subject. *Television & New Media*, *20*(4), 336–349. DOI: 10.1177/1527476418796632

Demaidi, M. N. (2025). Artificial intelligence national strategy in a developing country. *AI & Society*, *40*(2), 423–435. DOI: 10.1007/s00146-023-01779-x

Donatz-Fest, I. C. (2025). Values? Camera? Action! An ethnography of an AI camera system used by the Netherlands Police. *Policing and Society*, *35*(1), 50–67. DOI: 10.1080/10439463.2024.2370939

Duarte, L. L., & Baptista, G. (2025). Artificial intelligence: Understanding the determinants of adoption and intention to recommend the technology in the financial sector in Portugal. *International Journal of Bank Marketing*, •••, 1–26. DOI: 10.1108/MD-09-2025-2668

El Jaouhari, A., Arif, J., Samadhiya, A., Dhayal, K. S., Luthra, S., & Singh, S. (2025). From automation to augmentation: Exploring the role of industry 5.0 advanced technologies for developing countries. *Engineering Management Journal*, *37*(3), 306–323. DOI: 10.1080/10429247.2024.2407254

European Union. (2024). *Regulation (EU) 2024/1689 of the European Parliament and of the Council of 13 June 2024 on artificial intelligence*. Official Journal of the European Union, L 339, 1–144. Retrieved from https://eur-lex.europa.eu/legal-content/EN/TXT/PDF/?uri=OJ:L_202401689

Food and Drug Administration. (2023). *Artificial intelligence in drug manufacturing: Discussion paper.* U.S. Food and Drug Administration. https://www.fda.gov/media/165743/download

Freeman, R. E. (1984). *Strategic management: A stakeholder approach*. Pitman.

Gstrein, O. J., Haleem, N., & Zwitter, A. (2024). General-purpose AI regulation and the European Union AI Act. *Internet Policy Review*, *13*(3), 1–26. DOI: 10.14763/2024.3.1790

Hartley, J., Alford, J., Knies, E., & Douglas, S. (2017). Towards an empirical research agenda for public value theory. *Public Management Review*, *19*(5), 670–685. DOI: 10.1080/14719037.2016.1192166

Hermann, E., & Puntoni, S. (2025). Generative AI in marketing and principles for ethical design and deployment. *Journal of Public Policy & Marketing*, *44*(3), 332–349. DOI: 10.1177/07439156241309874

Hermann, I. (2023). Artificial intelligence in fiction: Between narratives and metaphors. *AI & Society*, *38*(1), 319–329. DOI: 10.1007/s00146-021-01299-6

Impito, P. F., Nwizu, U. N., Nallapareddy, C., & Mansoor, I. K. (2025). *Strategic responses to disruptive technologies*. Cari Journals USA LLC.

Janssen, M. (2025). Responsible governance of generative AI: Conceptualizing GenAI as complex adaptive systems. *Policy and Society*, *44*(1), 38–51. DOI: 10.1093/polsoc/puae040

Jasanoff, S. (2004). Ordering knowledge, ordering society. In *States of knowledge* (pp. 13–45). Routledge. DOI: 10.4324/9780203413845-7

Jecker, N. S., & Nakazawa, E. (2022). Bridging East-West differences in ethics guidance for AI and robotics. *AI*, *3*(3), 764–777. DOI: 10.3390/ai3030045

Keddell, E. (2019). Algorithmic justice in child protection: Statistical fairness, social justice and the implications for practice. *Social Sciences (Basel, Switzerland)*, *8*(10), 1–16. DOI: 10.3390/socsci8100281

Kulothungan, V., & Gupta, D. (2025, May). Towards adaptive AI governance: Comparative insights from the US, EU, and Asia. In *2025 IEEE 11th Conference on Big Data Security on Cloud (BigDataSecurity)* (pp. 32-38). IEEE.

Laato, S., Birkstedt, T., Mäantymäki, M., Minkkinen, M., & Mikkonen, T. (2022, May). AI governance in the system development life cycle: Insights on responsible machine learning engineering. In *Proceedings of the 1st International Conference on AI Engineering: Software Engineering for AI* (pp. 113-123). DOI: 10.1145/3522664.3528598

Lachmann, N. (2025). Chasing the Elusive Bird? The Technological Development of the Digital Economy and International Trade Law's Susceptibility to a Pacing Problem. *The Journal of World Investment & Trade*, *26*(3), 479–511. DOI: 10.1163/22119000-12340368

Lawniczak, L., & Benzmüller, C. (2025). Logical Modalities within the European AI Act: An Analysis. *arXiv preprint arXiv:2501.19112*. DOI: 10.1145/3769126.3769202

Lessig, L. (2000). Code is law. *Harvard Magazine*, *1*, •••. http://cyber.law.harvard.edu/

Liu, Y., Zhang, Z., & Wu, Y. (2025). Will generative AI create a new social divide? Investigating the impacts of generative AI use on social capital in China. *International Journal of Human-Computer Interaction*, *41*(18), 11324–11340. DOI: 10.1080/10447318.2024.2443242

Mahama, J. D. (2025, September). *Message to the 80th session of the United Nations General Assembly*. Government of Ghana. https://gadebate.un.org/sites/default/files/gastatements/80/gh_en_0.pdf (Accessed: September 2025)

Majone, G. (1999). Regulation in comparative perspective. *Journal of Comparative Policy Analysis*, *1*(3), 309–324. DOI: 10.1080/13876989908412630

Maras, M. H., & Logie, K. (2024). Countering the complex, multifaceted nature of nude and sexually explicit deepfakes: An Augean task? *Crime Science*, *13*(1), 1–17. DOI: 10.1186/s40163-024-00226-6

Maslej, N., Fattorini, L., Brynjolfsson, E., Etchemendy, J., Ligett, K., Lyons, T., Manyika, J., Ngo, H., Niebles, J. C., Parli, V., Shoham, Y., Wald, R., Clark, J., & Perrault, R. (2023). *The AI Index 2023 annual report.* AI Index Steering Committee, Institute for Human-Centered AI, Stanford University. https://hai.stanford.edu/assets/files/hai_ai_index_report_2025.pdf

Mazzucato, M. (2018). Mission-oriented innovation policies: Challenges and opportunities. *Industrial and Corporate Change, 27*(5), 803–815. DOI: 10.1093/icc/dty034

Mensah, J. O., & Van Wynsberghe, A. (2025). Where are the missing values: An exploration of the need to incorporate Ubuntu values into African AI policy. *AI and Ethics, 5*(4), 1–11. DOI: 10.1007/s43681-025-00746-0

Mishra, N., & Kugler, K. (2024). International community in the global digital economy: A case study on the African digital trade framework. *The International and Comparative Law Quarterly, 73*(4), 853–889. DOI: 10.1017/S0020589324000290

Misra, S. K., Das, S., Gupta, S., & Sharma, S. K. (2020, December). Public policy and regulatory challenges of artificial intelligence (AI). In *International Working Conference on Transfer and Diffusion of IT* (pp. 100-111). Cham: Springer International Publishing. DOI: 10.1007/978-3-030-64849-7_10

Moore, M. H. (1997). *Creating public value: Strategic management in government.* Harvard university press.

Moses, L. B. (2011). Agents of change: How the law "Copes" with technological change. *Griffith Law Review, 20*(4), 763–794. https://search.informit.org/doi/10.3316/informit.795045545574361. DOI: 10.1080/10383441.2011.10854720

Mügge, D. (2024). EU AI sovereignty: For whom, to what end, and to whose benefit? *Journal of European Public Policy, 31*(8), 2200–2225. DOI: 10.1080/13501763.2024.2318475

Naudé, W. (2021). Artificial intelligence: Neither Utopian nor apocalyptic impacts soon. *Economics of Innovation and New Technology, 30*(1), 1–23. DOI: 10.1080/10438599.2020.1839173

Obermeyer, Z., Powers, B., Vogeli, C., & Mullainathan, S. (2019). Dissecting racial bias in an algorithm used to manage the health of populations. *Science, 366*(6464), 447–453. DOI: 10.1126/science.aax2342 PMID: 31649194

Olujobi, O. J., & Ape, N. T. (2025). Reforming legal frameworks for combating gas flaring in Nigeria: A comparative legal analysis and sustainable solutions. *Resources Policy, 109*, 1–17. DOI: 10.1016/j.resourpol.2025.105721

Organisation for Economic Co-operation and Development (OECD). (2019). *OECD principles on artificial intelligence.* OECD Publishing. https://archive.epic.org/algorithmic-transparency/OECD-AI-Principles-flyer.pdf

Orwat, C., Bareis, J., Folberth, A., Jahnel, J., & Wadephul, C. (2024). Normative challenges of risk regulation of artificial intelligence. *NanoEthics*, *18*(2), 1–29. DOI: 10.1007/s11569-024-00454-9

Oxford Insights. (2024, December). *Government AI readiness index 2024* (7th ed.). Oxford Insights. https://www.oxfordinsights.com/government-ai-readiness-index-2024. (Accessed: October 2025).

Pantanowitz, L., Hanna, M., Pantanowitz, J., Lennerz, J., Henricks, W. H., Shen, P., Quinn, B., Bennet, S., & Rashidi, H. H. (2024). Regulatory aspects of artificial intelligence and machine learning. *Modern Pathology*, *37*(12), 1–8. DOI: 10.1016/j.modpat.2024.100609 PMID: 39260776

Paraman, P., & Anamalah, S. (2023). Ethical artificial intelligence framework for a good AI society: Principles, opportunities and perils. *AI & Society*, *38*(2), 595–611. DOI: 10.1007/s00146-022-01458-3

Peltzman, S. (1976). Toward a more general theory of regulation. *The Journal of Law & Economics*, *19*(2), 211–240. DOI: 10.1086/466865

. Rajan. (2025). Philosophy of technology for the lost age of freedom: a critical treatise on human essence and uncertain future. *AI & SOCIETY, 40*(3), 1705-1722. DOI: 10.1007/s00146-024-01942-y

Raji, I. D., Xu, P., Honigsberg, C., & Ho, D. (2022, July). Outsider oversight: Designing a third party audit ecosystem for ai governance. In *Proceedings of the 2022 AAAI/ACM Conference on AI, Ethics, and Society* (pp. 557-571). DOI: 10.1145/3514094.3534181

. Raso, F. A., Hilligoss, H., Krishnamurthy, V., Bavitz, C., & Kim, L. (2018). Artificial intelligence & human rights: Opportunities & risks. *Berkman Klein Center Research Publication*, (2018-6).

Reale, R., & Biasin, E. (2023, March). The Design and Implementation of a National AI Platform for Public Healthcare in Italy: Implications for Semantics. *In the European Data Conference on Reference Data and Semantics* (ENDORSE).

Reuel, A., & Undheim, T. A. (2024). Generative AI needs adaptive governance. arXiv:2406.04554. /arXiv.2406.04554DOI: 10.48550

Roberts, H., Hine, E., & Floridi, L. (2023). Digital sovereignty, digital expansionism, and the prospects for global AI Governance. In *Quo Vadis, Sovereignty? New Conceptual and Regulatory Boundaries in the Age of Digital China* (pp. 51–75). Springer Nature Switzerland. DOI: 10.1007/978-3-031-41566-1_4

Schertel Mendes, L., & Kira, B. (2023, Dec 21). The road to regulation of artificial intelligence: the Brazilian experience. *Internet Policy Review.* https://policyreview.info/articles/news/road-regulation-artificial-intelligence-brazilian-experience/1737

Sinha, S., & Lee, Y. M. (2024). Challenges with developing and deploying AI models and applications in industrial systems. *Discover Artificial Intelligence*, *4*(1), 55. DOI: 10.1007/s44163-024-00151-2

Stefanini, R., & Vignali, G. (2024). The influence of Industry 4.0 enabling technologies on social, economic and environmental sustainability of the food sector. *International Journal of Production Research*, *62*(10), 3800–3817. DOI: 10.1080/00207543.2023.2248523

Stigler, G. J. (2021). The theory of economic regulation. In *The political economy: Readings in the politics and economics of American public policy* (pp. 67–81). Routledge. DOI: 10.4324/9781315495811-8

Stilgoe, J., Lock, S. J., & Wilsdon, J. (2014). Why should we promote public engagement with science? *Public Understanding of Science (Bristol, England)*, *23*(1), 4–15. DOI: 10.1177/0963662513518154 PMID: 24434705

Swaminathan, N., & Danks, D. (2024). Governing ethical gaps in distributed AI development. *Digital Society : Ethics, Socio-Legal and Governance of Digital Technology*, *3*(1), 1–11. DOI: 10.1007/s44206-024-00088-0

Taeihagh, A. (2025). Governance of generative AI. *Policy and Society*, *44*(1), 1–22. DOI: 10.1093/polsoc/puaf001

Topol, E. J. (2019). High-performance medicine: The convergence of human and artificial intelligence. *Nature Medicine*, *25*(1), 44–56. DOI: 10.1038/s41591-018-0300-7 PMID: 30617339

Van Bekkum, M., & Borgesius, F. Z. (2023). Using sensitive data to prevent discrimination by artificial intelligence: Does the GDPR need a new exception? *Computer Law & Security Review*, *48*, 105770. DOI: 10.1016/j.clsr.2022.105770

Van Kolfschooten, H., & Van Oirschot, J. (2024). The EU artificial intelligence act (2024): Implications for healthcare. *Health Policy (Amsterdam)*, *149*, 105152. DOI: 10.1016/j.healthpol.2024.105152 PMID: 39244818

Veale, M., & Borgesius, F. Z. (2021). Demystifying the draft EU artificial intelligence act. *arXiv preprint arXiv:2107.03721*. DOI: 10.31235/osf.io/38p5f

Winner, L. (1980). Do Artifacts Have Politics? In *Computer Ethics* (pp. 177-192). Routledge.

World Economic Forum. (2024, June). *Annual report 2023-2024*. World Economic Forum. https://www3.weforum.org/docs/WEF_Annual_Report_2023_2024.pdf (Accessed: October 2025)

Wu, H., Liu, J., & Liang, B. (2025). AI-driven supply chain transformation in Industry 5.0: Enhancing resilience and sustainability. *Journal of the Knowledge Economy*, *16*(1), 3826–3868. DOI: 10.1007/s13132-024-01999-6

Züger, T., & Asghari, H. (2023). AI for the public. How public interest theory shifts the discourse on AI. *AI & Society*, *38*(2), 815–828. DOI: 10.1007/s00146-022-01480-5

Zwitter, A., Gstrein, O. J., & Haleem, N. (2024). General-purpose AI regulation and the European Union AI Act. *Available at SSRN* 4916400. DOI: 10.2139/ssrn.4916400

KEY TERMS AND DEFINITIONS

AI Governance: includes systems, policies, laws, and institutions designed to guide AI development, deployment, and use in line with societal values, ethics, principles, and public good.

Pacing Problem: is the term used to describe the pace at which AI innovates, which will outstrip governments' ability to regulate it effectively.

Policy: encompasses guiding principles, objectives, and structured frameworks that articulate the societal objectives AI should serve.

Policy–Regulation–Governance Compass: sets the pace and guides the world towards an AI governance framework for the public good.

Public AI: an institutional framework or platform that governs the set of algorithms in the collective interest of the public, rather than being controlled by proprietaries, ensuring equitable access and accountability.

Public Good: is described as using AI in diverse public values to ensure that these technologies promote collective well-being rather than private interests.

Regulation: refers to the legal statutes, standards, and enforcement mechanisms that define permissible and impermissible uses of AI.

About the Contributors

Raul Villamarin Rodriguez is the Vice President of Woxsen University, India, and a globally recognized cognitive technologist and AI strategist. Holding a PhD in Cognitive Science & Technology, his work bridges AI, behavioral science, and human-centered design. He is the inventor of patented technologies such as Behavioral GPS™ and the Neural Personalized Academic Classification Engine, applied across sectors like healthcare, education, and aviation. A regular speaker at global platforms including Web Summit and World Summit AI, Dr. Rodriguez also contributes to international advisory boards and thought leadership in ethical and inclusive AI.

Hemachandran K is the Director of the AI Research Centre, Vice Dean of the School of Business, and Area Chair of Analytics at Woxsen University, India. With over 17 years of experience spanning academia, research, and industry, his expertise lies in Artificial Intelligence, Machine Learning, NLP, Computer Vision, and Robotics. He has led several high-impact AI projects across healthcare, agriculture, education, and smart infrastructure and holds patents for innovations such as the Unified Healthcare Intelligence System and Quantum AI-Driven Smart Airport Management System. Dr. Kannan also serves on international advisory boards and actively contributes as an ambassador for DeepLearning.AI and a UNESCO Expert.

Mark Esposito, Professor of Strategy and Technology Policy, Faculty Fellow at the Center for Emerging Markets, and Affiliate Scientist of the Global Resilience Institute at Northeastern University, is a globally recognized scholar, educator, and practitioner at the forefront of technology, economics, and public policy. He specializes in the Fourth Industrial Revolution and artificial intelligence, with expertise spanning academia, international organizations, and the private sector. Mark Esposito equally holds appointments at Harvard University, where his affiliations span the Berkman Klein Center for Internet and Society at Harvard Law School, the Center for International Development at Harvard Kennedy School, and the Institute for Quantitative Social Science. He is also a Senior Associate at the University of Cambridge, an institution with which he has maintained deep ties since he co-founded and directed the Circular Economy Research Center at Cambridge Judge Business School in 2016, a role he held through 2020. Mark Esposito is closely connected to policy and innovation initiatives worldwide. He is a member of the World Economic Forum's Global AI Alliance and Fostering Converging Technologies group and has been a Fellow at the Mohammed Bin Rashid School of Government in Dubai since 2017. He has advised government agencies across the Gulf Cooperation Council and Eurasia, and for over a decade co-led the Institute's Council on Microeconomics of Competitiveness at Harvard Business School under Professor Michael Porter. An accomplished entrepreneur, he co-founded Nexus FrontierTech, an influential machine learning research firm, as well as The Chart ThinkTank and the AI Native Foundation. He currently serves as Chief Economist of micro1, a Silicon Valley AI lab developing human intelligence platforms for the training of frontier AI models. Recognized by Thinkers50 Radar in 2016 as one of the world's top 30 business thinkers, Dr. Esposito is a prolific author with over 150 publications, both peer-reviewed and non-peer-reviewed, and 13 books to date. Among them, Understanding How the Future Unfolds (2017) and The AI Republic (2019) are Amazon bestsellers. His recent titles include The Emerging Economies under the Dome of the Fourth Industrial Revolution (Cambridge University Press, 2022), The Great Remobilization: Strategies and Designs for a Global Smarter World (MIT Press, 2023), and Digitizing the Emerging Economies (Cambridge University Press, 2024). His forthcoming books include Becoming AI Native: Charting the Next AI Frontier (Routledge, 2026), Tectonic Shifts: How Technology is Remaking Global Power Dynamics(Penguin Random House, 2026), Global Visions (Bloomsbury, 2026), and Who Rules the Internet? Polycentric Governance and the Political Economy of the Digital Commons (Routledge, 2027). He holds a doctoral degree from École des Ponts ParisTech in France, the world's oldest civil engineering school.

***

Saifuddin Ahmad, Assistant Professor at the Institute of Management Sciences, University of Lucknow, is a committed researcher in management studies with a keen focus on emerging paradigms in banking, insurance, and digital transformation. His scholarly contributions include several Scopus-indexed publications and a patent, reflecting novelty and practical relevance. His research interests include finance, digital transformation, AI in management, and sustainable business practices. A Ph.D. in Management, he also authored a textbook and actively presents at national and international forums.

Ashraf Alam is currently serving as an Assistant Professor in the Department of Education at NIT Calicut, Kerala, India. He has been recognised among the *World's Top 2% Scientists* by Stanford University and Elsevier. His academic contributions include more than 75 Scopus-indexed publications and over 100 conference presentations delivered at national and international academic forums. His work has appeared in reputed research publication platforms including Elsevier, Springer, Taylor & Francis, CRC Press, Routledge, Atlantis Press, Springer Nature, Palgrave Macmillan, and IEEE. He earned his Ph.D. from IIT Kharagpur, where his research engaged with foundational concerns in educational policy, curriculum design, educational technology, philosophy of education, innovation in education, educational administration, governance, and pedagogical transformation. His work is grounded in a transdisciplinary orientation that draws from educational sciences, digital pedagogy, curriculum theory, learning analytics, sustainability education, instructional design, artificial intelligence, ethics of technology, and socio-technical studies. Dr. Ashraf's scholarship strives to interrogate and expand the boundaries of educational thought by examining the epistemological, ethical, and social dimensions of digital and post-digital pedagogies. His work critically engages with the changing nature of knowledge production, learner agency, pedagogical mediation, and institutional transformation in technologically saturated educational environments. He is particularly invested in advancing inclusive digital learning, gamified instructional strategies, AI-mediated educational ecosystems, and ethically governed technology integration in education. His research agenda is animated by the pursuit of design-oriented, ethically attuned, and contextually responsive frameworks that can support meaningful educational reform. Through his work, he seeks to contribute to the development of pedagogical models that are not only technologically innovative but also socially just, epistemically inclusive, developmentally sensitive, and aligned with the broader goals of sustainable and human-centred education.

Duggirala Aravind is currently associated with the Department of Aviation at ASET Institute of Science and Technology, Chennai, India. He previously completed his Bachelor's degree in Electrical and Electronics Engineering at MITS University. His research interests include aircraft maintenance engineering, aviation safety management, renewable energy technologies in air transport, and sustainable energy systems. (aravindaset.edu@gmail.com).

Anil Chourasia is a Chief General Manager at Power Grid Corporation of India Ltd., popularly known as POWERGRID. He has 32 years of diverse experience in Construction & Project Management, Liaison, Operation & Maintenance, Tariff and Regulatory Affairs, Contract & Procurement, Industrial Engineering, and Recruitment in the central power sector 'Maharatna' company, Power Grid Corporation of India Ltd, making him a committed power sector professional to hold challenging techno-managerial roles in a highly competent business environment.

Veena Christy completed her graduation in Commerce from the renowned Madras Christian College, affiliated to University of Madras. Her urge for higher studies led her to complete her MBA with laurels. She was awarded the University rank in her PG program which was also affiliated to University of Madras. Lured by passion for research, she completed her doctoral studies in organisational Behaviour and has many Scopus-indexed publications to her credit. With ten years of teaching experience in different institutions of repute and six years of research expertise, she is also the recipient of Innovative Researcher and Dedicated Academician Award conferred by The Innovative Scientific Research Professional Institute, Chennai. She is an Associate member of Madras Management Association (MMA) (Mem No: AM863) and Affiliate member of Association of Behavior Analysis International (Mem No: 102469). Presently, She is serving as Assistant Professor in SRM University, KTR Campus, Chennai.

Samuel Etse Dzakpasu is a goal-driven academic with strong research, teaching, and leadership abilities. He holds an MPhil in Accounting Education and is pursuing a PhD in the same field at the University of Cape Coast. His interests include accounting education, quality teaching, and teacher personality. He has experience in tutoring, supervision, and mentorship, as well as in active leadership roles.

Sude Erdağ is a final-year Industrial Design student at Istanbul Aydın University. She started her education in 2022. Her program skills include Rhinoceros, AutoCAD, and Adobe Photoshop for 3D and visual design. She also focuses on project management and design thinking. Having completed her industrial design internship, she is now looking to step into better opportunities as a junior industrial designer. She speaks Turkish (native) and English.

Mark Esposito is a public policy scholar and social scientist affiliated with Harvard's Berkman Klein Center for Internet and Society; the Center for International Development at Harvard Kennedy School and the Institute for Quantitative Social Science. He leads policy clinics on the governance of technology worldwide. He is a Professor at Hult Int'l Business School and Adjunct Professor at Georgetown University. He has co-founded several AI ventures, including Nexus FrontierTech, the AI Native Foundation, and The Chart ThinkTank, and serves as Chief Economist at micro1, a Silicon Valley–based AI lab. He has been affiliate faculty of the Microeconomics of Competitiveness at Harvard Business School, under the mentorship of Prof. Michael E. Porter for over a decade and he served as Founding Fellow of the Circular Economy Research Center at the Judge Business School, at the University of Cambridge, where he retains a Senior Associate role. He advises governments in the GCC and Eurasia regions and is a global expert of the World Economic Forum, working across the Global AI Alliance and the Converging Technology group. He has been working as a Resident Fellow as well as Professor of public policy at Mohammed Bin Rashid School of Government, where he authored several policy briefs and appointed Research Fellow at the New School in NYC, focusing on AI and Cooperatives. Dr. Esposito has written or co-written over 150 publications, both peer-reviewed and non-peer-reviewed, and 13 books, two of which are Amazon bestsellers: "Understanding how the Future Unfolds" (2017) and "The AI Republic" (2019). His most recent books include "The Emerging Economies under the Dome of the Fourth Industrial Revolution" (Cambridge University Press, 2022), "The Great Remobilization: Strategies and Designs for a Global Smarter World" (MIT University Press, 2023), and "Digitizing the Emerging Economies" (Cambridge University Press, 2024). His forthcoming books are and "Becoming AI Native: Charting the Next AI Frontier" (Routledge, 2026); "Tectonic Shifts: How Technology is Remaking Global Power Dynamics" (Penguin Random House, 2026), Global Visions,(Bloomsbury, 2026) and Who Rules the Internet? Polycentric Governance and the Political Economy of the Digital Commons (Routledge, 2027) He has a doctorate from Ecole des Ponts Paris Tech and he lives across Boston, Geneva and Dubai.

Nur Muhammad Fadilah is a dedicated Information Technology professional and a passionate technology enthusiast. With a strong commitment to continuous learning, He focus on understanding emerging digital solutions and applying technical expertise to solve real-world challenges.

Robert Ipiin Gnankob is an Assistant Professor at the School of Business, Woxsen University, India, where he contributes to teaching, research, and academic development in public policy, leadership, and management. Prof Gnankob played a distinguished role in student governance in Ghana, serving as President of the Graduate Students' Association of Ghana (GRASAG) at the University of Cape Coast (UCC), and served on the governing Council of UCC. He has authored and co-authored scholarly works on public policy, management, and leadership in emerging economies. Known for his analytical approach and passion for academic excellence, Prof. Gnankob continues to engage in collaborative research, present at conferences, and mentor young scholars.

Yusuf Kamal is working as an assistant professor at the Institute of Management, University of Lucknow. He has more than 18 years of Academic and Industrial experience, 5 Patents granted by the Government of India, and authored 1 Book and 13 Articles in Journals of International repute. His earned Ph.D. degree is from the Faculty of Commerce (Major- Marketing & Minor- Digital Marketing). He has also invented many Industrial Designs & Patentable Utilities for Social Causes, free from royalty claims for social upliftment. His research statement includes regard to the changing scenario of the business world and methodology, and he has committed to providing technology-intensive business models by 2030. His 2030 vision includes: 1- Exploring the possible untapped sources of energy. 2- Machine automation and Digital incorporations of novel electronic systems for business efficiency. 3- Reveal the new domains of Intelligent systems. 4- Innovation in Business Management.

Sadaf Khurshid is an academic and researcher with a passion for entrepreneurship, culture, and the power of networks. She earned her PhD from the University of Wollongong, Australia, where her work focused on the intricate relationships between these areas, particularly in the context of social enterprises. Since 2014, Sadaf has published numerous case studies highlighting the role of social enterprises across the Middle East and North Africa (MENA) region. With an MBA from the University of Wollongong in Dubai and a Bachelor's in Technology (Electronics and Communication) from Jamia Millia Islamia, New Delhi, she brings a unique blend of technical and business expertise to her research. She currently serves as Vice Chair (Marketing and Outreach) of the Academy of International Business (AIB) MENA Chapter and actively contributes to academic conferences, scholarly reviewing, and regional research collaborations.

Sanjeev Kumar Is a Professor in Lovely Professional University Punjab India. He received Undergraduate in Hotel Management from Osmania University, Postgraduate in Tourism Management from Madurai Kamaraj University. He completed Doctor of Philosophy in Hospitality Management in 2019 from Amity University Jaipur India. His research areas lies in food Service Industry, his research focuses on Alcoholic beverages, Event management and Sustainable Management Practices, Metaverse, AI Machine Learning and Artificial Intelligence. He published more than 80 research papers, articles and chapters in Scopus Indexed, UGC Approved and peer reviewed Journals and books. Dr. Sanjeev Kumar participated and acted as resource person in various National and International conferences, seminars, session chair, research workshops and industry talks and his work has been widely cited. Dr. Sanjeev also edited more than 20 books with various International Publishers such as IGI Global, Cambridge Scholars, Emerald, CABI, Bentham Science, Taylor & Francis. Dr Sanjeev is an active researcher and has published numerous research articles in prestigious National and International journals and books.

Chandan Maheshkar is one of the founders of the East Nimar Society for Education (2019), dedicated to quality improvement in higher education and the development of educator competencies. As a Senior Consultant, he has served the Centre for Internal Quality Assurance (CIQA), Madhya Pradesh Bhoj (Open) University, Bhopal, India. He is associated with several management institutes in central India, including the University of Indore, India, in various academic roles. Dr. Maheshkar earned his MBA and Ph.D. from the University of Indore, India. In 2014, the University of Indore awarded him the Golden Jubilee Research Fellowship on the occasion of its successful 50th anniversary. Business education, HRD, cross-cultural business, and organizational behavior are his core areas of research interests. His research papers and book chapters have been published in the journals of international repute and edited collections by Springer, Taylor & Francis, Emerald, Sage, IGI Global, Vernon Press, and others, respectively.

Raj Kumar Mishra is an Assistant Professor at the Institute of Management Sciences, University of Lucknow, with 14 years of teaching experience. He has also gained valuable experience working at prestigious institutions like PSIT Kanpur and BBD University. He has authored one book, holds one patent, and has published 21 research papers in esteemed national and international journals. With a Ph.D. in Management specializing in Marketing, his current research focuses on the influence of emerging technologies such as AI and Big Data on consumer behaviour and brand engagement. He also explores the effectiveness of personalized and healthcare marketing, as well as the role of ethical advertising in fostering consumer trust.

Jayanta Kumar Mohapatra has more than two decades of teaching and research experience. His research interest includes business strategy and mergers and acquisitions. His teaching interest includes strategic management, mergers and acquisitions, business ethics and leadership.

Raghu N has Ten years of experience in Academic and Research. He received the Ph.D. (2020) in Electronics Engineering from Jain University, and B.E. in Electronics and Telecommunications Engineering from VTU University, Karnataka. His research interests were in RF Communication and Image Processing. He has published his research findings in various Scopus indexed Journals and International conferences. He is a member of various professional bodies.

Nityasundar Nanda is an accomplished finance scholar and IIM Kashipur alumnus with a PhD from Ravenshaw University. An Economics graduate from

Utkal University and MBA Finance holder from Indian Institute of Finance, Delhi, he has transitioned from IndusInd Bank to academia. With 20+ Scopus-indexed publications, three doctoral scholars under supervision, and expertise in stock market analytics and corporate valuation, he combines academic excellence with industry experience and active research contributions.

Durgesh Nandan is an associate professor at SR University, Warangal, Telangana, India. Additionally, he is serving the role of Director, Center of Embedded Systems at the same university. He is a result-oriented leader with 16 years of professional experience, out of which 10 years of teaching experience and 3 years of industry experience. He possesses effective communication skills, exceptional presentation skills, and the ability to lead cross-cultural teams and establish relationships.

Tamrisha Patnaik is an accomplished Assistant Professor at Woxsden university Hyderabad and alumna of IIM Kozhikode's Executive Faculty Development Program, with over 10 years of teaching and research experience in Human Resource Management. She holds a PhD in Business Management and specializes in employee attrition, retention, and HR analytics. With multiple Scopus-indexed publications and expertise in strategic HRM, she combines academic excellence with innovative teaching methodologies and groundbreaking research contributions.

Ravikumar R. N. is currently a Postdoctoral researcher at Universiti Malaysia Perlis, Malaysia. Also, working as an Associate Professor in the Department of Computer Engineering at Marwadi University, Rajkot, Gujarat. He has 14+ years of national and international teaching experience. He completed his PhD at Amity University, Rajasthan, M.E at VMRF Deemed University, B.E at Anna University, Chennai. He is the co-founder of the Research Activities Club at Marwadi University and trains students on research activities, which leads to the various publications. He has published numerous research articles in Scopus, SCI journals, Book chapters, and Conference papers. Also, he holds several patents and authored books. His specialization is ML, DL, Federated Learning, Recommendation Systems, and Optimization Techniques.

Jyoti Rana is a distinguished academician and researcher with a strong background in marketing and commerce. She holds a Ph.D. in AI applied in customer journey studies and has authored over 20 research publications in Scopus and ABDC indexed journals. Dr. Rana has participated in and presented papers at many national and international conferences. Her keen research areas are marketing, retail, artificial intelligence and customer journey.

Ali Rizvi is working as Fellow with NITI Ayog . Published work in various journals.

Göran Roos is a globally recognised expert in intellectual capital management, manufacturing competitiveness, industry policy, and innovation systems. With around 19,000 Google citations and more than 300 publications, he has established himself as a leading authority on strategic management, economic complexity, and sustainable industrial development. Academic Credentials and Appointments Professor Roos holds a PhD from the University of South Australia (2019), an MBA from INSEAD (1989), and an MSc in Engineering Physics from Chalmers University of Technology (1982). His current academic positions and fellowships include: • Adjunct Professor at the Institute for Sustainability, Energy and Resources (ISER), The University of Adelaide (2023-2025) • Fellow of the Australian Academy of Technological Sciences and Engineering (ATSE) since 2014 • Fellow of the Royal Swedish Academy of Engineering Sciences (IVA) since 2015 Areas of Expertise Intellectual Capital and Knowledge Management Professor Roos is a world-renowned expert in intellectual capital, having developed theories and implemented practical solutions that have shaped the

Amitabh Roy is currently working as Associate Professor at Institute of Management Sciences, University of Lucknow.

Parthasarathy Seethapathy is a Plant Pathologist currently serving as a Scientist at CSIR-Institute of Himalayan Bioresource Technology, Palampur. My research involves developing biological formulations and evaluating their effectiveness against plant diseases, exploring microbial diversity, and molecular plant pathology. Recently, I have been focusing on the early detection of plant diseases using AI and machine learning tools.

Binastya Anggara Sekti has been a lecturer at the Faculty of Computer Science, Esa Unggul University, Jakarta, since 2016. He is actively involved in the Artificial Intelligence and Information Security Study Group. He holds a Bachelor's degree in Informatics Engineering from the Bandung Institute of Technology and a Master's degree in Information Systems Management from Trisakti University, Jakarta. Prior to entering academia, he accumulated over 25 years of professional experience as an IT practitioner in various companies across Jakarta. His work focused on developing data centers for internet service providers and delivering IT solutions for the manufacturing and oil & gas sectors. His research interests include artificial intelligence, cybersecurity, and the development of IT infrastructure.

Mohammad Talha Siddiqui is an academic and researcher specializing in marketing, supply chain management, and artificial intelligence applications in education. He holds a Ph.D. in Logistics, Materials, and Supply Chain Management from the University of Lucknow, where he currently serves as an Assistant Professor. His academic background also includes a Master of Business Administration (MBA) in Marketing from LUMBA and a Bachelor of Science (B.Sc.) in Electronics from the University of Lucknow. Dr. Siddiqui began his professional career in higher education as an Assistant Professor at the Bora Institute of Management Sciences, where he taught business and marketing courses and developed a keen interest in educational innovation. He later joined the University of Lucknow as a Junior Research Fellow, deepening his expertise in logistics systems, research methodologies, and data-driven decision-making before transitioning to a full-time faculty position. With nearly a decade of academic experience, Dr. Siddiqui has developed a strong portfolio in AI-driven learning systems, marketing analytics, and supply-chain optimization. His recent work explores the transformative impact of Agentic AI and RAG systems in both business strategy and higher education, emphasizing the integration of intelligent systems for better learning outcomes and institutional efficiency. He holds multiple professional certifications, including IBM RAG and Agentic AI, Python for Data Science, R Essentials, and Agentic AI and AI Agents for Leaders, reflecting his commitment to bridging the gap between research and technological advancement. Dr. Siddiqui is also an active participant in national academic forums and contributes to initiatives promoting innovation and digital transformation in education.

Ayşe Sirel, Ph.D. (Turkey), Prof. Dr, Istanbul Aydın University, Istanbul. Prof. Dr. Sirel completed her undergraduate education at Mimar Sinan University, Department of Architecture in 1980. In 1982, she received her master's degree from Yıldız Technical University. In 1993, she received her PhD degree from Mimar Sinan University, Department of Urban and Regional Planning. She worked as an assistant at the same university between 1983-1993 and as an assistant professor at Trakya University Faculty of Architecture between 1994-2008. Since 2009, she has been continuing her academic studies at Istanbul Aydın University. Sirel became a Professor in 2022. She has written many articles, symposium papers and book chapters in the field of architectural design and urban design in national and international publications. Between 2017 and 2024, she served as editor of "A+Arch Design," the journal of the IAU Faculty of Architecture and Design. She is the recipient of numerous awards in architectural and urban design competitions.

Melodena Stephens is Professor of Innovation & Technology Governance, at the Mohammed Bin Rashid School of Government (MBRSG) in Dubai, UAE. She has held several senior leadership and board positions (India, UAE and Germany) and consults and researches in strategy. She is an expert in AI ethics and tech policy working with IEEE SA, Agile Nations, Council of Europe, Meta, WEF, and the Dubai Future Foundation. Currently, she is the Co-Chair of the policy group at think tank - The Digital Economist and a Team Leader for the Spring 2025 cohort of the Center for AI & Digital Policy - AI Policy (Advanced) Clinic. Melodena is an well recognized author having published 14 books, the most recent being 'Anticipatory Governance: Shaping a Responsible Future' and 'AI Enabled Business: A Smart Decision Kit'. She is active in the nonprofit sector, founding the Academy of International Business MENA chapter, which promotes awareness, research and teaching in the region, for which the organization won the recognition of the Dubai government. Melodena is an IEEE CertifAIEd™ Authorized Assessor for AI.

N.V. Suresh, renowned academician from Chennai, Tamil Nadu. He completed his schooling in Chennai. He obtained his BA degree (2012) from D.G.Vaishnav College, Madras University. He joined the Bharathiar University received an MBA degree and PGDM. He Awarded Ph.D. in Management from University of Madras. Dr.N.V.Suresh joined then IIKM in June 2014 as an Assistant Professor (2014 to 2016). He moved to Remo International College in June, 2019 as an Assistant Professor and rose through the ranks to the position of Head of the Department in Aug 2020 to July 2021 then Shifted to Working Academy of Maritime Education and Training (Deemed to be University) Presently Working as Associate Professor and Vice-Principal in ASET College of Science and Technology from March 2022 to Till Date. He is the Visiting Faculty for Institute of Distance Education, University of Madras, Tamil Nadu Open University, Anna University and Directorate of Online Education SRM University for MBA from 2015 to till date. Dr.N.V. Suresh has carried out all his research work entirely in India and has published 30 research papers in international journals.

Pınar Tabak completed her undergraduate education at the Faculty of Architecture at Trakya University in Edirne in 2003. After completing her undergraduate education, she continued her graduate education at Istanbul Technical University's Department of Urban Planning. During this period, she worked on historical environment and heritage projects at various firms in Istanbul and participated in international projects such as the Kosovo Conservation Plan, led by the ITU Application and Research Center, and the Inventory of Historic Structures (ISMEP), conducted by the Ministry of Culture in collaboration with the World Bank. Between 2009 and 2010, she studied at Minho and UPC Universities and completed the "Advanced Master's Program in Structural Analysis of Monuments and Historic Buildings." After working on various projects over the past decade, she began his doctoral studies at Istanbul Aydın University in 2019. She started working as a lecturer in the Department of Architecture and Interior Architecture and became an assistant professor after completing his doctorate. The focus of her studies is “historical environment”, “historical context”, “new building in historic city”, and “architectural design.”

Index

A

C

D

E

F

G

H

I

L

M

O

P

R

S

T

U

W